Proceedings in Adaptation, Learning and Optimization

Volume 6

About this Series

The role of adaptation, learning and optimization are becoming increasingly essential and intertwined. The capability of a system to adapt either through modification of its physiological structure or via some revalidation process of internal mechanisms that directly dictate the response or behavior is crucial in many real world applications. Optimization lies at the heart of most machine learning approaches while learning and optimization are two primary means to effect adaptation in various forms. They usually involve computational processes incorporated within the system that trigger parametric updating and knowledge or model enhancement, giving rise to progressive improvement. This book series serves as a channel to consolidate work related to topics linked to adaptation, learning and optimization in systems and structures. Topics covered under this series include:

- complex adaptive systems including evolutionary computation, memetic computing, swarm intelligence, neural networks, fuzzy systems, tabu search, simulated annealing, etc.
- machine learning, data mining & mathematical programming
- hybridization of techniques that span across artificial intelligence and computational intelligence for synergistic alliance of strategies for problem-solving
- aspects of adaptation in robotics
- agent-based computing
- autonomic/pervasive computing
- dynamic optimization/learning in noisy and uncertain environment
- systemic alliance of stochastic and conventional search techniques
- all aspects of adaptations in man-machine systems.

This book series bridges the dichotomy of modern and conventional mathematical and heuristic/meta-heuristics approaches to bring about effective adaptation, learning and optimization. It propels the maxim that the old and the new can come together and be combined synergistically to scale new heights in problem-solving. To reach such a level, numerous research issues will emerge and researchers will find the book series a convenient medium to track the progresses made.

More information about this series at http://www.springer.com/series/13543

Jiuwen Cao · Kezhi Mao
Jonathan Wu · Amaury Lendasse

Editors

Proceedings of ELM-2015 Volume 1

Theory, Algorithms and Applications (I)

 Springer

Editors
Jiuwen Cao
Institute of Information and Control
Hangzhou Dianzi University
Hangzhou, Zhejiang
China

Kezhi Mao
School of Electrical and Electronic
    Engineering
Nanyang Technological University
Singapore
Singapore

Jonathan Wu
Department of Electrical and Computer
    Engineering
University of Windsor
Windsor, ON
Canada

Amaury Lendasse
Department of Mechanical and Industrial
    Engineering
University of Iowa
Iowa City, IA
USA

ISSN 2363-6084   ISSN 2363-6092   (electronic)
Proceedings in Adaptation, Learning and Optimization
ISBN 978-3-319-28396-8   ISBN 978-3-319-28397-5   (eBook)
DOI 10.1007/978-3-319-28397-5

Library of Congress Control Number: 2015958845

Printed on acid-free paper

This Springer imprint is published by SpringerNature
The registered company is Springer International Publishing AG Switzerland

Contents

Extreme Learning Machine for Multi-class Sentiment Classification of Tweets 1
Zhaoxia Wang and Yogesh Parth

Efficient Batch Parallel Online Sequential Extreme Learning Machine Algorithm Based on MapReduce 13
Shan Huang, Botao Wang, Yuemei Chen, Guoren Wang and Ge Yu

Fixed-Point Evaluation of Extreme Learning Machine for Classification 27
Yingnan Xu, Jingfei Jiang, Juping Jiang, Zhiqiang Liu and Jinwei Xu

Multi-layer Online Sequential Extreme Learning Machine for Image Classification 39
Bilal Mirza, Stanley Kok and Fei Dong

ELM Meets Urban Computing: Ensemble Urban Data for Smart City Application 51
Ningyu Zhang, Huajun Chen, Xi Chen and Jiaoyan Chen

Local and Global Unsupervised Kernel Extreme Learning Machine and Its Application in Nonlinear Process Fault Detection 65
Hanyuan Zhang, Xuemin Tian, Xiaohui Wang and Yuping Cao

Parallel Multi-graph Classification Using Extreme Learning Machine and MapReduce 77
Jun Pang, Yu Gu, Jia Xu, Xiaowang Kong and Ge Yu

Extreme Learning Machine for Large-Scale Graph Classification Based on MapReduce 93
Zhanghui Wang, Yuhai Zhao and Guoren Wang

**The Distance-Based Representative Skyline Calculation
Using Unsupervised Extreme Learning Machines**   107
Mei Bai, Junchang Xin, Guoren Wang and Xite Wang

**Multi-label Text Categorization Using L_{21}-norm Minimization
Extreme Learning Machine**   121
Mingchu Jiang, Na Li and Zhisong Pan

**Cluster-Based Outlier Detection Using Unsupervised Extreme
Learning Machines**   135
Xite Wang, Derong Shen, Mei Bai, Tiezheng Nie,
Yue Kou and Ge Yu

**Segmentation of the Left Ventricle in Cardiac MRI
Using an ELM Model**   147
Yang Luo, Benqiang Yang, Lisheng Xu, Liling Hao, Jun Liu,
Yang Yao and Frans van de Vosse

**Channel Estimation Based on Extreme Learning Machine
for High Speed Environments**   159
Fang Dong, Junbiao Liu, Liang He, Xiaohui Hu and Hong Liu

MIMO Modeling Based on Extreme Learning Machine.   169
Junbiao Liu, Fang Dong, Jiuwen Cao and Xinyu Jin

**Graph Classification Based on Sparse Graph Feature Selection
and Extreme Learning Machine**.   179
Yajun Yu, Zhisong Pan and Guyu Hu

**Time Series Prediction Based on Online Sequential Improved
Error Minimized Extreme Learning Machine**   193
Jiao Xue, Zeshen Liu, Yong Gong and Zhisong Pan

**Adaptive Input Shaping for Flexible Systems Using an Extreme
Learning Machine Algorithm Identification**   211
Jun Hu and Zhongyi Chu

**Kernel Based Semi-supervised Extreme Learning Machine
and the Application in Traffic Congestion Evaluation**   227
Qing Shen, Xiaojuan Ban, Chong Guo and Cong Wang

**Improvement of ELM Algorithm for Multi-object Identification
in Gesture Interaction**   237
Liang Diao, Liguo Shuai, Huiling Chen and Weihang Zhu

**SVM and ELM: Who Wins? Object Recognition with Deep
Convolutional Features from ImageNet**.   249
Lei Zhang, David Zhang and Fengchun Tian

**Learning with Similarity Functions: A Novel Design
for the Extreme Learning Machine**.   265
Federica Bisio, Paolo Gastaldo, Rodolfo Zunino,
Christian Gianoglio and Edoardo Ragusa

**A Semi-supervised Low Rank Kernel Learning Algorithm
via Extreme Learning Machine**   279
Bing Liu, Mingming Liu, Chen Zhang and Weidong Wang

**Application of Extreme Learning Machine on Large Scale
Traffic Congestion Prediction**.   293
Xiaojuan Ban, Chong Guo and Guohui Li

**Extreme Learning Machine-Guided Collaborative Coding
for Remote Sensing Image Classification**.   307
Chunwei Yang, Huaping Liu, Shouyi Liao and Shicheng Wang

**Distributed Weighted Extreme Learning Machine for Big
Imbalanced Data Learning**.   319
Zhiqiong Wang, Junchang Xin, Shuo Tian and Ge Yu

**NMR Image Segmentation Based on Unsupervised Extreme
Learning Machine**.   333
Junchang Xin, Zhongyang Wang, Shuo Tian and Zhiqiong Wang

**Annotating Location Semantic Tags in LBSN Using Extreme
Learning Machine**   347
Xiangguo Zhao, Zhen Zhang, Xin Bi, Xin Yu and Jingtao Long

**Feature Extraction of Motor Imagery EEG Based on Extreme
Learning Machine Auto-encoder**   361
Lijuan Duan, Yanhui Xu, Song Cui, Juncheng Chen
and Menghu Bao

**Multimodal Fusion Using Kernel-Based ELM for Video
Emotion Recognition**   371
Lijuan Duan, Hui Ge, Zhen Yang and Juncheng Chen

**Equality Constrained-Optimization-Based Semi-supervised ELM
for Modeling Signal Strength Temporal Variation in Indoor
Location Estimation**   383
Felis Dwiyasa, Meng-Hiot Lim, Yew-Soon Ong and Bijaya Panigrahi

**Extreme Learning Machine with Gaussian Kernel Based
Relevance Feedback Scheme for Image Retrieval**   397
Lijuan Duan, Shuai Dong, Song Cui and Wei Ma

**Routing Tree Maintenance Based on Trajectory Prediction
in Mobile Sensor Networks**   409
Junchang Xin, Teng Li, Pei Wang and Zhiqiong Wang

**Two-Stage Hybrid Extreme Learning Machine
for Sequential Imbalanced Data**   423
Wentao Mao, Jinwan Wang, Ling He and Yangyang Tian

**Feature Selection and Modelling of a Steam Turbine
from a Combined Heat and Power Plant Using ELM**   435
Sandra Seijo, Victoria Martínez, Inés del Campo,
Javier Echanobe and Javier García-Sedano

**On the Construction of Extreme Learning Machine
for One Class Classifier**   447
Chandan Gautam and Aruna Tiwari

**Record Linkage for Event Identification in XML Feeds Stream
Using ELM**   463
Xin Bi, Xiangguo Zhao, Wenhui Ma, Zhen Zhang and Heng Zhan

Timeliness Online Regularized Extreme Learning Machine   477
Xiong Luo, Xiaona Yang, Changwei Jiang and Xiaojuan Ban

**An Efficient High-Dimensional Big Data Storage Structure
Based on US-ELM**   489
Linlin Ding, Yu Liu, Baoyan Song and Junchang Xin

**An Enhanced Extreme Learning Machine for Efficient
Small Sample Classification**   501
Ying Yin, Yuhai Zhao, Ming Li and Bin Zhang

Code Generation Technology of Digital Satellite   511
Ren Min, Dong Yunfeng and Li Chang

Class-Constrained Extreme Learning Machine   521
Xiao Liu, Jun Miao, Laiyun Qing and Baoxiang Cao

Author Index   531

Extreme Learning Machine for Multi-class Sentiment Classification of Tweets

Zhaoxia Wang and Yogesh Parth

Abstract The increasing popularity of social media in recent years has created new opportunities to study and evaluate public opinions and sentiments for use in marketing and social behavioural studies. However, binary classification into positive and negative sentiments may not reveal too much information about a product or service. This research paper explores the multi-class sentiment classification using machine learning methods. Three machine learning methods are investigated in this paper to examine their respective performance in multi-class sentiment classification of tweets. Experimental results show that Extreme Learning Machine (ELM) achieves better performance than other machine learning methods.

Keywords Extreme learning machine · Machine learning · Multi-class classification · Sentiment analysis · Social media · Tweets

1 Introduction

In recent years, the increasing popularity of social media, including the use of tweets, has created new opportunities to study and evaluate public opinions and sentiments for use in marketing and social behavioural studies.

Sentiment analysis, or opinion mining, can be defined as a computational study of consumer opinions, sentiments and emotions, particularly towards specific products or services [1]. Sentiment classification can be thought of as a pattern-recognition

Z. Wang (✉)
Social and Cognitive Computing (SCC) Department, Institute of High Performance
Computing (IHPC), Agency for Science, Technology and Research (ASTAR),
Singapore 138632, Singapore
e-mail: wangz@ihpc.a-star.edu.sg

Y. Parth
Department of Space, Indian Institute of Space Science and Technology (IIST),
Thiruvananthapuram 695547, India
e-mail: yogeshparthiist@gmail.com

© Springer International Publishing Switzerland 2016
J. Cao et al. (eds.), *Proceedings of ELM-2015 Volume 1*,
Proceedings in Adaptation, Learning and Optimization 6,
DOI 10.1007/978-3-319-28397-5_1

and classification task analysing unstructured data for purposes towards improving product or service quality [2].

Sentiment classification can be broadly categorized into two main groups: machine learning-based and non-machine learning-based methods [3, 4]. In general, machine learning-based methods can achieve better classification results compared to non-learning based method (such as simple lexicon-based method), and are widely used [4]. Extreme Learning Machine (ELM) is one of the more recent and popular machine learning-based methods [5]. It is a kind of feedforward networks, which considers multi-hidden-layer of networks as a white box and trained layer-by-layer [6]. In general, ELM tends to perform better compared to other gradient-based learning algorithms [5]. It has been successfully implemented in many real-world applications [7–11].

Other machine learning methods like Support Vector Machine (SVM), Naïve Bayes (NB) and Maximum Entropy have also been used in classification applications [12–16]. Most of these machine learning methods are applied to binary-classification problems and their performance in handling multi-class sentiment classification has not been well researched.

In this paper, we investigate the performances of different machine learning methods such as ELM, Multi-Class SVM, and Multinomial Naïve Bayes in multi-class sentiment classification. The rest of the paper is organized as follows. Section 2 reviews the relevant work of the machine learning methods in classification problems. This is followed by the implementation of these methods in sentiment analysis of tweets in Sect. 3. Performance of different machine learning-based methods is evaluated with case studies in Sect. 4. Section 5 concludes this paper with recommendations for future studies.

2 Relevant Work of Machine Learning Methods

The machine-learning methods reviewed herein include Multinomial Naïve Bayes, Multi-Class support vector machine and ELM.

Naïve Bayes classifier is a probabilistic classifier based on the Bayes theorem [17, 18]. Relaxing the conditional independence assumption for each of the features in binary classification, the Multinomial Naïve Bayes classifier can be used to deal with multi classifications [19]. Given a set of objects, each of which belonging to a known class and having a known vector of variables, the algorithm attempts to construct a rule which will assign future objects to a class, while being given only the vectors of variables describing the future objects. Let x_i be the feature vector in multinomial model for the ith document D_i. Let $n_i = \sum_t x_{it}$ be the total number of words in D_i, where x_{it} is the t th element of x_i, and let $P(w_t|C)$ be the probability of words w_t occurring in class C. Then, by Naïve Bayes assumption of independence, the document likelihood $P(D_i|C)$ can be written as:

$$P(D_i|C) = \frac{n_i!}{\prod\limits_{t=1}^{|v|} x_{it}!} \prod_{t=1}^{|v|} P(w_t|C)^{x_{it}} \tag{1}$$

The probability, $P(w_t|C)$, of each word in a given the document class, C, can be written as,

$$\hat{P}(w_t|C = k) = \frac{\sum\limits_{i=1}^{N} x_{it} z_{ik}}{\sum\limits_{s=1}^{|v|} \sum\limits_{i=1}^{N} x_{is} z_{ik}} \tag{2}$$

where N is the total number of documents with the condition, z_{ik} equals to 1 when D_i contain class $C = k$, otherwise equals 0.

Each $P(w_i|c)$ term in Multinomial Naïve Bayes is assumed to a multinomial distribution. Multinomial distribution works well for data which can easily be turned into counts, and in this case, word counts in the text.

SVM is a non-probabilistic classifier that constructs a hyperplane in a high-dimensional space through the classification training process [4, 20]. The decision function can be defined as follows:

$$f(x) = sign(\sum_i \alpha_i K(x_i^v, x) + b) \tag{3}$$

where α_i is Lagrange multiplier determined during SVM training. The parameter b representing the shift of the hyperplane is determined during SVM training with the $K(x_i^v, x)$ as the kernel function [21].

ELM which was initially proposed by Huang [22] is different from BP and SVM which consider multi-layer of networks as a black box. It is also different from Deep Learning which requires intensive tuning in hidden layers and hidden neurons, ELM theories show that hidden neurons do not need tuning because its hidden nodes parameters (c_i, a_i) are randomly assigned [23]. For N arbitrary distinct samples $(x_k, t_k) \in R^n \times R^m$, the single ELM classifier with $\tilde{N}$ hidden nodes becomes a linear system as,

$$\sum_{i=1}^{\tilde{N}} \beta_i G(x_k; c_i; a_i) = t_k, k = 1, \ldots, N. \tag{4}$$

where $c_i \in R^n$ and $a_i \in R$ are the learning parameters of hidden nodes and randomly assigned weight β_i connecting the ith hidden node to the output node, x_k are the training examples, t_k is the target output for $k = 1, \ldots, N$, and $G(x_k; c_i; a_i)$ is the output of the ith hidden node with respect to the input x_k. The output weights can be described in matrix form as,

$$\beta = \lfloor \beta_1^T \ldots \beta_{\tilde{N}}^T \rfloor_{m \times \tilde{N}}^T \tag{5}$$

Equation (4) can be rewritten as:

$$H\beta = T \tag{6}$$

where $H(c_1, \ldots, c_{\tilde{N}}, a_1, \ldots, a_{\tilde{N}}, x_1, \ldots, x_N) =$

$$\begin{bmatrix} G(x_1; c_1, a_1) & \cdots & G(x_1; c_{\tilde{N}}, a_{\tilde{N}}) \\ \vdots & \ddots & \vdots \\ G(x_N; c_1, a_1) & \cdots & G(x_N; c_{\tilde{N}}, a_{\tilde{N}}) \end{bmatrix}_{N \times \tilde{N}} \tag{7}$$

$$T = \lfloor t_1^T \ldots t_N^T \rfloor_{m \times N}^T \tag{8}$$

The output weights β can be determined by finding the least-square solution as,

$$\hat{\beta} = H^\dagger T \tag{9}$$

where $H^\dagger$ is the Moore-Penrose generalized inverse [24] of the hidden layer output matrix H.

An ELM classifier (single) implements multi-class classification problem using a network architecture of multi-output nodes equal to the number of pattern classes n. The network output can be written as $y = (y_1, y_2, \ldots, y_n)^T$. For each training example say x, the target output t is coded into n bits: $(t_1, \ldots, t_n)^T$. For a pattern of class i, only the target output t_i is "1" and the rest is "−1".

3   Proposed Implementation of ELM and Other Machine Learning Methods for Sentiment Classifications

The task of applying machine learning methods for classification requires several steps.

Pre-processing of unstructured tweet data is the first step in implementing machine learning methods for sentiment classification. The collected data have to undergo cleaning, tokenization [25], and stemming [26] to convert them into structured text data. Cleaning involves the removal of url links, usernames ("@username"), punctuations, whitespaces, hashtags etc. The structured texts are then tokenized with labels to create a word features list. Using chi-square (χ_N^2) distribution, the word features are assigned an intermediate score. The scores are subsequently updated to find collocation and bigrams. After pre-processing, the top n features in the list with the best bigrams and collocations are extracted to train the machine learning classifier. The latest advanced enhancement methods [4] have been used to obtain the best possible results for SVM and Naïve Bayes classifier in this paper.

In the case of ELM, the optimized parameters have been chosen to minimize the root mean square error (RMSE). Tables 1 and 2 show the list of parameter sets for

Table 1 ELM optimized parameter set for tweet data without any regressor

RMSE	Hidden node	Alpha	RBF_width	Activation_Function
5.53667074826311	96	0.524301588135944	0.00192745905713552	inv_tribas
0.117694994097736	**23**	**0.362227734140632**	**7.05900482368913**	**tanh**
0.119176722319975	14	0.0862999041065237	0.0000154232670105994	multiquadric
0.121964679113459	360	0.267976516333136	0.0000335082533900404	softlim
0.128636656277498	27	0.8158426513257	0.0533106256560502	sigmoid
0.116351185300275	**728**	**0.8209323214078**	**0.0153925477267926**	**hardlim**
0.129012115416406	147	0.399149654069297	22.5662635770545	inv_multiquadric
0.117714432502741	734	0.301426069647834	3.634415143724	hardlim
0.119782506893025	570	0.4309900031149163	0.000191208358888382	tanh
0.119752243475673	10	0.869747469515247	0.0174366766856871	softlim
0.123482000618614	17	0.999118291073565	1.9406508280759	hardlim
0.15629835641910	405	0.854263388991001	0.00341937930632422	gaussian
0.193590715031852	192	0.319027484499291	0.0644530059543435	inv_tribas
0.117524241972222	**850**	**0.373112162422129**	**1.15291865399892**	**hardlim**
0.120402689130944	13	0.6223561171948	0.0029834065089396	multiquadric
0.130039185254618	17	0.60576214037852	0.0165939924916861	inv_tribas
0.146980644034541	30	0.1947764934742	0.0000137035341999231	tribas
0.126685956956185	23	0.8133601307840	0.923405718456997	sigmoid
0.15177520202209191	81	0.478881694502156	27.3035063972805	inv_multiquadric

Table 2 ELM optimized parameter set for tweet data with ridge as regressor

RMSE	Hidden node	Alpha	RBF_width	Activation_Function	Ridge_alpha
0.11721977272709406	12	0.0973901333207798	1.48500198321801	gaussian	0.000240288138976629
0.17111518171 9981	87	0.894978744640209	48.345992769188	multiquadric	0.00518583033923811
0.112284877403487	**965**	**0.1340791976002**	**0.0984951895364843**	**multiquadric**	**0.298806226947076**
0.125655943802341	284	0.29186818541362	0.09855690804485006	softlim	0.541801573766452
0.122583509721702	196	0.018282092561289	0.927267663139151	sine	0.000561830406504939
0.112803169495448	**675**	**0.0638767992050295**	**0.0778255414360264**	**multiquadric**	**0.719153883848897**
0.1179177952119	77	0.516497025526854	0.190687099943598	inv_multiquadric	31.118255688534
0.122992691773984	18	0.651179646424338	0.0125897731660419	sigmoid	0.0000180218698166426
0.131352330647929	17	0.596875638531792	2.90337806004152	gaussian	0.00000638733172727182
0.15669449022685	60	0.459878608804425	0.0003599990252415064	sigmoid	0.0000085619264740847
0.117270497665036	26	0.182715737779089	6.22597637491161	tanh	0.0000288300935812876
0.117270497665036	109	0.381036038270451	7.75350974738416	hardlim	0.00000104973023516875
0.1131814807259	**652**	**0.0769244046713881**	**0.0890436054223181**	**multiquadric**	**0.841414102931452**
0.1356524004731523	46	0.351763852755516	1.042073355382151	inv_multiquadric	0.0195730715339201

a particular type of data which are optimized using the ridge regression and linear regression respectively. The selected parameter sets which produce the least RMSE (shown in bold) are used to train the classifier. The number of iterations for training and testing the classifier is limited to a maximum of 20.

Algorithm 1 shows the pseudo code for minimizing the RMSE, which is coded in hyper-parameter optimization library in python "Hyperopt" [27]. The optimization is performed over the given parameters or search space such as hidden node, alpha, ridge alpha, and radial basis function (RBF) width. The Tree-structured Parzen Estimator (TPE) [28], is used for optimization over the given conditions for training an ELM classifier.

Algorithm 1 Minimize Root Mean Square Error

```
 1: procedure REQUIRE(numpy, hyperopt, sklearn, elm)
 2:     input ← training file
 3:     iteration_max ← 20
 4: test_run:
 5:     test(params, ridge)
 6: test(params, ridge):
 7:     hidden_n, alpha, rbf_width, activation_func = params
 8:     layer ← RandomLayer(params)
 9:     ridge ← Ridge(ridge_alpha)
10:     elm = pipeline([layer, ridge])
11:     elm.train
12:     p ← elm.predict
13:     rmse = sqrt(mse(test, p))
14:     return rmse
15: parameters:
16:     hidden_n ← ploguniform(a,b)
17:     alpha ← uniform(a',b')
18:     rbf_width ← loguniform(c,d)
19:     ridge_alpha ← uniform(c',d')
20:     activation_func ← choice(tanh,sine, ...,gaussian)
21: main():
22:     best ← min(test_run, parameters, algo=tpe, iteration_max).
23:     print [rmse],[params],[ridge_alpha]
24:     print best
```

---

4 Performance Evaluation with Discussion

4.1 Data Collection and Preparation

The datasets used in this study are tweets data obtained from two different sources. In case 1, the data are downloaded from the "twitter-sentiment-analyzer" (https://github.com/ravikiranj/twitter-sentiment-analyzer/tree/master/data), which contained 1.6 million pre-classified tweets reported previously [4]. We downloaded ds_5k, ds_10k, ds_20k, ds_40k, which consisted of 5k, 10k, 20k, and 40k of pre-classified tweets respectively. However, this set of data only contains binary sentiment tweets i.e. positive and negative tweets, so we extracted neutral tweets data from our tweet data collections which were collected by using twitter wrapper API (application program interface), and appended to the downloaded tweet datasets.

In case 2, the data were collected through twitter API by using the keyword "MRT" (Mass Rapid Transit) over the region of Singapore. Location-constraining geo codes were used to ensure that the tweets were collected within the region in and around Singapore. We performed this data collection with the aim to investigate the public attitudes towards Singapore public transportation services. The collected tweets were annotated manually with the help of field experts in order to obtain ground-truth data to be used in machine-learning-based methods.

In both cases 1 and 2, the training dataset contains 75 % of the data, while the test set consists of the remaining 25 %.

4.2 Performance Evaluation

We trained each classifier, namely Multinomial Naïve Bayes, Multi-class SVM, and ELM using the training set and tested its accuracy using the test sets. The performance metric is measured by the accuracy of classification. This defines how close the performance is to the idle or benchmark value. For binary classification problems, this is calculated using the formula:

$$Accuracy = (T_p + T_n)/(T_p + T_n + F_p + F_n) \tag{10}$$

where T_p is the number of correctly identified positives, F_p is the number of incorrectly identified positives while, T_n is the number of correctly identified negatives and F_n the number of incorrectly identified negatives.

For multi-classification problems, accuracy is calculated as:

$$Accuracy = N_c/N_t \tag{11}$$

where N_c is the number of correctly identified samples, N_t is the number of total samples.

4.3 Case Studies

4.3.1 Case Study 1

In this case study, we used tweets data downloaded from the web and extracted from our tweet collections. Table 3 shows the comparison of the classification accuracy among the different machine learning methods.

Table 3 Comparison of machine-learning algorithms for classifying the downloaded tweets data-sets

Datasets	Number of class	Number of feature	ELM (%)	Multi-class SVM (%)	Multinomial NB (%)
ds_5k	3	1000	**83.50**	83.07	71.07
ds_10k	3	4000	**68.53**	68.18	43.44
ds_20k	3	2000	**91.12**	84.73	78.92
ds_40k	3	8000	**78.73**	77.49	74.67
tweets*	3	2000	**99.27**	93.71	92.98

Table 4 Comparison of machine-learning algorithms for classifying the MRT tweets data-sets

Datasets	Number of class	Number of feature	ELM (%)	Multi-class SVM (%)	Multinomial NB (%)
MRT DATA	3	700	**89.09**	83.64	76.36

The accuracy in ELM ranges from 68 to 99 % but is higher the that of Multi-class SVM and Multinomial Naïve Bayes for all the datasets. ELM is significantly better than Multinomial Naïve Bayes and marginally better than Multi-class SVM. In general, ELM outperforms the others in larger datasets. This indicates the efficiency of the ELM for multi-class classification.

4.3.2 Case Study 2

Tweets collected and extracted using the twitter API, having search queries related to MRT services over the region of Singapore. The accuracy of the different classifiers is compared and shown in Table 4. ELM achieves an accuracy of nearly 90 %, outperforming SVM (84 %) and Multinomial Naïve Bayes (76 %).

5 Conclusions and Future Work

In this paper, we have investigated the performance of different machine learning classifiers, including ELM, Multi-Class SVM, and Multinomial Naïve Bayes for multi-class sentiment analysis. The experimental results show that ELM achieves better performance than other machine learning methods for multi-class sentiment classification of tweet data.

As the performance of machine learning methods is dependent on how the features are selected, the machine-learning based multi-class sentiment classifiers may be improved if enhanced feature selection can be incorporated. Further studies on ELM with sophisticated feature selection techniques are currently being explored.

Acknowledgments This work is supported by the A*STAR Joint Council Office Development Programme "Social Technologies+ Programme".

References

1. Pang, B., Lee, L.: Opinion mining and sentiment analysis. Found. Trends Inf. Retr. **2**(1–2), 1–135 (2008)
2. Liu, B., Zhang, L.: A survey of opinion mining and sentiment analysis. In: Mining Text Data, pp. 415–463. Springer (2012)
3. Wang, Z., Tong, V.J.C., Chan, D.: Issues of social data analytics with a new method for sentiment analysis of social media data. In: 2014 IEEE 6th International Conference on Cloud Computing Technology and Science (CloudCom 2014), pp. 899–904. IEEE (2014)
4. Wang, Z., Tong, V.J.C., Chin, H.C.: Enhancing machine-learning methods for sentiment classification of web data. In: Information Retrieval Technology, pp. 394–405. Springer (2014)
5. Huang, G.-B.: Extreme learning machine: theory and applications. Neurocomputing **70**(1), 489–501 (2006)
6. Huang, G.-B., Bai, Z., Kasun, L.L.C., Vong, C.M.: Local receptive fields based extreme learning machine. IEEE Comput. Intell. Mag. **10**(2), 18–29 (2015)
7. Liang, N.-Y., Saratchandran, P., Huang, G.-B., Sundararajan, N.: Classification of mental tasks from eeg signals using extreme learning machine. Int. J. Neural Syst. **16**(01), 29–38 (2006)
8. Handoko, S.D., Keong, K.C., Soon, O.Y., Zhang, G.L., Brusic, V.: Extreme learning machine for predicting hla-peptide binding. In: Advances in Neural Networks-ISNN 2006, pp. 716–721. Springer (2006)
9. Yeu, C.-W., Lim, M.-H., Huang, G.-B., Agarwal, A., Ong, Y.-S.: A new machine learning paradigm for terrain reconstruction. IEEE Geosci. Remote Sens. Lett. **3**(3), 382–386 (2006)
10. Kim, J., Shin, H., Lee, Y., Lee, M.: Algorithm for classifying arrhythmia using extreme learning machine and principal component analysis. In: Engineering in Medicine and Biology Society: EMBS 2007. 29th Annual International Conference of the IEEE, pp. 3257–3260. IEEE (2007)
11. Wang, G., Zhao, Y., Wang, D.: A protein secondary structure prediction framework based on the extreme learning machine. Neurocomputing **72**(1), 262–268 (2008)
12. Chaovalit, P., Zhou, L.: Movie review mining: a comparison between supervised and unsupervised classification approaches. In: Proceedings of the 38th Annual Hawaii International Conference on System Sciences. HICSS'05, pp. 112c–112c. IEEE (2005)
13. Galitsky, B., McKenna, E.W.: Sentiment extraction from consumer reviews for providing product recommendations, May 12, 2008, US Patent App. 12/119,465
14. Hu, M., Liu, B.: Mining and summarizing customer reviews. In: Proceedings of the tenth ACM SIGKDD International Conference on Knowledge Discovery and Data Mining, pp. 168–177. ACM (2004)
15. Si, J., Mukherjee, A., Liu, B., Li, Q., Li, H., Deng, X.: Exploiting topic based twitter sentiment for stock prediction. In: ACL (2), pp. 24–29 (2013)
16. Chi, L., Zhuang, X., Song, D.: Investor sentiment in the chinese stock market: an empirical analysis. Appl. Econ. Lett. **19**(4), 345–348 (2012)
17. Dalton, L.A., Dougherty, E.R.: Optimal classifiers with minimum expected error within a bayesian framework part ii: properties and performance analysis. Pattern Recognit. **46**(5), 1288–1300 (2013)
18. Muralidharan, V., Sugumaran, V.: A comparative study of naïve bayes classifier and bayes net classifier for fault diagnosis of monoblock centrifugal pump using wavelet analysis. Appl. Soft Comput. **12**(8), 2023–2029 (2012)
19. Pappas, E., Kotsiantis, S.: Integrating global and local application of discriminative multinomial bayesian classifier for text classification. In: Intelligent Informatics, pp. 49–55. Springer (2013)

20. Chang, C.-C., Lin, C.-J.: Libsvm: a library for support vector machines. ACM Trans. Intell. Syst. Techn. (TIST) **2**(3), 27 (2011)
21. Byvatov, E., Fechner, U., Sadowski, J., Schneider, G.: Comparison of support vector machine and artificial neural network systems for drug/nondrug classification. J. Chem. Inf. Comput. Sci. **43**(6), 1882–1889 (2003)
22. Huang, G.-B., Zhu, Q.-Y., Siew, C.-K.: Extreme learning machine: a new learning scheme of feedforward neural networks. In: 2004 IEEE International Joint Conference on Neural Networks. Proceedings, vol. 2, pp. 985–990. IEEE (2004)
23. Huang, G.-B.: An insight into extreme learning machines: random neurons, random features and kernels. Cogn. Comput. **6**(3), 376–390 (2014)
24. Rao, C.R., Mitra, S.K.: Generalized Inverse of Matrices and Its Applications, vol. 7. Wiley, New York (1971)
25. Maršík, J., Bojar, O.: Trtok: a fast and trainable tokenizer for natural languages. Prague Bull. Math. Linguist. **98**, 75–85 (2012)
26. Willett, P.: The porter stemming algorithm: then and now. Program **40**(3), 219–223 (2006)
27. Bergstra, J., Yamins, D., Cox, D.D.: Hyperopt: a python library for optimizing the hyperparameters of machine learning algorithms (2013)
28. Bergstra, J.S., Bardenet, R., Bengio, Y., Kégl, B.: Algorithms for hyper-parameter optimization. In: Advances in Neural Information Processing Systems, pp. 2546–2554 (2011)

Efficient Batch Parallel Online Sequential Extreme Learning Machine Algorithm Based on MapReduce

Shan Huang, Botao Wang, Yuemei Chen, Guoren Wang and Ge Yu

Abstract With the development of technology and the widespread use of machine learning, more and more models need to be trained to mine useful knowledge from large scale data. It has become a challenging problem to train multiple models accurately and efficiently so as to make full use of limited computing resources. As one of ELM variants, online sequential extreme learning machine (OS-ELM) provides a method to learn from incremental data. MapReduce, which provides a simple, scalable and fault-tolerant framework, can be utilized for large scale learning. In this paper, we propose an efficient batch parallel online sequential extreme learning machine (BPOS-ELM) algorithm for the training of multiple models. BPOS-ELM estimates the Map execution time and Reduce execution time with historical statistics and generates execution plan. BPOS-ELM launches one MapReduce job to train multiple OS-ELM models according to the generated execution plan. BPOS-ELM is evaluated with real and synthetic data. The accuracy of BPOS-ELM is at the same level as those of OS-ELM and POS-ELM. The speedup of BPOS-ELM reaches 10 on a cluster with maximum 32 cores.

Keywords Parallel learning · Extreme learning machine · MapReduce · Sequential learning

S. Huang (✉) · B. Wang · Y. Chen · G. Wang · G. Yu
College of Information Science and Engineering, Northeastern University,
Shenyang 110819, Liaoning, China
e-mail: huangshan.neu@gmail.com

B. Wang
e-mail: wangbotao@ise.neu.edu.cn

Y. Chen
e-mail: 1044210092@qq.com

G. Wang
e-mail: wanggr@ise.neu.edu.cn

G. Yu
e-mail: yuge@ise.neu.edu.cn

© Springer International Publishing Switzerland 2016
J. Cao et al. (eds.), *Proceedings of ELM-2015 Volume 1*,
Proceedings in Adaptation, Learning and Optimization 6,
DOI 10.1007/978-3-319-28397-5_2

13

1 Introduction

With the development of technology and the widespread use of machine learning, more and more models need to be trained to mine useful knowledge from large scale data. It has become a challenging problem to train multiple models accurately and efficiently so as to make full use of limited computing resources. For example, in a machine learning organization where high performance computing cluster is a limited resource, researchers must schedule the jobs on the cluster legitimately to make full use of the cluster. For another example, resizable cloud hosting services such as Amazon Elastic Compute Cloud (EC2) [1], which become more and more popular, make it possible to rent large amount of virtual machines by the hour at lower costs than operating a data center year-round. It is important for users to schedule multiple jobs running on this kind of environment as the rented virtual machines are charged by the used time.

Extreme learning machine (ELM) was proposed based on single-hidden layer feed-forward neural networks (SLFNs) [7], and it has been verified to have high learning speed as well as high accuracy [5]. It has also been proved that ELM has universal approximation capability and classification capability [6]. As one of ELM variants, online sequential extreme learning machine (OS-ELM) [8] supports incremental learning.

MapReduce [3] is a well-known framework which supports large scale data processing and analyzing on a large cluster of commodity machines. Recent research has studied on parallelizing ELM [4, 11, 12], however the strategies are not suitable to parallelize OS-ELM. POS-ELM [10] supports training one single OS-ELM model in parallel with MapReduce, but it does not support training multiple OS-ELM models efficiently.

In this paper, we propose an efficient batch parallel online sequential extreme learning machine (BPOS-ELM) algorithm for the training of multiple OS-ELM models on MapReduce. BPOS-ELM first estimates the execution time of Map and Reduce tasks of each OS-ELM based on historical statistics. Then it generates a Map execution plan and a Reduce execution plan with greedy strategy based on the estimations. After that, BPOS-ELM launches a MapReduce job to train multiple OS-ELM models. At the same time, BPOS-ELM collects execution information of selected Map tasks and Reduce tasks, and merges them to historical statistics to improve the accuracy of time estimation. The algorithm is evaluated with real and synthetic data. The accuracy is at the same level as those of OS-ELM and POS-ELM. The speedup reaches 10 on a cluster with maximum 32 cores.

The remainder of this paper is organized as follows. Section 2 describes the batch parallel online sequential learning machine algorithm in detail. An extensive experimental evaluation of BPOS-ELM is presented in Sect. 3. A brief conclusion is presented in Sect. 4.

2 BPOS-ELM

As shown in Fig. 1, BPOS-ELM (1) assigns each model with a unique ID which is
used to specify it from the other training models. Then (2) the Map execution time
and Reduce execution time are estimated according to historical statistics described
with parameters shown in Table 1. After that, (3) A job execution plan is generated
according to the estimations. Finally, (4) the generated execution plan is executed to
train the models and (5) the actual execution information of selected tasks is collected
for future time estimation. The details of execution time estimation, execution plan
generation and job execution are described in Sects. 2.1–2.3, respectively.

2.1 Execution Time Estimation

Map execution time is estimated with Inverse Distance Weighted (IDW) [9] interpolation method. First, each parameter of job execution information is mapped to
one dimension at a multi-dimensional space, so the historical statistics are mapped
to a set of points in the space. Then k nearest neighbour points of the point to be
estimated in the space are selected and used to estimate Map execution time. After
that, Inverse Distance Weighted (IDW) [9] interpolation method shown as Formula
(1) is used to estimate Map execution time.

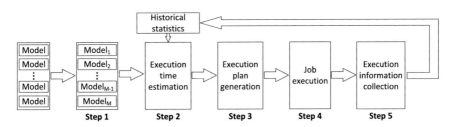

Fig. 1 Execution framework of BPOS-ELM

Table 1 Notations used in
BPOS-ELM

Parameter	Description
B	Block size
N	Number of training data
D	Number of attributes in training data set
M	Number of training models
$\tilde{N}$	Number of hidden layer nodes
C	Number of classifications

$$t_{map}(\mathbf{x}) \approx \begin{cases} \dfrac{\sum_{i=1}^{k} w_i(\mathbf{x})t_i}{\sum_{i=1}^{k} w_i(\mathbf{x})}, & \text{if } d(\mathbf{x}, \mathbf{x}_i) \neq 0 \text{ for all } i \\ \\ t_i, & \text{if } d(\mathbf{x}, \mathbf{x}_i) = 0 \text{ for some } i \end{cases} \qquad (1)$$

where $w_i(\mathbf{x}) = \frac{1}{d(\mathbf{x},\mathbf{x}_i)^p}$ is a simple IDW weighting function, as defined by Shepard [9], $\mathbf{x}$ denotes the parameter vector of point to be predict, $\mathbf{x_i}$ is the selected k nearest neighbour points, d is a given distance from the point $\mathbf{x_i}$ to point $\mathbf{x}$ and p is a positive real number, called the power parameter. Euclidean distance is used to measure the distance between two points. According to the the complexity analysis of POS-ELM reduce phase algorithm in [10], the Reduce execution time is estimated by Formula (2).

$$t_{red} \approx \frac{N}{B}(\alpha_1 B^3 + \alpha_2 B^2 \tilde{N} + \alpha_3 B \tilde{N}^2 + \alpha_4 B \tilde{N} C + \alpha_5 (B^2 + BC + \tilde{N}^2 + \tilde{N}C))$$

$$= N(\alpha_1 B^2 + \alpha_2 B\tilde{N} + \alpha_3 \tilde{N}^2 + \alpha_4 \tilde{N}C + \alpha_5 (B + C + \frac{\tilde{N}^2}{B} + \frac{\tilde{N}C}{B})) \qquad (2)$$

where $\alpha_n (1 \leq n \leq 5)$ are the factors that need to be determined using historical statistics. Then, the Reduce execution time estimation transforms to a multi-parameter regression problem. In this paper, OS-ELM is used as the regression model to solve the problem.

2.2 Execution Plan Generation

2.2.1 Map Execution Plan Generation

The execution plan generation algorithm of Map phase is shown in Algorithm 1. It first calculates the predictable average execution time of Map tasks (lines 1–3). The models whose estimated Map execution time is less than average time are treated as unit executions during the execution plan generation (lines 5–6). The models whose estimated Map execution time is more than average time are split to multiple unit executions (lines 7–11). After generating the list of unit executions, heuristic algorithm GeneratePlan is executed to generate Map execution plan.

The GeneratePlan algorithm is shown in Algorithm 2, which is used in both Map execution plan generation and Reduce execution plan generation. When the number of unit executions in the list is less than that of tasks, each of the unit execution is assigned to each task (lines 1–3). Otherwise, greedy strategy is used to generate execution plan. *Unassigned* is initialized and used to count the number of unassigned unit executions in the list (line 5). First, the list of unit executions is sorted by

---

Algorithm 1: Map execution plan generation

Input: *models* [ ] : array of OS-ELM models.

           MapNum : the maximum number of Map tasks in the cluster.

Result: *MapPlan* < *List* < *ID*, *start*, *end* >>[ ]: the Map execution plan.

1 **for** *m = 1 to sizeof(models)* **do**

2     $TimeSum = TimeSum + model[i].EstimatedMapTime$;

3 $AvgTime = \dfrac{TimeSum}{MapNum}$;

4 **for** *m= 1 to sizeof(models)* **do**

5   **if** $models[m].EstimatedMapTime \leq AvgTime * \alpha$ **then**

6   list.add(< *models[m].id*, 0, *models[m].InputSize*,
        models[m].EstimatedMapTime >);

7   **else**

8        $splits = \dfrac{models[m].EstimatedMapTime}{AvgTime}$;

9        $splitsize = \dfrac{models[m].InputSize}{splits}$;

10   **for** *i = 1 to splits* **do**

11   list.add(< *models[m].id*, *i* ∗ *splitsize*, (*i* + 1) ∗ *splitsize*,
          $\dfrac{models[m].EstimatedMapTime}{splits}$ >);

12 *MapPlan* = **GeneratePlan**(*list.toArray()*, *MapNum* , *AvgTime*);

---

estimated execution time in descending order (line 6). Then the sorted list is scanned
and the unit executions in it are added to the execution plan (lines 7–16). The assigned
unit executions are skipped (lines 8–9) and the loop is broken when *Count* exceeds
the number of tasks (lines 10–11). After that, the unassigned unit execution which
has the longest execution time is added to execution plan (line 12) and the algorithm
scans the remaining list to find the suitable unit execution and add it to execution plan
recursively (lines 14–15). Finally, the algorithm scans the list of unit executions again
and adds the unassigned unit execution to the expected shortest task (lines 17–20).

2.2.2 Reduce Execution Plan Generation

The execution plan generation algorithm of Reduce phase is shown in Algorithm 3.
The algorithm first calculates the expected average execution time of Reduce tasks
(lines 1–3). Then it scans the OS-ELM models and adds them to the list of unit
executions (lines 4–5). As the calculations of POS-ELM algorithm in Reduce phase
is indivisible, each Reduce task is treated as a unit execution. Since *start* and *end*
are not used in Reduce execution plan generation, they are set to 0 to be compatible
with GeneratePlan algorithm. After that, GeneratePlan introduced in Map execution
plan generation is executed to generate Reduce execution plan (line 6). At last, the
algorithm scans the execution plan and assigns the OS-ELM models in the plan with

---

Algorithm 2: GeneratePlan()

Input: $Tasks < ID, start, end, time > [\,]$: array of quadruples, in which each quadruple represents task information of OS-ELM model.
    $TaskNum$: the maximum number of tasks that the cluster can hold.
    $AvgTime$: the expected average execution time for each task.
Result: $Plan < List < ID, start, end >>[\,]$: the execution plan.

1 **if** $sizeof(Tasks) \leq TaskNum$ **then**
2 | **for** $m = 1$ to $sizeof(Tasks)$ **do**
3 | | $Plan[m].add(< Tasks[m].id, Tasks[m].start, Tasks[m].end >)$;

4 **else**
5 | $Unassigned = Size = sizeof(Tasks)$;
6 | SortByTimeInDescendingOrder($Tasks$);
7 | **for** $i = 1$ to $Size$ **do**
8 | | **if** $used[i] == true$ **then**
9 | | | continue;
10 | | **if** $Count \geq TaskNum$ **then**
11 | | | break;
12 | | **addToPlan**($Count, i$)
13 | | $Start = i+1$;
14 | | **for** $Start \leq Size$ **do**
15 | | | $Start = $ **FindAndAdd**($Start$);
16 | | $Count = Count+1$;
17 | **for** $i = 1$ to $Size$ **do**
18 | | **if** $Unassigned > 0$ && $used[i] == false$ **then**
19 | | | $insertIndex = $ **findMinTimeIndex**($Time$);
20 | | | **AddToMapPlan**($insertIndex, i$);

21 **FindMinTimeIndex**($Time$)
22 | **for** $i = 1$ to $Size$ **do**
23 | | **if** $Time[i] < MinTime$ **then**
24 | | | $MinTime = Time[i]$;
25 | | | $MinIndex = i$;
26 | **return** $MinIndex$;

27 **FindAndAdd**($Start$)
28 | **for** $j = Start$ to $Size$ **do**
29 | | **if** $used[j] == false$ && $Tasks[j].time + Time[Count] \leq AvgTime * \alpha$ **then**
30 | | | addToMapPlan($Count, j$);
31 | | | **return** j;
32 | **return** j;

33 **AddToPlan**(P_I, T_I)
34 | $used[T\_I] = true$;
35 | $Time[P\_I] = Time[P\_I] + Tasks[T\_I].time$
36 | $Plan[P\_I].add(< Tasks[T\_I].id, Tasks[T\_I].start, Tasks[T\_I].end >)$
37 | $Unassigned = Unassigned-1$;

Algorithm 3: Reduce execution plan generation

Input: *models* [ ] : array of OS-ELM models.
       ReduceNum : the maximum number of Reduce tasks in the cluster.
Result: *ReducePlan < List < ID, start, end >>*[ ]: the Reduce execution plan.
1 **for** *m = 1 to sizeof(models)* **do**
2   TimeSum = TimeSum+model[i].EstimatedReduceTime;
3 $AvgTime = \dfrac{TimeSum}{ReduceNum};$
4 **for** *m = 1 to sizeof(models)* **do**
5   list.add(*< models[m].id, 0, 0, models[m].EstimatedReduceTime >*);

6 *ReducePlan* = GeneratePlan(*list.toArray()*, *ReduceNum, AvgTime*);
7 **for** *i = 1 to sizeof(ReducePlan)* **do**
8   **for** *j = 1 to sizeof (ReducePlan[i])* **do**
9   Index = FindByID(models, ReducePlan [i][j].ID);
10   models [Index].ReduceKey = i ;

11 **FindByID(***list, ID***)**
12   **for** *i = 1 to sizeof(list)* **do**
13   **if** *list[i].ID == ID* **then**
14   **return** *i*;

---

correct *ReduceKey*s (lines 7–10). The *ReduceKey*s are used to mark which Reduce task that intermediate results should pass to.

2.3 Job Execution

Figure 2 shows the execution procedure of BPOS-ELM. Each Map task is responsible for calculating **H** for one OS-ELM model, part of one OS-ELM model or multiple OS-ELM models according to the Map execution plan generated. Each Reduce task

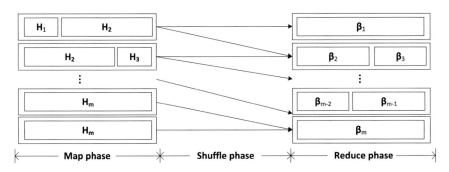

Fig. 2 Job execution of BPOS-ELM

Algorithm 4: BPOS-ELM map()

Input: (Key, Value): Key is the OS-ELM model ID, Value is a sample pair $(x_i, t_i) \in (X_k, T_k)$
where $0 \le i \le |(X_k, T_k)|$ for the model.

Result: m : OS-ELM model ID;

k : blockID;

$ReduceKey_m$: Key that marks which Reduce task trains the model;

$\mathbf{H}_{m,k}$: Output weight;

$\mathbf{T}_{m,k}$: Observation value vector;

1   $m = Key$;

2   add to $block_m$;

3   $count_m + +$;

4   **if** $count \ge BLOCK_m$ **then**

5      $\mathbf{H}_{m,k}$=calcH($block_m$);

6      $\mathbf{T}_{m,k}$=calcT($block_m$);

7   output(($m, k_m, ReduceKey_m$), ($\mathbf{H}_{m,k}, \mathbf{T}_{m,k}$));

8      $count_m = 0$;

9   k_m++;

Algorithm 5: BPOS-ELM reduce()

Input: Set of (*key*, *value*): *key* is a combination of OS-ELM model ID m, blockID k and
ReduceKey. *value* is a vector pair (H_{kb}, T_{kb});

Result: β_m: output weight vector (corresponding to $\beta_{m,k}$).

1   $m = getm(key)$;

2   $\mathbf{H}_{m,k+1} = getH(value)$;

3   $\mathbf{T}_{m,k+1} = getT(value)$;

4   $\mathbf{P}_{m,k+1} = \mathbf{P}_{m,k} - \mathbf{P}_{m,k}\mathbf{H}_{m,k+1}^T(\mathbf{I} + \mathbf{H}_{m,k+1}\mathbf{P}_{m,k}\mathbf{H}_{m,k+1}^T)^{-1}\mathbf{H}_{m,k+1}\mathbf{P}_{m,k}$;

5   $\beta_{m,k+1} = \beta_{m,k} + \mathbf{P}_{m,k+1}\mathbf{H}_{m,k+1}^T(\mathbf{T}_{m,k+1} - \mathbf{H}_{m,k+1}\beta_{m,k})$;

---

is responsible for calculating β for one OS-ELM model or multiple OS-ELM models according to the Reduce execution plan.

The pseudo codes of Map procedure are shown in Algorithm 4. Firs it collects $BLOCK_m$ training instances into a buffer $block_m$ (lines 1–3). After $BLOCK_m$ training instances are collected (line 4), matrix $\mathbf{H}_{m,k}$ is calculated (line 5) and $\mathbf{T}_{m,k}$ is also generated (line 6). After that, a key-value pair is generated as output (line 7). *key* is composed with OS-ELM model ID m, block ID k and $ReduceKey_m$ while *value* includes $\mathbf{H}_{m,k}$ and $\mathbf{T}_{m,k}$. Finally, the counter is cleared (line 8) and the block ID k is increased by one (line 9).

The pseudo codes of Reduce procedure are shown in Algorithm 5. The output results of Map phase which have the same *ReduceKey* are partitioned to the same Reducer and then sorted by m and k. The OS-ELM model ID m is first resolved (line 1). Then $\mathbf{H}_{m,k}$ and $\mathbf{T}_{m,k}$ included in *value* are resolved (lines 2–3). Finally, the $\mathbf{P}_{m,k}$ and $\beta_{m,k}$ are updated according to the formulas (lines 4–5).

3 Experimental Evaluation

3.1 Experimental Setup

In this section, POS-ELM indicates parallel online sequential learning machine algorithm in [10] that trains each OS-ELM model one by one. BPOS-ELM is compared with POS-ELM and OS-ELM algorithms. All the three algorithms are implemented in Java 1.6. Universal java matrix package (UJMP) [2] with version 0.2.5 is used for matrix storage and processing. The activation function of OS-ELM, POS-ELM and BPOS-ELM algorithm is $g(x) = \frac{1}{1+e^{-x}}$. The number of hidden layer node is set to 128 in accuracy evaluation and it is set to 64 in training speed evaluation and scalability evaluation.

Hadoop-0.20.2-cdh3u3 is used as our evaluation platform. The Hadoop cluster is deployed on 9 commodity PCs in a high speed Gigabit network, with one PC as the Master node and the others as the Slave nodes. Each PC has an Intel Quad Core 2.66 GHZ CPU, 4 GB memory and CentOS Linux 5.6 operating system. Each PC is set to hold maximum 4 Map or Reduce tasks running in parallel and the cluster is set to hold maximum 32 tasks running in parallel. Each task is configured with 1024M java heap. Other parameters are using the default values of Hadoop.

BPOS-ELM algorithm is evaluated with real data and synthetic data. The real data sets (MNIST,[1] DNA (see footnote 1) and KDDCup99[2]) are mainly used to evaluate training accuracy and testing accuracy. Some attributes of KDDCup99 data set are symbolic-valued attributes which cannot be directly used for BPOS-ELM, POS-ELM or OS-ELM, so we preprocess the data set by mapping symbolic-valued attributes to numeric-valued attributes with the method in [11]. For testing data, we use the KDDcup99 (corrected) evaluation data set by excluding those attack instances which do not belong to the set of attack types in the training data set. The specifications of real data are shown in Table 2.

The synthetic data sets are used for training speed evaluation and scalability evaluation, which are generated by extending based on Flower.[3] The volume and attributes of training data are extended by duplicating the original data in a round-robin way. In one training model group, there are 11 OS-ELM models training with synthetic data sets with N varies from $2^0 \times 10^4$ to $2^{10} \times 10^4$. The parameters used in scalability evaluation are summarized in Table 3. In the experiments, all the parameters use default values unless otherwise specified.

---

[1] Downloaded from http://www.csie.ntu.edu.tw/~cjlin/libsvmtools/datasets/.

[2] Downloaded from http://kdd.ics.uci.edu/databases/kddcup99/kddcup99.html.

[3] Downloaded from http://www.datatang.com/data/13152.

Table 2 Specifications of real data

Data set	#attributes	#class	#training data	#testing data	Size of test data (KB)
MNIST	780	10	60000	10000	176001.138
DNA	180	3	2000	1186	1126.301
KDDCup99	41	2	4898431	292300	708197.916

Table 3 Specifications of synthetic data and running parameters for scalability evaluation

Parameter	Value range	Default value
#training data	10k, 20k, 40k, 80k, 160k,320k, 640k, 1280k, 2560k, 5120k, 10240k	640k
#attributes	64, 128, 256, 512	64
#cores	1, 2, 4, 8, 16, 32	32
#data per block	1, 2, 4, 8, 16, 32, 64, 128, 256, 512, 1024	64
#neurons	1, 2, 4, 8, 16, 32, 64, 128, 256, 512, 1024	64
#classifications	1, 2, 4, 8, 16, 32, 64, 128, 256, 512, 1024	2
#model groups	3, 6, 9, 12, 15, 18, 21, 24, 27, 30	6

3.2   Evaluation Results

3.2.1   Accuracy Evaluation

Table 4 shows the results of accuracy evaluations with real data. It can be found that the training accuracy and testing accuracy of BPOS-ELM algorithm are at the same level with those of POS-ELM and OS-ELM. The reason for this is that BPOS-ELM algorithm does not change the computation sequence of matrices calculation of OS-ELM.

3.2.2   Training Speed Evaluation

Table 5 shows the results of training speed evaluation with real data and synthetic data. As shown in Table 5, the training speed of BPOS-ELM is faster than the training speed of POS-ELM and OS-ELM. For the models training with real data sets, the training speed of BPOS-ELM is only a little faster than that of POS-ELM. The reason for this is that most of the cores are idle in Reduce phase of BPOS-ELM since the number of the training models is less than that of cores and the Reduce tasks are indivisible. For the models training with synthetic data sets, the training speed of BPOS-ELM is much faster than that of POS-ELM and OS-ELM. This is because the cores of the cluster are fully utilized in the Reduce phase of BPOS-ELM algorithm.

Table 4 Accuracy evaluation with real data

Data set	Algorithm	Training accuracy	Testing accuracy
MNIST	OS-ELM	0.824	0.831
	POS-ELM	0.823	0.830
	BPOS-ELM	0.825	0.831
DNA	OS-ELM	0.845	0.779
	POS-ELM	0.846	0.781
	BPOS-ELM	0.844	0.780
KDDCup99	OS-ELM	0.992	0.856
	POS-ELM	0.991	0.856
	BPOS-ELM	0.992	0.855

Table 5 Execution time evaluation with real data and synthetic data

Data set	Algorithm	Training time (s)
Real data	OS-ELM	1146
	POS-ELM	1025
	BPOS-ELM	1021
Synthetic data	OS-ELM	20701
	POS-ELM	10651
	BPOS-ELM	2130

This result also shows that BPOS-ELM trains large scale multiple OS-ELM models efficiently.

3.2.3 Scalability Evaluation

Figure 3a shows the scalability (speedup) of BPOS-ELM compared with that of POS-ELM. The speedup of BPOS-ELM reaches 10 when the number of cores increases to 32. It means that BPOS-ELM has good scalability. It benefits from accurate estimations of Map and Reduce execution time and the execution plan which is suitable for parallel processing. It can also be found that the speedup of BPOS-ELM reaches 10 whereas the speedup of POS-ELM only reaches 1.96. The reason is that BPOS-ELM calculates $\beta_{m,k}$ for different models in parallel instead of calculating them sequentially.

There are several reasons for the changing trend of speedup decreases as the number of cores increases. First, since the Reduce tasks cannot be further split into smaller ones, the execution time of the OS-ELM model which has the longest Reduce execution time does not decrease as the number of cores increases. In this case, the MapReduce job has to wait for the completion of the slowest task. Second, the cost of

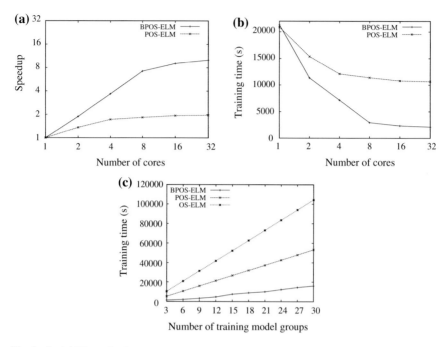

Fig. 3 Scalability evaluation

scheduling tasks among multiple cores increases as the number of cores increases. Third, the memory and the number of I/Os become bottlenecks as the number of cores increases since all the Map tasks and Reduce tasks running on a physical machine share the same memory and disks.

Figure 3b shows the training time of BPOS-ELM compared with that of POS-ELM. The training time of BPOS-ELM is a little longer than that of POS-ELM on one core due to the overhead derived from task scheduling. However, as the number of cores increases, the training time drops significantly and becomes much shorter than that of POS-ELM. It means that BPOS-ELM is more efficient than POS-ELM for training multiple models for the reason that BPOS-ELM trains multiple models in parallel in Reduce phase.

Figure 3c shows the training time of BPOS-ELM with different number of model groups compared with that of POS-ELM and OS-ELM. As shown in Fig. 3c, the training time increases much more slowly than that of POS-ELM and OS-ELM. The reason for this is that BPOS-ELM trains multiple OS-ELM models in parallel in both Map phase and Reduce phase whereas POS-ELM only parallelizes the training in Map phase and OS-ELM does not parallelize the training. It means that BPOS-ELM utilizes computing resources efficiently.

4 Conclusions

In this paper, a batch parallel online sequential extreme learning machine (BPOS-ELM) algorithm has been proposed for large scale batch learning. It estimates the execution time of Map and Reduce tasks with historical statistics, and generates a Map execution plan and a Reduce execution plan with greedy strategy based on the estimation. It launches a MapReduce job to train multiple OS-ELM models. The algorithm also collects information of selected tasks in the job and merges it to historical statistics to help to improve the estimation accuracy. BPOS-ELM algorithm is evaluated with real and synthetic data. The experimental results show that the accuracy of BPOS-ELM is at the same level as those of POS-ELM and OS-ELM. The speedup of BPOS-ELM reaches 10 on a cluster with maximum 32 cores. Compared with OS-ELM and POS-ELM, BPOS-ELM trains multiple OS-ELM models more efficiently.

Acknowledgments This research was partially supported by the National Natural Science Foundation of China under Grant nos. 61173030, 61272181, 61272182, 61173029, 61332014; and the National Basic Research Program of China under Grant no. 2011CB302200-G.

References

1. Amazon elastic compute cloud (2015). http://aws.amazon.com/cn/ec2/
2. Arndt, H., Bundschus, M., Naegele, A.: Towards a next-generation matrix library for java. In: Computer Software and Applications Conference, 2009. COMPSAC'09. 33rd Annual IEEE International. vol. 1, pp. 460–467. IEEE (2009)
3. Dean, J., Ghemawat, S.: Mapreduce: simplified data processing on large clusters. Commun. ACM **51**(1), 107–113 (2008)
4. He, Q., Shang, T., Zhuang, F., Shi, Z.: Parallel extreme learning machine for regression based on mapreduce. Neurocomputing **102**, 52–58 (2013)
5. Huang, G.B., Chen, L.: Convex incremental extreme learning machine. Neurocomputing **70**, 3056–3062 (2007)
6. Huang, G.B., Zhou, H., Ding, X., Zhang, R.: Extreme learning machine for regression and multiclass classification. IEEE Trans. Syst. Man Cybern. Part B: Cybern. **42**(2), 513–529 (2012)
7. Huang, G.B., Zhu, Q.Y., Siew, C.K.: Extreme learning machine. In: Technical Report ICIS/03/2004. School of Electrical and Electronic Engineering, Nanyang Technological University, Singapore (2004)
8. Liang, N.Y., Huang, G.B., Saratchandran, P., Sundararajan, N.: A fast and accurate online sequential learning algorithm for feedforward networks. IEEE Trans. Neural Netw. **17**(6), 1411–1423 (2006)
9. Shepard, D.: A two-dimensional interpolation function for irregularly-spaced data. In: Proceedings of the 1968 23rd ACM National Conference, pp. 517–524. ACM'68 (1968)
10. Wang, B., Huang, S., Qiu, J., Liu, Y., Wang, G.: Parallel online sequential extreme learning machine based on mapreduce. Neurocomputing **149**(Part A), 224–232 (2015)
11. Xiang, J., Westerlund, M., Sovilj, D., Pulkkis, G.: Using extreme learning machine for intrusion detection in a big data environment. In: Proceedings of the 2014 Workshop on Artificial Intelligent and Security Workshop, pp. 73–82. AISec'14 (2014)
12. Xin, J., Wang, Z., Chen, C., Ding, L., Wang, G., Zhao, Y.: Elm*: distributed extreme learning machine with mapreduce. World Wide Web pp. 1–16 (2013)

Fixed-Point Evaluation of Extreme Learning Machine for Classification

Yingnan Xu, Jingfei Jiang, Juping Jiang, Zhiqiang Liu and Jinwei Xu

Abstract With growth of data sets, the efficiency of Extreme Learning Machine (ELM) model combined with accustomed hardware implementation such as Field-programmable gate array (FPGA) became attractive for many real-time learning tasks. In order to reduce resource occupation in eventual trained model on FPGA, it is more efficient to store fixed-point data rather than double-floating data in the on-chip RAMs. This paper conducts the fixed-point evaluation of ELM for classification. We converted the ELM algorithm into a fixed-point version by changing the operation type, approximating the complex function and blocking the large-scale matrixes, according to the architecture ELM would be implemented on FPGA. The performance of classification with single bit-width and mixed bit-width were evaluated respectively. Experimental results show that the fixed-point representation used on ELM does work for some application, while the performance could be better if we adopt mixed bit-width.

Keywords Extreme Learning Machine (ELM) · Fixed-point evaluation · Classification

1 Introduction

Due to advances in technology, the size and dimensionality of data sets used in machine learning tasks have grown very large and continue to grow by the day. For this reason, it is important to have efficient computational methods and algorithms that can be applied on very large data sets, such that it is still possible to complete the machine learning tasks in reasonable time [8].

Extreme Learning Machine (ELM) is well known for its computational efficiency, making it well-suited for large data processing. However, it is still worth speeding

Y. Xu (✉) · J. Jiang · J. Jiang · Z. Liu · J. Xu
Science and Technology on Parallel and Distributed Processing Laboratory,
National University of Defense Technology, ChangSha 410073, Hunan, China
e-mail: yingnanpearl@163.com

© Springer International Publishing Switzerland 2016
J. Cao et al. (eds.), *Proceedings of ELM-2015 Volume 1*,
Proceedings in Adaptation, Learning and Optimization 6,
DOI 10.1007/978-3-319-28397-5_3

up its implementation in many real-time learning tasks. Hardware implementation is one of the most popular approaches, and two types of reconfigurable digital hardware have been adopted, i.e., Field-programmable gate array device (FPGA) and complex programmable logic device (CPLD) [3]. As rival application-specific integrated circuit, FPGA can attain performances and logic densities at lower development costs and privilege computational optimization over area optimization, thus many of previous works [6] implementing other machine learning algorithms have applied FPGA. The parameters, such as weights and Bias of hidden nodes, are stored in on-chip RAM during processing and are swapped out to off-chip memory after processing. Since it is too expensive to support a large number of floating-point units on chip and store values using the standard double precision floating-point representations in on-chip RAMs, most of these works has adopted fixed-point data. Bit-widths with integral multiple of bytes are convenient to align with other components (such as IP cores and user interfaces) and easier to design. Recent work [2] found that very low precision storage is sufficient not just for running trained networks but also for training them by training the Maxout networks with three distinct storing formats: floating point, fixed point and dynamic fixed point. However, the efficiency of ELM model combined with FPGA adopting fixed bit-width is not very clearly.

Motivated by this gap between workloads and state-of-the-art computing platforms, we evaluate fixed-point implementation of ELM for classification. At first, we present the architecture of FPGA. Then with this in mind, we converted the ELM algorithm into a fixed-point version by changing the operator type, approximating the complex function and blocking the large-scale matrixes. Finally, we evaluated the performance of classification with single bit-width and mixed bit-width respectively.

Experiments are performed on a large data setSatImage. Results of the experiments show it does work for some application to use the fixed-point representation on ELM, however, considering resource occupation, the performance of implementation adopting single bit-width is not so optimistic, and it could be improved if we adopt mixed bit-width.

The organization of this paper is as follows. Section 2 introduces the algorithm of ELM. Section 3 shows the specific procedure of fixed-point conversion for ELM including changing the operator type, approximating the complex function and blocking the large-scale matrixes. Section 4 presents our experiment and the results of simulations adopting single bit-width and mixed bit-width respectively. Finally, the results are discussed and an overview of the work in progress is given.

2 Extreme Learning Machine (ELM)

ELM was proposed for generalized single-hidden layer feedforward networks where the hidden layer need not be neuron alike. It offers three main advantages: low training complexity, the minimization of a convex cost that avoids the presence of local minima, and notable representation ability. The output function of ELM for generalized SLFNs is

$$f_L(x) = \sum_{i=1}^{L} \beta_i h_i(x) = h(x)\beta \ , \tag{1}$$

where $\beta = [\beta_1, \dots, \beta_L]^T$, is the output weight vector between the hidden layer of L nodes to the $m1$ output nodes, and $h(x) = [h_1(x), \dots, h_L(x)]$ is ELM nonlinear feature mapping, $h_i(x)$ is the output of the ith hidden node output. In particular, in real applications $h_i(x)$ can be

$$h_i(x) = G(a_i, b_i, x), a_i \in R^d, b_i \in R \ . \tag{2}$$

Basically, ELM trains an SLFN in two main stages: (1) random feature mapping and (2) linear parameters solving. In the first stage, ELM randomly initializes the hidden node parameters (a, b) to map the input data into a feature space by nonlinear piecewise continuous activation function. The most often used activation function is sigmoid function, the formula is

$$G(a, b, x) = \frac{1}{1 + \exp\left(-(ax + b)\right)} \ . \tag{3}$$

In the second stage of ELM learning, the weights connecting the hidden layer and the output layer, denoted by β, are solved by minimizing the approximation error in the squared error sense:

$$\min_{\beta \in R^{L \times m}} \|H\beta - T\|^2 \ , \tag{4}$$

where H is the hidden layer output matrix (randomized matrix):

$$H = \begin{bmatrix} h(x_1) \\ \vdots \\ h(x_N) \end{bmatrix} = \begin{bmatrix} h_1(x_1) & \cdots & h_L(x_1) \\ \vdots & \vdots & \vdots \\ h_1(x_N) & \cdots & h_L(x_N) \end{bmatrix} \ , \tag{5}$$

and T is the training data target matrix:

$$T = \begin{bmatrix} t_1^T \\ \vdots \\ t_N^T \end{bmatrix} = \begin{bmatrix} t_{11} & \cdots & t_{1m} \\ \vdots & \vdots & \vdots \\ t_{N1} & \cdots & t_{Nm} \end{bmatrix} \ , \tag{6}$$

where $\|\cdot\|$ denotes the Frobenius norm.

The optimal solution to (4) is given by

$$\beta^* = H^{\dagger}T \ , \tag{7}$$

where $H^{\dagger}$ denotes the MoorePenrose generalized inverse of matrix H.

A positive value C can be added to the diagonal of $H^T H$ or HH^T of the Moore-Penrose generalized inverse H the resultant solution is more stable and tends to have better generalization performance [4]. Thus

$$\beta^* = (H^T H + 1/C)^{-1} H^T T \,, \tag{8}$$

where I is an identity matrix of dimension L.

Overall, the ELM algorithm is then:

ELM Algorithm: Given a training set $\aleph = \{(x_i, t_i) \,|\, x_i \in R^n, \, t_i \in R^m, \, i = 1, \dots, N\}$, hidden node output function $G(a, b, x)$, and the number of hidden nodes L,

1. generate random hidden nodes (random hidden node parameters) $(a_i, b_i), i = 1, \dots, L$.
2. Calculate the hidden layer output matrix H.
3. Calculate the output weight vector $\beta^* = (H^T H + 1/C)^{-1} H^T T$.

3 Fixed-Point Conversion for ELM

Since FPGAs attain performances and logic densities at lower development costs and the resource consumption can be further reduced with fixed-point data, we attempt to implement the ELM algorithm for classification on FPGA using fixed-point format, the overall FPGA architecture of ELM algorithm for classification is shown in Fig. 1. According to previous works [7], the QR decomposition adopts float-point format while the multiplication of matrix adopts fixed-point format. In this work, we simulate the behaviour of FPGA adopting fixed bit-width on Matlab environment. Figure 2 shows the execution flow.

In the beginning of simulation, we adjust the radix point position shown in Fig. 3 [6] according to the corresponding range of data. For example, since the range of InputWeight matrix shown in Fig. 2 is approximately $[-1, 1]$, the optimal bit-width of integer part is 1 and cannot be allowed any further reduction or increment. This way can make the precision better when considering a fixed-point representation for real numbers, the integer part of a number mainly influences the representation scope while the fractional part mainly decides the precision.

In the procedure of fixed-point conversion, we choose Piecewise Linear Approximation of nonlinearity algorithm (PLA) [1] as the method of sigmoid function approximation, this method uses linear functions and can be implemented on FPGA easily. PLA has uniform structures like Table 1. For the main operations like matrix multiplication in ELM, parallel multiply-accumulators are often used on FPGA. The operands are stored on distributed block RAM, which bit-width is n bits. A 2n bits partial product can be produced by the n bits multiplier. An accumulator with larger bit-width can be used to accumulate the partial product, avoiding the precision lost and not increasing much logic cost at the same time. So, we often chose a bit-width

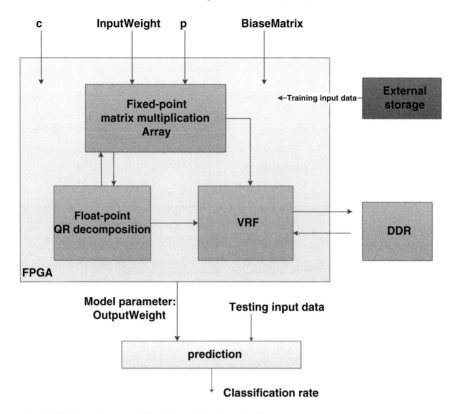

Fig. 1 FPGA architecture of ELM algorithm for classification

in the range of n bits to 2n bits for the adder and the accumulator. Only the bit-width of the final result which needs to store back to on-chip RAM is constrained to n bits. The partition of the integer part and fractional part for the result depends on the representation range of the data, which must be studied when converting to the fixed-point hardware.

Under the implementation assumption above, it is more reasonable that maintaining the precision of a block matrix multiplication instead of converting the partial product for each element. Assuming that we can chose enough wide bit-width for the accumulation operation, thus we only need to cut down the bit-width to n bits for the result of a block multiplication when simulating the fixed-point operations. From this observation, we converted all matrix operations in ELM to a loop code of block matrix operations and converted each element of the block result to a fixed-point format. The size of block matrix is set 64. The flow diagram of computation is shown in Fig. 4.

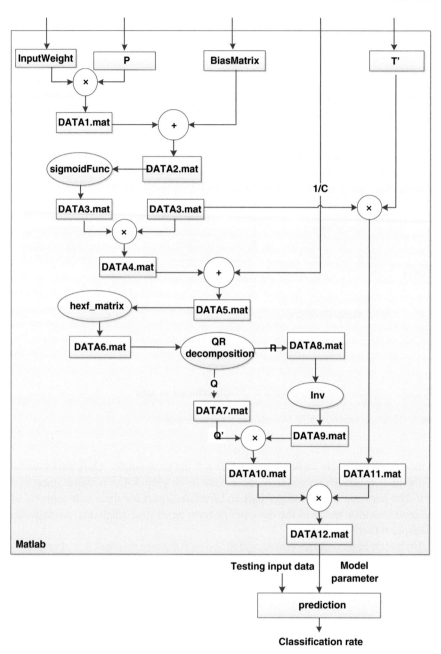

Fig. 2 Execution flow of ELM for classification

Fig. 3 Fixed-point data format

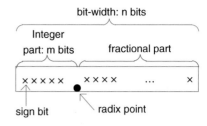

Table 1 Piecewise linear approximation algorithm

x	y
$0 \leq \lvert x \rvert < 1$	$y = (\lvert x \rvert + 2)/4$
$1 \leq \lvert x \rvert < 19/8$	$y = (\lvert x \rvert + 5)/8$
$19/8 \leq \lvert x \rvert < 5$	$y = (\lvert x \rvert + 27)/32$
$\lvert x \rvert \geq 5$	$y = 1$
$x < 0$	$y = 1 - y$

Fig. 4 Flow diagram of block matrix multiplication

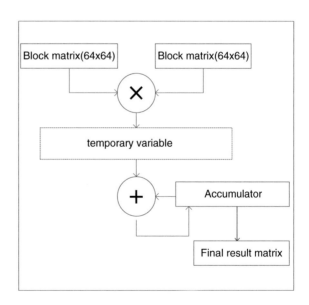

4 Experiment and Results

All the simulations are conducted in MATLAB R2009a environment on an ordinary PC with Intel(R) Core(TM) i3-2120 and 4 GB RAM.

The SatImage dataset with 36 input attributes and 6 class label is chosen as experimental Dataset. It can be downloaded from the official ELM website with pre-scaled values [6]. 3217 instances are used as training data and the rest 3218 of the instances are used for testing.

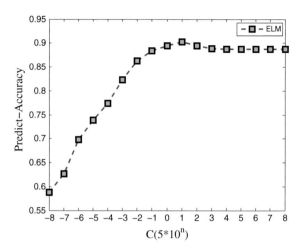

Fig. 5 Classification accuracy with different C

In ELM, the number of output nodes is equal to the number of classes. For the SatImage data set used in this paper, there are 6 classes and, thus, ELM has 6 output nodes. The activation function used in ELM in our experiment is sigmoid function.

In the implementation of ELM, it is found that the generalization performance of ELM is not sensitive to the dimensionality of the feature space (L) and good performance can be reached as long as L is large enough. In our simulations, $L = 1000$ is set for all tested cases no matter whatever size of the training data sets. And since training data sets are very large N ≫ L, we apply solutions (11) in Sect. 2 to reduce computational costs [5].

The hidden node parameters a_i and b_i are not only independent of the training data but also of each other. Unlike conventional learning methods which MUST see the training data before generating the hidden node parameters, ELM could generate the hidden node parameters before seeing the training data. Thus, a set of random values are produced to be applied in all of our experiment.

And the value C can affect the performance to a large extent. In our experiment, we first trained the ELM classification problem with different C in float-point algorithm. And the Fig. 5 shows the classification accuracy with different value C. It can be seen that the accuracy with C being set 50 is much better than other chosen value. And in order to make C expressed in fixed-point algorithm more accurate, we ignored the possibility that the absolute value of C can be set too large. So, in the following experiment, we applied the fixed value with $C = 50$.

4.1 Single Bit-Width

The Fig. 6 shows the classification accuracy applying different fixed bit-width. It can be seen that the accuracies with the bit-width set 16 bits, 24 bits and 32 bits are all bad, however the performance in the bit-width of 16 bits is better than 24

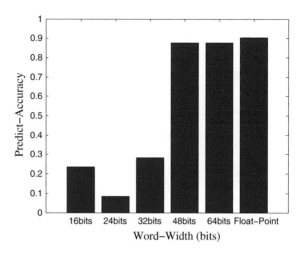

Fig. 6 Prediction accuracy with different bit-widths

bits, it means that the representation domain constraint throws away the redundant and useless information of the high-dimensional input data. The performance in the bit-widths of 48 bits and 64 bits indicate that fixed-point representation used on ELM does work for some application. In order to balance classification accuracy and resource occupation in the eventual trained model, we can only choose 48 bits as the optimal bit-width on FPGAs if we adopt single bit-width. It is obvious that the result is not optimistic.

To solve this problem, we analyzed the result of each operation and tracked the source where error comes from. In this subsection, we computed the Forbenius Norms (FN)

$$\|A - B\|_F = \sqrt{\sum_{i=1}^{n} \sum_{j=1}^{n} \left| a_{ij} - b_{ij} \right|^2} \tag{9}$$

of error matrixes which can weigh the degree of error, the error matrixes are subtraction between float-point and fixed-point data from the output generated by each execution stage shown in Fig. 2, the result of computation is presented in Fig. 7. It can be seen that the error mainly begins with the operation of large scale matrix multiplication generating DATA4.mat and is propagated in latter operations. Because of the linear nature of the operations and the dynamic range compression of the sigmoid generating DATA7.mat, quantization errors tend to propagate sub-linearly and not cause numerical instability [9].

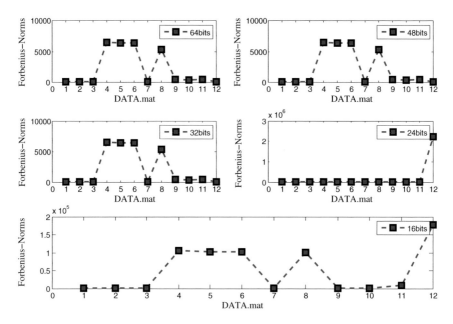

Fig. 7 FN with different bit-widths

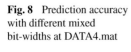

Fig. 8 Prediction accuracy
with different mixed
bit-widths at DATA4.mat

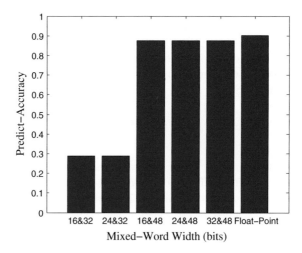

4.2 Mixed Bit-Widths

In order to improve the performance, we re-trained the ELM by adopting mixed bit-widths which can change bit-width at a special point. The prediction accuracy of training with mixed bit-widths applied at the point computing DATA4.mat is shown in Fig. 8, it can be seen that we can also get attractive result even we adopt mixed

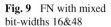

Fig. 9 FN with mixed bit-widths 16&48

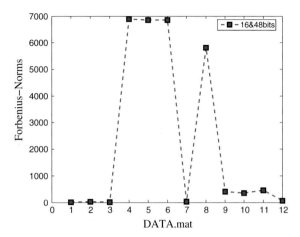

bit-widths which can decrease the occupation of memory resource. According to the FN of the optimal mixed bit-widths (16&48) shown in Fig. 9, the propagated error produced by bit-width of 16 bits can be improved through changing the bit-width into 48 bits and would not affect the performance.

5 Conclusion

This research has tackled the fixed-point evaluation of ELM for classification. We conduct the conversion of fixed-point for ELM and then make simulations on Matlab. Experimental results show that the resource occupation of implementation adopting single bit-width is too large, and the performance could be improved if we adopt mixed bit-width. Our results can act as a guide to inform the design choices on bit-widths when implementing ELM in FPGA documenting clearly the trade-off in accuracy. However, the use of mixed bit-width makes the computing resource rise, we need to conduct the further evaluation of resource occupation and then implement the ELM for classification on FPGA with the parameter discussed in this work.

Acknowledgments This work is funded by National Science Foundation of China(number 61303070).

References

1. Amin, H., Curtis, K., Hayes Gill, B.: Piecewise linear approximation applied to nonlinear function of a neural network. IEE Proc. Circuits Devices Syst. **144**(6), 313–317 (1997)
2. Courbariaux, M. Bengio, Y. David, J.P.: Low precision storage for deep learning. C. Eprint Arxiv (2014)

3. Decherchi, S., Gastaldo, P., Leoncini, A., et al.: Efficient digital implementation of extreme learning machines for classification. IEEE Trans. J. Circuits Syst. II: Express. Briefs **59**(8), 496–500 (2012)
4. Hoerl, A.E., Kennard, R.W.: Ridge regression: biased estimation for nonorthogonal problems. J. Technometrics **12**(1), 55–67 (2012)
5. Huang, G.B., Zhou, H., Ding, X., et al.: Identification of common molecular subsequences. IEEE Trans. J. Syst. Man Cybern. Part B: Cybern. **42**(2), 513–529 (2012)
6. Jiang, J., Hu, R., Zhang, F., et al.: Experimental demonstration of the fixed-point sparse coding performance. J. Cybern. Inform. Technol. **14**(5), 40–50 (2014)
7. Jie, Z.: Fine-grained algorithm and architecture for data processing in SAR applications (in Chinese). University of Defense and Technology (2010)
8. van Heeswijk, M., Miche, Y., Oja, E., et al.: GPU-accelerated and parallelized ELM ensembles for large-scale regression. J. Neurocomput. **74**(16), 2430–2437 (2011)
9. Vanhoucke, V., Senior, A., Mao, M.Z.: Improving the speed of neural networks on CPUs. In: Proceedings of Deep Learning and Unsupervised Feature Learning Workshop Nips (2011)

Multi-layer Online Sequential Extreme Learning Machine for Image Classification

Bilal Mirza, Stanley Kok and Fei Dong

Abstract In this paper, a multi-layer online sequential extreme learning machine (ML-OSELM) is proposed for image classification. ML-OSELM is an online sequential version of a recently proposed multi-layer extreme learning machine (ML-ELM) method for batch learning. Existing ELM-based sequential learning methods, such as state-of-the-art online sequential extreme learning machine (OS-ELM), were proposed only for single-hidden-layer networks. A distinctive feature of the new method is that it can sequentially train a multi-hidden-layer ELM network. Auto-encoders are used to perform layer-by-layer unsupervised sequential learning in ML-OSELM. We used four image classification datasets in our experiments and ML-OSELM performs better than the OS-ELM method on all of them.

Keywords Deep learning · Extreme learning machine · Feature learning · Image classification · Sequential learning

1 Introduction

Deep learning has attracted much attention in the past decade or so due to its successes in various research domains such as pattern recognition, computer vision and automatic speech recognition [1, 2]. Neural-network-based deep architectures or deep neural networks, convolutional neural networks, and deep belief networks are some of the commonly used deep learning architectures. Multiple layers in deep architectures provide multiple non-linear transformations of the original raw

B. Mirza (✉) · S. Kok · F. Dong
Information Systems Technology & Design Pillar, Singapore University
of Technology & Design, Singapore, Singapore
e-mail: bilal2@e.ntu.edu.sg

S. Kok
e-mail: stanleykok@sutd.edu.sg

F. Dong
e-mail: fei_dong@mymail.sutd.edu.sg

© Springer International Publishing Switzerland 2016
J. Cao et al. (eds.), *Proceedings of ELM-2015 Volume 1*,
Proceedings in Adaptation, Learning and Optimization 6,
DOI 10.1007/978-3-319-28397-5_4

input data for better representation learning. Deep networks can capture higher level abstractions which most single layer networks are unable to achieve. Every layer in deep networks may learn a different representation of the input by performing feature or representation learning using efficient unsupervised or semi-supervised learning algorithms.

Restricted Boltzmann machine (RBM) [3] and auto-encoders (AE) [4] have been successfully applied for feature learning in deep networks. Examples of RBM-based deep networks include deep belief networks (DBNs) [3] and deep Boltzmann machines (DBMs) [5] while examples of auto-encoder-based deep networks are stacked auto-encoders (SAEs) [4] and stacked denoising auto-encoders (SDAEs) [5]. Multiple RBMs are stacked to create DBN and DBM networks, whereas multiple AEs are stacked to create SAE and SDAE networks. These methods have outperformed support vector machines, single-hidden-layer feed-forward neural networks and traditional multi-layer neural networks on image classification, automatic speech recognition, and other tasks. However, the time taken for learning deep networks on these big dataset applications is generally long. Recently, multi-layer extreme learning machine (ML-ELM) [6], which is based on extreme learning machine theory [7, 8], has been proposed as a computationally efficient alternative to existing state-of-the-art deep networks. ML-ELM learns significantly faster than existing deep networks, and obtains better or similar generalization performance to DBNs, SAEs, SDAEs and DBMs. In ML-ELM, extreme learning machine auto-encoders (ELM-AEs) are used for unsupervised layer-by-layer feature learning in the hidden layers.

The training of ELM-AEs is similar to that of regular ELMs except that the output in ELM-AEs is the same as the input. The extreme learning machine (ELM) algorithm [7, 8] is becoming popular in large datasets and online learning applications due to its fast learning speed. ELM provides a single step least square estimation (LSE) method for training single-hidden-layer feedforward network (SLFN) instead of using iterative gradient descent methods such as backpropagation.

ML-ELM has been proposed for image classification and pattern recognition applications, and the dimensions of the corresponding datasets are generally very high. When constrained with limited memory, both single layer ELM and ML-ELM are not suitable for big datasets. Therefore, batch learning or learning with complete datasets in one step becomes challenging. Also, graphical processing units (GPUs), widely used in scientific computing, generally have limited memory that cannot accommodate big datasets. To overcome this problem, online sequential extreme learning machine (OS-ELM) has been proposed for sequential learning from big data streams [9]. OS-ELM has become one of the standard algorithms for sequential learning due to its fast learning speed. Sequential or incremental learning algorithms store the previously learned information and update themselves only with a chunk of new data [10–13]. Once the chunk of data has been used for training, it may be discarded. OS-ELM is preferable over batch ELM not only for their reduced computational time on large datasets, but also for their ability to adapt to online learning applications. Note that OS-ELM is an online sequential version of the original single-hidden-layer extreme learning machine.

In this paper, to imbue ML-ELM with the online sequential learning advantages of OS-ELM, we propose a sequential version of ML-ELM, which we term multi-layer online sequential extreme learning machine (ML-OSELM). To our knowledge, we are the first to develop an ELM that is both multi-layered and can learn from data in an online sequential manner. In order to perform layer-by-layer unsupervised sequential learning in ML-OSELM, an online sequential extreme learning machine auto-encoder (OS-ELM-AE) is also proposed. ML-OSELM resembles deep networks since it stacks on top of OS-ELM-AE to create a multi-layer neural network. This paper is organized as follows. Section 2 discusses the preliminaries. Section 3 presents the details of the new multi-layer online sequential learning method. This is followed by experiments for validating the performance of the proposed framework in Sect. 4. Finally, Sect. 5 concludes the paper.

2 Preliminaries

2.1 *Extreme Learning Machine (ELM) and Online Sequential ELM (OS-ELM) Methods*

Extreme learning machine [7] provides a single step least squares error (LSE) estimate solution for a single-hidden-layer feedforward network (SLFN). With ELM there is no need to tune the hidden layer of the SLFN as in traditional gradient-based algorithms. ELM randomly assigns weights and biases in the hidden layer while the output weights connecting the hidden layer and the output layer are determined using the LSE method.

Consider a q class training dataset $\{x_i, y_i\}$, $i = 1, \dots, N$ and $y_i \in R^q$. $x_i \in R^d$ is a d-dimensional data point. The SLFN output with L hidden nodes is given by

$$o_i = \Sigma_{j=1}^{L} \beta_j G(a_j, b_j, x_i), i = 1, \dots, N \tag{1}$$

where a_j and b_j, $j = 1, \dots, L$ are the jth hidden node's weights and biases respectively, and they are assigned randomly, independent of the training data. $\beta_j \in R^q$ is the output weight vector connecting the jth hidden node to the output nodes and $G(x)$ can be any infinitely differentiable activation function such as the sigmoidal function or radial basis function in the hidden layer.

The N equations in (1) can be written in a compact form as below.

$$O = H\beta \tag{2}$$

where $H = \begin{bmatrix} G(a_1, b_1, x_1) & \dots & G(a_L, b_L, x_1) \\ \vdots & \dots & \vdots \\ G(a_1, b_1, x_N) & \dots & G(a_L, b_L, x_N) \end{bmatrix}_{N \times L}$

is called the hidden layer output matrix, H_{ij} represents the jth hidden node output corresponding to the input x_i, $\beta = [\beta_1, \beta_2, \ldots, \beta_L]^T$ and $O = [o_1, o_2, \ldots, o_N]^T$.

In order to find the output weight matrix β which minimizes the cost function $\|O - Y\|$, a LSE solution of (2) is obtained as below

$$\beta = H^\dagger Y = (H^T H)^{-1} H^T Y \tag{3}$$

where $Y = [y_1, y_2, \ldots, y_N]^T$ and $H^\dagger$ is the Moore-Penrose generalized inverse of matrix H. This closed-form single step LSE solution is referred to as an extreme learning machine [7].

Online sequential extreme learning machine (OS-ELM) [9] has been proposed as an incremental version of the batch ELM. OS-ELM achieves better generalization performance than the previous algorithms proposed for SLFN and at a much faster learning speed. It can learn from data one example at a time as well as chunk-by-chunk (with a fixed or varying chunk size). Only the newly arrived samples are used at any given time so the examples that have already been used in the learning procedure can be discarded. The learning in OS-ELM consists of two phases.

Step 1: Initialization
A small portion of training data $n_0 = \{x_i, y_i\}$, $i = 1, \ldots, N_0$ with $N_0 \in N$ is considered for initializing the network. The initial output weight matrix is calculated according to the ELM algorithm by randomly assigning weights a_j and bias b_j, $j = 1, \ldots, L$ as follows

$$\beta^{(0)} = P_0 H_0^T Y_0 \tag{4}$$

where $P_0 = (H_0^T H_0)^{-1}$ and H_0 is the initial hidden layer output matrix.

It is recommended that the number of initial training samples should be greater than or equal to the number of hidden neurons. With this setting the generalization performance of online sequential ELM reaches that of the batch ELM.

Step 2: Sequential Learning
Upon the arrival of a new set of observations $n_{k+1} = \{x_i, y_i\}$, $i = (\Sigma_{l=0}^{k} N_l) + 1, \ldots,$ $\Sigma_{l=0}^{k+1} N_l$, i.e., the $(k + 1)$th chunk of data, we first compute the partial hidden layer output matrix H_{k+1}. N_{k+1} is the number of samples in the $(k + 1)$th chunk. Then by using the output weight update equation shown below, we calculate the output weight matrix β^{k+1} with $Y_{k+1} = [y_{(\Sigma_{l=0}^{k} N_l)+1}, \ldots, y_{\Sigma_{l=0}^{k+1} N_l}]^T$.

$$P_{k+1} = P_k - P_k H_{k+1}^T (I + H_{k+1} P_k H_{k+1}^T)^{-1} H_{k+1} P_k$$
$$\beta^{(k+1)} = \beta^{(k)} + P_{k+1} H_{k+1}^T (Y_{k+1} - H_{k+1} \beta^{(k)}) \tag{5}$$

Each time a new chunk of data arrives, the output weight matrix is updated according to (5). Note the one-by-one learning can be considered a special case of chunk-by-chunk learning when the chunk size is set to 1 and the matrices in (5) become vectors.

2.2 Multi-layer Extreme Learning Machine (ML-ELM)

In multi-layer neural networks (ML-NN), the hidden layer weights are initialized by layer-by-layer unsupervised learning and then the whole network is fine-tuned using backpropagation. ML-NN performs better with layer-by-layer unsupervised learning as compared to only using backpropagation. However, fine-tuning is avoided in such deep networks using the recently proposed multi-layer extreme learning machine (ML-ELM) method.

ML-ELM hidden layer weights are initialized randomly using extreme learning machine auto-encoders (ELM-AEs). ELM-AE performs layer-by-layer unsupervised learning. ELM-AE is trained differently from ELM in that the output is set to be equal to the input, *i.e.*, $Y = X$ in (3) (see Fig. 1), and hidden layer weights and biases are chosen to be orthogonal to the random weights in ELM. Orthogonalization of these weights tends to result in better generalization performance. Note that ELM-AE output weights are obtained analytically unlike RBMs and other auto-encoders, which require iterative algorithms.

ELM-AE's main objective is to transform features from input data space to lower or higher dimensional feature space. Since ELM is a universal approximator [8], ELM-AE is also a universal approximator.

Once the hidden layer weights are learnt using ELM-AE, the output weights connecting the last hidden layer to the output layer of ML-ELM are determined analytically using (3).

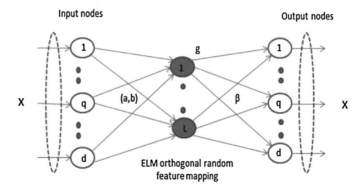

Fig. 1 ELM-AE uses the same architecture as original ELM with the exception that the target output 'x' is the same as input. Here 'g' is the activation function, (a, b) represents random weights and biases, and β represents output weights

3 Proposed Method

In this section, we propose an online sequential version of the multi-layer extreme learning machine, which we term multi-layer online sequential extreme learning machine (ML-OSELM). In Sect. 3.1, we propose an online sequential extreme learning machine auto-encoder (OS-ELM-AE) for feature or representation learning from sequential data streams. In Sect. 3.2, the proposed OS-ELM-AE is used to perform layer-by-layer unsupervised sequential learning in multi-layer extreme learning machines. The deep sequential learning in ML-OSELM is performed by stacking several OS-ELM-AEs. Note that the network architectures of OS-ELM-AE and ML-OSELM are identical to ELM-AE (Fig. 1) and ML-ELM (Fig. 2) respectively. In the following, we will discuss how these architectures are used for sequential learning.

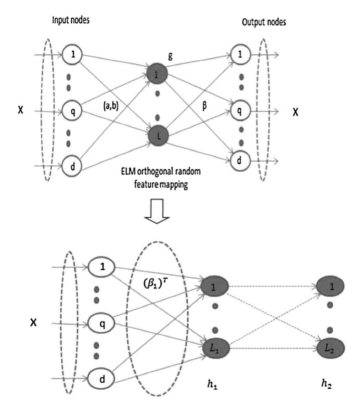

Fig. 2 The weights of each hidden layer in ML-ELM are determined using ELM-AE. $(\beta_1)^T$ is responsible for feature learning at the first hidden layer of ML-ELM

3.1 Online Sequential Extreme Learning Machine Auto-Encoder (OS-ELM-AE)

OS-ELM-AE is a special case of OS-ELM where output is the same as the input at every time step. The hidden layer weights are randomly generated, as in OS-ELM, but in OS-ELM-AE orthogonal of random weights and biases are used. Learning in OS-ELM-AE is done in two phases as described below.

Step 1: Initialization

The OS-ELM-AE is initialized with a portion of training data $n_0 = \{x_i, y_i\}$, $i = 1, \ldots, N_0$ with $N_0 \in N$. The initial output weights $\beta^{(0)}$ of OS-ELM-AE is given as

$$\beta^{(0)} = P_0 H_0^T X_0 \qquad (6)$$

where $X_0 = [x_1, x_2, \ldots, x_{N_0}]^T$, $P_0 = (H_0^T H_0)^{-1}$ and H_0 is the hidden layer output matrix.

Step 2: Sequential Learning

With the arrival of a new chunk of training data, the recursive least square equation in (5) is used, but with input set equal to output as below

$$
\begin{aligned}
P_{k+1} &= P_k - P_k H_{k+1}^T (I + H_{k+1} P_k H_{k+1}^T)^{-1} H_{k+1} P_k \\
\beta^{(k+1)} &= \beta^{(k)} + P_{k+1} H_{k+1}^T (X_{k+1} - H_{k+1} \beta^{(k)})
\end{aligned}
\qquad (7)
$$

The output weight β in OS-ELM-AE is responsible of learning the transformation from input space to feature space. Note that (7) is different from (5) since Y_{k+1} is replaced with X_{k+1}.

3.2 Multi-layer Online Sequential Extreme Learning Machine (ML-OSELM)

The OS-ELM-AE method proposed in the previous subsection is now applied to layer-by-layer unsupervised sequential learning in a multi-layer online sequential extreme learning machine. All the hidden layer are initialized using (6) and then sequentially trained with the arrival of new data using (7). The hidden layer output matrix corresponding to hidden layer m at time step k is given as

$$H_k^m = g((\beta_m^{(k)})^T H_k^{m-1})$$

where g can be any activation function that can be used with ELMs [7, 8] and $\beta_m^{(k)}$ is the output weight matrix obtained using OS-ELM-AE for layer m at time step k.

The input or data layer can be considered as the *zero*th hidden layer where $m = 0$. Assuming a total of p hidden layers in the network, the output weight matrix in ML-OSELM connecting the last hidden layer to the output layer is obtained as follows,

Step 1: Initialization

$$\beta^{(0)} = P_0 H_0^T Y_0 \tag{8}$$

where $H_0 = H_0^p = g((\beta_p^{(0)})^T H_0^{p-1})$.

Step 2: Sequential Learning

$$P_{k+1} = P_k - P_k H_{k+1}^T (I + H_{k+1} P_k H_{k+1}^T)^{-1} H_{k+1} P_k$$
$$\beta^{(k+1)} = \beta^{(k)} + P_{k+1} H_{k+1}^T (Y_{k+1} - H_{k+1} \beta^{(k)}) \tag{9}$$

where $H_{k+1} = H_{k+1}^p = g((\beta_p^{(k+1)})^T H_{k+1}^{p-1})$.
The training algorithm of ML-OSELM is summarized in Algorithm 1.

Algorithm 1 Multi-layer online sequential extreme learning machine (ML-OSELM) algorithm

---

Input: $\{x_i, y_i\}, x_i \in R^d$ and $y_i \in R^q, i = 1, \ldots, N$.
Output: $\beta^{(k+1)}$

Initialization:
Initial data set: $n_o = \{x_i, y_i\}, x_i \in R^d$ and $y_i \in R^q, i = 1, \ldots, N_0$ with $N_0 \in N$.

for $m = 1 \to p$ **do**   % number of hidden layers %
   $H_0^m = g((\beta_m^{(0)})^T H_0^{m-1})$
   where $\beta_m^{(0)}$ is obtained using (6)
end for

   $\beta^{(0)} = P_0 H_0^T Y_0$
   where $H_0 = H_0^p$

Sequential Learning:
for $k = 0 \to K$ **do**   % number of time steps %
   for $m = 1$ to p **do**
      $H_{k+1}^m = g((\beta_m^{(k+1)})^T H_{k+1}^{m-1})$
      where $\beta_m^{(k+1)}$ is obtained using (7)
   end for

   $\beta^{(k+1)} = \beta^{(k)} + P_{k+1} H_{k+1}^T (Y_{k+1} - H_{k+1} \beta^{(k)})$
   where $H_{k+1} = H_{k+1}^p$
end for

---

4 Experiments

In this section, we use four image classification datasets for the performance evaluation of ML-OSELM. The first dataset is CIFAR10 [14], the second CALTECH [15], the third Olivetti faces [16] and the last UMIST [17]. CIFAR10 consists of 50,000 training and 10,000 testing samples belonging to 10 different categories. We use a feature extraction pipeline as described in [18, 19]. The pipeline extracts 6×6 pixel patches from the training set images, performs ZCA whitening of those patches, runs K-means for 50 rounds, and then normalizes its dictionary to have zero mean and unit variance. The final feature vector is of dimension $G \times G \times K$, where K is the dictionary size and $G \times G$ represents the size of the pooling grid. We set $K = 400$ and $G = 4$ in our experiments. We compare the results obtained by multi-layer OS-ELM with that of single layer OS-ELM and also ML-ELM (batch mode) under identical settings.

For the CALTECH dataset, we used 5 classes representing faces, butterfly, crocodile, camera and cell phone. For each class, we set aside one third of the images (up to 50) for testing and used the rest for training. The pixel values represent gray scale intensities which are normalized to have zero mean and unit variance. Similar to [20], we used fixed-size square images in all the categories even though the original sizes may vary in different categories. The original images are rescaled so that the longer side is of length 100, and then we used the inner 64×64 portions for our experiments.

Out of 400 samples in Olivetti faces, we used 300 for training and 100 for testing. Each sample is a 64×64 gray scale image. It represents 40 unique people with 10 images each, all frontal and with a slight tilt of the head.

The UMIST database has 565 total images of 20 different people with 19–36 images per person. 400 samples are used for training while 175 are used for testing. Each sample is a 112×92 gray scale image. The subjects differ in race, gender and appearances. The dataset covers various angles from left profile to right profile.

We conducted all the experiments on a high performance computer with Xeon-E7-4870 2.4 GHz processors, 256 GBytes of RAM, and running Matlab 2013b. For consistency, we used a three-hidden-layer (3000-4000-5000) network structure for multi-layer ELM in all the datasets. For OS-ELM, the number of hidden neurons is set to 5000, *i.e.*, the same as the last layer in the multi-layer network. The sizes of each initialization set and chunk of samples at each time step in sequential learning are respectively set to 15000 and 500 for CIFAR10, 200 and 50 for CALTECH, 100 and 50 for Olivetti faces, and 150 and 50 for UMIST faces. Sigmoid is used as the activation function throughout the experiments. The classification accuracy comparison between ML-OSELM and OS-ELM on the four datasets is given in Table 1. The results represent average accuracy over 20 runs with different initialization, training and testing sets.

We use the two-sample t-test to see whether the results obtained by ML-OSELM are statistically significantly better than those obtained by single layer OS-ELM. The null hypothesis is rejected at $\alpha = 0.01$ for UMIST and CIFAR10 with p values of

Table 1 Accuracy comparisons between multi-layer OS-ELM (ML-OSELM) and single-layer OS-ELM

Dataset	ML-OSELM	OS-ELM
UMIST	**98.03**	96.51
Olivetti	**92.81**	90.23
CIFAR10	**64.96**	64.01
CALTECH	**59.22**	59.07

Table 2 Accuracy comparisons between multi-layer OS-ELM (ML-OSELM) and batch multi-layer ML-ELM

Dataset	ML-OSELM	ML-ELM
UMIST	98.03	**98.83**
Olivetti	92.81	**94.91**
CIFAR10	64.96	**66.41**
CALTECH	59.22	**59.57**

0.0054 and 1.35E-5 respectively. For Olivetti faces, the null hypothesis is rejected at $\alpha = 0.05$ with a p value of 0.043. For CALTECH, p value is 0.735 and the null hypothesis is not rejected.

With these statistical significances test, ML-OSELM is found to be superior to OS-ELM on the UMIST, CIFAR10, CALTECH datasets.

The aim of sequential learning methods is to achieve a performance similar to that of batch learning methods when constrained with limited memory. It can be observed from Table 2 that ML-OSELM results are competitive against those obtained by ML-ELM (batch mode).

5 Summary and Future Work

We propose the multi-layer online sequential extreme learning machine (ML-OSELM). Our empirical results show that by using multiple layers, our proposed ML-OSELM outperforms the state-of-the-art single-layer online sequential ELM. Further, our ML-OSELM achieves competitive results against a batch multi-layer ELM that has the advantage of having the full dataset available for training.

As future work, we want to come up with a method to find the optimal number of hidden layers in ML-OSELM, incorporate concept drift learning into ML-OSELM, and compare it against the recently proposed concept drift learning method for single-hidden-layer ELMs [21].

References

1. Bengio, Y., Courville, A., Vincent, P.: Representation learning: a review and new perspectives. IEEE Trans. Pattern Anal. Mach. Intell. **35**(8), 1798–1828 (2013)
2. Bengio, Y.: Learning deep architectures for AI. Found. Trends Mach. Learn. **2**(1), 1–127 (2009)
3. Hinton, G.E., Salakhutdinov, R.R.: Reducing the dimensionality of data with neural networks. Science **313**(5786), 504–507 (2006)
4. Vincent, P., et al.: Stacked Denoising Autoencoders, Learning Useful representations in a deep network with a local denoising criterion. J. Mach. Learn. Res. **11**, 3371–3408 (2010)
5. Salakhutdinov, R., Larochelle, H.: Efficient learning of deep Boltzmann machines. J. Mach. Learn. Res. **9**, 693–700 (2010)
6. Kasun, L.L.C., Zhou, H., Huang, G.B., Vong, C.M.: Representational learning with ELMs for big data. IEEE intell. Syst. **28**(6), 31–34 (2013)
7. Huang, G.-B., Zhu, Q.-Y., Siew, C.-K.: Extreme learning machine: theory and applications. Neurocomputing **70**, 489–501 (2006)
8. Huang, G.-B., et al.: Extreme learning machine for regression and multiclass classification. IEEE Trans. Syst. Man Cybern. **42**(2), 513–529 (2012)
9. Lian, N.Y., Huang, G.B., Saratchandran, P., Sundararajan, N.: A fast and accurate online sequential learning algorithm for feedforward networks. IEEE Trans. Neural Netw. **17**(6), 1411–1423 (2006)
10. Mirza, B., Lin, Z., Toh, K.A.: Weighted online sequential extreme learning machine for class imbalance learning. Neural Process. Lett. **38**(3), 465–486 (2013)
11. Kim, Y., Toh, K.A., Teoh, A.B.J., Eng, H.L., Yau, W.Y.: An online learning network for biometric scores fusion. Neurocomputing **102**, 65–77 (2013)
12. Gu, Y., Liu, J., Chen, Y., Jiang, X., Yu, H.: TOSELM: timeliness online sequential extreme learning machine. Neurocomputing **128**, 119–127 (2014)
13. Zhao, J., Wang, Z., Park, D.S.: Online sequential extreme learning machine with forgetting mechanism. Neurocomputing **87**, 79–89 (2012)
14. Krizhevsky, A., Hinton, G.: Learning multiple layers of features from tiny images. Computer Science Department, University of Toronto, Techinical Report, vol. 1, no. 4, 2009
15. Fei-Fei, L., Fergus, R., Perona, P.: Learning generative visual models from few training examples: an incremental Bayesian approach tested on 101 object categorie, IEEE. CVPR 2004, Workshop on Generative-Model Based Vision, 2004
16. Samaria, F., Harter, A.: Parameterisation of a stochastic model for human face identification. In: Proceedings of 2nd IEEE Workshop on Applications of Computer Vision, Sarasota FL, Dec 1994
17. Graham, D.B., Allinson, N.M.: Characterizing Virtual Eigensignatures for General Purpose Face Recognition. Face Recognition: From Theory to Applications, NATO ASI Series F, Computer and Systems Sciences, vol. 163, pp. 446–456 (1998)
18. Coates, A., Ng, A.Y., Lee, H.: An analysis of single-layer networks in unsupervised feature learning. In: International Conference on Artificial Intelligence and Statistics, 2011
19. Jia, Y., Huang, C., Darrell, T.: Beyond spatial pyramids: receptive field learning for pooled image features. CVPR, 2012
20. Poon, H., Domingos, P.: Sum-product networks: a new deep architecture. In: IEEE Computer Vision Workshops (ICCV Workshops), pp. 689–690 (2011)
21. Mirza, B., Lin, Z., Nan, L.: Ensemble of subset online sequential extreme learning machine for class imbalance and concept drift. Neurocomputing **149**(A), 316–329 (2015)

ELM Meets Urban Computing: Ensemble Urban Data for Smart City Application

Ningyu Zhang, Huajun Chen, Xi Chen and Jiaoyan Chen

Abstract In recent years, big data analysis has been applied to the design and development of smart cities, which creates opportunities as well as challenges. It is necessary to retrieve a large amount of social media data and physical sensor data for this purpose. However, different cities have different infrastructures and populations, resulting in the sparsity of some types of data, such as social media data. In this paper, we propose ELM based method for smart cities and apply it to optimal retail store placement owing to its importance in the success of a business. Traditional approaches to the problem have considered demographics, revenue, and aggregated human flow statistics from nearby or remote areas; however, the acquisition of relevant data is usually expensive. The rapid growth of location-based social networks in recent years has led to the availability of fine-grained data describing the mobility of users and popularity of places. However, circumstances vary from one city to another. Furthermore, the number of sensors may not be sufficient to cover all the relevant areas of a particular city. In such cases, it would be useful to transfer knowledge to small cities. We study the predictive power of various machine-learning features with regard to the popularity of retail stores in a city by using datasets collected from open data sources in several big cities. In addition, we adopt a ELM based method to transfer knowledge to small cities. The results of experiments involving cities in China confirm the effectiveness of the proposed framework.

Keywords Urban computing · Optimal retail location · Smart city · Extreme learning machine

N. Zhang (✉) · H. Chen · X. Chen · J. Chen
College of Computer Science and Technology, Zhejiang University,
Hangzhou 310027, China
e-mail: zhangningyu@zju.edu.cn

H. Chen
e-mail: huajunsir@zju.edu.cn

X. Chen
e-mail: xichen@zju.edu.cn

J. Chen
e-mail: jiaoyanchen@zju.edu.cn

© Springer International Publishing Switzerland 2016
J. Cao et al. (eds.), *Proceedings of ELM-2015 Volume 1*,
Proceedings in Adaptation, Learning and Optimization 6,
DOI 10.1007/978-3-319-28397-5_5

1 Introduction

Urban computing, which aims to tackle urban problems by using city-generated data (e.g., traffic flow, human mobility, and geographical data), connects urban sensing, data management, data analytics, and service provision into a recurrent process to continuously and unobtrusively improve the quality of life, city operation systems, and the environment. For instance, the geographical placement of a retail store or new business has been of prime importance since the establishment of the first urban settlements in ancient times, and it assumes the same importance from the viewpoint of modern trading and commercial ecosystems in today's cities. A new coffee shop that is set up in a street corner may thrive with hundreds of customers, but it may close in a matter of months if it is set up a few hundred meters down the road. Nevertheless, infrastructure statistics are not sufficient for evaluating investment values. For example, the noise and pollution associated with train/bus systems can lower the value of a coffee shop.

In contrast, from the perspective of urban computing, more dynamic and information-rich data can be accumulated with the development of mobile, internet, and sensor technologies. For example, people may post comments and ratings for places of interest (POIs; e.g., schools, restaurants, and shopping centers) via mobile apps after consumption. Moreover, mobility data such as smart card transactions and taxi GPS traces consist of both trajectories and consumption records of residents' daily commutes. If properly analyzed, these data (e.g., user reviews and location traces) can serve as a rich source of intelligence for determining optimal retail store placement.

Indeed, these retail-related dynamic data generated by users could better reflect values of placement than urban infrastructure statistics. In general, if people have good opinions of a store, the demand for this store as well as its investment value will be high. The challenge is how to uncover people's opinions of a store. In fact, the opinions of users can be mined from (1) online user reviews and (2) offline urban regional data. Specifically, online reviews (e.g., Dianping ratings) contain explicit opinions regarding places surrounding a store. For example, the quality of a neighborhood can be partially approximated by the ratings of business venues, such as overall rating, service rating, and environment rating. On the other hand, offline urban regional data near a store not only encode the static statistics of urban infrastructure but also reflect residents' implicit opinions of the neighborhood. All these indications provided by dynamic user-generated data reveal important facets of a store that are of great concern to customers and convey the implicit user opinions of a neighborhood. Therefore, we consider and mine both the explicit opinions from user reviews and the implicit opinions from urban regional data to enhance the evaluation of optimal retail store placement.

However, different cities have different infrastructures and populations, resulting in the sparsity of some types of data for smart cities. For example, it is relatively easy to obtain heterogeneous data such as online user reviews in a metropolis because of its large population and infrastructure. However, small towns have small populations, and hence, relatively low social media activity. Therefore, it is difficult to assess optimal retail store placement on the basis of such data from small cities. On the other hand, large cities have been extensively modeled for numerous applications through big data analysis. In this paper, we propose ELM based method to transfer knowledge between smart cities and apply it to optimal retail store placement.

Specifically, transfer learning aims to extract common knowledge across domains such that a model trained on one domain can be adapted effectively to other domains. In reality, different cities are equivalent to different domains, and online and offline data can be regarded as two different views. Given a set of candidate areas in a city for opening a store, our aim is to identify the best ones in terms of their potential to attract a large number of users (i.e., to become popular). We formulate this problem as a rank problem, where, by extracting a set of features, we seek to exploit them to assess the retail quality of a geographic area. More specifically, we adopt ELM with domain adaptation to train a classifier for predictions on the target domain [7]. Based on the framework, a particular solution is proposed to learn the classifier simultaneously.

The major contributions of this paper are as follows:

(1) We propose a ELM based method for urban computing between smart cities. We apply this method to optimal retail store placement in order to transfer knowledge from large cities to small ones to improve accuracy.
(2) We study the factors affecting results in order to select the candidates. We analyze the changes in different cities, and we formulate rules to select candidates.
(3) We evaluate our approach using various data sources from the Web, including traffic data, bus data, user comments in China, in order to verify the effectiveness of the proposed framework.

2   Related Work

2.1   Urban Computing

The dynamics of a city (e.g., human mobility and the number of changes in a POI category) may indicate trends in the city's economy. For instance, the number of movie theaters in Beijing kept increasing from 2008 to 2012 [6, 8]. This could mean that an increasing number of Beijing's residents preferred to watch movies in movie theaters. In contrast, some categories of POIs may vanish in a city, signifying a downturn in business. Likewise, human mobility could indicate the unemployment rate in some major cities and therefore facilitate the prediction of stock market trends [5]. Thus, human mobility combined with POIs can help determine the placement of some businesses.

With respect to previous work in the general area, in this paper, we examine how the problem can be framed through Domain Adaption Transfer ELM. The richness of information provided by the above-mentioned services in big cities could enable us to study the retail quality of an area in a fine-grained manner: various types of geographic, semantic, and mobility information can not only complement traditional techniques but also form the basis for a new generation of business analytics driven by online data.

2.2 ELM

Given N samples $[x_1, x_2, \ldots, x_N]$ and their corresponding target$[y_1, y_2, \ldots, y_N]$, where $x_i = [x_{i1}, x_{i2}, x_{i3}, x_{in}]^T \in R^n$ and $y_i = [y_{i1}, y_{i2}, y_{i3}, y_{im}]^T \in R^m$. The output of the hidden layer is denoted as $h(x_i) \in R^{1 \times L}$, where L is the number of hidden nodes and h(.) is the activation function. The output weights between the hidden layer and the output layer being learned is denoted as $\beta \in R^{L \times m}$, where m is the number of output nodes.

Regularized ELM aims to solve the output weights by minimizing the squared loss summation of prediction errors and the norm of the output weights for over-fitting control, which results in the following formulation

$$\begin{cases} min_\beta L_{ELM} = \frac{1}{2}||\beta||^2 + C \cdot \frac{1}{2} \cdot \sum_{i=1}^{N} \xi_i^2 \\ s.t. \quad h(x_i)\beta = y_i - \xi_i, \quad i = 1, \ldots, N \end{cases} \qquad (1)$$

There are two cases when solving β, i.e. if the number N of training patterns is larger than L, the gradient equation is over-determined, and the closed form solution can be obtained as

$$\beta^* = (H^T H + \frac{I_L}{c})^{-1} H^T T \qquad (2)$$

If the number N of training patterns is smaller than L, an under-determined least square problem would be handled. In this case, the solution of (2) can be obtained as

$$\beta^* = H^T (HH^T + \frac{I_N}{c})^{-1} T \qquad (3)$$

Therefore, in classifier training of ELM, the output weights can be computed by using (3) or (4) which depends on the number of training instances and the number of hidden nodes.

3 Overview

3.1 Preliminaries

Definition 1 (*City Block*): We divide a city into blocks, assuming that placement in a block g is uniform; each block has several data samples and only one label that denotes whether it contains only one store (if not, we will try to use a store with more data as a label).

Definition 2 (*Social View*): Information aggregation index *svi* obtained by the analyses of online user review data of smart cities.

Definition 3 (*Physical View*): Information aggregation index *pvi* obtained through offline urban region data from various physical sensors and satellite data from smart cities.

3.2 Framework

As shown in Fig. 1, our framework consists of two major components: feature extraction of the source and target cities, and domain adaptation transfer ELM, which involves the analysis of optimal retail store placement in other cities. We retrieved massive amounts of online user review data from big cities. Through proper feature learning from social and physical views, we fed these data into our framework. Then, through domain adaptation transfer ELM, we transferred knowledge to other cities with sparse online user review data.

   Problem statement: Formally, by considering the existence of a candidate set of areas L in which a commercial enterprise is interested in placing its business, we wish to identify the optimal area $l \in L$ such that a newly opened store in l will potentially attract the largest number of visitors. An area l is derived from block g_i.

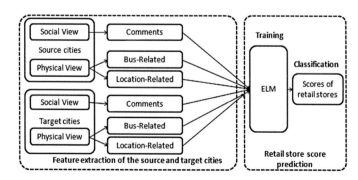

Fig. 1 Domain adaptation transfer ELM framework

We compute a score y_i for every candidate area l: the top-ranked area in terms of this score will be the optimal area for placing the new store. Our main assumption in the formulation of this task is that the Dianping score empirically observed by users can be used as a proxy for the relative popularity of a place.

Suppose that we are given a group of source cities $U_{s1}, U_{s2}, \ldots, U_{sn}$ and a target city U_t. Each source city has a set of blocks $U_{si} = \{D_s\}$, each block has labeled source-domain data $D_s = \{(s^{(i)}, p^{(j)}, y^{(k)})\}$, and the target city has labeled target-domain data $D_t = \{(s^{(m)}, p^{(n)}, y^{(p)})\}$, $m \ll i$ and $n \approx j$, consisting of two views, where s_i and p_i are column vectors of the ith instance from the social and physical views, respectively, and y_i is its class label, $y_i \in \{0, 1, 2, \ldots, k\}$ ($k = 3$ in this paper). The different class label (0, 1, 2, 3) corresponds to the store's score. The source and target domain data follow different distributions.

Our goal is to assign appropriate class labels to the instances in the target domain. We adopt domain adaptation transfer ELM to train a classifier for predictions on the target domain. Eventually, we generate a final score y_i for a region g.

4   Approach

4.1   Model Social View

Prosperity and users' opinions of a neighborhood are two important factors determining property investment value. Recent studies have shown that a strong regional economy usually indicates high demand [4]. Thus, we decided to mine online user reviews collected from dianping.com.

Overall Satisfaction: For each block g, we access the overall satisfaction of users over the neighborhood r_i. Since the overall rating of a business venue p represents the satisfaction of users, we extract the average of the overall ratings of all business venues located in r_i as a numeric score of overall satisfaction. Formally, we have

$$f_i^{OS} = \frac{\sum_{p:p,q \in P \& p \in r_i} OverallRating_p}{\{p : p \in P \& p \in r_i\}}. \tag{4}$$

Service Quality: Similarly, we compute the average service rating of business venues in r_i and express the service quality of the neighborhood as

$$f_i^{SQ} = \frac{\sum_{p:p,q \in P \& p \in r_i} ServiceRating_p}{\{p : p \in P \& p \in r_i\}}. \tag{5}$$

Environment Class: The environment class of business venues could reflect whether the neighborhood is high-class. Therefore, we extract the average environment rating as

$$f_i^{EC} = \frac{\sum_{p:p,q\in P\&p\in r_i} EnvironmentRating_p}{\{p : p \in P\&p \in r_i\}}.$$ (6)

Consumption Cost: The average cost of consumption behaviors in business venues can partially reflect the income and neighborhood class. We calculate the average consumption cost of business venues of a targeted neighborhood as a feature:

$$f_i^{CC} = \frac{\sum_{p:p,q\in P\&p\in r_i} AverageCost_p}{\{p : p \in P\&p \in r_i\}}.$$ (7)

Dianping Comments. We extend the existing word-embedding learning algorithm and develop five-layer neural networks for learning. Assuming that there are K labels, we modify the dimension of the top layer in the C&W model [1] as K, and add a *softmax* layer on the top layer. The *softmax* layer is suitable for this scenario because its outputs are interpreted as conditional probabilities. Unlike the C&W model, our model does not generate any corrupted n-gram. Let $f^g(t)$, where K denotes the number of polarity labels, be the gold K-dimensional multinomial distribution of input t and $\sum_k f_k^g(t) = 1$. The cross-entropy error of the *softmax* layer is given by

$$loss_h(t) = - \sum_{k=0,1} f_k^g(t) \cdot log(f_k^h(t)),$$ (8)

where $f^g(t)$ is the gold event distribution and $f^h(t)$ is the predicted event distribution.

4.2 Model Physical View

Recent studies have reported that different types of transit systems have different impacts on a region owing to the differences in fares, frequencies, speeds, and scopes of service. Economic information of a region can also reflect its pulse.

Bus-Related Features: Most moderate-income residents choose buses, which are cheaper and travel at acceptable speeds, instead of taxies, which are expensive and travel at faster speeds. Let BT denote the set of all bus trajectories of a city, p is a pickup bus stop and d is a drop-off bus stop.

Bus Arriving, Departing, and Transition Volume: We extract the arriving, departing, and transition volumes of buses from smart card transactions. Formally,

$$F_i^{BAV} = |< p, d >\in BT : p \notin r_i \& d \in r_i|$$ (9)

$$F_i^{BLV} = |< p, d >\in BT : p \in r_i \& d \notin r_i|$$ (10)

$$F_i^{BTV} = |< p, d >\in BT : p \in r_i \& d \in r_i|.$$ (11)

Real Estate Features: Recent studies report that real estate prices reflect the purchasing power and economic index of this region. Formally, we have

$$F_i^{RE} = \frac{\sum_{p:p,q\in P \& p\in r_i} RealEstate_p}{\{p : p \in P \& p \in r_i\}}.$$ (12)

Traffic Index Features: Increased travel velocity and reduced traffic congestion should be reflected by values. We investigate the traffic index from nitrafficindex.com, which gives us as value to evaluate local traffic conditions in each block. Formally, we have

$$F_i^{RE} = \frac{\sum_{p:p,q\in P \& p\in r_i} TrafficIndex_p}{\{p : p \in P \& p \in r_i\}}.$$ (13)

Competitiveness Features: We devise a feature to factor in the competitiveness of the surrounding area. Given the type of the place under prediction γ_l (e.g., Coffee Shop for Starbucks), we measure the proportion of neighboring places of the same type γ_l with respect to the total number of nearby places. Then, we rank areas in reverse order, assuming that the least competitive area is the most promising one:

$$\hat{X}_l(r) = -\frac{N_{\gamma l}(l,r)}{N(l,r)}.$$ (14)

Quality by Jensen Features: To consider spatial interactions between different place categories, we exploit the metrics defined by Jensen et al. [3]. To this end, we use the inter-category coefficients described to weight the desirability of the places observed in the area around the object, i.e., the greater the number of the places in the area that attract the object, the better is the quality of the location. More formally, we define the quality of location for a venue of type γ_l as

$$\hat{X}_l(r) = \sum_{\gamma_p\in\Gamma} log(\chi_{\gamma_p->\gamma_l}) \times (N_{\gamma_p(l,r)} - \overline{N_{\gamma_p(l,r)}}),$$ (15)

POIs: The category of POIs and their density in a region indicate land use as well as patterns in the region, thereby contributing to optimal placement. A POI category may even have a direct causal relation to it. Let $\sharp(i, c)$ denote the number of POIs of category $c \in C$ located in g_i, and let $\sharp(i)$ be the total number of POIs of all categories located in g_i. The entropy is defined as

$$f_i^{POI} = -\sum_{c\in C} \frac{\sharp(i, c)}{\sharp(i)} \times log \frac{\sharp(i, c)}{\sharp(i)}.$$ (16)

4.3 Domain Adaptation Transfer ELM

Suppose that the source domain and target domain are represented D_S and D_T. In this paper, we assume that all the samples in the source domain are labeled data. The proposed method aims to learn a classifier β_S using a number of labeled instances from the source domain, and set the few labeled data from the target domain as an appropriate regularizer for adapting to the source domain as shown in Fig. 2, which can be formulated as

$$min_{\beta_S, \xi_S^i, \xi_T^i} \frac{1}{2}||\beta||^2 + C_S \frac{1}{2} \sum_{i=1}^{N_S} (\xi_S^i)^2 + C_T \frac{1}{2} \sum_{j=1}^{N_T} (\xi_T^j)^2 \qquad (17)$$

$$s.t. \begin{cases} H_S^i \beta_S = t_S^i - \xi_S^i, i = 1, \ldots, N_S \\ H_T^j \beta_T = t_T^j - \xi_T^j, j = 1, \ldots, N_T \end{cases}$$

We can find that the very few labeled guide samples from target domain can assist the learning of β_S and realize the knowledge transfer between source domain and target domain by introducing the third term as regularization with the second constraint, which makes the feature mapping of the guide samples from target domain approximate the labels with the output weights β_S learned by the training data from the source domain.

In fact, according to [7] the optimization (18) can be reformulated an equivalent unconstrained optimization problem in a matrix form by substituting the constraints into the objective function as

$$min_{\beta_S} L(\beta_S) = \frac{1}{2}||\beta||^2 + C_S \frac{1}{2} \sum_{i=1}^{N_S} ||t_S - H_S \beta_S||^2 + C_T \frac{1}{2} \sum_{j=1}^{N_T} ||t_T - H_T \beta_T||^2 \qquad (18)$$

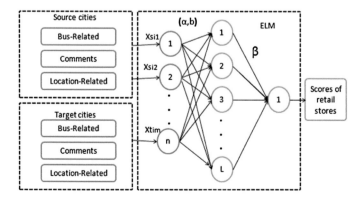

Fig. 2 The proposed retail store classification model

By setting the gradient of L with respect to β_S as zero, Then, we can easily solve the β_S.

For the case that the number of training samples N_S is smaller than L ($N_S < L$), we can obtain the output weights as:

$$\beta_S = H_S^T (CB^{-1}A - D)^{-1}(CB^{-1}t_T - t_S) +$$
$$H_T^T[B^{-1}t_T - B^{-1}A(CB^{-1}A - D)^{-1}(CB^{-1})t_T - t_S)] \tag{19}$$

For the case that the number of training samples N_S is larger than L($N_S > L$),

$$\beta_S = (I + C_S H_S^T t_S + C_T H_T^T t_T)^{-1}(C_S H_S^T t_S + C_T H_T^T t_T) \tag{20}$$

For recognition of the numerous unlabeled data in target domain, we calculate the final output using the following

$$\hat{y}_{Tp}^k = H_{Tp}^k \cdot \beta_S, k = 1, \dots, N_{Tp} \tag{21}$$

In terms of the above discussion, the algorithm is summarized as

---

Algorithm 1 Domain adaption transfer ELM algorithm

Input:
The source dataset $D_s = \{(s^{(i)}, p^{(j)}, y^{(k)})\}$
The target dataset $D_t = \{(s^{(m)}, p^{(n)}, y^{(p)})\}$
trade-off parameters C_S and C_T.
Output: The output weights β_S; The predicted output y_{Tp} of unlabeled data in target domain.
1. Initialize the ELM network of L hidden neurons with random input weights W and hidden bias B.
2. Calculate the output matrix H_S and H_T of hidden layer with source and target domains as $H_S = h(W \cdot X_S + B)$ and $H_T = h(W \cdot X_T + B)$.
3. Compute the output weights β_S using (21)(22). s 4. Calculate the predicted output y_{Tp} using (23).

---

5 Experiments

5.1 Datasets

We extract features from smart card transactions in five cities. Each bus trip has an associated card id, time, expense, balance, route name, and pick-up and drop-off stop information (names, longitudes, and latitudes). In addition, we crawl the traffic index from nitrafficindex.com, which is open to the public. Furthermore, we crawl online business reviews from www.dianping.com, which is a site for reviewing business establishments in China. Each review includes the shop ID, name, address, latitude, longitude, consumption cost, star rating, poi category, city, environment, service, overall ratings, and comments. Finally, we crawl the estate data from www.soufun. com, which is the largest real-estate online system in China.

5.2 Evaluation Metrics

To verify the effectiveness of our method, we compared our method with the following algorithms: (1) **MART**, (2) **RankBoost**, (3) **TrAdaBoost** [2]

Normalized Discounted Cumulative Gain. The discounted cumulative gain (DCG@N) is given by

$$
DCG[n] = \begin{cases} rel_1 & if \quad n = \quad 1 \\ DCG[n-1] + \dfrac{rel_n}{log_2 n} & if \quad n >= \quad 2 \end{cases}
$$

Precision and Recall. Because we use a four-level rating system ($3 > 2 > 1 > 0$) instead of binary rating, we treat the rating 3 as a high value and ratings less than 2 as low values. Given a top-N block list E_N sorted in descending order of the prediction values, the precision and recall are defined as $\text{Precision@N} = \frac{E_N \bigcap E_{>=2}}{N}$ and Recall@N $= = \frac{E_N \bigcap E_{>=2}}{E_{>=2}}$, where $E_{>=2}$ are blocks whose ratings are greater than or equal to 2.

5.3 Model Evaluation

We use data for a single city as the baseline for our experiments. The dataset contains data from five cities in China. Each city's block that has a retail store is annotated with a score of $\{0, 1, 2, 3\}$ based on the dianping.com score empirically observed by users. Each city is considered as a domain, and each domain contains hundreds of blocks. Each block is represented as a vector of features. We randomly select one of the five domains as the target domain, and all the domains serve as the source domains. Therefore, we can formulate four multi-source classification problems.

Figure 3 shows the NDCG, precision, and recall of the social view, physical view, both views, and Domain Adaption Transfer ELM for Starbucks in Beijing. In all cases, we observe the performance of Domain Adaption Transfer ELM outperformed the other methods with single cites. For cites of Beijing, Shanghai and Shenzhen our method mostly outperformed the others.

In Table 1, we present the results obtained for the NDCG@10 metric for all features across the three chains. In all cases, we observe a significant improvement with respect to the baseline.

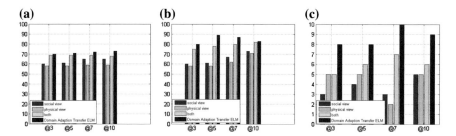

Fig. 3 NDCG, precision, and recall of @N for Starbucks in Beijing

Table 1 The best average NDCG@10 results of baseline and Domain Adaption Transfer ELM

Cities	Starbucks	TrueKungFu	YongheKing
MART (Single City)			
Beijing	0.743	0.643	0.725
Shanghai	0.712	0.689	0.712
Hangzhou	0.576	0.611	0.691
Guangzhou	0.783	0.691	0.721
Shenzhen	0.781	0.711	0.722
RankBoost (Single City)			
Beijing	0.752	0.678	0.712
Shanghai	0.725	0.667	0.783
Hangzhou	0.723	0.575	0.724
Guangzhou	0.812	0.782	0.812
Shenzhen	0.724	0.784	0.712
TrAdaBoost			
Beijing	0.772	0.678	0.732
Shanghai	0.755	0.667	0.753
Hangzhou	0.743	0.575	0.734
Guangzhou	0.712	0.782	0.712
Shenzhen	0.754	0.774	0.752
Domain Adaption Transfer ELM			
Beijing	**0.781**	**0.721**	**0.756**
Shanghai	**0.735**	**0.760**	**0.801**
Hangzhou	**0.744**	**0.710**	0.731
Guangzhou	0.798	**0.805**	**0.823**
Shenzhen	**0.781**	**0.784**	0.722

6 Conclusions

In this paper, from the perspective of a smart city, we analyzed retail store placement using four datasets observed in cities in general. Using the proposed Domain Adaption Transfer ELM method, we transferred knowledge from some cities to other cities with sparse data. In addition, we tested our approach for five cities in China. The results showed that our approach is applicable to different city environments.

In the future, we plan to apply our approach to other cities. Moreover, the sparsity of labeled data for machine learning remains a problem. To overcome this problem, it is necessary to use Domain Adaption Transfer ELM to train sparse labeled data with abundant labeled data.

References

1. Collobert, R., Weston, J., Bottou, L., Karlen, M., Kavukcuoglu, K., Kuksa, P.: Natural language processing (almost) from scratch. J. Mach. Learn. Res. **12**, 2493–2537 (2011)
2. Dai, W., Yang, Q., Xue, G.-R., Yu, Y.: Boosting for transfer learning. In: Proceedings of the 24th International Conference on Machine Learning, pp. 193–200. ACM (2007)
3. Jensen, P.: Network-based predictions of retail store commercial categories and optimal locations. Phys. Rev. E **74**(3), 035101 (2006)
4. Karamshuk, D., Noulas, A., Scellato, S., Nicosia, V., Mascolo, C.: Geo-spotting: Mining online location-based services for optimal retail store placement. In: Proceedings of the 19th ACM SIGKDD International Conference on Knowledge Discovery and Data Mining, pp. 793–801. ACM (2013)
5. Vaca Ruiz, C., Quercia, D., Aiello, L.M., Fraternali, P.: Taking brazil's pulse: tracking growing urban economies from online attention. In: Proceedings of the Companion Publication of the 23rd International Conference on World Wide Web Companion, pp. 451–456. International World Wide Web Conferences Steering Committee (2014)
6. Yuan, J., Zheng, Y., Xie, X.: Discovering regions of different functions in a city using human mobility and pois. In: Proceedings of the 18th ACM SIGKDD International Conference on Knowledge Discovery and Data Mining, pp. 186–194. ACM (2012)
7. Zhang, L., Zhang, D.: Domain adaptation transfer extreme learning machines. In: Proceedings of ELM-2014, vol. 1, pp. 103–119. Springer (2015)
8. Zheng, Y., Capra, L., Wolfson, O., Yang, H.: Urban computing: concepts, methodologies, and applications. ACM Trans. Intell. Syst. Technol. **5**(3), 38 (2014)

Local and Global Unsupervised Kernel Extreme Learning Machine and Its Application in Nonlinear Process Fault Detection

Hanyuan Zhang, Xuemin Tian, Xiaohui Wang and Yuping Cao

Abstract Unsupervised extreme learning machine (UELM) only concentrates on the local structure analysis of the observation data and thus may perform unsatisfactorily for process monitoring. In this paper, local and global unsupervised kernel extreme learning machine and support vector data description (LGUKELM-SVDD) is proposed for effective fault detection. LGUKELM model incorporates the global structure analysis into the standard UELM model. Meanwhile, the kernel trick is adopted to avoid the problem of explicitly selecting the number of the hidden layer nodes. After the nonlinear data features are extracted from LGUKELM model, SVDD is applied to build a monitoring index to detect fault. The simulation results on the continuous stirred tank reactor system show that the proposed method can detect process fault more effectively than conventional methods.

Keywords Unsupervised extreme learning machines · Support vector data description · Global structure analysis · Fault detection · Nonlinear process

H. Zhang · X. Tian (✉) · X. Wang · Y. Cao
College of Information and Control Engineering,
China University of Petroleum (East China), Qingdao 266580, China
e-mail: tianxm@upc.edu.cn

H. Zhang
e-mail: zhanghanyuan1@yeah.net

X. Wang
e-mail: qduwxh@163.com

Y. Cao
e-mail: cao_yp@163.com

X. Wang
College of Applied Technology, Qingdao University, Qingdao 266061, China

© Springer International Publishing Switzerland 2016
J. Cao et al. (eds.), *Proceedings of ELM-2015 Volume 1*,
Proceedings in Adaptation, Learning and Optimization 6,
DOI 10.1007/978-3-319-28397-5_6

1 Introduction

As large amounts of data are available in industrial processes, data-driven based fault detection methods have gained great attentions. In the past decades, many data-driven based methods have been proposed including principal component analysis (PCA) and partial least squares (PLS) [1]. In order to capture nonlinear feature existing in real industrial processes, many nonlinear extensions of the traditional PCA have been developed [2, 3].

Extreme learning machine (ELM) is a special type of single layer feedforward neural networks (SFLNs) without an iterative calculation process. By mapping the observation data from input space to a high-dimensional feature space via nonlinear activation function, ELM can be seen as a nonlinear method [4]. In the nonlinear mapping process of ELM, however, the number of the hidden layer nodes is usually empirically chosen according to the learning tasks. In order to avoid the application of time-consuming methods to identify the number of the hidden layer nodes, kernel versions of the ELM have been recently proposed [5, 6], which are expected to achieve better generalization performance than basic ELM.

To overcome the disadvantage of ELM based algorithms that cannot make use of unlabeled data, semi-supervised ELM versions have been proposed [7–9]. In order to explore the underlying structure of the data under the case where no labeled data are available, Huang et al. [10] discussed unsupervised ELM (UELM) by introducing the manifold regularization framework. However, UELM only focuses on the detailed local structure information and ignores the global structure information in a dataset which is very important for data mining and feature extraction.

In this paper, we develop a novel nonlinear fault detection method based on local and global unsupervised kernel extreme leaning machine and support vector data description (LGUKELM-SVDD). The optimization objective of the LGUKELM method is constructed to preserve both the local and global structure information of input data simultaneously. With the application of kernel trick, the challenging problem of selecting the number of hidden layer nodes is avoided. After the data features are extracted from LGUKELM model, support vector data description (SVDD) is used to build a monitoring index for. Simulation results obtained on the continuous stirred tank reactor (CSTR) system illustrate the effectiveness of the proposed fault detection method.

The remaining sections are organized as follows. A brief review of regularized ELM and standard UELM is provided in Sect. 2. The proposed LGUKELM method is detailed presented in Sect. 3. Section 4 introduces the LGUKELM-SVDD based fault detection strategy. The simulation study using the CSTR system is discussed in Sect. 5. Finally, we offer conclusions in Sect. 6.

2 Regularized ELM and Unsupervised ELM

2.1 Regularized ELM

Given a training dataset with N samples, $\{X, Y\} = \{x_i, y_i\}_{i=1}^{N}$, where $x_i \in R^{n \times 1}$ is a n-dimensional data point and $y_i \in R^{n_o \times 1}$ is a binary vector, and the number of hidden nodes L. Standard SLFNs are modeled as

$$H\beta = O \tag{1}$$

where $O = [o_1, o_2, \ldots, o_N]^T \in R^{N \times n_o}$ is the output matrix of SLFNs and $\beta = [\beta_1, \beta_2, \ldots, \beta_L]^T \in R^{L \times n_o}$ is the output weight matrix. H is the random feature mapping matrix:

$$H = \begin{bmatrix} h(x_1) \\ \vdots \\ h(x_N) \end{bmatrix} = \begin{bmatrix} G(w_1 \cdot x_1 + b_1) & \cdots & G(w_L \cdot x_1 + b_L) \\ \vdots & \cdots & \vdots \\ G(w_1 \cdot x_N + b_1) & \cdots & G(w_L \cdot x_N + b_L) \end{bmatrix}_{N \times L} \tag{2}$$

Due to the fact that the smaller the norm of output weights are, the better generalization performance the network tends to have [11], so the trained network of regularized ELM [12] not only aims to reach the smallest training error $\|O - Y\|^2$, but also aims to reach the smallest output weights norm. Therefore, the optimization model of regularized ELM can be formulated as

$$\min_{\beta \in R^{L \times n_o}} \frac{1}{2}\|\beta\|^2 + \frac{C}{2}\sum_{i=1}^{N}\|e_i\|^2 \tag{3}$$
$$s.t. \quad h(x_i)\beta = y_i^T - e_i^T, \quad i = 1, 2, \ldots, N$$

where $e_i \in R^{n_o \times 1}$ is the error vector with respect to the ith training sample and C is a penalty coefficient on the training errors.

2.2 Unsupervised ELM

To enforce the assumption used in unsupervised ELM (UELM): if two points x_i and x_j are close to each other, then the conditional probabilities $P(y|x_i)$ and $P(y|x_j)$ should be similar as well [10], the following cost function is minimized:

$$J_m = \min \frac{1}{2}\sum_{i,j} w_{ij}\|P(y|x_i) - P(y|x_j)\|^2 \tag{4}$$

where w_{ij} is a weight parameter that represents the local information structure in the unlabeled training data.

After the weight matrix $W = [w_{ij}]$ is calculated, Eq. (4) is simplified into a matrix form approximately.

$$\hat{J}_m = \min Tr(\hat{Y}^T L \hat{Y}) \tag{5}$$

where $\hat{Y}$ is the prediction matrix with respect to X. $L = D - W$ is the Laplacian matrix and D is a diagonal matrix with its diagonal element $D_{ii} = \sum_{j=1}^{N} w_{ij}$.

By modifying the regularized ELM objective function in Eq. (3), the optimization model of UELM is formulated as

$$\min_{\beta \in R^{L \times n_o}} \frac{1}{2} \|\beta\|^2 + \frac{\lambda}{2} Tr(F^T L F) \tag{6}$$
$$s.t. \quad f_i = h(x_i)\beta, \quad i = 1, 2, \ldots, N$$

where F is the output matrix of the network with its ith row equal to f_i, and λ is the tradeoff parameter.

3   Local and Global Unsupervised Kernel ELM

Minimizing the optimization of UELM can pledge that if x_i and x_j are neighboring data points, then the corresponding outputs of the network f_i and f_j are also neighboring. However, due to lack of the constraint for the faraway points, UELM may project the distant input data points in a small region in output space. In this section, LGUKELM is developed by integrating the global structure analysis into the objective function of the standard UELM method. Furthermore, kernel trick is used to overcome the tough problem of selecting the number of hidden layer nodes.

Given the scaled dataset $X \in R^{n \times N}$, the global structure analysis is imposed on UELM to find an output weight matrix $\beta \in R^{L \times n_o}$, so that the mean square of the Euclidean distance between all pairs of the network output points is maximized:

$$J_G = \max \frac{1}{N} \sum_{i=1}^{N} (f_i - \bar{f})^2 \tag{7}$$

where $\bar{f} = (1/N) \sum_{i}^{N} f_i$, $i = 1, 2, \ldots, N$.

We substitute $f_i = h(x_i)\beta$ into the objective function (7) and rewrite it as:

$$J_G(\beta) = \max \frac{1}{N} \sum_{i=1}^{N} \left(h(x_i)\beta - \frac{1}{N} \sum_{i=1}^{N} h(x_i)\beta \right)^2$$
$$= \max \frac{1}{N} \sum_{i=1}^{N} \beta^T \left(h(x_i) - \frac{1}{N} \sum_{i=1}^{N} h(x_i) \right)^T \left(h(x_i) - \frac{1}{N} \sum_{i=1}^{N} h(x_i) \right)\beta \tag{8}$$

On the assumption $\sum_{i=1}^{N} h(x_i) = 0$, Eq. (8) is reduced to

$$J_G(\beta) = \max \frac{1}{N} \beta^T \left(\sum_{i=1}^{N} h(x_i)^T h(x_i) \right) \beta = \max \frac{1}{N} \beta^T H^T H \beta \qquad (9)$$

The optimization of standard UELM in Eq. (6) can be reformulated as follows by substituting the constraint into the objective function

$$J_L(\beta) = \min \|\beta\|^2 + \lambda \, Tr(\beta^T H^T L H \beta) = \min Tr(\beta^T (I_L + \lambda H^T L H) \beta) \qquad (10)$$

where $I_L \in R^{L \times L}$ is an identity matrix.

The optimization objective $J_{LG}(\beta)$ of the LGUKELM is to minimize the $J_L(\beta)$ and to maximize the $J_G(\beta)$ simultaneously.

$$J_{LG}(\beta) = \frac{\min J_L(\beta)}{\max J_G(\beta)} = \frac{\min Tr(\beta^T (I_L + \lambda H^T L H) \beta)}{\max \frac{1}{N} \beta^T H^T H \beta} = \min \frac{\beta^T (I_L + \lambda H^T L H) \beta}{\frac{1}{N} \beta^T H^T H \beta} \qquad (11)$$

When $N < L$, we restrict β to be a linear combination of the rows of H [12]:

$$\beta = H^T A \qquad (12)$$

where $A \in R^{N \times n_o}$ is defined as the loading matrix in this paper.

Substituting Eq. (12) into Eq. (11), we can obtain

$$J_{LG}(\alpha) = \min \frac{A^T H (I_L + \lambda H^T L H) H^T A}{\frac{1}{N} A^T H H^T H H^T A} = \min \frac{A^T (H H^T + \lambda H H^T L H H^T) A}{A^T \frac{1}{N} H H^T H H^T A} \qquad (13)$$

The minimization problem in Eq. (13) can be converted to the following generalized eigenvalue problem.

$$(H H^T + \lambda H H^T L H H^T) \alpha_j = \gamma_j \frac{1}{N} H H^T H H^T \alpha_j \qquad (14)$$

where $\alpha_j, j = 1, 2, \ldots, n_o$ is the jth eigenvector corresponding to the jth eigenvalue γ_j.

Multiplying both side of Eq. (14) by $(H H^T)^{-1}$, we get

$$(I_N + \lambda L H H^T) \alpha_j = \gamma_j \frac{1}{N} H H^T \alpha_j \qquad (15)$$

where $I_N \in R^{N \times N}$ is an identity matrix.

With the introduction of the kernel function $k(x_i, x_j) = \langle h(x_i), h(x_j) \rangle$, kernel matrix of LGUKELM is defined as $K = H H^T$, where $K_{i,j} = k(x_i, x_j)$,

$i, j = 1, 2, \ldots, N$. The kernel function is selected as Gaussian kernel: $k(\boldsymbol{x}_i, \boldsymbol{x}_j) = exp\left(-\left\|\boldsymbol{x}_i - \boldsymbol{x}_j\right\|^2 / \sigma\right)$.

Then Eq. (15) can be expressed as

$$(\boldsymbol{I}_N + \lambda \boldsymbol{L} \boldsymbol{K}) \boldsymbol{\alpha}_j = \gamma_j \frac{1}{N} \boldsymbol{K} \boldsymbol{\alpha}_j \tag{16}$$

After resolving the generalized eigenvalue problem in Eq. (16), the final solution to the output weights $\boldsymbol{\beta}$ is given by

$$\boldsymbol{\beta}^* = \boldsymbol{H}^T \boldsymbol{A} = \boldsymbol{H}^T [\tilde{\boldsymbol{\alpha}}_2, \tilde{\boldsymbol{\alpha}}_3, \ldots, \tilde{\boldsymbol{\alpha}}_{n_o+1}] \tag{17}$$

where $\tilde{\boldsymbol{\alpha}}_i = \boldsymbol{\alpha}_i / \|\boldsymbol{K} \boldsymbol{\alpha}_i\|, i = 2, 3, \ldots, n_o + 1$ and $\boldsymbol{A} = [\tilde{\boldsymbol{\alpha}}_2, \tilde{\boldsymbol{\alpha}}_3, \ldots, \tilde{\boldsymbol{\alpha}}_{n_o+1}]$.

The lower-dimensional representation of the input dataset $\boldsymbol{X}$ is calculated as

$$\boldsymbol{T} = \boldsymbol{H} \boldsymbol{\beta}^* = \boldsymbol{H} \boldsymbol{H}^T \boldsymbol{A} = \boldsymbol{K} \boldsymbol{A}. \tag{18}$$

For a test data point $\boldsymbol{x}_t$, its corresponding projection vector $\boldsymbol{t}_t$ is calculated as

$$\boldsymbol{t}_t = \boldsymbol{h}(\boldsymbol{x}_t) \boldsymbol{\beta}^* = \boldsymbol{h}(\boldsymbol{x}_t) \boldsymbol{H}^T \boldsymbol{A} = \boldsymbol{k}_t \boldsymbol{A} \tag{19}$$

where $\boldsymbol{k}_t \in R^{1 \times N}$ is kernel vector and $\boldsymbol{k}_{t,i} = k(\boldsymbol{x}_t, \boldsymbol{x}_i), i = 1, 2, \ldots, N$.

In order to ensure $\sum_{i=1}^{N} \boldsymbol{h}(\boldsymbol{x}_i) = 0$, kernel matrix $\boldsymbol{K}$ should be mean centered using Eq. (20) before solving Eqs. (16) and (18) and test kernel vector $\boldsymbol{k}_t$ should also be mean centered using Eq. (21) before calculating $\boldsymbol{t}_t$.

$$\widetilde{\boldsymbol{K}} = \boldsymbol{K} - \boldsymbol{I}_K \boldsymbol{K} - \boldsymbol{K} \boldsymbol{I}_K + \boldsymbol{I}_K \boldsymbol{K} \boldsymbol{I}_K \tag{20}$$

$$\widetilde{\boldsymbol{k}}_t = \boldsymbol{k}_t - \boldsymbol{I}_t \boldsymbol{K} - \boldsymbol{k}_t \boldsymbol{I}_K + \boldsymbol{I}_t \boldsymbol{K} \boldsymbol{I}_K \tag{21}$$

where $\boldsymbol{I}_K$ is the $N \times N$ matrix whose elements are all equal to $1/N$, and $\boldsymbol{I}_t = 1/N[1, \ldots, 1] \in R^{1 \times N}$.

4 The LGUKELM-SVDD Based Fault Detection

4.1 Support Vector Data Description

In this paper, support vector data description (SVDD) [13, 14] is employed to monitor the output data of the LGUKELM for fault detection. For the output dataset $\boldsymbol{T} = [\boldsymbol{t}_1, \boldsymbol{t}_2, \ldots, \boldsymbol{t}_N]$ of the LGUKELM model. The dual form of the SVDD optimization problem can be obtained as

$$\max \sum_{i=1}^{N} \beta_i k(t_i, t_i) - \sum_{i=1}^{N} \sum_{j=1}^{N} \beta_i \beta_j k(t_i, t_j)$$

$$s.t. \quad 0 \le \beta_i \le C_s, \quad \sum_{i=1}^{N} \beta_i = 1 \tag{22}$$

where β_i is the Lagrange multiplier, C_s gives the trade-off between the volume of the hypersphere and the number of errors, $k(t_i, t_j) = \langle \Phi(t_i), \Phi(t_j) \rangle$ is used as Gaussian kernel function. Solving the above optimization problem, we can get the centre of the hypersphere b.

To judge whether a test point x_t is a fault sample, the distance D from t_t to the centre of the hypersphere is calculated as a monitoring index:

$$D = \|\Phi(t_t) - b\| = \sqrt{k(t_t, t_t) - 2 \sum_{i=1}^{N} \beta_i k(t_t, t_i) + \sum_{i=1}^{N} \sum_{j=1}^{N} \beta_i \beta_j k(t_i, t_j)} \tag{23}$$

The confidence limit D_{limit} of the monitoring index D is determined as the distance from the centre of the hypersphere b to any support vector on the boundary.

4.2 Fault Detection Strategy Based on LGUKELM-SVDD

The detailed fault detection procedure of LGUKELM-SVDD based method can be summarized as follows.

- The off-line modelling stage

  1. Construct the weight matrix W using scaled normal operating dataset X, and then calculate the diagonal matrix D and the Laplacian matrix L.
  2. After computing the kernel matrix K, carry out mean centering operation in Eq. (20) to get centered kernel matrix $\widetilde{K}$.
  3. Solve the generalized eigenvalue problem by replacing K with $\widetilde{K}$ in Eq. (16) to obtain the loading matrix A.
  4. Calculate the output matrix T according to Eq. (18) by replacing K with K and compute the confidence limit D_{limit} using SVDD.

- The on-line detection stag

  1. After obtaining the scaled new observed data x_t, compute the test kernel vector k_t and mean center it according to Eq. (21).

2. Calculate the projection vector t_t based on Eq. (19) by replacing k_t with $\tilde{k}_t$ and obtain the monitoring index D according to Eq. (23).
3. Compare the monitoring index D of x_t with its confidence limit D_{limit} to detect fault.

5 Simulation Study

We apply the LGUKELM-SVDD, KPCA and standard UELM based method to detect faults in CSTR system [15]. SVDD is also applied to construct monitoring index after obtaining the output data of the standard UELM, and the fault detection method is referred to as UELM-SVDD. 900 observations are produced in normal operation condition to build statistical models. The simulated seven kinds of fault pattern are shown in Table 1. For each kind of fault pattern, fault dataset with 900 samples are simulated and the fault is introduced at the 201th sample.

For the KPCA, the number of principal components that describes at least 90 % of the variance in dataset is selected and kernel function is chosen as Gaussian kernel. For the UELM-SVDD, the number of hidden layer nodes L is chosen as 1000 and Sigmoid function is selected as activation function. For a fair comparison, the same kernel functiona used in KPCA and the same dimension of output space ($n_o = 10$) used in UELM-SVDD are chosen in the LGUKELM-SVDD. A fault is considered to be detected if 5 continuous samples exceed confidence limit.

The fault detection results for fault F6 are plotted in Figs. 1, 2 and 3. As is shown in Fig. 1, the T^2 statistic off KPCA detects the fault at the 381th sample and its *SPE* statistic detects the faut at the 266th sample. The fault detection rate of T^2 and *SPE* statistic are 83.89 and 87.91 % respectively. In the monitoring chart of the UELM-SVDD, the fault is detected at the 218th sample with the fault detection rate of 94.09 %. However, the D statistic of UELM-SVDD brings some fales alarming samples under normal operation condition, which deteriorates the fault detection performance. With the application of the LGUKELM-SVDD based method, the D statistic exceeds its threshold from the 204th sample and the fault detection rate is 97.55 %. From the detection results under fault F6, we can see that the proposed

Fault	Description
F1	The activation energy ramps up
F2	The heat transfer coefficient ramps down
F3	The reactor temperature measurement has a bias
F4	Step change in feed flow rate
F5	The feed concentration ramps up
F6	The feed temperature ramps up
F7	The coolant feed temperature ramps up

Table 1 Fault pattern for CSTR sustem

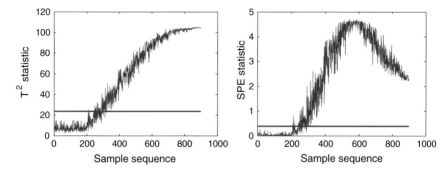

Fig. 1 KPCA monitoring charts for fault F6

Fig. 2 UELM-SVDD
monitoring chart for fault F6

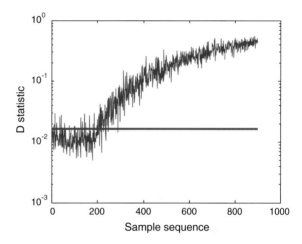

Fig. 3 LGUKELM-SVDD
monitoring chart for fault F6

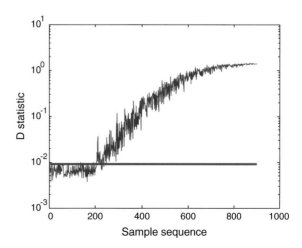

Table 2 Comparison of fault detection rate (%)

Fault	KPCA		UELM-SVDD	LGUKELM-SVDD
	T^2	SPE	D	D
F1	75.72	63.68	78.68	88.26
F2	82.39	80.98	84.81	90.13
F3	100.00	100.00	100.00	100.00
F4	100.00	100.00	100.00	100.00
F5	73.59	65.28	81.12	86.97
F6	83.89	87.91	94.09	97.55
F7	72.37	73.53	83.14	90.66

LGUKELM-SVDD based monitoring method can detect fault F6 faster and more effective than KPCA based and UELM-SVDD based monitoring methods.

The fault detection performance of the three methods for all seven kinds of fault pattern is compared in Table 2. For the step-change faults F3 and F4, all the three methods are seen to achieve 100 % fault detection rates. However, for the challenging ramp fault (F1, F2 and F5 to F8) detection problem, our proposed LGUKELM-SVDD method achieves much higher fault detection rates than both KPCA and UELM-SVDD, which confirms the superior monitoring performance of LGUKELM-SVDD.

6 Conclusions

In this paper, we present a novel nonlinear process monitoring method based on LGUKELM-SVDD. LGUKELM model is developed by integrating the global structure analysis into the local optimization of the standard UELM menthod and utilizing the kernel trick to avert the tough problem of determining the number of hadden layer nodes. After the local and global structure feature information is extracted, SVDD is applied to derive monitoring index. The simulation results obtained on CSTR demonstrate that the proposed LGUKELM-SVDD method outperforms both KPCA method and UELM-SVDD method significantly in terms of fault detection performance.

Acknowledgments This work was supported by the National Natural Science Foundation of PR China (Grant Nos. 61273160 and 61403418), the Natural Science Foundation of Shandong Province, PR China (Grant No. ZR2014FL016) and the Fundamental Research Funds for the Central Universities, PR China (Grant No. 14CX06132A).

References

1. Qin, S.J.: Surney on data-driven industrial process monitoring and diagnosis. Ann. Rev. Control **36**(2), 220–234 (2012)
2. Liu, X., Li, K., McAfee, M., Irwin, G.W.: Improved nonlinear PCA for process monitoring using support vector data description. J. Process Control **21**(9), 1306–1317 (2001)

3. Lee, J.M., Yoo, C.K., Choi, S.W., Vanrolleghem, P.A., Lee, I.B.: Nonlinear process monitoring using kernel principal component analysis. Chem. Eng. Sci. **59**(1), 223–234 (2004)
4. Iosifidis, A., Tefas, A., Pitas, I.: Minimum class variance extreme learning machine for human action recognition. IEEE Trans. Circuits Syst. Video Technol. **23**(11), 1968–1979 (2013)
5. Iosifidis, A., Tefas, A., Pitas, I.: On the kernel extreme learning machine classifier. Pattern Recogn. Lett. **54**, 11–17 (2015)
6. Liu, X.W., Wang, L., Huang, G.B., Zhang, J., Yin, J.P.: Multiple kernel extreme learning machine. Neurocomputing **149**, 253–264 (2015)
7. Liu, S.L., Feng, L., Xiao, Y., Wang, H.B.: Robust activation function and its application: semi-supervised kernel extreme learning method. Neurocomputing **144**(1), 318–328 (2014)
8. Luo, X.Z., Liu, F., Yang, S.Y., Wang, X.D., Zhou, Z.G.: Joint sparse regularization based sparse semi-supervised extreme learning machine (S3ELM) for classification. Knowl. Based Syst. **73**, 149–160 (2015)
9. Ayerdi, B., Marques, I., Grana, M.: Spatially regularized semisupervised ensembles of extreme learning machines for hyperspectral image segmentation. Neurocomputing **149**, 373–386 (2015)
10. Huang, G., Song, S.J., Gupta, J.N.D., Wu, C.: Semi-supervised and unsupervised extreme learning machines. IEEE Trans. Cybern. **44**(12), 2405–2417 (2014)
11. Bartlett, P.L.: The sample complexity of pattern classification with neural networks: the size of the weights is more important than the size of the network. IEEE Trans. Inf. Theor. **44**(2), 525–536 (1998)
12. Huang, G.B., Zhou, H., Ding, X., Zhang, R.: Extreme learning machine for regression and multiclass classification. IEEE Trans. Syst. Man Cybern. Part B Cybern. **42**(2), 513–529 (2012)
13. Tax, D.M.J., Duin, R.P.W.: Support vector data description. Mach. Learn. **54**(1), 45–66 (2004)
14. Ge, Z.Q., Song, Z.H.: Bagging support vector data description model for batch process monitoring. J. Process Control **23**(8), 1090–1096 (2013)
15. Cai, L.F., Tian, X.M., Chen, S.: A process monitoring method based on noisy independent component analysis. Neurocomputing **127**, 231–246 (2014)

Parallel Multi-graph Classification Using Extreme Learning Machine and MapReduce

Jun Pang, Yu Gu, Jia Xu, Xiaowang Kong and Ge Yu

Abstract A multi-graph is represented by a bag of graphs and modelled as a generalization of a multi-instance. Multi-graph classification is a supervised learning problem for multi-graph, which has a wide range of applications, such as scientific publication categorization, bio-pharmaceu-tical activity tests and online product recommendation. However, existing algorithms are limited to process small datasets due to high computation complexity of multi-graph classification. Specially, the precision is not high enough for a large dataset. In this paper, we propose a scalable and high-precision parallel algorithm to handle the multi-graph classification problem on massive datasets using MapReduce and extreme learning machine. Extensive experiments on real-world and synthetic graph datasets show that the proposed algorithm is effective and efficient.

Keywords Multi-graph · Classification · Extreme learning machine · MapReduce

J. Pang (✉) · Y. Gu · X. Kong · G. Yu
College of Information Science and Engineering, Northeastern University,
Liaoning 110819, China
e-mail: pangjun@research.neu.edu.cn

Y. Gu
e-mail: guyu@ise.neu.edu.cn

X. Kong
e-mail: kongxiaowang.neu@gmail.com

G. Yu
e-mail: yugu@ise.neu.edu.cn

J. Xu
School of Compute, Electronics and Information, Guangxi University, Guangxi
530004, China
e-mail: xujia@gxu.edu.cn

© Springer International Publishing Switzerland 2016
J. Cao et al. (eds.), *Proceedings of ELM-2015 Volume 1*,
Proceedings in Adaptation, Learning and Optimization 6,
DOI 10.1007/978-3-319-28397-5_7

1   Introduction

Multi-graph learning is a generalization of multi-instance learning in which instances
are organized as graphs instead of feature vectors. Multi-graph representation maintains rich structure information which makes it outperforms other multi-instance
representation. Nowadays, multi-graph learning has many successful application
scenarios. We give two typical examples as follows. (1) Scientific publication categorization [1]: a paper is represented as a multi-graph, i.e., the abstract of the paper is
modelled as a graph and the abstract of every reference is also modelled as a graph. If
the paper or one of its references is related with the topic, this paper is positive. Otherwise, it is negative. Given training papers with classification labels, we can predict
the unseen papers' labels. (2) Bio-pharmaceutical activity tests [2]: a molecule has
a lot of forms. If one of its forms resists the disease, the molecule be used to man-
ufacture drugs. Otherwise, it cannot be applied. A specific form of a molecule can
be described as a graph and a multi-graph demotes different forms of the molecule.
Multi-graph learning can predict the molecules' activities.

   Although multi-graph learning has important practical applications, existing
multi-instance learning algorithms can not be directly used to solve this problem.
Because these multi-instance learning algorithms are designed to process tabular
instances which are represented in a common vectorial feature space. To the best
of our knowledge, few work focuses on exploring it. Wu et al. propose the *gMGFL*
approach, which mines informative feature subgraphs and has higher accuracy than
the extended multi-instance learning algorithms. Nowadays, the quantity of information is very large and fast growing. And more and more multi-graphs are modelled
from these increasing information. It is a non-trivial task to mine valued knowledge from so large scale multi-graphs. Specifically, the following challenges need
to be tackled. (1) *gMGFL* is not suited to deal with large-scale datasets because
gMGFL adopts in-memory frequent subgraph mining and classification algorithms.
(2) To support high-quality exploratory analysis and decision making, the precision
and recall are desired to be improved. Our experimental results show the traditional
parallel Bayes algorithm does not obtain high precisions and recalls on large-scale
multi-graph datasets.

   For the first challenge, we adopt popular MapReduce framework to execute large-
scale multi-graph binary classification. MapReduce is a popular parallel programming framework to handle big data owning to its good properties of fault tolerance,
high scalability and low deployment cost [3]. We propose a parallel approach based
on MapReduce, named *ME-MGC*, to solve multi-graph binary classification problem
on massive multi-graph dataset. For the second challenge, we adapt a parallel ELM
approach to improve the classification performance. Experimental results display
that our approaches obtain higher precisions and recalls on both real and synthetic
datasets than extended approaches adopting NBayes or SVM.

   Specifically, the major contributions we have made in this paper are summarized
as follows. (1) We propose a parallel approach based on MapReduce to solve the
massive multi-graph binary classification problem. (2) We adapt extreme learning

machine (ELM) to process multi-graph classification for improving the performance of classification. Moreover, we study the variation of the precision with a different number of hidden nodes for ELM algorithm. (3) We have conducted extensive experiments on both real and simulated data sets and the results demonstrate that our approaches are effective and efficient.

The remainder of this paper is organized as follows. Related works are introduced in Sect. 2. Problem definition and backgrounds are discussed in Sect. 3. Our approach is provided in Sect. 4. Experimental results and discussions are presented in Sect. 5. We conclude the paper in Sect. 6.

2 Related Works

Related works with our study include multi-graph classification, extreme learning machine based on MapReduce and frequent subgraph mining based on MapReduce.

2.1 Multi-graph Learning

Although multi-graph learning is very valuable in real applications, the researches on it are still quite limited. Wu et al. [1] propose *gMGFL* approach to solve multi-graph classification problem. Inspired by multi-instance learning, *gMGFL* mines feature subgraph from overall graph dataset and converts multi-graphs into feature-value vectors which fit conventional classification models, such as naive Bayes, kNN classifier, decision tree and support vector machine. Feature subgraphs are top-k frequent subgraphs with a score function *score(g)*. *gMGFL* is not suitable to process large-scale datasets because it is an in-memory algorithm.

2.2 Extreme Learning Machine Based on MapReduce

Extreme learning machine(ELM) is a type of artificial neural network [4, 5]. Parallel ELM based on MapReduce has attracted attention of many researchers. He et al. [6] propose a MapReduce version of ELM, named *PELM*, to implement regression for large-scale datasets. *PELM* consists of two MapReduce jobs. Xin et al. [7] design and implement *ELM** which combines the previous two MapReduce jobs into one MapReduce job. Xin et al. [8] propose an incremental algorithm E^2LM to process large-scale updating training datasets. Bi et al. [9] propose a distributed extreme learning machine with kernels based on MapReduce. Wang et al. [10] design a parallel online sequential extreme learning machine based on MapReduce. To the best of our knowledge, *ELM** is the-state-of-the-art parallel ELM algorithm processing large static training datasets. In this paper, we adapt *ELM** into the scenario of massive multi-graph classification.

2.3 Frequent Subgraphs Mining Based on MapReduce

A typical problem of large-scale frequent subgraphs mining can use two settings: (1) one single big graph: its target is to mine subgraphs from one single big graph such that supports of these subgraphs are not smaller than a given support threshold [11, 12]; (2) a large collection of graphs: its target is to mine frequent subgraphs from a large collection of graphs [13, 14]. In this paper, we consider the second setting. Hill et al. [13] propose iterative *MapReduce-FSG* algorithm which is an incremental approach to mine frequent subgraphs from a large collection of graphs. Lin et al. [14] design and implement *MRFSM* approach which contains only three MapReduce jobs. To the best of our knowledge, *MRFSM* is the state-of-the-art frequent subgraphs mining approach for a large collection of graphs. In this paper, *MRFSM* is leveraged during the process of frequent subgraphs generation.

3 Preliminaries

In this section, we firstly define related basic concepts. Then, we give a simple overview of the extreme learning machine.

3.1 Problem Definition

Definition 1 (*Labeled multi-graph*) A multi-graph is a bag of graphs. A labeled multi-graph mg is a multi-graph with binary class label $l(mg) \in \{positive, negative\}$. If the class label for one graph of multi-graph is positive, the multi-graph has a positive class label $l(mg) = positive$. Otherwise, the multi-graph has a negative class label.

Definition 2 (*Feature subgraph representation of multi-graph*) Given a multi-graph set $MG = \{mg_1, mg_2, \ldots, mg_n\}$ and k feature subgraphs $F = \{f_1, \ldots, f_k\}$, $mg_i \in MG$ is represented as a feature vector $v(mg_i)$ of k dimensions. The weight w_i of the *ith* dimension is 1 if the *ith* feature subgraph $f_i \in F$ is a subgraph of one graph for mg_i. Otherwise, w_i is set to 0.

Definition 3 (*Score function of the frequent subgraphs*) The score function of the frequent subgraphs is used to mine feature subgraphs [1]. The score function, named $score(g)$, is as follow.

$$score(g) = 1/2(s(A)/A - s(B)/B + s(C)/C - s(D)/D)$$

Definition 4 (*Multi-graph classification*) Given a labeled multi-graph dataset, we aim to construct prediction model from labeled multi-graphs to predict unseen multi-graphs with maximum precision.

Definition 5 (*Massive multi-graph classification*) Massive multi-graph classification is a special multi-graph classification with large-scale training dataset and large-scale test dataset.

Based on *gMGFL*, we propose a parallel algorithm *ME-MGC* to solve massive multi-graph classification problem. Next, we simply introduce ELM.

3.2 Extreme Learning Machine

Huang proposes ELM for single hidden-layer feedforward neural networks (SLFNs) and then extends it to the "generalized" SLFNs [4, 5, 15–22]. Compared to traditional feedforward neural networks, ELM has better generalization performance, faster learning speed and less training error. The training process of ELM approach is described in Algorithm 1. ELM approach has a wide range of applications, such as protein secondary structure prediction [23], XML document classification [24], classification in P2P networks [25] and graph classification [26].

Algorithm 1: *training process of ELM*

Input   : a training set $V = (x_i, t_i) | x_i \in R^n, t_i \in R^m, i = 1, \dots, N$, the number of hidden node L
              and activation function g(v)
Output: an ELM instance
1   1)randomly generate every input weight w_i and bias b_i, i = 1,...,L;
2   2)calculate hidden node output matrix H;
3   3)calculate output weight $\beta = H^\dagger T$, where $H^\dagger$ is the Moore-Penrose generalized inverse of
     matrix H, $T = [t_1, \dots, t_N]^T$.

---

4 ME-MGC Algorithm

In this section, we propose a *ME-MGC* approach.

4.1 Overview of ME-MGC

In this section, an overview of *ME-MGC* algorithm based on MapReduce is provided. Given a multi-graph set $MG = \{mg_1, mg_2, \dots, mg_n\}$ and the graph set $G = \{g | g \in mg_i, mg_i \in MG\}$ consisting of overall graphs of *MG*. *ME-MGC* contains three steps: (1) mining frequent subgraphs *FG* of *G*, (2) mining feature subgraphs *F* and (3) constructing the predict model. The first step is implemented based on *MRFSM* [14] which has three-round MapReduce jobs: *getCFS* (i.e. getting candidate

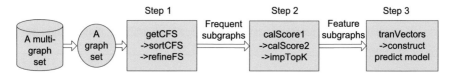

Fig. 1 An overview of *ME-MGC*

frequent subgraphs); *sortCFS* (i.e. sorting candidate frequent subgraphs); *refineFS* (i.e. refining and obtaining frequent subgraphs). Also, the second step needs three-round MapReduce jobs: *calScore*1 (i.e. calculating partial scores *ps* of frequent subgraphs produced in the first step); *calScore*2 (i.e. aggregating partial results output by calScore1 to get final scores); *impTopK* (i.e. obtaining top-*k* subgraphs, namely feature subgraphs). The last step is completed based on *ELM**. Figure 1 shows the overview processing framework of *ME-MGC*. Next, the aforementioned three steps of *ME-MGC* are represented in detail.

4.2   Mining Frequent Subgraphs

In order to mine frequent subgraphs of a large graph dataset, a MapReduce job chain is implemented consisting of *getCFS*, *sortCFS* and *refineFS*.

4.2.1   Getting Candidate Frequent Subgraphs

GetCFS retrievals candidate frequent subgraphs described as Algorithm 2. In the map phase, each map task outputs the local frequent subgraphs which are candidate subgraphs. In the reduce phase, the upper bounds of frequency of candidate subgraphs are estimated. A candidate frequent subgraph is eliminated whose upper bound is less than minimum frequency threshold.

4.2.2   Sorting Candidate Frequent Subgraphs

SortCFS sorts the candidate subgraphs produced by *getCFS* according to edge size shown as Algorithm 3. We utilize the sort function of MapReduce to improve the performance. The size of every candidate subgraph is used as the key for a map task. Meanwhile, only one reduce task is adopted. The records received by the same reduce task are then sorted by the sort function of MapReduce. In addition, inclusion relations of candidate frequent subgraphs are calculated and output by this reduce task.

Algorithm 2: *getCFS*

---

1   //map task
2   List *graphPartition*;// store a subset of graphs set
3   estimate f; //local frequent threshold
4   **Map** (< *Offset, a multi − graph* >)
5   add into *graphPartition* all graphs of this multi-graph;

6   **Cleanup** ()
7   calculate local frequent subgraphs $LFS = \{lfs_1, lfs_2, \dots, lfs_i\}$ for *graphPartition* with frequency *fre* $\geq f$;
8   encode frequent subgraphs $EFS = \{v(lfs_1), v(lfs_2), \dots, v(lfs_i)\}$;
9   //$v(lfs_i)$ is the code of lfs_i
10   emit(< $v(lfs)$, (*partitionId, fre*) >);
11   //$v(lfs) \in EFS$, *partitionId* is id of *graphPartition*, *fre* is local frequency of v

12 //reduce task
13 **Reduce** (v, *list* < *partitionId, fre* >)
14   calculate the frequency upper bound *fub(v)* of v;
15   **if** *fub(v)* $\geq f$ **then**
16   emit(< v, efs >);
17   //*efs* means the sum of exact frequent for v

4.2.3 Refining and Obtaining Frequent Subgraphs

RefineFS refines the candidate subgraphs and gets the final results. A map task reads a subset S_i of graph data set and sorted candidate subgraphs(*SCS*). After that, we calculate the local frequency $f_i(cg)$ of candidate subgraph $cg \in SCS$ for the subset S_i in the map phase which outputs key-value pair $< cg, f_i(cg) >$. If the local frequencies of candidate subgraphs have been calculated in the *getCFS*, they do not need to be recalculated. After that, exact global frequency of every candidate subgraph is calculated in the reduce tasks. If the global frequency of a subgraph is no less than the minimum frequency threshold, this subgraph is a desirable frequent subgraph and is output.

After mining frequent subgraphs of G, we mine feature subgraphs from the derived frequent subgraphs.

4.3 Mining Feature Subgraphs

Feature subgraphs $F = \{f_1, f_2, \dots, f_k\}$ are top-k frequent subgraphs with score function *score(fg)*, $fg \in FG$. Different from the traditional top-k query problem, the score calculation does not depend on one record but the overall dataset. So, we mine feature subgraphs using the following two steps instead of adopting traditional top-k query techniques. At first, we calculate scores of frequent subgraphs. Then, we answer top-k query. A MapReduce job chain is designed to complete this. *calScore*1 gets partial scores, which are aggregated by *calScore2* to get scores $\{score(fg)|fg \in FG\}$. After that, *impTopK* answers top-k query to get F. In the following, we discuss the details of these MapReduce jobs.

Algorithm 3: *sortCFS*

---

1 //map task
2 **Map** (*v, efs*)
3   emit(< *s*, (*v, efs*) >);// *s* is the edge size of *v*

4 //reduce task
5 List *canSubGraph*=empty; //candidate subgraphs in current layer.
6 List *Id*=empty;//ids of candidate subgraphs in canSubGraph
7 *layer*=0;
8 *currentId*=0;
9 **Reduce** (< *s, list(v, efs)* >)
10   **if** *layer==0* **then**
11   layer++;
12   **for** *each element* (*v, efs*) $\in$ *list(v, efs)* **do**
13   add *v* into *canSubGraph*;
14   add *currentId* into *Id*;
15   *currentId*++;
16   emit(< (*v, efs*), *NULL* >)//*NULL* means having no subgraphs;

17   **else**
18   List *subGraph*=empty;
19   **for** *each element* (*v, efs*) $\in$ *list(v, efs)* **do**
20   **for** *each element* $v' \in$ *canSubGraph* **do**
21   **if** *subgraphIsomorphismTest(v',v)* **then**
22   add *id(v')* into *subGraph*;
23   emit(< (*v, efs*), *subGraph* >);
24   update *canSubGraph* and *Id*;

4.3.1 Calculating Partial Scores of Frequent Subgraphs

We define some concepts before we introduce *CalScore1*.

Definition 6 (*Partition*) Given a dataset D, a *partition* p_i is a subset of D which meets the following two conditions: (1) $\cup p_i = D$, where $i \geq 0$ and $i \leq m - 1$; (2) $p_i \cap p_j = \emptyset$, where $i/j \geq 0$, $i/j \leq m - 1$ and $i \neq j$.

Definition 7 (*Fragment*) Given a dataset D and its *partition* set $P = \{p_0, p_1, \ldots, p_{m-1}\}$, a fragment is a partition pair $< p_i, p_j >$, $p_i, p_j \in P$. In total, there are $m * (m + 1)/2$ fragments for D and P.

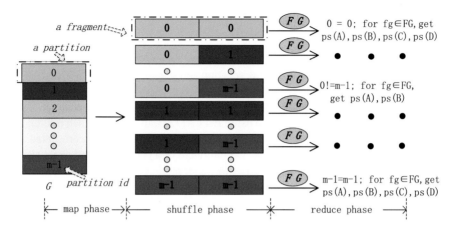

Fig. 2 overview of *calScore*1

*calScore*1 gets the partial results. In map phase, with hash and copy techniques, the multi-graph dataset *MG* is divided into a *partition* set $P = \{p_0, p_1, \ldots, p_{m-1}\}$. If $hash(mg_j) = i$, $mg_j \in p_i (i \geq 0$ and $i \leq m - 1)$. In reduce phase, overall fragments for *MG* and *P* are produced to calculate the partial scores of subgraphs showed in Fig. 2.

4.3.2 Getting Final Scores

*CalScore*2 aggregates the results of *calScore*1 to get the final scores of frequent subgraphs. After reading outputs of *canScore*1, map tasks output $< v(fg), partial\ result >$ pairs. Overall partial results of a frequent subgraph are shuffled to the same reduce task and are aggregated to the final score.

4.3.3 Obtaining Feature Subgraphs

ImpTopK selects as feature subgraphs *k* frequent subgraphs from *FG* whose scores are larger than others. Every map task computes the local top-*k* frequent subgraphs in the corresponding input split. A reduce task is launched to aggregate overall local top-*k* frequent subgraphs and to calculate the global top-*k* frequent subgraphs.

4.3.4 Building Prediction Model

After getting feature subgraphs, multi-graphs are preprocessed to vectors according to *feature subgraph representation of multi − graph*. Then, we build prediction model based on *ELM**.

5   Performance Evaluation

In this section, we compare the precision and recall for ELM and other classification models, and compare the training time, speedup and scalability for *gMGFL* and *ME-MGC* over both real and synthetic datasets.

DBLP dataset. Every paper p_i is regarded as a multi-graph mg_i. The abstract of p_i is a graph $g \in mg_i$, which is obtained using E-FCM [27]. In addition, each reference is modelled as a graph. For example, p_i has m references. So p_i can be represented as a multi-graph including $m + 1$ graphs. Domain field of a paper is treated as its class label. Two domain fields are selected in this paper namely artificial intelligence AI and computer vision CV. After preprocessing, there are 7661 AI multi-graphs and 1817 CV multi-graphs. We randomly select 1817 AI multi-graphs and using all 1817 CV multi-graphs to test.[1]

Synthetic dataset (SYN). Each National Cancer Institute (NCI) data set belongs to a bio-assay task for anti-cancer activity prediction [28]. If a chemical compound is active against the corresponding cancer, it is positive. Otherwise, it is negative. We generate a synthetic multi-graph dataset with a graph data set(with ID 1) [1]. We randomly select one to four positive graphs and several negative graphs to build a positive multi-graph. A negative multi-graph is build by randomly selecting a number of negative graphs. The number of graphs in each multi-graph varies from 1 to 10. In total, we built 500,000 positive and 500,000 negative multi-graphs. The total number of graphs is 4,997,537.

A 31-node (1 master and 30 slaves) cluster is used to test. Every machine is collocated with two 3.1 HZ CPUs, 8 GB Memory, 500 GB hard disk, Redhat 4.4.4-13 operation system and Hadoop-1.2.1. 10-fold cross-validation is adopted. Mean precision and recall are reported in this paper.

5.1   Precision and Recall

In this section, we compare precision and recall over variable real and synthetic datasets.[2]

Figure 3 shows the precision on variable DBLP datasets. Figure 3 displays precisions of *gMGFL + ELM* is highest among all algorithms. Because the precision of *ELM* is higher than *NBayes* and *SVM* on the same DBLP dataset.

---

[1]DBLP dataset can be downloaded from http://arnetminer.org/citation.

[2]*gMGFL + NBayes(orSVM, orELM)* denotes *gMGFL* using NBayes, SVM and ELM classification model, respectively. *ME-MGC + PNBayes(ELM)* represents *ME-MGC* using parallel NBayes and parallel ELM prediction model, respectively. In the case of without causing ambiguity, *ME-MGC* represents *ME-MGC + ELM*.

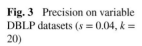

Fig. 3 Precision on variable DBLP datasets ($s = 0.04$, $k = 20$)

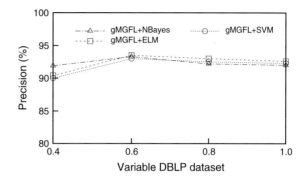

Fig. 4 Recall on variable DBLP datasets ($s = 0.04$, $k = 20$)

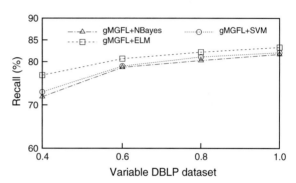

Figure 4 shows the recalls on variable DBLP datasets. The recalls of *gMGFL + ELM* are highest among all algorithms because *gMGFL + ELM* adopt *ELM* whose recall is higher than those algorithms embedded with *NBayes* and *SVM* on the same DBLP dataset.

Figure 5 shows the precisions of different methods on variable SYN datasets. Figure 6 shows the recalls of different methods on variable SYN datasets. The experimental results on variable synthetic datasets are similar with variable DBLP datasets. The reasons can also be referred to DBLP datasets.

Fig. 5 Precision on variable SYN datasets ($s = 0.12$, $k = 15$)

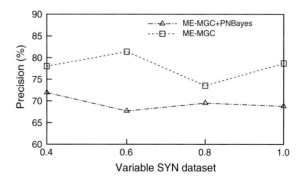

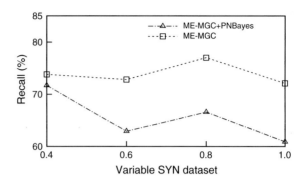

Fig. 6 Recall on variable SYN datasets ($s = 0.12$, $k = 15$)

5.2 Training Time

In this section, we compare the classification model constructing time(training time) of *gMGFL* and *ME-MGC* on variable DBLP and synthetic datasets. *gMGFL* runs in stand-alone. *ME-MGC* runs in 31-node cluster.

Figure 7 exhibits the training time on variable DBLP datasets. *gMGFL* is faster than *ME-MGC*. Because *ME-MGC* consists of several MapReduce jobs. The launch of these jobs costs much time. In addition, *gMGFL* is in-memory algorithm suitable for small datasets.

Figure 8 exhibits the training time on variable synthetic datasets. *gMGFL* can not run on these synthetic datasets successfully.

5.3 Speedup and Scaleup

We evaluate the speedup and scaleup of *ME-MGC + PNBayes* and *ME-MGC + ELM* on synthetic dataset.

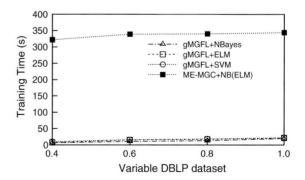

Fig. 7 Training time on variable DBLP datasets ($s = 0.04$, $k = 20$)

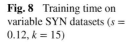

Fig. 8 Training time on variable SYN datasets ($s = 0.12, k = 15$)

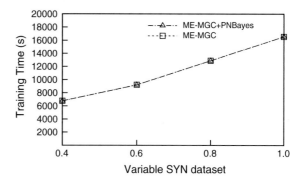

Fig. 9 Training time for processing *SYN* dataset on m-node cluster (where m = 12, 18, 24 and 30)

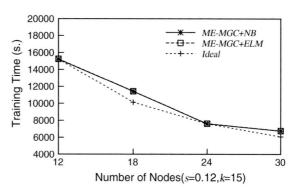

Fig. 10 Training time for processing *SYN**n datasets (where n = 0.4, 0.6, 0.8 and 1.0) on a 31-node cluster

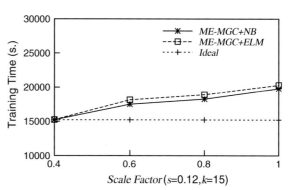

Figure 9 exhibits that both *ME-MGC + PNBayes* and *ME-MGC + ELM* have a good speedup on synthetic dataset. Figure 10 exhibits that both *ME-MGC + PNBayes* and *ME-MGC + ELM* have a good scalability on synthetic dataset.

5.4 Performance with Different Hidden Node Number

In this section, we evaluate the precisions and recalls of *gMGFL* and *ME-MGC* for different hidden node number on DBLP and SYN dataset.

Figure 11 shows comparisons of precision and recall for *gMGFL* with variable hidden node number on DBLP dataset. Figure 12 shows comparisons of precision and recall for *ME-MGC* with a variable number of hidden nodes on SYN dataset. With the increment of the hidden node number, precision and recall for *gMGFL* and *ME-MGC* are stable.

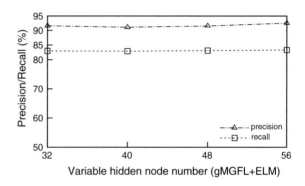

Fig. 11 Precision and recall with variable number of hidden nodes on DBLP dataset ($s = 0.04$, $k = 20$)

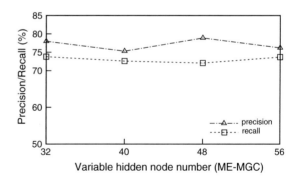

Fig. 12 Precision and recall with variable number of hidden nodes on SYN dataset ($s = 0.12$, $k = 15$)

6 Conclusions

In this paper, we propose a parallel approach *ME-MGC* based on MapReduce to resolve massive multi-graph classification problem. Meanwhile, ELM prediction model is applied to predict multi-graph data type for improving the algorithm performance. Extensive experimental results on both real and synthetic datasets display that our algorithm apparently outperforms *gMGFL* and the extended algorithm with parallel Bayes because of its high precision and good scalability.

Acknowledgments The work is partially supported by the National Basic Research Program of China (973 Program) (No. 2012CB316201), the National Natural Science Foundation of China (No. 61272179, No. 61472071).

References

1. Wu, J., Zhu, X., Zhang, C., et al.: Bag constrained structure pattern mining for multi-graph classification. TKDE **26**(10), 2382–2396 (2014)
2. Wu, J., Pan, S., Zhu, X., et al.: Boosting for multi-graph classification. T. Cybern. **45**(3), 430–443 (2015)
3. MapReduce. http://en.wikipedia.org/wiki/MapReduce
4. Huang, G., Zhu, Q., Siew, C.K.: Extreme learning machine: a new learning scheme of feedforward neural networks. In: IJCNN, pp. 985–990 (2004)
5. Huang, G., Liang, N., Rong, H., et al.: On-line sequential extreme learning machine. In: IASTED, pp. 232–237 (2005)
6. He, Q., Shang, T., Zhuang, F., et al.: Parallel extreme learning manchine for regression based on MapReduce. Neurocomputing **102**(2), 52–58 (2013)
7. Xin, J., Wang, Z., Chen, C., et al.: *ELM**: distributed extreme learning machine with MapReduce. World Wide Web **17**(5), 1189–1204 (2014)
8. Xin, J., Wang, Z., Qu, L., et al.: Elastic extreme learning machine for big data classification. Neurocomputing, **149**(Part A), 464–471 (2015)
9. Bi, X., Zhao, X., Wang, G., et al.: Distributed extreme learning machine with kernels based on MapReduce. Neurocomputing **149**, 456–463 (2015)
10. Wang, B., Huang, S., Qiu, J., et al.: Parallel online sequential extreme learning machine based on MapReduce. Neurocomputing **149**, 224–232 (2015)
11. Kuramochi, M., Karypis, G.: Finding frequent patterns in a large sparse graph. In: SDM, pp. 345–356 (2004)
12. Kuramochi, M., Karypis, G.: Grew-a-scalable frequent subgraph discovery algorithm. In: ICDM, pp. 439–442 (2004)
13. Hill, S., Srichandan, B., Sunderraman, R.: An Iterative mapreduce approach to frequent subgraph mining in biological datasets. In: BCB, pp. 661–666 (2012)
14. Lin, W., Xiao, X., Ghinita, G.: Large-scale frequent subgraph mining in MapReduce. In: ICDE, pp. 844–855 (2014)
15. Huang, G., Zhu, Q., Siew, C.K., et al.: Extreme learning machine: theory and applications. Neurocomputing **70**(1–3), 489–501 (2006)
16. Huang, G., Chen, L.: Enhanced random search based incremental extreme learning machine. Neurocomputing **71**(16–18), 3460–3468 (2008)
17. Huang, G., Ding, X., Zhou, H.: Optimization method based extreme learning machine for classification. Neurocomputing **74**(1–3), 155–163 (2010)

18. Huang, G., Wang, D., Lan, Y.: Extreme learning machines: a survey. Int. J. Mach. Learn. Cybern **2**(2), 107–122 (2011)
19. Huang, G., Zhou, H., Ding, X., et al.: Extreme learning machine for regression and multiclass classification. In: TSMC. Part B Cybern. **42**(2), 513–529 (2012)
20. Huang, G., Wang, D.: Advances in extreme learning machines (ELM2011). Neurocomputing **102**, 1–2 (2013)
21. Huang, G.: An insight into extreme learning machines: random neurons, random features and kernels. Cogn. Comput. **6**(3), 376–390 (2014)
22. Huang, G., Bai, X., Kasun, L.L.C., et al.: Local receptive fields based extreme learning machine. In: Comp. Int. Mag. **10**(2), 18–29 (2015)
23. Wang, G., Zhao, Y., Wang, D.: A protein secondary structure prediction framework based on the extreme learning machine. Neurocomputing **72**(1–3), 262–268 (2008)
24. Zhao, X., Wang, G., Bin, X., et al.: XML document classification based on ELM. Neurocomputing **74**(16), 2444–2451 (2011)
25. Sun, Y., Yuan, Y., Wang, G.: An OS-ELM based distributed ensemble classification framework in P2P networks. Neurocomputing **74**(16), 2438–2443 (2011)
26. Wang, Z., Zhao, Y., Wang, G., et al.: On extending extreme learning machine to non-redundant synergy pattern based graph classification. Neurocomputing **149**, 330–339 (2015)
27. Perusich, K., Senior, M.: Using fuzzy connitive maps for knowledge management in a conflict environment. TSMC. Part C **36**(6), 810–821 (2006)
28. Pubchem. http://pubchem.ncbi.nlm.nih.gov

Extreme Learning Machine for Large-Scale Graph Classification Based on MapReduce

Zhanghui Wang, Yuhai Zhao and Guoren Wang

Abstract Discriminative subgraph mining from a large collection of graph objects is a crucial problem for graph classification. Extreme Learning Machine (ELM) is a simple and efficient Single-hidden Layer Feedforward neural Networks (SLFNs) algorithm with extremely fast learning capacity. In this paper, we propose a discriminative subgraph mining approach based on ELM-Filter strategy within the scalable MapReduce computing model. We randomly partition the collection of graphs among worker nodes, and each worker applies a fast pattern evolutionary method to mine a set of discriminative subgraphs with the help of ELM-Filter strategy in its partition. And, the set of discriminative subgraphs must produce higher ELM training accuracy. The union of all such discriminative subgraphs is the mining result for the input large-scale graphs. Extensive experimental results on both real and synthetic datasets show that our method obviously outperforms the other approaches in terms of both classification accuracy and runtime efficiency.

Keywords Discriminative subgraph pattern · MapReduce · Extreme Learning Machine · Graph classification

1 Introduction

The graph classification framework being widely used is first to select a set of subgraph features from graph databases, and then to build a generic classification model using the set of subgraph features selected. Discriminative subgraphs that are frequent in one class labeled graph set but infrequent in the other class labeled graph sets are more suitable for classification requirement.

Z. Wang (✉) · Y. Zhao · G. Wang
College of Information Science and Engineering, Northeastern University, Shenyang, China
e-mail: wzh_neu@163.com

Y. Zhao
Key Laboratory of Computer Network and Information Integration, Southeast University,
Ministry of Education, Nanjing, China
e-mail: zhaoyuhai@ise.neu.edu.cn

© Springer International Publishing Switzerland 2016
J. Cao et al. (eds.), *Proceedings of ELM-2015 Volume 1*,
Proceedings in Adaptation, Learning and Optimization 6,
DOI 10.1007/978-3-319-28397-5_8

Several main memory-based approaches [1, 2] have been proposed to mine discriminative subgraphs in small-scale graph databases, but they are both time and memory costly when applied to process large-scale graph databases. Cloud computing and the widespread MapReduce framework can be used to solve the scalability and computationally-intensive problems.

With a large amount of mined discriminative graph patterns, how to effectively build a graph classification model with these graph pattern features becomes the key problem. Previous work [1, 2] generally adopt Library Support Vector Machines (LIBSVM) [3] as the classification model. Although LIBSVM can get good classification accuracy, it faces the problems of slow learning speed and trivial human intervene in general. Extreme Learning Machine (ELM) is a simple and efficient Single-hidden Layer Feedforward neural Networks (SLFNs) algorithm with extremely fast learning capacity [4–9]. Furthermore, ELM for classification is less sensitive to user specified parameters and can be implemented easily [10].

In this paper, we employ the HaLoop MapReduce framework and evolutionary computation techniques based on ELM-Filter to find discriminative subgraphs efficiently and propose a large-scale discriminative subgraph mining algorithm with MapReduce, named MRGC.[1]

The rest of the paper is organized as follows. We briefly review the related work in Sect. 2. Section 3 describes the discriminative subgraph mining problem and introduces the MapReduce framework and ELM algorithm. In Sect. 4, we describe the discriminative subgraph mining approach based on evolutionary computation and ELM-Filter strategy. In Sect. 5, we evaluate our experiment results. Finally, we make a conclusion in Sect. 6.

2   Related Work

Discriminative graph pattern mining aims to mine the patterns that occur with disproportionate frequency in some classes versus others. Various efficient algorithms have been developed, such as LEAP [1], LTS [2].

Due to the computation and I/O intensive characteristic of graph pattern ming in large-sale graph database, more and more efforts are geared towards solving it with parallel techniques. MapReduce [11, 12] has emerged as a popular alternative for large-scale parallel data analysis and Hadoop is an open-source implementation of MapReduce. Evolutionary computation [13, 14] are usually used to obtain global solutions that can be used for discriminative subgraph pattern mining.

Extreme learning machine (ELM) has been originally developed based on single-hidden layer feed-forward neural networks(SLFNs) with the weights in hidden nodes randomly assigned and the output weights decided by the MoorePenorose pseudoinverse. It has been shown that ELM has an extremely high learning speed and a good

---

[1]*MapReduce Graph Classifcation.*

generalization performance [4–9]. In this paper, we integrate MapReduce with evolutionary computation based on ELM-Filter to mine discriminative subgraph patterns efficiently.

3 Preliminaries and Problem Definition

In this section, we first provides a brief MapReduce and ElM primer, and then
Sect. 3.3 describe the concept of discriminative subgraph and formalizes the problem
statement.

3.1 MapReduce

MapReduce which is a distributed framework for processing large-scale data contains three phases: map, shuffle, and reduce. With the MapReduce framework, users
can implement a map function and a reduce function to process their applications.
We only focus on the design of the map and reduce functions, as the shuffle phase is
automatically handled by the MapReduce infrastructure. Figure 1 shows our MapReduce framework for discriminative subgraph mining based on ELM-Filter. First, the
master node assign the map and the reduce tasks to each worker. Then, discriminative subgraphs was searched based on evolutionary computation and ELM-Filter
strategy. At last, the discriminative subgraph results was aggregated in each worker.

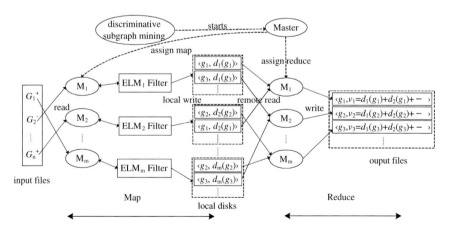

Fig. 1 The framework of MapReduce for discriminative subgraph mining based on ELM-Filter

3.2   Extreme Learning Machine

In this section, we present a brief overview of Extreme Learning Machine (ELM). ELM is a generalized Single Hidden-layer Feedforward Network. In ELM, the hidden-layer node parameters is mathematically calculated instead of being iteratively tuned, providing good generalization performance at thousands of times faster speed than traditional popular learning algorithms for feedforward neural networks. Given N arbitrary samples $(\mathbf{x}_i, \mathbf{t}_i) \in \mathbf{R}^{n \times m}$ and activation function $g(x)$, standard SLFNs are modeled mathematically as

$$\sum_{i=1}^{L} \beta_i g_i(\mathbf{x}_j) = \sum_{i=1}^{L} \beta_i \, g(\mathbf{w}_i \cdot \mathbf{x}_j + b_i) = \mathbf{o}_j, j = 1, \dots, N \qquad (1)$$

where L is the number of hidden layer nodes, $\mathbf{w}_i = [w_{i1}, w_{i2}, \dots, w_{in}]^T$ is the input weight vector, $\beta_i = [\beta_{i1}, \beta_{i2}, \dots, \beta_{im}]^T$ is the output weight vector, b_i is the bias of ith hidden node, and $\mathbf{o}_j$ is the output of the jth node.

To approximate these samples with zero errors means that $\sum_{j=1}^{L} ||\mathbf{o}_j - \mathbf{t}_j|| = 0$ [5], where exist β_i, $\mathbf{w}_i$ and b_i satisfying that

$$\sum_{i=1}^{L} \beta_i \, g(\mathbf{w}_i \mathbf{x}_j + b_i) = \mathbf{t}_j, j = 1, \dots, n \qquad (2)$$

which can be rewritten in terms as

$$\mathbf{H}\beta = \mathbf{T} \qquad (3)$$

where $\mathbf{T} = \left[\mathbf{t}_1^T, \dots, \mathbf{t}_L^T\right]^T_{m \times L}$, $\beta = [\beta_1^T, \dots, \beta_L^T]^T_{m \times L}$ and

$$\mathbf{H} = \begin{bmatrix} h(\mathbf{x}_1) \\ \vdots \\ h(\mathbf{x}_N) \end{bmatrix} = \begin{bmatrix} g(\mathbf{w}_1 \cdot \mathbf{x}_1 + b_1) & \cdots & g(\mathbf{w}_L \cdot \mathbf{x}_1 + b_L) \\ \vdots & \cdots & \vdots \\ g(\mathbf{w}_1 \cdot \mathbf{x}_N + b_1) & \cdots & g(\mathbf{w}_L \cdot \mathbf{x}_N + b_L) \end{bmatrix}_{n \times L} \qquad (4)$$

$\mathbf{H}$ is the ELM feature space to map the n-dimensional input data space into l-dimensional hidden nodes space. The ELM Algorithm [5] is described as Algorithm 1.

In ELM, the parameters of hidden layer nodes, namely $\mathbf{w}_i$ and b_i, is chosen randomly without acknowledging the training data sets. The output weight β is then calculated with matrix computation formula $\beta = \mathbf{H}^\dagger \mathbf{T}$, where $\mathbf{H}^\dagger$ is the Moore-Penrose Inverse of $\mathbf{H}$. In the binary classification case, ELM only uses a single-output node, and the class label closer to the output value of ELM is chosen as the predicted class label of the input data.

Algorithm 1: ELM

1 **for** $i = 1$ *to* n **do**
2 | randomly assign input weight $\mathbf{w}_i$;
3 | randomly assign bias b_i;
4 calculate $\mathbf{H}$;
5 calculate $\beta = \mathbf{H}^\dagger \mathbf{T}$;

3.3 Discriminative Subgraph

An undirected graph can be modeled as $G = (V, E, L)$ where V is a set of vertices and E is a set of edges connecting the vertices. As in Fig. 2, the set of positive graphs in D denoted as D^+ and the set of negative graphs denoted as D^-.

Definition 1 (*Frequency*) Given a graph database $D = \{G_1, G_2, \dots, G_n\}$ and a graph pattern G, $D = D^+ \cup D^-$, the supporting graph set of G is $D_G = \{G_i \mid G \subseteq G_i, G_i \in D\}$. The support of G in D is $|D_G|$, denoted as $sup(G, D)$, the support of G in D^+ and D^- denoted as $sup(G, D^+)$ and $sup(G, D^-)$, respectively; the frequency of G is $|D_G|/|D|$, denoted as $freq(G, D)$, meanwhile, the frequency of G in D^+ and D^- denoted as $freq(G, D^+)$ and $freq(G, D^-)$, respectively.

Definition 2 (*Discriminative Subgraph*) Discriminative subgraph G is a subgraph pattern that occur with disproportionate frequency in one class versus others. The discriminate power score $d(G)$ can be calculated by a given discrimination function.

Problem Statement: Let $D = \{G_1, G_2, \dots, G_n\}$ be a graph database that consists of graph G_i, for $1 \leq i \leq |D|$. Large-scale discriminative subgraph mining problem is to find a set of subgraph patterns with which are more discriminative and can be built for graph classifiers.

Fig. 2 A graph database D with 4 positive graphs and 4 negative graphs

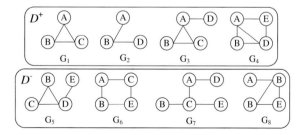

4 Discriminative Subgraph Mining Based on ELM-Filter with MapReduce

In this section, we present the solution for large-scale discriminative subgraph mining based on ELM-Filter with MapReduce.

Map Function: In the map step, each worker node $M_i(i = 1, \ldots, m)$ reads a subset D_i of D, and identifies a set of discriminative subgraphs R_i after pattern reproduction iterations based on ELM-Filter to represent the set of positive graphs in D_i that stored in M_i. Then, for each discriminative subgraph $G \in R_i$, M_i outputs a key-value pair where the key is G and the value indicates the corresponding discrimination score $d_i(G)$.

Reduce Function: With the key and value list obtained from the shuffle phase, M_i inspects the list of values with key G and computes the sum discrimination scores. Then, M_i outputs a key-value pair with key G and value equal to the sum of discrimination scores.

4.1 Pattern Evolutionary Computation

In the map step, we randomly divide graph database D into m disjoint subsets $D_i(i = 1, \ldots, m)$ and sent D_i to worker node M_i. The first goal of our work is to find a set of discriminative subgraphs in each worker node $M_i(i = 1, \ldots, m)$, each positive graph in D_i that stored in M_i must have at least one representative subgraph for classification. We achieve this goal in each worker node by exploring candidate subgraph patterns in a process resembling biological evolution (evolutionary commutation) which consist of two evolutionary mechanisms such as reproduction, and selection. Evolutionary computation begins with a set of sample points in the search space and gradually biases to the regions of high quality fitness [13, 14].

We use a discrimination score definition as the fitness function in Eq. (5) for MapReduce.

$$d_i(G) = sup(G, D_i^+) - sup(G, D_i^-) \qquad (5)$$

The discrimination score in Eq. (5) is used to measure the fitness of subgraph patterns in each worker node M_i, the bigger the score is, the more discriminative subgraph pattern is.

4.2 Pattern Reproduction

All the representative subgraph candidates should have a probability of being selected for subgraph pattern reproduction to generate a larger subgraph. During each iteration, we give a proportional threshold α ($0 < \alpha \leq 1$) to randomly select a subset of the representative subgraph candidates for reproduction.

The probability is always between 0 and 1. This reproduction strategy is commonly used in evolutionary algorithms [13]. The intuition here is that candidate subgraph patterns with higher discriminative scores are more likely be extended to larger subgraph patterns with high scores.

4.3　Pattern Selection

With the new representative subgraph candidates generated from subgraph pattern reproduction step in M_i, the goal of subgraph pattern selection is to find a subset of discriminative subgraph patterns among which each positive graph in D_i can have at least one representative pattern for graph classification. We should select a subset of representative subgraphs $R_i = \{g_1, g_2, \ldots, g_k\}$ to cover the set of positive graphs in D_i. R_i and the value k should simultaneously satisfy Eqs. (6) and (7). So, the values k can not be the same for different workers.

$$c_i(g_1, D_i) \cup c_i(g_2, D_i) \cup \cdots \cup c_i(g_k, D_i) = D_i^+ \tag{6}$$

We use a heuristic algorithm to select R_i from the set of representative subgraph candidates. First, the representative subgraph candidates should be sorted in descending order of their discrimination scores, and we choose some highest scores subgraph patterns which satisfy Eq. (6) as the select result R_i. Apparently, when we choose the top-k highest scores subgraph patterns, Eq. (7) is satisfied.

$$max\{\{d_i(g_1, D_i) + d_i(g_2, D_i) + \cdots + d_i(g_k, D_i)\}/k\} \tag{7}$$

4.4　Patterns for ELM-Filter

For the positive graph set D_i^+ that stored in M_i and a subset of selected representative subgraphs $R_i = \{g_1, g_2, \ldots, g_k\}$. we can use the ELM algorithm to evaluate the training accuracy of the selected representative subgraphs R_i. A high training accuracy of the selected representative subgraphs R_i indicate that R_i is viable and having high precision of prediction. We use 0.9 for the ELM training accuracy threshold in this paper. Otherwise we should reselect a new subset of discriminative subgraphs R_j which satisfy the high training accuracy for the positive graph set D_i^+.

With subgraph pattern evolutionary operations and the ELM-Filter strategy, we could quickly identify a set of locally optimal discriminative subgraph results R_i based on evolutionary computation and ELM-Filter to cover the set of positive graphs in D_i that stored in M_i. Then, Map function showed in algorithm 2 output the key-value pairs that the key is a subgraph pattern and the value is the corresponding discrimination score. Next, the key and value list got from the shuffle phase, M_i inspects the list of values with the key and computes the sum discrimination scores.

Algorithm 2: Map Function

Input: Graph dataset D_i, θ and α

Output: Representative subgraph set and corresponding discrimination scores

1 θ: maximum number of iterations;
2 α: reproduction threshold;
3 Worker node $M_i(i = 1, \cdots, m)$ reads a subset D_i of D;
4 $S = $ All the candidate subgraph patterns in D_i and corresponding discrimination scores;
5 **for** $j = 1$ *to* θ **do**
6   randomly select a subgraphs pattern g in the pattern candidate set S;
7   extend pattern g to g' with one more edge attached to g;
8   insert g' to the pattern candidate set S;
9   calculate the discrimination score of g' and insert it to S;

10 select $\mathcal{R}_i$ from S which should satisfy Eq. (6) and Eq. (7);
11 **if** $\mathcal{R}_i$ *satisfy the ELM-Filter strategy* **then**
12   output $\mathcal{R}_i$ and corresponding discrimination scores;

13 **else**
14   reselect a new $\mathcal{R}'_i$ that satisfy the ELM-Filter strategy;
15   output $\mathcal{R}'_i$ and corresponding discrimination scores

---

At last, M_i outputs a key-value pair with key equal to the subgraph pattern and the value equal to the sum of discrimination scores.

$$\mathcal{R} = \sum_{i=1}^{m} \mathcal{R}_i \tag{8}$$

We introduce an algorithm named MRGC for discriminative subgraph mining with MapReduce, and MRGC consists of two main steps. The fist step is the map step showed in algorithm 2. Through three pattern operation strategies, the map function outputs the representative subgraphs and their corresponding discrimination scores efficiently. The next reduce step showed in algorithm 3 aggregates the key-values pairs from the shuffle step, meanwhile, outputs the results $\mathcal{R}$ which defined in Eq. (8).

Algorithm 3: Reduce Function

Input: Representative subgraphs set $\mathcal{R}_i$ and corresponding discrimination scores

Output: Union of the representative subgraphs set $\mathcal{R}$

1 aggregate the key-value pairs;
2 output $\mathcal{R}$;

---

5 Experimental Evaluation

In this section, we first evaluate the efficiency of MRGC with ELM-Filter and without ELM-Filter on the synthetic dataset and next evaluate experimentally MRGC against two competitors on both synthetic and real-world graph datasets. And, then study the performance of ELM for classification compared with SVM based on the mined graph patterns.

5.1 Experimental Setup

We set up a cluster of 11 commodity PCs in a high speed Gigabit network, with one PC as the Master node, the others as the worker nodes. Each PC has an Intel Quad Core 2.66 GHZ CPU, 4 GB memory and GentOS Linux 5.6. We use HaLoop (a modified Hadoop 0.20.201) to run MRGC and the default configuration of Hadoop, i.e., $dfs.replication = 3$ and $fs.block.size = 64\,MB$. The MRGC algorithms were implemented in C++ using STL and the SVM and ELM for classification algorithms are carried out in MATLAB 2010.

As competitors, we consider two main memory-based discriminative subgraph mining algorithms LEAP [1] and LTS [2], and we adapt the two competitors namely MRLEAP and MRLTS to run on MapReduce, respectively. The parameter $\sigma = 0.1$ in the MRLEAP. The parameters $\theta = 1000$, $m = 10$ and $\alpha = 0.3$ are default values.

5.2 Data Description

(1) Synthetic: We use a synthetic graph data generator[2] to generate 100000 undirected, labeled and connected graphs, and randomly select half of the graphs to be positive graphs.
(2) NCI: The NCI cancer screen data sets are widely used for graph classification evaluation [1, 15]. We download ten NCI data sets from the PubChem database.[3] Table 1 lists the summary of the ten data sets. We randomly select a negative data set with the same size as the positive one (e.g., 2047 poistive and 2074 negative graphs in NCI 1). As a result, we obtain ten balanced data sets, which comprise 56000 graphs in total. We combine them to construct a new NCI-A dataset (balanced dataset). Meanwhile, we also use all the graphs from the ten data sets (400000 graphs) to construct a new NCI-B dataset (unbalanced dataset) in our experiment.

---

[2]http://www.cais.ntu.edu.sg/~jamescheng/graphgen1.0.zip.
[3]http://pubchem.ncbi.nlm.nih.gov/.

Table 1 Summary of the NCI data sets

Bioassay ID	Tumor description	Actives	Inactives
NCI 1	Lung cancer	2047	38410
NCI 33	Melanoma	1642	38456
NCI 41	Prostate cancer	1568	25967
NCI 47	Nervous sys. tumor	2018	38350
NCI 81	Colon cancer	2401	38236
NCI 83	Breast cancer	2287	25510
NCI 109	Ovarian tumor	2072	38551
NCI 123	Leukemia	3123	36741
NCI 145	Renal cancer	1948	38157
NCI 167	Yeast anticancer	8894	53622

5.3   Efficiency of MRGC

In this subsection, we study the efficiency of MRGC on the synthetic dataset with respect to the parameter: α (reproduction threshold) and θ (maximum number of iterations).

Figure 3 shows the running times vs the two parameters. From the result, we can discover that when the reproduction threshold α increases, the running times of the two methods increase and they have no large difference cost time. In addition, we also observe that the bigger the maximum number of iterations θ is, the more time will be cost.

The phenomena above demonstrate that the most time susceptible operation is the subgraph pattern reproduction in evolutionary computation, which is the same as the traditional subgraph pattern mining algorithms. The more subgraphs are explored, the more time will be cost. Also, the bigger number iterations, the more subgraph evolutionary operations to be need.

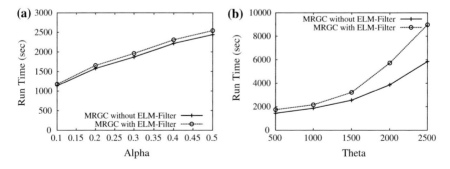

Fig. 3 Running time versus α and θ. **a** Varying α. **b** Varying θ

5.4 Comparison with Other Methods

We compare MRGC with two other representative discriminative subgraph mining methods: MRLEAP and MRLTS that modified to the MapReduce framework. We randomly select half of each dataset to be the corresponding training graph set, and the remainder half of graphs to be the testing graph set. We run the three algorithms on both synthetic and real-world datasets and show the efficiency and effectiveness in Fig. 4, Tables 2, 3 and 4.

Figure 4 shows that MRGC clearly outperforms MRLEAP and MRLTS in terms of all the datasets. The three methods all decrease their time cost when the worker nodes increase, and MRLEAP need more worker nodes to complete the mining task and have the worst performance in the three methods.

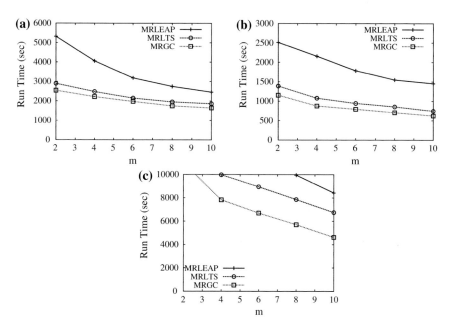

Fig. 4 Running time versus number of workers m. **a** Synthetic dataset. **b** NCI-A dataset. **c** NCI-B dataset

Table 2 Average training and testing time between ELM and SVM

Classifier	Average training time(s)	Average testing time(s)
ELM	0.128	0.032
SVM	9.425	0.943

Table 3 Average SVM testing accuracy between MRLEAP, MRLTS and MRGC

Graph datasets	MRLEAP+SVM	MRLTS+SVM	MRGC+SVM
Synthetic	0.624	0.724	0.754
NCI-A	0.736	0.758	0.784
NCI-B	0.686	0.695	0.725
Average	0.682	0.726	0.754

Table 4 Average ELM testing accuracy between MRLEAP, MRLTS and MRGC

Graph datasets	MRLEAP+ELM	MRLTS+ELM	MRGC+ELM
Synthetic	0.654	0.746	0.778
NCI-A	0.767	0.783	0.812
NCI-B	0.703	0.722	0.785
Average	0.708	0.75	0.792

We use the Support Vector Machine (SVM)[4] [3] with RBF kernel using default parameters values and Extreme Learning Machine (ELM) [5] wiht 1000 hidden nodes based on SGVM to build the classifiers. The training time, testing time are the average values of the 3 datasets. Table 2 shows that ELM needs much less training time and testing time compared to SVM.

As is shown in Tables 3 and 4, MRGC outperforms MRLEAP and MRLTS based on SVM and ELM. On average, MRGC achieves normalized accuracy of 0.042 and 0.084 higher than MRLTS and MRLEAP respectively in Table 4. And the results in Table 3 are the same obvious. Compared Table 4 to Table 3, the prediction accuracy of MRGC+ELM is usually obviously better than MRGC+SVM in all the datasets. From the aspect of average normalized classification accuracy, the contrast results intuitively demonstrate the quality of representative discriminative subgraphs mined by the three methods.

6 Conclusion

In this paper, we propose the problem for large-scale graph classification based on evolutionary computation and ELM-Filter strategy with MapReduce. We propose a fast pattern evolutionary method to mine discriminative subgraphs efficiently and explore candidate subgraph pattern space in a biological evolution way. Through graph pattern operations and ELM-Filter strategy, the method could obtain a set of discriminative subgraphs efficiently. Experiments on both real and synthetic datasets show that our method obviously outperforms the other approaches in terms of classification accuracy and runtime efficiency.

---

[4]http://www.csie.ntu.edu.tw/~cjlin/libsvm/.

Acknowledgments This work was supported by National Natural Science Foundation of China (61272182, 61100028, 61173030 and 61173029), 973 program (2011*CB*302200-*G*), State Key Program of National Natural Science of China (61332014, *U*1401256), 863 program (2012*AA*011004), New Century Excellent Talents (*NCET*-11-0085), Fundamental Research Funds for the Central Universities *N*130504001).

References

1. Yan, X., Cheng, H., Han, J., Yu, P.S.: Mining significant graph patterns by leap search. In: Proceedings of the 2008 ACM SIGMOD International Conference on Management of Data, pp. 433–444, ACM (2008)
2. Jin, N., Wang, W.: Lts: discriminative subgraph mining by learning from search history. In: 2011 IEEE 27th International Conference on Data Engineering (ICDE), pp. 207–218, IEEE (2011)
3. Chang, C.-C., Lin, C.-J.: Libsvm: a library for support vector machines. ACM Trans. Intell. Syst. Technol. (TIST) **2**(3), 27 (2001)
4. Huang, G.-B., Zhu, Q.-Y., Siew, C.-K.: Extreme learning machine: a new learning scheme of feedforward neural networks. In: Proceedings of the 2004 IEEE International Joint Conference on Neural Networks, vol. 2, pp. 985–990, July 2004
5. Huang, G.-B., Zhu, Q.-Y., Siew, C.-K.: Extreme learning machine: theory and applications. Neurocomputing **70**, 489–501 (2006)
6. Huang, G.-B., Zhu, Q.-Y., Mao, K.Z., Siew, C.-K., Saratchandran, P., Sundararajan, N.: Can threshold networks be trained directly? IEEE Trans. Circuits Syst. Ii: Analog Digit. Signal Process. **53**, 187–191 (2006)
7. Huang, G.-B., Chen, L., Siew, C.K.: Universal approximation using incremental constructive feedforward networks with random hidden nodes. IEEE Trans. Neural Netw. **17**, 879–892 (2006)
8. Huang, G.-B., Chen, L.: Enhanced random search based incremental extreme learning machine. Neurocomputing **71**, 3460–3468 (2008)
9. Huang, G.-B., Chen, L.: Convex incremental extreme learning machine. Neurocomputing **70**, 3056–3062 (2007)
10. Huang, G.-B., Ding, X., Zhou, H.: Optimization method based extreme learning machine for classification. Neurocomputing **74**, 155–163 (2010)
11. Tao, Y., Lin, W., Xiao, X.: Minimal mapreduce algorithms. In: Proceedings of the 2013 international conference on Management of data, pp. 529–540, ACM (2013)
12. Cui, B., Mei, H., Ooi, B.C.: Big data: the driver for innovation in databases. Natl. Sci. Rev. **1**(1), 27–30 (2014)
13. De Jong, K.: Evolutionary computation: a unified approach. In: Proceedings of the Fourteenth International Conference on Genetic and Evolutionary Computation Conference Companion, pp. 737–750, ACM (2012)
14. Storn, R., Price, K.: Differential evolution-a simple and efficient adaptive scheme for global optimization over continuous spaces. ICSI Berkeley (1995)
15. Ranu, S., Singh, A.K.: Graphsig: a scalable approach to mining significant subgraphs in large graph databases. In: IEEE 25th International Conference on Data Engineering, 2009. ICDE'09, pp. 844–855, IEEE (2009)

The Distance-Based Representative Skyline Calculation Using Unsupervised Extreme Learning Machines

Mei Bai, Junchang Xin, Guoren Wang and Xite Wang

Abstract *A representative skyline* contains k skyline points that can represent its full skyline, which is very useful in the multiple criteria decision making problems. In this paper, we focus on the distance-based representative skyline (k-DRS) query which can describe the tradeoffs among different dimensions offered by the full skyline. Since k-DRS is a NP-hard problem in d-dimensional ($d \geq 3$) space, it is impossible to calculate the exact k-DRS in d-dimensional space. By in-depth analyzing the properties of the k-DRS, we propose a new perspective to solve this problem and a k distance-based representative skyline algorithm based on US-ELM (DRSELM) is presented. In DRSELM, first we apply US-ELM to divide the full skyline set into k clusters. Second, in each cluster, we design a method to select a point as the representative point. Experimental results show that our DRSELM significantly outperforms its competitors in terms of both accuracy and efficiency.

Keywords Skyline · k representative skyline · k-DRS · US-ELM

1 Introduction

Given a large dataset, it is impracticable for a user to browse all the points in the dataset. Hence, obtaining a succinct representative subset of the dataset is crucial. A well-established approach to representing a dataset is with the *skyline* operator [1].

M. Bai (✉) · J. Xin · G. Wang · X. Wang
College of Information Science & Engineering, Northeastern University,
Shenyang, People's Republic of China
e-mail: baimei861221@163.com

J. Xin
e-mail: xinjunchang@ise.neu.edu.cn

G. Wang
e-mail: wanggr@mail.neu.edu.cn

X. Wang
e-mail: wangxite@research.neu.edu.cn

© Springer International Publishing Switzerland 2016
J. Cao et al. (eds.), *Proceedings of ELM-2015 Volume 1*,
Proceedings in Adaptation, Learning and Optimization 6,
DOI 10.1007/978-3-319-28397-5_9

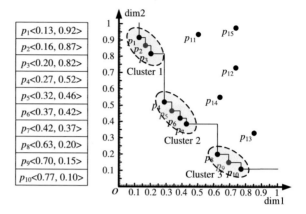

Fig. 1 Example of distance-based representative skyline

p_1<0.13, 0.92>
p_2<0.16, 0.87>
p_3<0.20, 0.82>
p_4<0.27, 0.52>
p_5<0.32, 0.46>
p_6<0.37, 0.42>
p_7<0.42, 0.37>
p_8<0.63, 0.20>
p_9<0.70, 0.15>
p_{10}<0.77, 0.10>

The skyline consists of the points which are not *dominated* by any other point. Given two points p_1 and p_2, if the values of p_1 are as good as or better than those of p_2 in any dimension, and better in at least one dimension. With loss of generality, we assume that a smaller value indicates a better performance in all dimensions. However, when the skyline size is large, the full skyline is helpless to the user. Detecting a subset of the full skyline set with fixed size (such as k points) is necessary. As investigated in [2], Tao et al. proposed a *distance-based representative skyline* (k-DRS for short) which can best describe the tradeoffs among different dimensions offered by the full skyline. They applied a *distance* metric to measure the "representativeness" of the chosen set. Given a subset $\mathcal{K}$ with k skyline points from the full skyline set S, $Er(\mathcal{K}, S) = \max_{p \in S-\mathcal{K}} \min_{p' \in \mathcal{K}} \| p,p' \|$, where $\| p,p' \|$ is the Euclidean distance between p and p'. The k-DRS is the set $\mathcal{K}$ with the minimum value $Er(\mathcal{K}, S)$. As illustrated in Fig. 1, given $k = 3$, the 3-DRS is $\{p_2, p_5, p_9\}$ with the corresponding value $Er(\mathcal{K}, S) = \| p_5, p_7 \| = 0.134$. Obviously, when the full skyline are divided into k clusters, k-DRS aims to select k skyline points from these k clusters and these k skyline points should come from different clusters.

In this paper, we deeply analyze the properties of k-DRS query, and solve the problem using the extreme learning machine (ELM for short) [3–5]. Compared with support vector machines (SVMs) [6, 7], ELM shows better predicting accuracy than that of SVMs [4, 8–10]. Moreover, various extensions have been made to the basic ELMs to make it more efficient and more suitable for special problems. such as ELMs for online sequential data [10–12], ELMs for distributed environments [13], and ELMs for semi-labeled data and unlabeled data [14]. As proposed in [14], US-ELM can be applied to unsupervised data which has more widely applications. Meanwhile, the experiments show that US-ELM gives favorable performance compared to the state-of-the-art clustering algorithms [15–18]. Therefore, we apply US-ELM to cluster the full skyline points, and then select the appropriate the skyline points from every cluster as the k-DRS.

As mentioned in Lemma 4 in [2], k-DRS is NP-hard when the dimensionality $d \geq 3$. Hence, it is impossible to calculate the exact k-DRS in d-dimensional space

$(d \geq 3)$. To solve the challenging issue, we attempt to solve k-DRS problem from another perspective.

The key contributions are summarized as follow. Through in-depth analysis of k-DRS properties, we propose a k distance-based representative skyline algorithm based on US-ELM methods (DRSELM for short). The calculation of the k-DRS is divided into two stages. Step 1: the full skyline set is divided into k clusters by using the US-ELM algorithm. Step 2: for each cluster, an appropriate skyline point is added to the k-DRS set. The chosen k skyline points are the k-DRS result. Then we test our algorithm on a variety of data sets, and comparisons with other related algorithms [2]. The results show that our algorithm is competitive in terms of both accuracy and efficiency.

The rest of paper is organized as follows. In Sect. 2, we give a brief overview of clustering data using US-ELM algorithm. In Sect. 3, we present our k distance-based representative skyline algorithm based on US-ELM. Experimental results and related work are given in Sects. 4 and 5, respectively. Section 6 concludes the paper.

2 Preliminaries

In this paper, we process k-DRS query using ELM to cluster the skyline points. Here, we introduce how to extend ELMs to cluster the data.

2.1 Brief Introduction to ELM

ELM is an algorithm for neural network, and is a single-hidden layer feed forward network. ELM aims to learn a decision rule or an approximation function based on a training set with N samples, $\{X, Y\} = \{x_i, y_i\}_{i=1}^{N}$, where $x_i \in \mathbb{R}^{n_i}$ and $y_i \in \mathbb{R}^{n_o}$, n_i and n_o are the dimensions of input and output, respectively.

As described in Fig. 2, the training of ELMs contains two phases. Step 1: a pair of parameters $\{a_j, b_j\}$ are randomly generated for the jth hidden layer node, where a_j is a n_i-dimensional vector and b_j is a random value. For an input vector x_i, its output on the jth hidden node can be obtained by the following mapping function (we use

Fig. 2 ELM framework

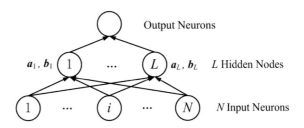

Output Neurons

L Hidden Nodes

N Input Neurons

the Sigmoid function in this paper).

$$g(\boldsymbol{x}_i, \boldsymbol{a}_j, b_j) = \frac{1}{1 + exp(-(\boldsymbol{a}_j^T \times \boldsymbol{x}_i + b_j))} \tag{1}$$

Hence, the output on the hidden layer nodes can be written as

$$\boldsymbol{H} = \begin{bmatrix} g(\boldsymbol{x}_1, \boldsymbol{a}_1, b_1) & \cdots & g(\boldsymbol{x}_1, \boldsymbol{a}_L, b_L) \\ \vdots & & \vdots \\ g(\boldsymbol{x}_N, \boldsymbol{a}_1, b_1) & \cdots & g(\boldsymbol{x}_N, \boldsymbol{a}_L, b_L) \end{bmatrix}_{N \times L} \tag{2}$$

Step 2: On the jth hidden node, an adjustment factor $\boldsymbol{\beta}_j$ is generated, where $\boldsymbol{\beta}_j$ is a n_o-dimensional vector. The output on the output neuron is $\boldsymbol{Y}$ which is the output of the N samples. Then we can obtain the following equation

$$\boldsymbol{H} \cdot \boldsymbol{\beta} = \boldsymbol{Y} \tag{3}$$

where

$$\boldsymbol{\beta} = \begin{bmatrix} \boldsymbol{\beta}_1 \\ \vdots \\ \boldsymbol{\beta}_L \end{bmatrix}_{L \times n_o} and \ \boldsymbol{Y} = \begin{bmatrix} \boldsymbol{y}_1 \\ \vdots \\ \boldsymbol{y}_N \end{bmatrix}_{N \times n_o} \tag{4}$$

According to Eq. 3, the values of $\boldsymbol{H}$ and $\boldsymbol{Y}$ haven been known, we can compute the values of $\boldsymbol{\beta}$ by the equation $\boldsymbol{\beta} = \boldsymbol{H}^\dagger \boldsymbol{Y}$ where $\boldsymbol{H}^\dagger$ is the Moore-Penrose [19] of $\boldsymbol{H}$. In order to avoid over-fitting, they introduced two parameters, $\boldsymbol{e}_i$ and C. $\boldsymbol{e}_i$ is the error vector with respect to the ith training sample, and C is a penalty coefficient on the training errors. Then the following equation is used to generate $\boldsymbol{\beta}$.

$$\min_{\boldsymbol{\beta} \in \mathbb{R}^{L \times n_o}} L_{ELM} = \frac{1}{2} \| \boldsymbol{\beta} \|^2 + \frac{C}{2} \| \boldsymbol{Y} - \boldsymbol{H}\boldsymbol{\beta} \|^2 \\ s.t. \ \boldsymbol{H}\boldsymbol{\beta} = \boldsymbol{Y} - \boldsymbol{e} \tag{5}$$

where $\| \cdot \|$ denotes the Euclidean norm and $\boldsymbol{e} = [\boldsymbol{e}_1^T, \ldots, \boldsymbol{e}_N^T] \in \mathbb{R}^{N \times n_o}$.

According to the ridge regression or regularized least squares principle, the gradient of L_{ELM} with respect to $\boldsymbol{\beta}$ is set to zero. We have

$$\nabla L_{ELM} = \boldsymbol{\beta} + C\boldsymbol{H}^T(\boldsymbol{Y} - \boldsymbol{H}\boldsymbol{\beta}) = 0 \tag{6}$$

If $\boldsymbol{H}$ has more rows than columns and is full of column rank, we use Eq. 7 to evaluate $\boldsymbol{\beta}$. If the number of training samples N is smaller than L, we restrict $\boldsymbol{\beta}$ to be a linear combination of the rows of H: $\beta = H^T \alpha (\alpha \in \mathbb{R}^{N \times n_o})$. Then β can be calculated by Eq. 8.

$$\beta^* = (H^T H + \frac{I_L}{C})^{-1} H^T Y \tag{7}$$

$$\beta^* = H^T \alpha^* = H^T (HH^T + \frac{I_N}{C})^{-1} Y \tag{8}$$

where I_L and I_N are the identity matrices of dimensions L and N, respectively.

2.2 Unsupervised ELM

In [14], they extended ELM to process unlabeled data and made ELM a wide applications. The unsupervised learning is built on the following assumption: (1) all the unlabeled data X_u is drawn from the same marginal distribution $\mathcal{P}_X$ and (2) if two points x_1 and x_2 are close to each other, then the probabilities $P(y|x_1)$ and $P(y|x_2)$ should be similar. The manifold regularization framework proposes to minimize the following cost function:

$$L_m = \frac{1}{2} \sum_{i,j} w_{ij} \parallel P(y|x_i) - P(y|x_j) \parallel^2 \tag{9}$$

where w_{ij} is the pair-wise similarity between x_i and x_j. w_{ij} is usually computed using Gaussian function $exp(- \parallel x_i - x_j \parallel^2 / 2\sigma^2)$.

Equation 9 can be simplified in a matrix form

$$\hat{L}_m = Tr(\hat{Y}^T L \hat{Y}) \tag{10}$$

where $Tr(\cdot)$ denotes the trace of a matrix, $\hat{Y}$ is the predictions of X_u, $L = D - W$ is known as graph Laplacian, and D ia a diagonal matrix with its diagonal elements $D_{ii} = \sum_{j=1}^{u} w_{ij}$.

Hence, in unsupervised setting, the entire data set $X = \{x_i\}_{i=1}^N$ are unlabeled. According to Eqs. 5 and 10, the formulation of US-ELM is reduced to

$$\min_{\beta \in \mathbb{R}^{L \times n_o}} \parallel \beta \parallel^2 + \lambda Tr(\beta^T H^T L H \beta) \tag{11}$$

where λ is an tradeoff parameter. Usually, Eq. 11 attains its minimum value at $\beta = \mathbf{0}$. As suggested in [16], a constraint $(H\beta)^T H\beta = I_{n_o}$ is introduced. According to the conclusion in [14], if $L \leq N$, the adjustment factor β is given by

$$\beta^* = [\tilde{v}_2, \tilde{v}_3, \dots, \tilde{v}_{n_o+1}] \tag{12}$$

where $\tilde{v}_i = v_i / \parallel Hv_i \parallel, i = 2, \ldots, n_o + 1$ is the normalized eigenvectors. γ_i is the ith smallest eigenvalues of Eq. 13 and v_i is the corresponding eigenvectors.

$$(I_L + \lambda H^T LH)v = \gamma H^T Hv \tag{13}$$

If $L > N$, Eq. 13 is underdetermined. In this case, the following alternative formulation is given by using the same trick

$$(I_N + \lambda LHH^T)u = \gamma HH^T u \tag{14}$$

Also, u_i is the generalized eigenvectors corresponding the ith smallest eigenvalues of Eq. 14. Then, the final solution is given by

$$\beta^* = H^T[\tilde{u}_2, \tilde{u}_3, \ldots, \tilde{u}_{n_o+1}] \tag{15}$$

where $\tilde{u}_i = \tilde{u}_i / \parallel HH^T \tilde{u}_i \parallel, i = 2, \ldots, n_o + 1$ are the normalized eigenvectors.
The US-ELM is described in Algorithm 1.

3 *k*-DRS Query Processing Based on US-ELM

First, we describe the formal definition of the k-DRS in Sect. 3.1. Then, our proposed algorithm DRSELM is presented in Sect. 3.2.

Algorithm 1: US-ELM Algorithm [14]

---

input : The training data: $X \in \mathbb{R}^{N \times n_i}$.
output: The label vector of cluster y_i corresponding to x_i

1 **Step 1**: Construct the graph Laplacian $L = D - W$ from X;
2 **Step 2**: For each hidden neuron, generate a pair of random values $\{a_i, b_i\}$; Calculate the output matrix $H \in \mathbb{R}^{N \times L}$;
3 **Step 3**:
4 **if** $L \leq N$ **then**
5    $\quad$ Find the generalized eigenvectors $v_2, \ldots, v_{n_o+1}$ of Eq. 13. Let $\beta = [\tilde{v}_2, \tilde{v}_3, \ldots, \tilde{v}_{n_o+1}]$.
6 **else**
7    $\quad$ Find the generalized eigenvectors $u_2, \ldots, u_{n_o+1}$ of Eq. 14. Let $\beta = H^T[\tilde{u}_2, \tilde{u}_3, \ldots, \tilde{u}_{n_o+1}]$;
8 **Step 4**: Calculate the embedding matrix: $E = H\beta$;
9 **Step 5**: Each row of E is treated as a point, and then cluster these N points into K clusters using the k-means algorithm. Let y_i be the label vector of cluster index for x_i.
10 **return** Y;

3.1 Problem Statement

Given a data set D in the d-dimensional space, and two points $p_i = \langle p_i[1], \dots, p_i[d] \rangle$ and $p_j = \langle p_j[1], \dots, p_j[d] \rangle$, then p_i *dominates* p_j (denoted as $p_i \prec p_j$) if $\forall m \in [1, d]$, $p_i[m] \leq p_j[m]$ and $\exists n \in [1, d], p_i[n] < p_j[n]$. The *skyline* of D consists of the points which are not dominated by others, denoted as $S = \{p_i | \nexists p_j \in D, p_j \prec p_i\}$. Next, we give the formal definition of the k-DRS.

Definition 1 (*Representation Error*) Given the full skyline set S and a subset $\mathcal{K}$ of S with k skyline points, the *representation error* $Er(\mathcal{K}, S)$ quantifies the representation quality as the maximum distance between a non-representative skyline point in $S - \mathcal{K}$ and its nearest representative in $\mathcal{K}$, which is formally denoted as:

$$Er(\mathcal{K}, S) = \max_{p \in S - \mathcal{K}} \{ \min_{p' \in \mathcal{K}} \| p, p' \| \} \tag{16}$$

Definition 2 (*Distance-based Representative Skyline*) The distance-based representative skyline (k-DRS) is the set $\mathcal{K}$ with the minimum representation error $Er(\mathcal{K}, S)$.

As shown in Fig. 1, the skyline set is $S = \{p_1, \dots, p_{10}\}$. Given $k = 3$, and two subsets $\mathcal{K}_1 = \{p_2, p_6, p_9\}$ and $\mathcal{K}_2 = \{p_2, p_5, p_9\}$, the representation errors $Er(\mathcal{K}_1, S) = \| p_4, p_6 \| = 0.141$ and $Er(\mathcal{K}_2, S) = \| p_5, p_7 \| = 0.135$. Consequently, $\mathcal{K}_2$ is the k-DRS because its representation error is the minimum.

3.2 DRSELM Algorithm

Reviewing the conclusion in [2], the k-DRS problem is NP-hard in d-dimensional ($d \geq 3$) space. Hence, it is impossible to calculate the exact k-DRS. In this paper, we answer the k-DRS problem from another perspective.

Since the initial objective of the k-DRS is to avoid selecting k points that appear in an arbitrarily tiny cluster, we first divide the full skyline points into k clusters using the US-ELM algorithm (introduced in Sect. 2.2). Specifically, given a data set D in d-dimensional space, each point $p_i \in D$ is considered as a d-dimensional vector. $p_i[j]$ denotes the jth dimension value of p_i. According to the Algorithm 1, there is a corresponding output y_i with regard to p_i. Because the full skyline needs to be divided into k clusters, each output y_i is a k-dimensional vector. Only one dimension value is 1, and the other dimension values are 0. As shown in Fig. 1, p_1, p_2, p_3 have the same outputs $y_1 = y_2 = y_3 = [1, 0, 0]$. p_4, p_5, p_6, p_7 have the same outputs $y_4 = y_5 = y_6 = y_7 = [0, 1, 0]$. p_8, p_9, p_{10} have the same outputs $y_8 = y_9 = y_{10} = [0, 0, 1]$.

Given a cluster $c_i = \{p_1, p_2, \dots, p_{|c_i|}\}$ with $|c_i|$ points, the centroid point m_i of c_i can be calculated by the formula below:

$$m_i[j] = \frac{\sum_{p \in c_i} p[j]}{|c_i|} \tag{17}$$

As shown in Fig. 1, given the cluster $c_1 = \{p_1, p_2, p_3\}$, its centroid point m_1 is calculated as: $m_1[1] = \frac{0.13+0.16+0.20}{3} \approx 0.16$ and $m_1[2] = \frac{0.92+0.87+0.82}{3} = 0.87$. Hence, p_2 is regarded as the centroid point in c_1. Similarly, the centroid points of c_2 and c_3 are $m_2 = \langle 0.345, 0.4425 \rangle$ and $m_3 = \langle 0.70, 0.15 \rangle$.

After the clustering, we have the following properties.

Observation 1 *Given a point p_1 comes from cluster c_1, and a point p_2 comes from cluster c_2, m_1 and m_2 are the centroid points of c_1 and c_2, respectively. Then $\| p_1, m_1 \| < \| p_1, m_2 \|$.*

We have divided the full skyline into k clusters. The target of the k-DRS wants to get the minimum representation error $Er(\mathcal{K}, S)$. In order to obtain this goal, all the points should come from different clusters. Given two clusters c_i and c_j, we should select any 2 points $\mathcal{K}^2$ from $C = c_i \bigcup c_j$, in order to obtain the minimum value $Er(\mathcal{K}^2, C) = \max_{p \in C - \mathcal{K}^2} \{ \min_{p' \in \mathcal{K}^2} \| p, p' \| \}$. The two points in $\mathcal{K}^2$ should come from c_i and c_j, respectively.

Theorem 1 *Give two clusters c_1 and c_2, and two sets $S_1 = \{m_1, m_2\}$ and $S_2 = \{p_m, p_n\}$, m_1 and m_2 are the centroid points of c_1 and c_2. p_m and p_n are any two points come from c_1. Then $Er(S_1, C) < Er(S_2, C)$.*

Proof Suppose the point in c_1 with the largest distance to m_1 is p_1, and the point in c_2 with the largest distance to m_2 is p_2, then $Er(S_1, C) = \max\{\| p_1, m_1 \|, \| p_2, m_2 \|\}$. Obviously, a good clustering method can ensure that $\| m_1, m_2 \| > \max\{\| p_1, m_1 \|, \| p_2, m_2 \|\}$. Hence, there must exist a point $p' \in c_2$, the distance between p' and any point $p_m \in c_1$ is larger than $Er(S_1, C)$. The theorem can be proven.

Lemma 1 *Given k clusters $c_1, \ldots, c_k$ of the full skyline S, in order to obtain the minimum representation error $Er(\mathcal{K}, S)$, the selected k skyline points should come from different k clusters.*

Proof This lemma can be obtained directly from Theorem 1.

According to Lemma 1, the selected points come from different clusters. As shown in Fig. 1, the full skyline is divided into 3 clusters. The selected 3 skyline points should come from different clusters. For each cluster $c_i, i \in [1, 3]$, we choose one point.

Next, we introduce how to select a point from a cluster. According to the objective function $Er(\mathcal{K}, S)$, the selected point p_i from c_i should have the minimum value $Er(p_i, c_i) = \max_{p_i \in c_i, p' \in c_i - \{p_i\}} \{\| p_i, p' \|\}$. The details to calculate the k-DRS based on ELM is described in Algorithm 2.

The calculation in a cluster needs to compute the distances between any two points in a cluster. Hence, the time cost of calculating an appropriate in a cluster is $O(|c_i|^2)$ where $|c_i|$ is the size of the largest cluster.

Algorithm 2: The DRSELM Algorithm

---

input : the data set D in d-dimensional space; the parameter k
output: The distance-base representative set k-$DRS(D)$;

1 Using BNL algorithm [1] to calculate the skyline S of D;
2 Using Algorithm 1 to divide the full skyline S into k clusters;
3 **for** *each cluster c_i* **do**
4 | From all the points in c_i, add the one with the minimum value $MaxDis(p, c_i)$ to the
     | k-DRS;
5 **return** k-DRS;

---

4 Experimental Evaluation

In this section, we demonstrate the efficiency and effectiveness of the DRSELM. We test 3 algorithms: 2d-opt, I-greedy, DRSELM. Specifically, 2d-opt and I-greedy are the algorithms in [2] for 2-dimensional dataset and d-dimensional ($d \geq 3$) dataset, respectively.

DataSets. We apply the same datasets in [2], a synthetic dataset *Island* and a real dataset *NBA*. *Island* follows a cluster distribution along the anti-diagonal, which is shown in Fig. 3. There are 27868 points in the *Island*, and the skyline of the *Island* consists of 110 points. *NBA* is downloadable at http://www.databasebasketball.com. It includes 17265 5-dimensional points and skyline of *NBA* consists of 494 points.

The distance-base representative skyline of *Island* is shown in Fig. 4 when k varies. As shown in Fig. 5, our DRSELM shows outstanding performances. Comparing with 2d-opt, DRSELM has more efficiency and good accuracy. In Fig. 5a,

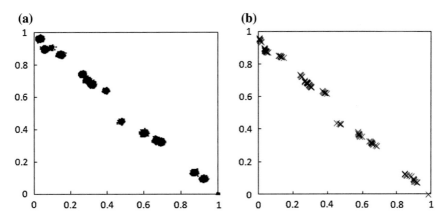

Fig. 3 The synthetic dataset *Island*. **a** The *Island* Dataset. **b** The Skyline of *Island*

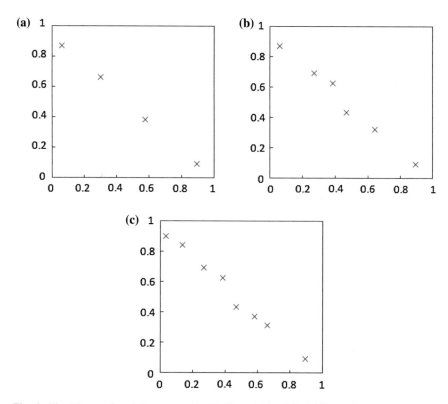

Fig. 4 The Distance-based Representative Skyline of *Island* for Different k. **a** $k = 4$. **b** $k = 6$. **c** $k = 8$.

with the increase of k, the running time of DRSELM and 2d-opt has little change. In Fig. 5b, as k grows, the representation error becomes smaller. DRSELM has the same representation errors with 2d-opt. Since 2d-opt is an exact algorithm, DRSELM has good accuracy in 2-dimensional datasets.

The experimental results of *NBA* is shown in Fig. 6. According to Fig. 6a, the running time of DRSELM is shorter than that of I-greedy. With the increase of k, the running time of DRSELM is stable, and the running time of I-greedy raises slightly. Hence, the efficiency of DRSELM is better than that of I-greedy. As illustrated in Fig. 6b, the representation errors of DRSELM and I-greedy are close. Therefore, comparing with I-greedy, the accuracy of DRSELM is competitive.

Comparing Fig. 5 with Fig. 6, with the increase of dimensionality, the running time of DRSELM has a little increment. Based on analysis above, it can be concluded that our DRSELM can process the k-DRS effectively.

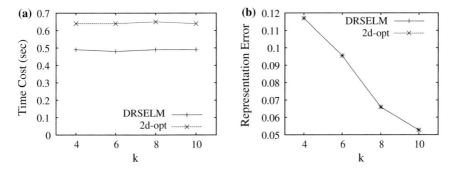

Fig. 5 The Experimental Results of *Island*. **a** Running time versus *k*. **b** Representation Error versus *k*

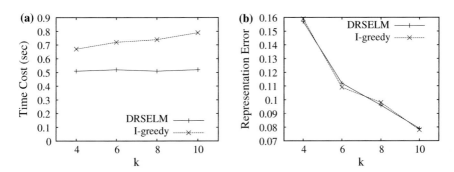

Fig. 6 The Experimental Results of *NBA*. **a** Running time versus *k*. **b** Representation Error versus *k*

5 Related Work

The skyline operator was first introduced by Börzönyi et al. [1]. Then many efficient skyline algorithms [20–23] have been proposed. Algorithms BNL and D&C [1], SFS [20], Bitmap [24] and NN [21] can process skyline query in the datasets without indexes. BBS [22] calculate the skyline query using R-tree index and ZINC [23] apply the Z-order index to process the skyline query. When the full skyline set is large, it is difficult to understand the full skyline. Thus, selecting *k* representative points is significant. There are some definitions [2, 25] about the representative skylines. In this paper, we focus on the distance-based representative skyline (*k*-DRS). *k*-DRS is NP-hard when $d \geq 3$. By in-depth analysis of the properties of the *k*-DRS, first, we use the clustering algorithms to cluster the full skyline set. Second, we calculate the representative point in each cluster. So far, there are some state-of-the-art clustering algorithms [15–18]. The experimental results show that US-ELM [14] is competitive in terms of both accuracy and efficiency. Hence, in this paper, we apply US-ELM to cluster the full skyline set.

6 Conclusion

As an important variant of skyline, the k representative skyline is a useful tool if the size of the full skyline is large. In this paper, we focus on the distance-based representative skyline (k-DRS). Since k-DRS is a NP-hard problem in d-dimensional ($d \geq 3$) space, we design a 2-step algorithm DRSELM to solve the k-DRS problem efficiently. Step 1 divides the full skyline set into k clusters using US-ELM algorithm. In step 2, a point in each cluster is selected as the representative skyline point. The k selected skyline points consist of the k-DRS. Comprehensive experimental results demonstrate that DRSELM is competitive with the state-of-the-art algorithm in terms of both accuracy and efficiency.

Acknowledgments This research was partially supported by the National Natural Science Foundation of China under Grant Nos. 61472069, 61402089, 61100022 and 61173029; the National High Technology Research and Development Plan (863 Plan) under Grant No. 2012AA011004; and the Fundamental Research Funds for the Central Universities under Grant No. N130404014.

References

1. Stephan, B.: Donald Kossmann, Konrad Stocker: The Skyline Operator. ICDE, pp. 421–430 (2001)
2. Tao, Y., Ding, L., Lin, X., Pei, J.: Distance-Based Representative Skyline. ICDE, pp. 892–903 (2009)
3. Huang, G., Zhu, Q., Siew, C.-K.: Extreme learning machine: a new learning scheme of feedforward neural networks. Proc. Int. Jt. Conf. Neural Netw. **2**, 985–990 (2004)
4. Huang, G., Zhu, Q., Siew, C.-K.: Extreme learning machine: theory and applications. Neurocomputing **70**, 489–501 (2006)
5. Huang, G.: What are extreme learning machines? Filling the gap between Frank Rosenblatt's Dream and John von Neumann's Puzzle. Cogn. Comput. **7**(3), 263–278 (2015)
6. Cherkassky, V.: The nature of statistical learning theory. IEEE Trans. Neural Netw. **8**(6), 1564 (1997)
7. Cortes, C., Vapnik, V.: Support-vector networks. Mach. Learn. **20**(3), 273–297 (1995)
8. Shi, L., Baoliang, L.: EEG-based vigilance estimation using extreme learning machines. Neurocomputing **102**, 135–143 (2013)
9. Huang, G., Zhou, H., Ding, X., Zhang, R.: Extreme learning machine for regression and multiclass classification. IEEE Trans. Syst. Man Cybern. **Part B 42**(2), 513–529 (2012)
10. Rong, H., Huang, G., Sundararajan, N., Saratchandran, P.: Online sequential fuzzy extreme learning machine for function approximation and classification problems. IEEE Trans. Syst. Man Cybern. **Part B 39**(4), 1067–1072 (2009)
11. Zhao, J., Wang, Z.: Dong Sun Park: online sequential extreme learning machine with forgetting mechanism. Neurocomputing **87**, 79–89 (2012)
12. Liang, N., Huang, G., Saratchandran, P., Sundararajan, N.: A fast and accurate online sequential learning algorithm for feedforward networks. IEEE Trans. Neural Netw. **17**(6), 1411–1423 (2006)
13. He, Q., Changying, D., Wang, Q., Zhuang, F., Shi, Z.: A parallel incremental extreme SVM classifier. Neurocomputing **74**(16), 2532–2540 (2011)
14. Huang, G., Song, S., Gupta, J.N.D., Wu, C.: Semi-Supervised and unsupervised extreme learning machines. IEEE Trans. Cybern. **44**(12), 2405–2417 (2014)

15. Kanungo, T., Mount, D.M., Netanyahu, N.S., Piatko, C.D., Silverman, R., Wu, A.Y.: An efficient k-means clustering algorithm: analysis and implementation. IEEE Trans. Pattern Anal. Mach. Intell. **24**(7), 881–892 (2001)
16. Belkin, M., Niyogi, P.: Laplacian eigenmaps for dimensionality reduction and data representation. Neural Comput. **15**(6), 1373–1396 (2003)
17. Ng, A.Y., Jordan, M.I., Weiss, Y.: On spectral clustering: analysis and an algorithm. Adv. Neural Inform. Process. Syst. **2**, 849C856 (2002)
18. Bengio, Y.: Learning deep architectures for AI. Found. Trends Mach. Learn. **2**(1), 1–127 (2009)
19. Serre, D.: TMatrices: Theory and Applications. Springer, Berlin (2002)
20. Chomicki, J., Godfrey, P., Gryz, J., Liang, D.: Skyline with presorting. ICDE 717–719 (2003)
21. Kossmann, D., Ramsak, F., Rost, S.: Shooting stars in the sky: an online algorithm for skyline queries. VLDB 275–286 (2002)
22. Papadias, D., Tao, Y., Fu, G., Seeger, B.: An optimal and progressive algorithm for skyline queries. SIGMOD Conference 467–478 (2003)
23. Liu, B., Chan, C.-Y.: ZINC: efficient indexing for skyline computation. PVLDB, **4**(3), 197–207 (2020)
24. Pei, J., Jin, W., Ester, M., Tao, Y.: Catching the best views of skyline: a semantic approach based on decisive subspaces. VLDB 253–264 (2005)
25. Lin, X., Yuan, Y., Zhang, Q., Zhang, Y.: Selecting stars: the k most representative skyline operator. ICDE 86–95 (2007)

Multi-label Text Categorization Using L_{21}-norm Minimization Extreme Learning Machine

Mingchu Jiang, Na Li and Zhisong Pan

Abstract Extreme learning machine (ELM) was extended from the generalized single hidden layer feedforward networks where the input weights of the hidden layer nodes can be assigned randomly. It has been widely used for its much faster learning speed and less manual works. Considering the field of multi-label text classification, in this paper, we propose an ELM based algorithm combined with L_{21}-norm minimization of the output weights matrix called L_{21}-norm Minimization ELM, which not only fully inherits the merits of ELM but also facilitates group sparsity and reduces complexity of the learning model. Extensive experiments on several benchmark data sets show a more desirable performance compared with other common multi-label classification algorithms.

Keywords Text categorization · Multi-label learning · L_{21}-norm minimization · Extreme learning machine

1 Introduction

Continued development of the Internet and information technology has spawned a large number of text data in various forms. How to organize, manage and analyze such a huge data, and find the user information quickly, accurately and comprehensively is a big challenge. Text automatic classification is an important research point in the field of information mining. Compared to the traditional single classification problem, multi-label text classification has more value of research and application.

In multi-label learning, the text data are always in high dimensionality and sparsity. e.g. In a large number of feature words, only a few are related to the topic of a

M. Jiang (✉) · N. Li · Z. Pan
College of Command Information System, PLA University of Science
and Technology, Nanjing 210007, China
e-mail: 735906675@qq.com

Z. Pan
e-mail: hotpzs@hotmail.com

© Springer International Publishing Switzerland 2016
J. Cao et al. (eds.), *Proceedings of ELM-2015 Volume 1*,
Proceedings in Adaptation, Learning and Optimization 6,
DOI 10.1007/978-3-319-28397-5_10

text and most of the rest are redundant. Therefore, introducing sparsity into machine learning has become a popular technology, which not only meet the need of practical problems but also can simplify the learning model. In resent years, extreme learning machine (ELM) [1–4] has attracted increasing attention and been widely used for its distinguishing characteristics: (1) fast learning speed, (2) good generalization performance on classification or regression, (3) less human intervention with randomly setted hidden layer parameters. For these reasons, the theoretical analysis and various improvement algorithms of ELM are put forward continuously.

In ELM network, the function of the random hidden layer nodes can be seen as feature mapping. It maps the data from the input feature space to the hidden layer feature space, which is called ELM feature space in literature [5]. In this ELM feature space, each instance may still remains the sparsity. Meantime, considering the characteristics of multi-label learning and the advantages of the classifier ELM, in this paper, we propose an embedded model for multi-label text classification, which is derived from a formulation based on ELM with L_{21}-norm minimization of the output weights matrix. Through the constraint of the L_{21}-norm regularization, the training model becomes simplified, also we can sufficiently preserve the intrinsic relation of different nodes in the ELM feature space and select them by joint multiple related labels, where the labels are not always independent to each other. Experimental results on several benchmark data sets verify the efficiency of our proposed algorithm.

The main contributions of this paper can be summarized below:

- According to the characteristics of the multi-label text data we introduce the sparsity model.
- Applying L_{21}-norm for joint hidden layer nodes selection and avoiding individual training for each label.
- Using ELM for multi-label text classification.

The remainder of this paper is organized as follows. After reviewing the related works in Sect. 2, we present the algorithm L_{21}-ELM in Sect. 3 and describe the evaluation measures of multi-label learning in Sect. 4. Experimental results are presented in Sect. 5 and we conclude this paper in Sect. 6.

2 Related Work

2.1 Multi-label Learning

Unlike traditional supervised learning, in multi-label learning each instance may belong to multiple classes and for a new instance we try to predict its associated set of labels. This is a generalized case of the prevalent multi-class problems where in multiple classes each instance has only one class restrictedly.

Let $\mathcal{X} \in \mathbb{R}^d$ denote the d-dimensional space of instances, $\mathcal{Y} = \{y_1, \ldots, y_k\}$ denote the label space with k possible class labels. Given the training data set $\{(x_1, Y_1), \ldots, (x_n, Y_n)\}$ where $x_i \in \mathcal{X}$ and $Y_i \subseteq \mathcal{Y}$. the task of multi-label learning is to learn a multi-label classifier $f : \mathcal{X} \rightarrow 2^k$ from the training data set. For any unknown instance $x \in \mathcal{X}$, the multi-label classifier $f(\cdot)$ predicts $f(x) \subseteq \mathcal{Y}$ as the set of proper labels. Existing multi-label learning algorithms can be divided into two main categories [6, 7].

Problem transformation methods. The main idea of most problem transformation methods is to transform the original multi-label learning problem into multiple single-label learning problems, which usually reconstructs the multi-label data sets and then existing classification algorithms can be applied directly.

The binary relevance (BR) [8] algorithm is a popular kind of this transformation method and has been widely used in many practical applications. This algorithm divides the multi-label classification problem into k independent binary classification problems, however, the assumption of label independence is too implicit and the label correlations are ignored. The label powerset (LP) [9] algorithm is another common transformation method. It considers each unique set of labels in a multi-label training data as one class in the new transformed data. While the computational complexity of LP is too big and it may pose class imbalance problem. The basic idea of the classifier chains (CC) [10] is to chain the transformed binary classifiers one by one, but the sequence of each classifier is a problem. The ensembles of classifier chains (ECC) [11] improved the CC algorithm and identify the sequence of each classifier effectively.

Algorithm adaptation methods. From another perspective, this method improves conventional algorithms to deal with multi-label data directly. Some representative algorithms include ML-kNN [12] adapting k-nearest neighbor techniques, which has the advantage of both lazy learning and Bayesian but ignores label correlations. ML-DT [13] adapting decision tree techniques, Rank-SVM [14] adapting kernel techniques, etc.

In this paper, the algorithm based on ELM we proposed is designed to deal with multi-label data directly, therefore, it can be considered as a kind of algorithm adaptation method.

2.2 L_{21}-norm Regularization for Parameter Estimation

In recent years, parameter estimation via sparsity-promoting regularization has be widely used in machine learning and statistics. Perhaps L_1-norm regularization is the most successful and common method to promote sparsity for the parameter vector (the lasso approach). Along with the development of multi-task learning, in 2006, Obozinski et al. [15, 16] proposed to constrain the sum of L_2-norms of the blocks of weights connected with each feature, and then leading to the L_{21}-norm regularized optimization problem (the group lasso).

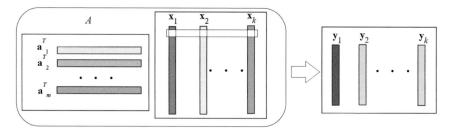

Fig. 1 Illustration of the data matrix A, Y, and the weights matrix X

In this section, we will briefly review the basics of this technique. Usually, the optimization problem can be described as following:

$$\min_{X} : loss(X) + \lambda \parallel X \parallel_{2,1} \qquad (1)$$

where $\lambda > 0$ is the regularization parameter, $X \in \mathbb{R}^{n \times k}$ is the weights matrix, $\parallel X \parallel_{2,1} = \sum_{i=1}^{n} \parallel X \parallel_{2}$ and $loss(X)$ is a smooth and convex loss function (such as the logistic loss, the least square loss and the hinge loss). Take the least squares problem as an example, the Eq.1 is expressed as:

$$\min_{X} : \frac{1}{2} \parallel AX - Y \parallel_{2}^{2} + \lambda \parallel X \parallel_{2,1} \qquad (2)$$

where $A \in \mathbb{R}^{m \times n}$, $Y \in \mathbb{R}^{m \times k}$ are the data matrices, each row of X forms a feature group. Figure 1 visualizes this optimization problem.

This optimization problem will be more challenging to solve due to the non-smoothness and non-differential of the L_{21}-norm regularization. In this paper, we apply the strategy proposed in literature [17] to solve this problem, which reformulates the non-smooth L_{21}-norm regularized problem to an equivalent smooth convex optimization problem and can be solved in linear time.

3 L_{21}-minimization ELM (L_{21}-ELM)

In this section, we propose L_{21}-ELM algorithm for multi-label learning problem, which takes the significant advantages of ELM like affording good generalization performance at extremely fast learning speed, meantime, offers us some additional characteristics. Firstly, we will review the theories of ELM, then, introduce the algorithm we proposed.

Fig. 2 Structure of ELM
network

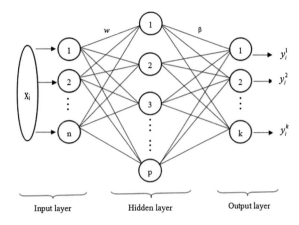

Input layer   Hidden layer   Output layer

3.1 Extreme Learning Machine

Extreme learning machine [2, 3] was originally proposed for single hidden layer feedforward neural networks and then extended to the generalized single hidden layer feedforward networks where the hidden layer need not be neuron alike [1]. Figure 2 shows the structure of ELM network. It contains an input layer, a hidden layer and an output layer.

In ELM, the hidden layer parameters are chosen randomly, and the output function can be represented as following (take the case of p hidden layer nodes and one output layer node as an example):

$$f_{output}(x) = \sum_{i=1}^{p} \beta_i h_i(x) = \mathbf{h}(x)\beta \tag{3}$$

where $x \in \mathbb{R}^n$ is the input variable, $\beta = (\beta_1, \beta_2, \dots, \beta_p)^T$ is the weights vector between the hidden layer nodes and the output layer nodes. $\mathbf{h}(x) = \left[h_1(x), h_2(x), \dots, h_p(x)\right]$ is the output vector of the hidden layer with respect to the input vector x. $h_i(x)$ is the ith activation function, its input weights vector and bias are w_i and b_i.

Figure 2 shows that $\mathbf{h}(x)$ actually maps the input variables from the n-dimension to the p-dimensional hidden layer space (ELM feature space), thus, it appears to be a feature mapping function.

The ELM reliably approximates m samples, $X = [x_1, \dots, x_m]$, with minimum error:

$$\min_{\beta} : \| H\beta - Y \|_2^2 \tag{4}$$

where H is hidden layer output matrix,

$$H = \begin{bmatrix} h(x_1) \\ \vdots \\ h(x_m) \end{bmatrix} = \begin{bmatrix} g(w_1 \cdot x_1 + b_1) & \cdots & g(w_p \cdot x_1 + b_p) \\ \vdots & \ddots & \vdots \\ g(w_1 \cdot x_m + b_1) & \cdots & g(w_p \cdot x_m + b_p) \end{bmatrix}_{m \times p} \qquad (5)$$

and $Y = [y_1, \ldots, y_m]^T$ is the target vector.

The analytical result of this least squares equation is:

$$\hat{\beta} = H^{\dagger} Y \qquad (6)$$

where $H^{\dagger}$ is called Moore-Penrose generalized inverse of matrix H.

3.2 L_{21}-norm Minimization ELM for Multi-label Learning

In this section, we consider adapting the ELM network to solve the multi-label learning problem. Given the multi-label training data with m samples (x_i, y_i), where $x_i = (x_{i1}, x_{i2}, \ldots, x_{in})^T \in \mathbb{R}^n$ and $y_i = (y_{i1}, y_{i2}, \ldots, y_{ik}) \in \mathbb{R}^k$. As shown in the Fig. 2, we set the number of output layer nodes k, which equals the number of labels, and set the number of hidden layer nodes p randomly.

Inspired by ELM, we consider combining the smallest training error of ELM with the L_{21}-norm minimization of output weights matrix. It is reformulated as following:

$$\min_{\beta} : \| H\beta - Y \|_2^2 + \lambda \| \beta \|_{2,1} \qquad (7)$$

where $\| \beta \|_{2,1} = \sum_{i=1}^{p} \| \beta_i \|_2$ is the L_{21}-norm of the matrix β, and $\beta_i = (\beta_{i1}, \beta_{i2}, \ldots, \beta_{ik})$, λ is the regularization parameter.

To solve the nonsmooth optimization problem in Eq. (7), the literature [17] proposed to employ the Nesterov's optimal method by optimizing its equivalent smooth convex reformulation. When using a constraint to replace the nonsmooth L_{21}-norm, the original problem can be equivalent to the L_{21}-ball constrained smooth convex optimization problem as following:

$$\min_{\beta} : \| H\beta - Y \|_2^2 \ s.t. \ \| \beta \|_{2,1} \le z \qquad (8)$$

When applying the Nesterov's optimal method to solve Eq. (8), one key building block of this method is Euclidean projection onto the L_{21}-ball. The Euclidean projection problem is defined as:

$$\pi_Z(U) = arg \min_{\beta \in Z} \frac{1}{2} \| \beta - U \|_2^2 \qquad (9)$$

where $Z = \{\beta \in \mathbb{R}^{p \times k} \,|\| \beta \|_{2,1} \leq z\}$ is the L_{21}-ball and $z \geq 0^1$ is the radius of L_{21}-ball. To solve the problem in Eq. (9), the Lagrangian variable α is introduced for the inequality constrain $\| \beta \|_{2,1} \leq z$, then we can lead to the Lagrangian function of Eq. (9) as:

$$\mathcal{L}(\beta, \alpha) = \frac{1}{2} \| \beta - U \|_2^2 + \alpha(\| \beta \|_{2,1} - z) \tag{10}$$

Let β^* be the primal optimal point, and α^* be the dual optimal point. This two points must satisfy the condition: $\| \beta^* \|_{2,1} \leq z$ and $\alpha^* \geq 0$. Since considering the strong duality holds of the Slater's condition, and values of the primal and dual optimal points are equal: $\alpha^*(\| \beta \|_{2,1} - z) = 0$. Therefore, the primal optimal point β^* can be given by Eq. (11) if the dual optimal point α^* is known.

$$\beta_i^* = \begin{cases} (1 - \frac{\alpha^*}{\|u^i\|})u^i, & \alpha^* > 0, \| u^i \| > \alpha^* \\ 0, & \alpha^* > 0, \| u^i \| \leq \alpha^* \\ u^i, & \alpha^* = 0 \end{cases} \tag{11}$$

where $u^i \in \mathbb{R}^{1 \times k}$ is the ith row of U.

According to Eq. (11), β^* can be computed as long as α^* is solved. Now, the key step is how to compute the unknown dual optimal point α^*. Liu et al. [17] gives the theorem : if $\| U \|_{2,1} \leq z$ the value of α^* is zero, otherwise, it can be solved as the unique root of the following auxiliary function.

$$\omega(\alpha) = \sum_{i=1}^p max(\| u^i \| - \alpha, 0) - z \tag{12}$$

The Eq. (12) can be solved in $O(p)$ flops by the bisection [18], and it costs $O(pk)$ flops to compute β^* by Eq. (11). Above all, for solving Eq. (7) the time complexity is $O(pk)$. When testing an unseen instance, we will use a threshold function $t(x)$ to determine its associated label set. For an actual outputs c_j, $Y = \{j \mid c_j > t(x)\}$. An usual solution is to set $t(x)$ to be zero. In this paper, we adopt the threshold category used in literature [19].

4 Evaluation Measures

Being different from the traditional single-label learning system, in multi-label learning an instance usually have one or more labels simultaneously, therefore those classical evaluation methods would be no longer applied in multi-label learning system. For this reason, a series of evaluation metrics for multi-label learning are proposed.

In order to compare our algorithm with other commonly used methods, we adopt five evaluation measures in multi-label learning in this section, including: hamming loss, one-error, coverage, ranking loss and average precision [6, 20, 21]. The following is a look at these measures based on a test data set $S = \{(x_i, Y_i) \mid 1 \leqslant i \leqslant n\}$ and a trained model $f(\cdot, \cdot)$ or $g(\cdot)$.

Hamming loss. This measure evaluates the error rate of all instances on all labels, e.g. a relevant label of an instance is not predicted or an irrelevant one is predicted. the smaller the value of hamming loss, the better the performance.

$$hloss_S(g) = \frac{1}{n} \sum_{i=1}^{n} \frac{1}{m} \mid g(x_i) \triangle Y_i \mid \tag{13}$$

where $\triangle$ stands for the symmetric difference between two sets, m is the total number of labels. It is worth noting that when most of these instances have little correlative labels, it can also get a small value of hamming loss even if all the labels of an instance are predicted in error. Therefore, we should integrate it with other measures.

One-error. This measure evaluates the times that the top-ranked label of an instance is not in its relevant label set. The smaller the value of $one - error_S(f)$, the better the performance.

$$one - error_S(f) = \frac{1}{n} \sum_{i=1}^{n} \left[\arg \max_{y \in y} f(x_i, y) \notin Y_i \right] \tag{14}$$

One-error mainly focuses on the most relevant label being correct or not, and it don't pay attention to other labels. Note that, it is equal to ordinary error identically in single-label classification problems.

Coverage. This measure evaluates the average steps we need to go down the ranked-label list for the sake of covering all the relevant labels. The smaller the value of coverage, the better the performance.

$$coverage_S(f) = \frac{1}{n} \sum_{i=1}^{n} \max_{\ell \in Y_i} rank_f(x_i, \ell) - 1 \tag{15}$$

where the $rank_f(\cdot, \cdot)$ is derived from the real-valued function $f(\cdot, \cdot)$, and the bigger the value of f, the smaller the $rank_f$ is. The performance is perfect when $coverage_S(f) = 0$.

Ranking loss. This measure evaluates the average fraction of the reversely ordered label pairs. The smaller the value of $rloss_S(f)$, the better the performance.

$$rloss_S(f) = \frac{1}{n} \sum_{i=1}^{n} \frac{1}{\mid Y_i \mid \mid \overline{Y_i} \mid} \mid \left\{ (y, \bar{y}) \mid f(x_i, y) \leq f(x_i, \bar{y}), (y, \bar{y}) \in Y_i \times \overline{Y_i} \right\} \mid \tag{16}$$

where Y_i and $\overline{Y_i}$ denote the possible and impossible label sets of the instance x_i. When the value is zero, it means that all impossible labels follow possible ones.

Average precision. This measure evaluates the average fraction of relevant labels ranked above a particular one $\ell \in Y_i$. It is typically used in information retrieval (IR) system to evaluate the document ranking performance query retrieval [22]. The bigger the value of $avgpec_S(f)$, the better the performance.

$$avgpec_S(f) = \frac{1}{n} \sum_{i=1}^{n} \frac{1}{|Y_i|} \sum_{y \in Y_i} \frac{|L_i|}{rank_f(x_i, y)} \tag{17}$$

where $L_i = \{y' \mid rank_f(x_i, y') \leq rank_f(x_i, y), y' \in Y_i\}$. Note that $avgpec_S(f) = 1$ ranks perfectly, that means there is no instance x_i for which a label not in Y_i is ranked above on a label in Y_i.

5 Experimental Results

In this section, L_{21}-ELM is compared with the performance of the original ELM as well as other common multi-label classification algorithms. The benchmark data sets we used are all in text areas, including: Enron for email analysis, Reuters for text categorization, BibTeX for tags of paper and Yahoo for web page categorization. Table 1 describes the datasets in detail. For Enron and Reuters without pre-separated training and testing sets, therefore, we decide to select 1,500 instances of them for

Table 1 Data sets

Items	Size	Train	Test	Features	Classes	Average labels
Enron	1702	–	–	1001	53	3.38
Reuters	2000	–	–	243	7	1.15
BibTeX	7395	4880	2515	1836	159	2.40
Arts	5000	2000	3000	462	26	1.64
Business	5000	2000	3000	438	30	1.59
Computers	5000	2000	3000	681	33	1.51
Education	5000	2000	3000	550	33	1.46
Entertainment	5000	2000	3000	640	21	1.42
Health	5000	2000	3000	612	32	1.66
Recreation	5000	2000	3000	606	22	1.42
Reference	5000	2000	3000	793	33	1.17
Science	5000	2000	3000	743	40	1.45
Social	5000	2000	3000	1047	39	1.28
Society	5000	2000	3000	636	27	1.69

Table 2 Results on data set Enron

Measure	Rank-SVM	ML-kNN	ECC	ELM	L_{21}-ELM
HL ↓	0.071 ± 0.0044	0.051 ± 0.002	0.055 ± 0.002	0.053 ± 0.002	$\mathbf{0.048 \pm 0.002}$
OE ↓	0.714 ± 0.087	0.299 ± 0.031	$\mathbf{0.224 \pm 0.036}$	0.281 ± 0.036	0.236 ± 0.0276
Co ↓	31.269 ± 2.233	12.959 ± 0.832	21.079 ± 1.265	17.118 ± 1.176	$\mathbf{12.809 \pm 0.906}$
RL ↓	0.338 ± 0.037	0.091 ± 0.008	0.249 ± 0.023	0.121 ± 0.012	$\mathbf{0.084 \pm 0.008}$
AP ↑	0.312 ± 0.045	0.639 ± 0.018	0.636 ± 0.023	0.649 ± 0.019	$\mathbf{0.683 \pm 0.015}$
Time	>100	16.1	61.7	0.6	3.4

Table 3 Results on data set Reuters

Measure	Rank-SVM	ML-kNN	ECC	ELM	L_{21}-ELM
HL↓	0.093 ± 0.007	0.049 ± 0.003	0.036 ± 0.003	0.044 ± 0.004	$\mathbf{0.033 \pm 0.003}$
OE↓	0.205 ± 0.056	0.126 ± 0.013	0.068 ± 0.009	0.091 ± 0.011	$\mathbf{0.062 \pm 0.011}$
Co↓	0.639 ± 0.163	0.439 ± 0.035	0.350 ± 0.036	0.380 ± 0.034	$\mathbf{0.276 \pm 0.029}$
RL↓	0.078 ± 0.027	0.045 ± 0.004	0.040 ± 0.006	0.034 ± 0.004	$\mathbf{0.019 \pm 0.003}$
AP↑	0.867 ± 0.037	0.920 ± 0.007	0.953 ± 0.006	0.940 ± 0.006	$\mathbf{0.962 \pm 0.006}$
Time	>100	3.4	2.8	1.8	2.6

Table 4 Results on data set Recreation

Measure	Rank-SVM	ML-kNN	ECC	ELM	L_{21}-ELM
HL↓	0.061	0.064	0.070 ± 0.001	0.084 ± 0.001	$\mathbf{0.058 \pm 0.001}$
OE↓	0.499	0.746	$\mathbf{0.485 \pm 0.005}$	0.577 ± 0.002	0.501 ± 0.023
Co↓	4.066	5.432	6.365 ± 0.128	6.169 ± 0.060	$\mathbf{3.955 \pm 0.012}$
RL↓	0.140	0.208	0.364 ± 0.008	0.228 ± 0.003	$\mathbf{0.136 \pm 0.001}$
AP↑	0.608	0.422	0.569 ± 0.006	0.528 ± 0.002	$\mathbf{0.611 \pm 0.014}$
Time	95	19	34	3.5	15

Table 5 Results on data set BibTeX

Measure	ML-kNN	ECC	ELM	L_{21}-ELM
HL↓	$\mathbf{0.014}$	0.017 ± 0.0001	0.014 ± 0.0001	0.015 ± 0.0002
OE↓	0.585	$\mathbf{0.371 \pm 0.007}$	0.409 ± 0.005	0.461 ± 0.018
Co↓	56.218	60.113 ± 0.369	37.266 ± 0.329	$\mathbf{23.041 \pm 0.436}$
RL↓	0.217	0.463 ± 0.002	0.128 ± 0.001	$\mathbf{0.081 \pm 0.001}$
AP↑	0.345	0.486 ± 0.003	0.516 ± 0.003	$\mathbf{0.528 \pm 0.015}$
Time	348	1007	40	94

training randomly and the rest data for testing. We repeat the data partition for thirty times randomly, and give the "average results" ± "standard deviations".

Table 2, 3, 4 and 5 shows the comparison results on a single data set. Among them, the symbol "↓" means the smaller the better, "↑" on the contrary. HL, OE, Co,

Table 6 Results on data set Yahoo

Measure	Rank-SVM	ML-kNN	ECC	ELM	L_{21}-ELM
HL↓	0.046 ± 0.014	0.043 ± 0.014	0.051 ± 0.021	0.050 ± 0.019	$\mathbf{0.042 \pm 0.014}$
OE↓	0.441 ± 0.118	0.471 ± 0.157	0.383 ± 0.123	0.437 ± 0.134	$\mathbf{0.379 \pm 0.125}$
Co↓	$\mathbf{3.564 \pm 1.043}$	4.098 ± 1.237	8.563 ± 1.867	6.362 ± 1.207	4.836 ± 1.080
RL↓	$\mathbf{0.083 \pm 0.031}$	0.102 ± 0.045	0.329 ± 0.080	0.154 ± 0.051	0.111 ± 0.034
AP↑	0.661 ± 0.089	0.625 ± 0.117	0.621 ± 0.085	0.631 ± 0.104	$\mathbf{0.685 \pm 0.095}$
Time	213	19	45	3	17

RL and AP are the abbreviations of hamming loss, one-error, coverage, ranking loss and average precision respectively, unit of Time (training) is seconds. The number of ELM hidden layer nodes is randomly setted but not more than the training samples and the best results are selected.

Overall, compared with other algorithms, L_{21}-ELM achieves the best performance in most case. Especially, it shows the absolute advantage on coverage, ranking loss and average precision in all datasets. On hamming loss it is worse than Rank-SVM only on BibTeX data set, and performs better on other cases. On one-error, ECC achieves comparable performance with other approaches. Without consideration of ECC, L_{21}-ELM outperforms other approaches by the metric of one-error.

Compared with the original ELM approach, L_{21}-ELM achieves obviously better performance on almost all datasets over all the 5 criteria. This validates the effectiveness of the L_{21}-norm regularization on the original ELM and eliminating redundant information.

On the training time, the ELM group has faster training time than other approaches. This validates that L_{21}-ELM could fully inherit the merits of ELM with extreme learning speed. Compared with original ELM, L_{21}-ELM consumes more training time, but considering its better performance it is worth.

Note that Yahoo is comprised of 11 independent data sets, including: Arts, Business, Computers, Education, Entertainment, Health, Recreation, Reference, Science, Social and Society. We just give the average results over the 11 data sets. From the results as Table 6 shows, our approach could also achieve a better performance relatively.

6 Conclusion

In this paper, we propose a L_{21}-norm Minimization ELM algorithm for multi-label learning problem, which not only inherits the advantage of ELM but also offers us additional characteristics. Through the constraint of the L_{21}-norm regularization on the original ELM, the output weights matrix of the hidden layer nodes become sparse and then leading to the simplification of the learning model. Experimental results validate that L_{21}-ELM has highly competition to state-of-the-art multi-label algorithms (e.g. Rank-SVM, ML-kNN and ECC) especially in training time. Our approach greatly improves the performance of the original ELM, although it sacrifices more time.

References

1. Huang, G.B., Chen, L.: Convex incremental extreme learning machine. Neurocomputing **70**, 3056–3062 (2007)
2. Huang, G.B., Zhu, Q.Y., Siew, C.K.: Extreme learning machine: a new learning scheme of feedforward neural networks. In: 2004 IEEE International Joint Conference on Neural Networks, 2004. Proceedings, vol. 2, pp. 985–990. IEEE (2004)
3. Huang, G.B., Chen, L., Siew, C.K.: Universal approximation using incremental constructive feedforward networks with random hidden nodes. IEEE Trans. Neural Netw. **17**, 879–892 (2006)
4. Huang, G.B., Siew, C.K.: Extreme learning machine with randomly assigned RBF kernels. Int. J. Inf. Technol. **11**(1), 16–24 (2005)
5. Huang, G.B., Ding, X., Zhou, H.: Optimization method based extreme learning machine for classification. Neurocomputing **74**(1–3), 155–163 (2010)
6. Zhang, M.L., Zhou, Z.H.: A review on multi-label learning algorithms. IEEE Trans. Knowl. Data Eng. **26**(8), 1819–1837 (2014)
7. Tsoumakas, G., Katakis, I.: Multi-label classification: an overview. Int. J. Data Warehousing Min. **2007**(3), 1–13 (2007)
8. Boutell, M.R., Luo, J., Shen, X., Brown, C.M.: Learning multi-label scene classification. Pattern Recogn. **37**(9), 1757–1771 (2004)
9. Tsoumakas, G., Vlahavas, I.: Random k-labelsets: an ensemble method for multilabel classification. Lecture Notes in Computer Science, pp. 406-417 (2007)
10. Read, J., Pfahringer, B., Holmes, G., Frank, E.: Classier chains for multi-label classification. In: Proceedings of ECML-KDD, vol. 22, no. 4, pp. 829–840 (2009)
11. Read, J., Pfahringer, B., Holmes, G., Frank, E.: Classifier chains for multi-label classification. Mach. Learn. **85**(3), 333–359 (2011)
12. Zhang, M.L., Zhou, Z.H.: ML-kNN: a lazy learning approach to multi-label learning. Pattern Recogn. **40**(7), 2038–2048 (2007)
13. Clare, A., King, R.D.: Knowledge Discovery in Multi-label Phenotype Data. Lecture Notes in Computer Science, pp. 42–53 (2001)
14. Elisseeff, A., Weston, J.: A Kernel Method for Multi-labelled Classification, pp. 681–687. MIT Press, USA (2002)
15. Obozinski, G., Taskar, B., Jordan, M.I.: Multi-task feature selection. Statistics Department, UC Berkeley, Technical Report, 1693–1696 (2006)
16. Obozinski, G., Taskar, B., Jordan, M.I.: Joint covariate selection for grouped classification. Statistics Department, UC Berkeley, Technical Report (2007)
17. Liu, J., Ji, S., Ye, J.: Multi-task feature learning via efficient L_{21}-norm minimization. In: Proceedings of the Twenty-Fifth Conference on Uncertainty in Artificial Intelligence, pp. 339–348 (2009)
18. Liu, J., Ye, J.: Efficient Euclidean projections in linear time. In: Proceedings of the Twenty-Sixth Annual International Conference on Machine Learning, pp. 657–664. ACM, (2009)
19. Zhang, M.L., Zhou, Z.H.: Multi-label neural networks with applications to functional genomics and text categorization. IEEE Trans. Knowl. Data Eng. **18**(10), 1338–1351 (2006)
20. Gjorgji, M.A., Dejan, G.A.: Two stage architecture for multi-label learning. Pattern Recogn. **45**(3), 1019–1034 (2012)

21. Schapire, R.E., Singer, Y.: Boostexter: a boosting-based system for text categorization. Mach. Learn. **39**(2–3), 135–168 (2000)
22. Salton, G.: Developments in automatic text retrieval. Science **253**(5023), 974–980 (1991)

Cluster-Based Outlier Detection Using Unsupervised Extreme Learning Machines

Xite Wang, Derong Shen, Mei Bai, Tiezheng Nie, Yue Kou and Ge Yu

Abstract Outlier detection is an important data mining task, whose target is to find the abnormal or atypical objects from a given data set. The techniques for detecting outliers have a lot of applications, such as credit card fraud detection, environment monitoring, etc. In this paper, we proposed a new definition of outlier, called cluster-based outlier. Comparing with the existing definitions, the cluster-based outlier is more suitable for the complicated data sets that consist of many clusters with different densities. To detect cluster-based outliers, we first split the given data set into a number of clusters using unsupervised extreme learning machines. Then, we further design a pruning method technique to efficiently compute outliers in each cluster. at last, the effectiveness and efficiency of the proposed approaches are verified through plenty of simulation experiments.

Keywords Outlier detection · Cluster-based · Unsupervised extreme learning machines

1 Introduction

Outlier detection is an important issue of data mining, and it has been widely studied by many scholars for years. According to the description in [1], "an outlier is an observation in a data set which appears to be inconsistent with the remainder of that set of data". The techniques for mining outliers can be applied to many fields, such as credit card fraud detection, network intrusion detection, environment monitoring, and so on.

There exist two primary missions in the outlier detection. First, we need to define what data are considered as outliers in a given set. Second, an efficient method to compute these outliers needs to be designed. The outlier problem is first studied

X. Wang (✉) · D. Shen · M. Bai · T. Nie · Y. Kou · G. Yu
College of Information Science & Engineering, Northeastern University,
Shenyang 110819, China
e-mail: xite-skywalker@163.com

© Springer International Publishing Switzerland 2016
J. Cao et al. (eds.), *Proceedings of ELM-2015 Volume 1*,
Proceedings in Adaptation, Learning and Optimization 6,
DOI 10.1007/978-3-319-28397-5_11

Fig. 1 Example of outliers

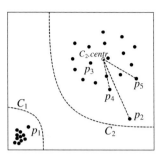

by the statistics community [2, 3]. They assume that the given data set follows a distribution, and an object is considered as an outlier if it shows distinct deviation from this distribution. However, it is almost an impossible task to find an appropriate distribution for high dimensional data. To overcome the drawback above, some model-free approaches are proposed by the data management community. Examples include distance-based outliers [4–6] and the density-based outlier [7]. Unfortunately, although these definitions do not need any assumption on the data set, some shortcomings still exist in them. Therefore, in this paper, we propose a new definition, the Cluster-Based (CB) outlier. The following example discusses the weaknesses of the existing model-free approaches and the motivation of our work.

As Fig. 1 shows, there are a denser cluster C_1 and a sparse cluster C_2 in a 2-dimensional data set. Intuitively, points p_1 and p_2 are the outliers because they show obvious differences from the other points. However, in the definitions of the distance-based outliers, a point p is marked as an outlier depending on the distances from p to its k-nearest neighbors. Then, most of the points in the sparse cluster C_2 are more likely to be outliers, whereas the real outlier p_1 will be missed. In fact, we must consider the **localization** of outliers. In other words, to determine whether a point p in a cluster C is an outlier, we should only consider the points in C, since the points in the same cluster usually have similar characters. Therefore, in Fig. 1, p_1 from C_1 and p_2 from C_2 can be selected correctly. The density-based outlier [7] also considers the localization of outliers. For each point p, they use the Local Outlier Factor (LOF) to measure the degree of being an outlier. To compute the LOF of p, we need to find the set of its k-nearest neighbors $nn_k(p)$ and all the k-nearest neighbors of each point in $nn_k(p)$. The expensive computational cost limits the practicability of the density-based outlier. Therefore, we propose the CB outlier to conquer the above deficiencies. The formal definition will be described in Sect. 3.1.

To detect CB-outliers in a given set, the data need to be clustered first. In this paper, we employ the Unsupervised Extreme Learning Machine (UELM) [8] for clustering. The Extreme Learning Machine (ELM) is a novel technique proposed by Huang et al. [9–11] for pattern classification, which shows better predicting accuracy than the traditional Support Vector Machines (SVMs) [12–15]. Thus far, the ELM techniques have attracted the attention of many scholars and various extensions of ELM have been proposed [16]. UELM [8] is designed for dealing with the unlabeled data, and it can efficiently handle clustering tasks. The authors show that

UELM provides favorable performance compared with the state-of-the art clustering algorithms [17–20].

This paper focuses on the problem of CB-outlier detection using UELM. The main contributions are summarized as follows. We propose the definition of CB outlier, which shows remarkable advantages than the existing definitions in complicated data sets (i.e., the data set consists of many clusters with different densities). We propose an efficient algorithm to detect CB outliers. The algorithm adopts a pruning strategy to improve the searching speed of k-nearest neighbors (kNNs). At last, we design a series of experiments to testify the effectiveness of the approaches proposed in this paper.

The rest of paper is organized as follows. Section 2 gives brief overviews of ELM and UELM. Section 3 formally defines the CB outlier and gives the method to detect CB-outliers. Section 4 analyzes the experimental results. Section 5 gives the related work of outlier detection. Section 6 concludes the paper.

2 Preliminaries

2.1 Brief Introduction to ELM

The target of ELM is to train a single layer feedforward network from a training set with N samples, $\{X, Y\} = \{x_i, y_i\}_{i=1}^{N}$. Here $x_i \in \mathbb{R}^d$, and y_i is a M-dimensional binary vector where only one entry is "1" to represent the class that x_i belongs to.

The training process of ELM includes two stages. In the first stage, we build the hidden layer with L nodes using a number of mapping neurons. In details, for the jth hidden layer node, a d-dimensional vector a_j and a parameter b_j are randomly generated. Then, for each input vector x_i, the relevant output value on the jth hidden layer node can be acquired using an activation function such as the Sigmoid function below.

$$g(x_i, a_j, b_j) = \frac{1}{1 + exp(-(a_j^T \times x_i + b_j))} \tag{1}$$

Then, the matrix outputted by the hidden layer is

$$H = \begin{bmatrix} g(x_1, a_1, b_1) & \cdots & g(x_1, a_L, b_L) \\ \vdots & & \vdots \\ g(x_N, a_1, b_1) & \cdots & g(x_N, a_L, b_L) \end{bmatrix}_{N \times L} \tag{2}$$

In the second stage, a M-dimensional vector β_j is the output weight that connects the jth hidden layer with the output node. The output matrix Y is acquired by Eq. 3.

$$H \cdot \beta = Y \tag{3}$$

where

$$\beta = \begin{bmatrix} \beta_1 \\ \vdots \\ \beta_L \end{bmatrix}_{L \times M} \quad and \quad Y = \begin{bmatrix} y_1 \\ \vdots \\ y_N \end{bmatrix}_{N \times M} \tag{4}$$

We have known the matrixes H and Y. the target of ELM is to solve the output weights β by minimizing the square losses of the prediction errors, leading to the following equation.

$$\min_{\beta \in \mathbb{R}^{L \times M}} L_{ELM} = \frac{1}{2} \parallel \beta \parallel^2 + \frac{C}{2} \parallel Y - H\beta \parallel^2$$
$$s.t. \; H\beta = Y - e \tag{5}$$

where $\parallel \cdot \parallel$ denotes the Euclidean norm, $e = Y - H\beta = [e_1^T, \dots, e_N^T] \in \mathbb{R}^{N \times M}$ is the error vector with respect to the training samples and C is a penalty coefficient on the training errors. The first term in the objective function is a regularization term against over-fitting.

If $N \geq L$, which means H has more rows than columns and it is full of column rank, Eq. 6 is the solution for Eq. 5.

$$\beta^* = (H^T H + \frac{I_L}{C})^{-1} H^T Y \tag{6}$$

If $N < L$, a restriction that β is a linear combination of the rows of H: $\beta = H^T \alpha (\alpha \in \mathbb{R}^{N \times M})$ is considered. Then β can be calculated by Eq. 7.

$$\beta^* = H^T \alpha^* = H^T (HH^T + \frac{I_N}{C})^{-1} Y \tag{7}$$

where I_L and I_N are the identity matrices of dimensions L and N, respectively.

2.2 Unsupervised ELM

Huang et al. [8] proposed UELM to process unsupervised data set, which shows good performance in clustering tasks. The unsupervised learning is based on the following assumption: if two points x_1 and x_2 are close to each other, their conditional probabilities $P(y|x_1)$ and $P(y|x_2)$ should be similar. To enforce this assumption on the data, we acquire the following equation:

$$L_m = \frac{1}{2} \sum_{i,j} w_{ij} \parallel P(y|x_i) - P(y|x_j) \parallel^2 \tag{8}$$

Algorithm 1: UELM Algorithm [8]

---

input : The training data: $X \in \mathbb{R}^{N \times d}$.
output: The label vector of cluster y_i corresponding to x_i

1   *a)* Construct the graph Laplacian L of X;
2   *b)* Generate a pair of random values $\{a_i, b_i\}$ for each hidden neuron, and calculate the output matrix $H \in \mathbb{R}^{N \times L}$;
3   *c)*
4   **if** $L \leq N$ **then**
5 | Find the generalized eigenvectors $v_2, \ldots, v_{M+1}$ of Eq. 11. Let $\beta = [\tilde{v}_2, \tilde{v}_3, \ldots, \tilde{v}_{M+1}]$.
6   **else**
7 | Find the generalized eigenvectors $u_2, \ldots, u_{M+1}$ of Eq. 13. Let $\beta = H^T[\tilde{u}_2, \tilde{u}_3, \ldots, \tilde{u}_{M+1}]$;
8   *d)* Calculate the embedding matrix: $E = H\beta$;
9   *e)* Treat each row of E as a point, and cluster the N points into K clusters using the k-means algorithm. Let Y be the label vector of cluster index for all the points.
10   **return** Y;

---

where w_{ij} is the pair-wise similarity between x_i and x_j, which can be calculated by Gaussian function $exp(- \| x_i - x_j \|^2 / 2\sigma^2)$.

Since it is difficult to calculate the conditional probabilities, the following Eq. 9 can approximate Eq. 8.

$$\hat{L}_m = Tr(\hat{Y}^T L \hat{Y}) \tag{9}$$

where $Tr(\cdot)$ denotes the trace of a matrix, $\hat{Y}$ is the predictions of the unlabeled dataset, $L = D - W$ is known as graph Laplacian, and D ia a diagonal matrix with its diagonal elements $D_{ii} = \sum_{j=1}^{u} w_{ij}$.

In the unsupervised learning, the data set $X = \{x_i\}_{i=1}^{N}$ is unlabeled. Substituting Eq. 9 to Eq. 5, the objective function of UELM is acquired.

$$\min_{\beta \in \mathbb{R}^{L \times M}} \| \beta \|^2 + \lambda Tr(\beta^T H^T L H \beta) \tag{10}$$

where λ is a tradeoff parameter. In most cases, Eq. 10 reaches its minimum value at $\beta = 0$. In [18], Belkin et al. introduced an additional constraint $(H\beta)^T H\beta = I_M$. On the base of the conclusion in [8], if $L \leq N$, we can obtain the following equation.

$$(I_L + \lambda H^T L H)v = \gamma H^T H v \tag{11}$$

Let γ_i be the ith smallest eigenvalues of Eq. 11 and v_i be the corresponding eigenvectors. Then the solution of the output weights β is given by

$$\beta^* = [\tilde{v}_2, \tilde{v}_3, \ldots, \tilde{v}_{M+1}] \tag{12}$$

where $\tilde{v}_i = v_i / \| H v_i \|, i = 2, \ldots, M + 1$ is the normalized eigenvectors.

If $L > N$, Eq. 11 is underdetermined. We obtain the alternative formulation below.

$$(I_N + \lambda LHH^T)u = \gamma HH^T u \tag{13}$$

Again, let u_i be the generalized eigenvectors corresponding the ith smallest eigenvalues of Eq. 13. Then, the final solution is

$$\beta^* = H^T[\tilde{u}_2, \tilde{u}_3, \dots, \tilde{u}_{M+1}] \tag{14}$$

where $\tilde{u}_i = \tilde{u}_i / \parallel HH^T\tilde{u}_i \parallel, i = 2, \dots, M+1$ are the normalized eigenvectors. Algorithm 1 shows the process of UELM.

3 CB Outlier Detection Using UELM

In this section, we first give the formal definition of the CB outlier. Then, we design an efficient algorithm to compute CB outliers from a given data set.

3.1 Defining CB Outliers

For a given data set P in a d-dimensional space, a point p is denoted by $p =< p[1], p[2], \dots, p[d] >$. The distance between two points p_1 and p_2 is $dis(p_1, p_2) = \sqrt{\sum_{i=1}^{d} (p_1[i] - p_2[i])^2}$. Suppose that there are m clusters $C_1, C_2, \dots, C_m$ in P outputted by UELM. For each cluster C_i, the centroid point $C_i.centr$ can be computed by the following equation.

$$C_i.centr[i] = \frac{\sum_{p \in C_i} p[i]}{|C_i|} \tag{15}$$

Intuitively, in a cluster C, most of the *normal* points are closely around the centroid point of C. In contrast, an *abnormal* point p (i.e., outlier) is usually far from the centroid point and the number of points close to p is quite small. Based on this observation, the weight of point is defined as follows.

Definition 1 (*Weight of a point*) Given an integer k, for a point p in cluster C, we use $nn_k(p)$ to denote the set of the k-nearest neighbors of p in C. Then, the weight of p is

$$w(p) = \frac{dis(p, C.centr) \times k}{\sum_{q \in nn_k(p)} dis(q, C.centr)} \tag{16}$$

Definition 2 (*Result set of CB outlier detection*) For a data set P, given two integers k and n, let R_{CB} be a subset of P with n points. If $\forall p \in R_{CB}$, there is no point $q \in P \backslash R_{CB}$ that $w(q) > w(p)$, R_{CB} is the result set of CB outlier detection.

For example in Fig. 1, in cluster C_2, the centroid point is marked in red. For $k = 2$, the k-nearest neighbors of p_2 are p_4, p_5. Because p_2 is an abnormal point and it is far from the centroid point, $dis(C_2.centr, p_2)$ is much larger than $dis(C_2.centr, p_4)$ and $dis(C_2.centr, p_5)$. Hence, the weight of p_2 is large. In contrast, for a normal point p_3 deep in the cluster, the distances from $C_2.centr$ to its k-nearest neighbors are similar to $dis(C_2.centr, p_3)$. The weight of p_3 is close to 1. Therefore, p_2 is more likely to be considered as a CB outlier.

3.2 The Algorithm for Detecting CB Outliers

According to Definitions 1 and 2, to determine whether a point p in a cluster C is an outlier, we need to search the k-nearest neighbors (kNNs) of p in C. In order to accelerate the kNN searching, we design an efficient method to prune the searching space.

For a cluster C, suppose that the points in C have been sorted according to the distances to the centroid point in ascending order. For a point p in C, we scan the points to search to its kNNs. Let $nn_k^{temp}(p)$ be the set of k points that are the nearest to p from the scanned points, and $kdis_{temp}(p)$ be the maximum value of the distances from the points in $nn_k^{temp}(p)$ to p. Then, the pruning method is described as follows.

Theorem 1 *For a point q in front of p, if $dis(q, C.centr) < dis(p, C.centr) - kdis_{temp}(p)$, the points in front of q and q itself cannot be the kNNs of p.*

Proof For a point q' in front of q, because the points in C have been sorted, $dis(q', C.centr) < dis(q, C.centr)$. Then, according to the triangle inequality, the distance from q' to p: $dis(q', p) > dis(p, C.centr) - dis(q', C.centr) > dis(p, C.centr) - dis(q, C.centr) > dis(p, C.centr) - (dis(p, C.centr) - kdis_{temp}(p)) = kdis_{temp}(p)$. Clearly, there exist k points closer to p than q', thus q' cannot be the kNN of p.

Similarly, for a point q at the back of p, if $dis(q, C.centr) > dis(p, C.centr) + kdis_{temp}(p)$, the points at the back of q and q itself cannot be the kNNs of p. For example, Fig. 2 shows a portion of points in a cluster C. First, we sort the points according to the distances to the centroid point, and obtain a point sequence $p_9, p_7, p_5, p_3, p_1, p, p_2, p_4, p_6, p_8$. For point p, we search its kNNs from p to both sides ($k = 2$). After p_1, p_2, p_3 are visited, p_1 and p_3 are the current top-k nearest neighbors for p, thus $kdis_{temp}(p) = dis(p, p_1)$. When we visit p_4, $dis(p_4, C.centr) > dis(p, C.centr) + kdis_{temp}(p)$. Hence, the points behind p_4 in the sequence (i.e., p_4, p_6, p_8) cannot be the kNNs of p. Similarly, when p_5 is visited, we do not need to further scan the points before p_5 in the sequence because $dis(p_5, C.centr) <$

Fig. 2 Example of *k*NN searching

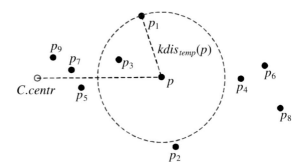

$dis(p, C.centr) - kdis_{temp}(p)$. Therefore, the *k*NN searching stops, the exact *k*NNs of p are p_1 and p_3.

4 Experimental Evaluation

In this section, we evaluate the performance of the proposed algorithm for CB outlier detection using a PC with an Intel Core i7-2600 @3.4 GHz CPU, 8 GB main memory and 1 TB hard disk. A synthetic data set is used for the experiments. In details, given the data size N, we generate $N/1000 - 1$ clusters, and randomly assign each of them a center point and a radius. In average, each cluster has 1000 points following Gaussian distribution. At last, the remaining 1000 points are scattered into the space. We implement the proposed method to detect CB outliers (CBOD) described in Sect. 3.2 using JAVA programming language. A naive method (naive) is also implemented as a comparing algorithm, where we simply search each point's *k*NNs and compute its weight. In the experiments, we mainly concern the runtime to represent the computational efficiency, and the Point Accessing Times (PAT) to indicate the disk IO cost. The parameters' default values and their variations are showed in Table 1.

As Fig. 3a shows, CBOD is much more efficient than the naive method because of the pruning strategy proposed in this paper. With the increase of k, we need to keep tracking more neighbors for a point, so the runtime of the naive method and the

Table 1 Parameter Settings

Parameter	Default	Range of variation
k	20	15–35
n	30	20–40
Data size $\|P\|$ ($\times 10^6$)	1	0.5–2.5
Dimensionality d	3	3–15

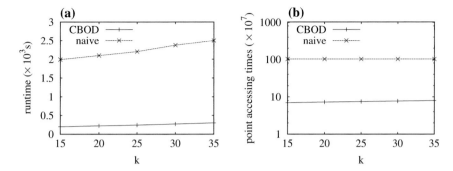

Fig. 3 The effect of parameter k. **a** Time cost versus k. **b** Point accessing times versus k

CBOD becomes larger. Figure 3b shows the effect of k on the PATs. For the naive method, each point needs to visit all the other points in the cluster to find its kNNs. Hence, PATs are large. In contrast, for CBOD, a point does not have to visit all the other points (Theorem 1). Therefore, the PATs are much smaller.

Figure 4 describes the effect of n. As n increases, more outliers are reported. Thus, the runtime of the naive method and the CBOD becomes larger. The effect on the PATs is showed in Fig. 4b, whose trend is similar to that in Fig. 3b. Note that the PATs of CBOD increase slightly with n, whereas the PATs of the naive method keep unchanged.

In Fig. 5, with the increase of the dimensionality, a number of operations (e.g. computing the distance of two points) become more time-consuming. Hence, the time cost of the two methods becomes larger. But, the variation of the dimensionality does not affect the PATs. The affect of the data size is described in Fig. 6. Clearly, with the increase of the data size, we need to scan more points to find the outliers. Therefore, both of the runtime and the PATs are liner to the data size.

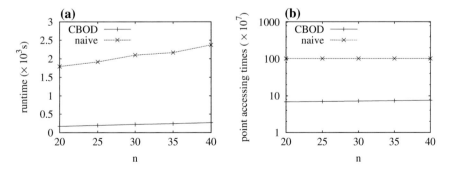

Fig. 4 The effect of parameter n. **a** Time cost versus n. **b** Point accessing times versus n

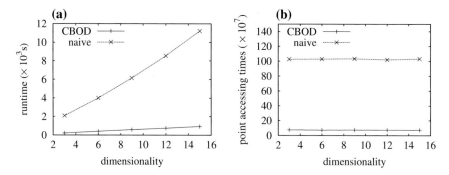

Fig. 5 The effect of dimensionality. **a** Time cost versus d. **b** Point accessing times versus d

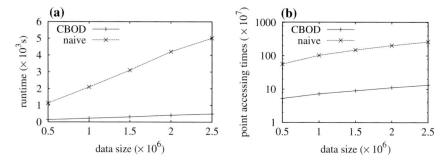

Fig. 6 The effect of data size. **a** Time cost versus $|P|$. **b** Point accessing times versus $|P|$

5 Related Work

Outlier detection is an important task in the area of data management, whose target is to find the abnormal objects in a given data set. The statistics community [2, 3] proposed the model-based outliers. The data set are assumed to follow a distribution. An outlier is the object that shows obvious deviation from the assumed distribution. Later, the data management community pointed out that building a reasonable distribution is almost an impossible task for high-dimensional data set. To overcome this weakness, they proposed several model-free approaches [21], including distance-based outliers [4–6], density-based outliers [7] and etc.

A number of studies focus on developing efficient methods to detect outliers. Knorr and Ng [4] proposes the well-known Nested-Loop (NL) algorithm to compute distance-based outliers. Bay et al. [22] proposed an improved nested loop approach, called ORCA. The approach efficiently prunes the searching space by randomizing the data set before outlier detection. Angiulli et al. [23] proposed DOLPHIN, which can reduce the disk IO cost through maintaining a small subset of the input data in main memory. Several researchers adopt the spatial indexes to further improve the computing efficiency. Examples include R-tree [24], M-tree [25], grids. However, the performance of these methods are quite sensitive to the dimensionality.

6 Conclusion

In this paper, we study on the outlier detection problem and proposed a new definition of outlier, called cluster-based outlier, which shows significant advantages than the existing definitions for complicated data sets. To detect cluster-based outliers, we first use unsupervised extreme learning machines to cluster the data in the given set. Then, we design a pruning method to reduce the kNN searching space. Finally, we evaluate the performance of the proposed approaches through a series of simulation experiments. The experimental results show that our method can effectively reduce the runtime and the disk IO cost for CB outlier detection.

Acknowledgments This work is supported by the National Basic Research 973 Program of China under Grant No.2012CB316201, the National Natural Science Foundation of China under Grant Nos. 61033007, 61472070.

References

1. Hawkins, D.M.: Identification of Outliers. Springer, New York (1980)
2. Barnett, V., Lewis, T.: Outliers in Statistical Data. Wiley, New York (1994)
3. Rousseeuw, P.J., Leroy, A.M.: Robust Regression and Outlier Detection. Wiley, New York (2005)
4. Knorr, E.M., Ng, R.T.: Algorithms for mining distancebased outliers in large datasets. In: Proceedings of the International Conference on Very Large Data Bases, pp. 392–403 (1998)
5. Ramaswamy, S., Rastogi, R., Shim, K.: Efficient algorithms for mining outliers from large data sets. ACM SIGMOD Rec. **29**(2), 427–438 (2000)
6. Angiulli, F., Pizzuti, C.: Outlier mining in large high-dimensional data sets. IEEE Trans. Knowl. Data Eng. **17**(2), 203–215 (2005)
7. Breunig, M.M., Kriegel, H.P., Ng, R.T., Sander, J.: Lof: identifying density-based local outliers. ACM Sigmod Rec. **29**(2), 93–104 (2000)
8. Huang, G., Song, S., Gupta, J.N.D., Wu, C.: Semi-supervised and unsupervised extreme learning machines. IEEE Trans. Cybern. **44**(12), 2405–2417 (2014)
9. Huang, G., Zhu, Q., Siew, C.-K.: Extreme learning machine: a new learning scheme of feedforward neural networks. Proc. Int. Joint Conf. Neural Netw. **2**, 985–990 (2004)
10. Huang, G., Zhu, Q., Siew, C.-K.: Extreme learning machine: theory and applications. Neurocomputing **70**, 489–501 (2006)
11. Huang, G.: What are extreme learning machines? Filling the gap between Frank Rosenblatt's Dream and John von Neumann's Puzzle. Cogn. Comput. **7**(3), 263–278 (2015)
12. Cherkassky, V.: The nature of statistical learning theory. IEEE Trans. Neural Netw. **8**(6), 1564 (1997)

13. Cortes, C., Vapnik, V.: Support-vector networks. Mach. Learn. **20**(3), 273–297 (1995)
14. Huang, G., Zhou, H., Ding, X., Zhang, R.: Extreme learning machine for regression and multiclass classification. IEEE Trans. Syst. Man Cybern. **Part B 42**(2), 513–529 (2012)
15. Rong, H., Huang, G., Sundararajan, N., Saratchandran, P.: Online sequential fuzzy extreme learning machine for function approximation and classification problems. IEEE Trans. Syst. Man Cybern. **Part B 39**(4), 1067–1072 (2009)
16. Liang, N., Huang, G., Saratchandran, P., Sundararajan, N.: A Fast and accurate online sequential learning algorithm for feedforward networks. IEEE Trans. Neural Netw. **17**(6), 1411–1423 (2006)
17. Kanungo, T., Mount, D.M., Netanyahu, N.S., Piatko, C.D., Silverman, R., Wu, A.Y.: An Efficient k-means clustering algorithm: analysis and implementation. IEEE Trans. Pattern Anal. Mach. Intell. **24**(7), 881–892 (2001)
18. Belkin, M., Niyogi, P.: Laplacian eigenmaps for dimensionality reduction and data representation. Neural Comput. **15**(6), 1373–1396 (2003)
19. Andrew, Y., Ng, M.I., Jordan, Y.W.: On spectral clustering: analysis and an algorithm. Adv. Neural Inform. Process. Syst. **2**, 849C856 (2002)
20. Bengio, Yoshua: Learning deep architectures for AI. Found. Trends Mach. Learn. **2**(1), 1–127 (2009)
21. He, Z., Xu, X., Deng, S.: Discovering cluster-based local outliers. Pattern Recog. Lett. **24**(9), 1641–1650 (2003)
22. Bay, S.D, Schwabacher, M.: Mining distance-based outliers in near linear time with randomization and a simple pruning rule. In: Proceedings of the Ninth ACM SIGKDD International Conference on Knowledge Discovery and Data Mining, pp. 29–38 (2003)
23. Angiulli, F., Fassetti, F.: Very efficient mining of distance-based outliers. In: Proceedings of the Sixteenth ACM Conference on Information and Knowledge Management, pp. 791–800 (2007)
24. Guttman, A.: R-trees: a dynamic index structure for spatial searching. ACM (1984)
25. Patella, M., Ciaccia, P., Zezula, P.: M-tree: an efficient access method for similarity search in metric spaces. In: Proceedings of the International Conference on Very Large Databases (VLDB). Athens, Greece (1997)

Segmentation of the Left Ventricle in Cardiac MRI Using an ELM Model

Yang Luo, Benqiang Yang, Lisheng Xu, Liling Hao, Jun Liu, Yang Yao and Frans van de Vosse

Abstract In this paper, an automatic left ventricle (LV) segmentation method based on an Extreme Learning Machine (ELM) is presented. Firstly, according to background and foreground, all sample pixels of Magnetic Resonance Imaging (MRI) images are divided into two types, and then 23-dimensional features of each pixel are extracted to generate a feature matrix. Secondly, the feature matrix is input into the ELM to train the ELM. Finally, the LV is segmented by the trained ELM. Experimental results show that the mean speed of LV segmentation based on the ELM is about 25 times faster than that of the level set, about 7 times faster than that of the SVM. The mean values of *mad* and *maxd* of image segmentation based on the ELM is about 80 and 83.1 % of that of the level set and the SVM, respectively. The mean value of *dice* of image segmentation based on the ELM is about 8 and 2 % higher than that of the level set and the SVM, respectively. The standard deviation of the proposed method is the lowest among all three methods. The results prove that the proposed method is efficient and satisfactory for the LV segmentation.

Yang Luo and Benqiang Yang contributed equally to this work and are co-first authors.

Y. Luo · L. Xu (✉) · L. Hao · J. Liu · Y. Yao
Sino-Dutch Biomedical and Information Engineering School,
Northeastern University, Shenyang 110167, Liaoning, China
e-mail: xuls@bmie.neu.edu.cn

Y. Luo
Anshan Normal University, Anshan 114005, Liaoning, China

B. Yang
General Hospital of Shenyang Military, Shenyang 110016, Liaoning, China

L. Xu
Ministry of Education, Key Laboratory of Medical Image Computing,
Hangzhou 110819, China

F. van de Vosse
Department of Biomedical Engineering, Eindhoven University of Technology,
5600 MB, Eindhoven, The Netherlands

© Springer International Publishing Switzerland 2016
J. Cao et al. (eds.), *Proceedings of ELM-2015 Volume 1*,
Proceedings in Adaptation, Learning and Optimization 6,
DOI 10.1007/978-3-319-28397-5_12

Keywords Extreme learning machine · Image segmentation · Left ventricle · Magnetic resonance imaging

1 Introduction

Cardiovascular disease is one of the main reasons of death over the past decades in the world. Cardiac magnetic resonance imaging (MRI) has proven to be a versatile and noninvasive imaging modality. Some techniques are used for the diagnoses of heart diseases. The MRI of the left ventricle (LV) is very important for the assessment of stroke volume, ejection fraction, and myocardial mass, as well as regional function parameters such as wall motion and wall thickening [1]. To perform a quantitative analysis of a LV, doctors need an accurate segmentation of the LV which can provide the anatomical and functional information of a heart, so it can be widely applied in clinical diagnoses [2, 3]. The segmentation of cardiac MRI images is one of the most critical prerequisites for quantitative study of the LV.

So far, in clinical practice, the segmentation of the LV is almost completed manually. This workload, however, is too heavy and time-consuming, subjective and irreproducible. Therefore, it is attractive to develop accurate and automatic segmentation algorithms for clinical applications.

This paper is organized as follows. Section 2 briefly reviews the related works on the segmentation of LV. In Sect. 3, this paper introduces the basic theory of ELM. In Sect. 4, the image segmentation methods are introduced in detail. The experimental results of the segmentation of LV based on the ELM are presented in Sect. 5. In Sect. 6, the conclusions are offered.

2 Related Works

In recent years, many methods have been proposed for LV segmentation. They can be classified into two types: deformable models and image-based methods. A complete review of recent literature is given in [3].

Deformable models include snakes [4–6], level set [7, 8], and their variants [9–11]. Deforming curves are derived iteratively in accordance with the minimization of an energy function which is composed of a data-driven term. A random active contour scheme for automatic image segmentation was proposed in [12].

Image-based approaches include thresholding [13], pixel-based classification [14–16], region-based and edge-based approaches. Otsu approaches [17] were employed by Lu [18] and Huang [19] in LV segmentation of cardiac MRI. However, the Otsu approach, which is based on the histogram of objects in the image, could be biased from the optimal threshold [20]. Dynamic programming

(DP) approach is used to find boundaries of the LV in MRI images [21, 22]. However, due to the complex boundaries of the myocardium [23], the performance of the DP approach sometimes is poor in boundary extraction.

3 A Brief Introduction to the Extreme Learning Machine

The extreme learning machine (ELM) is a learning algorithm, whose speed can be thousands of times faster than traditional feedforward network learning algorithms, and which has better generalization performance [24].

Given N arbitrary different samples (X_i, t_i), where $X_i = [x_{i1}, x_{i2}, \ldots, x_{in}]^T \in R^n$, and $t_i = [t_{i1}, t_{i2}, \ldots, t_{im}]^T \in R^m$, standard SLFNs with M hidden nodes and activation function $g(x)$ are modeled as

$$\sum_{i=1}^{M} \beta_i g_i(x_j) = \sum_{i=1}^{M} \beta_i g(w_i \cdot x_j + b_i) = o_j \quad (j = 1, \ldots, N) \tag{1}$$

where M is the number of the hidden layer nodes, $w_i = [w_{i1}, w_{i2}, \ldots, w_{in}]^T$ is the input weight vector, $\beta_i = [\beta_{i1}, \beta_{i2}, \ldots, \beta_{im}]^T$ is the output weight vector, and b_i is the threshold of the ith hidden node. $w_i \cdot x_j$ is the inner product of w_i and x_j [24].

If only the activation function is infinitely differentiable, the input weights and hidden layer biases can be randomly generated [24]. All the parameters of SLFNs need to be adjusted; training an SLFN is simply equivalent to finding a least squares solution $\widehat{\beta}$ of the linear system $H\beta = T$:

$$\left\| H(w_1, \ldots, w_M, b_1, \ldots, b_M)\widehat{\beta} - T \right\| = \min \left\| H(w_1, \ldots, w_{\tilde{N}}, b_1, \ldots, b_{\tilde{N}})\beta - T \right\|. \tag{2}$$

If the number M of the hidden nodes equals the number N of distinct training samples, matrix H is square and invertible, and SLFNs can approximate these training samples with zero error. However, in most cases the number of hidden nodes is much less than the number of distinct training samples, $M \ll N$, H is a non-square matrix and there may not exist w_i, b_i, β_i such that $H\beta = T$ where

$$H(w_1, \ldots, w_M, b_1, \ldots, b_M, x_1, \ldots, x_N) = \begin{bmatrix} g(w_1 \cdot x_1 + b_1) & \cdots & g(w_M \cdot x_1 + b_M) \\ \vdots & \cdots & \vdots \\ g(w_1 \cdot x_N + b_1) & \cdots & g(w_M \cdot x_N + b_M) \end{bmatrix}_{N \times M}, \tag{3}$$

$$\beta = \begin{bmatrix} \beta_1^T \\ \vdots \\ \beta_M^T \end{bmatrix}_{M \times m} \quad \text{and} \quad T = \begin{bmatrix} t_1^T \\ \vdots \\ t_N^T \end{bmatrix}_{N \times m} \tag{4}$$

$$\widehat{\beta} = H^{\dagger}T, \tag{5}$$

where $H^{\dagger}$ is the Moore–Penrose generalized inverse of matrix H. The ELM algorithm [24] can be described as follows [25].

Algorithm: ELM
1: for i = 1 to M do
2: randomly assign input weight w_i
3: randomly assign bias b_i
4: calculate H
5: calculate $\widehat{\beta} = H^{\dagger}T$

---

4   Methods

The whole algorithm of this image segmentation includes pre-processing techniques, training ELM, classification and post-processing as shown in Fig. 1.

4.1   Pre-processing Training Data and Training ELM

The pre-processing procedure of training data consists of the following steps:

1. Three typical images in cardiac MRI were selected as sample images.
2. All the pixels were selected from the LV of the ground truth, which were labeled as 1. The region of the LV was extended, and then all the pixels were selected from the extended region, which were labeled as 0 [26].
3. In addition to gray level value, gray mean value and gray median, representative features related to gray level co-occurrence matrix(GLCM) [27] such as energy, contrast, correlation entropy and inverse from 4 directions via a 5 × 5 window were used to represent a pixel, amounting to 23-dimentional features. Feature vectors of all pixels of an image were concentrated to generate a feature matrix.
4. Pre-processing procedure of each image includes steps 2–4, three feature matrices were merged into a feature matrix at last, whose values are normalized to [0, 1].

The training ELM is to find optimal parameters. The ELM kernel used in the pro-posed algorithm is Sigmoid function and the number of hidden nodes is 100, which are selected through multiple trials and means the optimal performance. Owing to the randomness of the input weights and hidden layer biases, in order to find the optimal input weights and hidden layer biases to gain optimal segmentation performance, therefore, the segmentations of images were performed over and over.

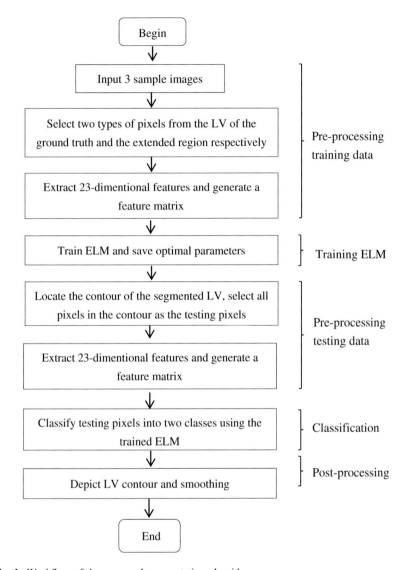

Fig. 1 Workflow of the proposed segmentation algorithm

4.2 Pre-processing Testing Data

The pre-processing procedure of testing data consists of the following steps:

1. A fitting threshold of the image was found using the Otsu method, and then the original image was converted into a binary image.
2. Little objects in the binary image were removed, whose areas are less than 300 pixels. The contours of the remaining objects were depicted.

(a)   (b)   (c)   (d)

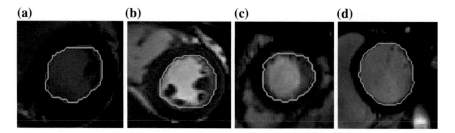

Fig. 2 ELM segmentation results of images (**a–d**). *Green* and *red* contours are obtained with manual and automatic segmentation, respectively

3. Due to the shape of LV approximating a circle, the roundness of each remaining object was calculated, and then the object which owned the biggest roundness is the LV, the LV was located approximately.
4. Usually, the errors of the contour of the LV aren't satisfactory, therefore, the morphological methods were used to process the contour, and as a result, the processed LV con tour included almost all the pixels of the LV regarded as the testing pixels set.
5. By the same methods (introduced in Sect. 4.1) 23-dimentional features of each pixel of the testing pixels set were extracted to generate a feature matrix.

4.3 Classification and Post-processing

All pixels were classified into two classes by using the trained ELM, namely one class belongs to the LV and the other one belongs to the non-LV. The LV contour was depicted and smoothed using the open-close operation in mathematical morphology, the segmentation results are shown in Fig. 2.

5 Results

5.1 Evaluation Measures

In this paper, a convenient tool for researchers is adopted to test and compare the segmentation results of the proposed method; the level set method and the SVM method easily and objectively. From the point of view of accuracy, several measures are used in our experiments, including mean absolute deviation (MAD), maximum absolute deviation (MAXD) and Dice Similarity Coefficient (DSC).

5.2 Performance

In this section, the performance of LV segmentation based on an ELM is studied through evaluating its efficiency and effectiveness. The algorithms are coded in MATLAB. All experiments are conducted on a 3.2-GHz PC with 4G memory running window 7. Real datasets are used in the experiments. The dataset includes a total of 1000 images whose sizes are 216 × 256 or 256 × 216.

Figures 2, 3 and 4 illustrate the segmentation results of the proposed method, the level set method and the SVM method of images (a), (b), (c), (d), respectively. The segmented LVs are delineated by red pixels. Green contours are obtained by manual segmentation. Many experiments were conducted on image segmentation based on the ELM. Among these experiments, the optimal performance is obtained, when the number of hidden layer nodes equals 100, and activate function is Sigmoid function.

Table 1 lists *mad*, *maxd*, *dice*, and time of our proposed method, the level set method and the SVM method about 24 images, respectively. From Table 1, it can be seen that the mean *mad* of proposed method is about 0.64 pixels, its standard deviation is about 0.18 pixels, the mean *maxd* of the proposed method is about 2.25 pixels, its standard deviation is about 0.71 pixels and the mean *dice* of the proposed

(a)   **(b)**   **(c)**   **(d)**

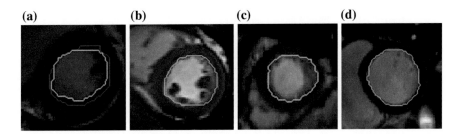

Fig. 3 Level set segmentation results of images (**a–d**). *Green* and *red* contours are obtained with manual and automatic segmentation, respectively

(a)   **(b)**   **(c)**   **(d)**

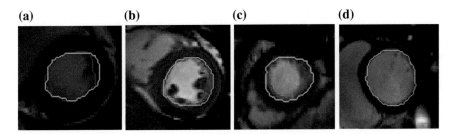

Fig. 4 SVM segmentation results of images (**a–d**). *Green* and *red* contours are obtained with manual and automatic segmentation, respectively

Table 1 Segmentation accuracy and speed on the ELM, the level set, the SVM

Id	Mad			Maxd			Dice (%)			Time		
	#1	#2	#3	#1	#2	#3	#1	#2	#3	#1	#2	#3
1	0.50	0.66	0.80	1.41	2.00	2.24	95	88	91	0.73	9.30	3.67
2	0.50	0.51	0.70	1.41	2.00	2.00	95	89	91	0.73	8.41	2.42
3	0.99	0.57	0.60	3.16	2.24	2.24	87	88	92	0.19	7.94	2.11
4	0.63	0.98	0.77	2.00	2.83	3.00	91	83	89	0.16	8.72	1.64
5	0.65	0.97	0.65	2.24	2.83	2.00	92	82	92	0.27	7.89	2.17
6	0.62	0.85	0.64	2.00	2.24	1.41	92	84	92	0.25	8.14	2.14
7	0.35	0.65	0.53	1.41	2.00	1.41	96	91	95	0.28	8.53	3.23
8	0.24	0.67	0.77	1.41	2.24	2.00	96	85	88	0.31	8.56	1.76
9	0.54	1.41	0.77	2.00	3.00	3.00	92	81	88	0.25	9.30	1.70
10	0.47	0.49	0.32	2.83	1.41	1.00	95	90	97	0.33	9.11	3.01
11	0.53	0.62	0.44	1.41	2.00	1.41	92	85	93	0.31	9.33	1.78
12	0.67	0.63	0.75	3.61	3.61	4.00	90	84	89	0.31	8.85	1.86
13	0.70	1.12	0.96	2.00	8.54	4.12	93	84	90	0.41	8.88	3.15
14	0.58	0.48	1.14	2.00	1.41	7.00	93	89	85	0.34	9.63	2.36
15	0.74	0.77	0.83	2.24	2.24	2.00	92	87	91	0.39	8.75	3.17
16	0.52	0.77	0.53	2.00	2.24	2.00	95	87	94	0.44	8.33	3.24
17	0.89	0.63	0.74	3.00	2.00	2.24	90	89	91	0.42	9.41	2.81
18	0.95	0.49	0.54	3.00	1.41	3.00	88	89	93	0.44	8.81	2.61
19	0.80	1.30	1.08	2.83	4.00	3.16	91	79	87	0.45	9.22	2.59
20	0.66	0.79	0.89	2.24	2.24	2.83	93	86	90	0.48	8.63	2.93
21	0.71	0.91	0.75	2.24	2.24	2.24	92	84	91	0.56	8.89	2.90
22	0.50	0.61	0.49	1.41	1.00	1.41	94	88	94	0.58	8.97	2.67
23	0.80	0.78	1.34	2.24	2.24	3.00	86	79	77	0.56	9.11	1.73
24	0.81	1.41	1.35	4.00	5.39	5.39	89	74	82	0.53	9.42	2.45
Mean	0.64	0.80	0.77	2.25	2.64	2.67	92	85	90	0.36	8.84	2.50
Std	0.18	0.27	0.26	0.71	1.53	1.34	2.67	4.05	4			

#1 denotes the ELM method, *#2* denotes the level set method, *#3* denotes the SVM method

method can reach up to 92 %, its standard deviation is about 2.67 %. The mean computation time of the proposed method is 0.36 s.The proposed method runs about 25 times faster than the level set method, about 7 times faster than the SVM method. The *dice* of image segmentation based on the ELM is about 8 and 2 % higher than that of the image segmentation based on the level set and the SVM, respectively. The standard deviation of our proposed method is the lowest in all three methods.

In order to further evaluate the performance of the proposed method, the local distributions of segmentation errors and the similarity between the segmentation and the ground truth have been illustrated in Fig. 5, respectively. The boxplots indicate the median, lower and upper quartiles of *mad*, *maxd*, *dice* of the above

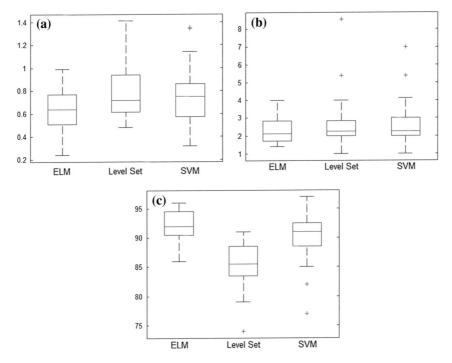

Fig. 5 Box plots of *mad* (**a**), *maxd* (**b**), and *dice* (**c**) between the ELM method, the level set method and the SVM method

three methods directly. It can be seen from Fig. 5 that the proposed method out-performs the other two methods since it obtains the higher *dice* and the lower *mad* and *maxd* than the other two methods.

6   Conclusions

This paper describes a new automatic LV segmentation method based on an ELM in cardiac MRI images in short-axis MRI view. This method takes into account the intensity inhomogeneity which often occurs in the left ventricular cavity and may cause many difficulties in image segmentation. Some images processing methods such as the morphological dilatation and the open-close operation in mathematical morphology have been used in the procedures of segmenting MRI images. Firstly, in accordance with background and foreground all the sample pixels were divided into two classes and then 23-dimensional features of each pixel were extracted to generate a feature matrix. Secondly, an ELM was trained using the feature matrix. Finally, the LV image was segmented using the trained ELM. Experimental results show that the mean speed of LV segmentation based on the ELM is about 25 times faster than that of the level set method, about 7 times faster than that of the SVM method, and in terms of the *mad* metric, *maxd* metric and *dice* metric, the image segmentation based on the proposed method is slightly better than those of the level set method and the SVM method. The results of this study prove that the proposed method is efficient and satisfactory for segmentation of LV.

Acknowledgments This work is supported by the National Natural Science Foundation of China (No. 61374015, No. 61202258), and the Fundamental Research Funds for the Central Universities (No. N130404016, N110219001).

References

1. Hu, H., Gao, Z., Liu, L., Liu, H., Gao, J., et al.: Automatic segmentation of the left ventricle in cardiac MRI using local binary fitting model and dynamic programming techniques. PLoS One 9(12), e114760. doi:10.1371/journal.pone.0114760
2. Frangi, A., Niessen, W., Viergever, M.: Three-dimensional modeling for functional analysis of cardiac images. a review. IEEE Trans. Med. Imaging 20(4), 2–5 (2001)
3. Petitjean, C., Dacher, J.N.: A review of segmentation methods in short axis cardiac MR images. Med. Image Anal. 15, 169–184 (2011)
4. Kaus, M.R., Berg, J., Weese, J., Niessen, W., Pekar, V.: Automated segmentation of the left ventricle in cardiac MRI. Med. Image Anal. 8, 245–254 (2004)
5. Lee, H.Y., Codella, N.C.F., Cham, M.D., Weinsaft, J.W., Wang, Y.: Automatic left ventricle segmentation using iterative thresholding and an active contour model with adaptation on short-axis cardiac MRI. IEEE Trans. Biomed. Eng. 57, 905–913 (2010)
6. Grosgeorge, D., Petitjean, C., Caudron, J., Fares, J., Dacher, J.-N.: Automatic cardiac ventricle segmentation in MR images: a validation study. Int. J. Comput. Assist. Radiol. Surg. 6, 573–581 (2011)
7. Chen, T., Babb, J., Kellman, P., Axel, L., Kim, D.: Semiautomated segmentation of myocardial contours for fast strain analysis in cine displacement-encoded MRI. IEEE Trans. Med. Imaging 27, 1084–1094 (2008)
8. Ammar, M., Mahmoudi, S., Chikh, M.A., Abbou, A.: Endocardial border detection in cardiac magnetic resonance images using level set method. J. Digit. Imaging 25, 294–306 (2012)
9. Pednekar, A., Kurkure, U., Muthupillai, R., Flamm, S., Kakadiaris, I.A.: Automated left ventricular segmentation in cardiac MRI. IEEE Trans. Biomed. Eng. 53, 1425–1428 (2006)
10. Zhang, H., Wahle, A., Johnson, R.K., Scholz, T.D., Sonka, M.: 4-D cardiac MR image analysis: left. and right ventricular morphology and function. IEEE Trans. Med. Imaging 29, 350–364 (2010)
11. O'Brien, S.P., Ghita, O., Whelan, P.F.: A novel model-based 3D + time left ventricular segmen-tation technique. IEEE Trans. Med. Imaging 30, 461–474 (2011)
12. Pluempitiwiriyawej, C., Moura, J.M.F., Wu, Y.J.L.: Ho C STACS: new active contour scheme for cardiac MR image segmentation. IEEE Trans. Med. Imaging 24, 593–603 (2005)
13. Lee, H.-Y., Codella, N.C., Cham, M.D., Weinsaft, J.W., Wang, Y.: Automatic left ventricle segmentation using iterative thresholding and an active contour model with adaptation on short-axis cardiac MRI. IEEE Trans. Biomed. Eng. 57, 905–913 (2010)
14. Nambakhsh, C., Yuan, J., Punithakumar K., Goela, A., Rajchl, M, et al.: Left ventricle segmentation in MRI via convex relaxed distribution matching. Med. Image Anal. 1010–1024 (2013)
15. Hadhoud, M.M., Eladawy, M.I., Farag, A., Montevecchi, F.M., Morbiducci, U.: Left Ventricle Segmentation in Cardiac MRI Images. Am. J. Biomed. Eng. 2, 131–135 (2012)

16. Eslami, A., Karamalis, A., Katouzian, A., Navab, N.: Segmentation by retrieval with guided random walks: application to left ventricle segmentation in MRI. Med. Image Anal. 236–253 (2012)
17. Otsu, N.: A threshold selection method from gray-level histograms. Automatica **11**, 285–296 (1975)
18. Lu, Y., Radau, P., Connelly, K., Dick, A,, Wright, G.: Automatic image-driven segmentation of left ventricle in cardiac cine MRI. The MIDAS J. 49 (2009)
19. Huang, S., Liu, J., Lee, L.C., Venkatesh, S.K., Teo, L.L.S., et al.: An image-based comprehensive approach for automatic segmentation of left ventricle from cardiac short axis cine MR images. J. Digital Imaging 1–11 (2010)
20. Moumena, A.-B., Ali, E.-Z.: Mammogram images thresholding for breast cancer detection using different thresholding methods. Adv. Breast Cancer Res. **2**, 72 (2013)
21. Yeh, J., Fu, J., Wu, C., Lin, H., Chai, J.: Myocardial border detection by branch-and-bound dynamic programming in magnetic resonance images. Comput. Methods Programs Biomed. **79**, 19–29 (2005)
22. Üzümcü, M., van der Geest, R.J., Swingen, C., Reiber, J.H.C., Lelieveldt, B.P.F.: Time continuous tracking and segmentation of cardiovascular magnetic resonance images using multidimensional dynamic programming. Invest. Radiol. 41–52 (2006)
23. Liu, H., Hu, H., Xu, X., Song, E.: Automatic left ventricle segmentation in cardiac MRI using topological stable-state thresholding and region restricted dynamic programming. Acad. Radiol. **19**, 723–731 (2012)
24. Huang, G.B., Zhu, Q.Y., Siew, C.K.: Extreme learning machine: theory and applications. Neurocomputing **70**, 489–501 (2006)
25. Zhao, Y., Wang, G., Yin, Y., et al.: Improving ELM-based microarray data classifcation by diversified sequence features selection. Neural Comput. Appl. doi:10.1007/s00521-014-1571-7
26. Zhao, Y., Xu Yu, J., Wang, G., Chen, L., Wang, B., Yu, G.: Maximal subspace coregulated gene clustering. IEEE Trans. Knowl. Data Eng. (TKDE) **20**(1), 83–98 (2008)
27. Wan, S.Y., William, H.: Symmetric region growing. IEEE Trans. Image Process. **12**(9), 1007–1015 (2003)

Channel Estimation Based on Extreme Learning Machine for High Speed Environments

Fang Dong, Junbiao Liu, Liang He, Xiaohui Hu and Hong Liu

Abstract Due to the complexity and extensive application of wireless systems, channel estimation has been a hot research issue, especially for high speed environments. High mobility challenges the speed of channel estimation and model optimization. Unlike conventional estimation implementations, this paper proposes a new channel estimation method based on extreme learning machine (ELM) algorithm. Simulation results of path loss estimation and channel type estimation show that the ability of ELM to provide extremely fast learning make it very suitable for estimating wireless channel for high speed environments. The results also show that channel estimation based on ELM can produce good generalization performance. Thus, ELM is an effective tool in channel estimation.

Keywords Channel estimation · Extreme learning machine · High speed · Path loss · COST 207

F. Dong (✉) · X. Hu · H. Liu
School of Information and Electrical Engineering, Zhejiang University City College,
HuZhou Street 50, Hangzhou, China
e-mail: dongf@zucc.edu.cn

X. Hu
e-mail: huxh@zucc.edu.cn

H. Liu
e-mail: liuhong@zucc.edu.cn

J. Liu
Institution of Information Science and Electrical Engineering, Zhejiang University,
Yugu Street 38, Hangzhou, China
e-mail: liujunbiao@zju.com

L. He
Department of Electronic Engineer, Tsinghua University,
Qinghua Street 1, Beijing, China
e-mail: heliang@mail.tsinghua.edu.cn

© Springer International Publishing Switzerland 2016
J. Cao et al. (eds.), *Proceedings of ELM-2015 Volume 1*,
Proceedings in Adaptation, Learning and Optimization 6,
DOI 10.1007/978-3-319-28397-5_13

1 Introduction

High-speed railways (HSR) and highway networks have developed rapidly for nearly ten years to meet people's travel needs. This explosive increase of high speed transportation raises higher requirements for wireless communication systems, including train ground communication (TGC) system [1], communication based train control (CBTC) system, vehicle ad hoc network (VAN) [2], vehicle to vehicle (V2V) communication, etc.

However, the speed of trains can reach 350 km/h and the speed of vehicles is up to 120 km/h, which make the users can not enjoy the smooth and high quality wireless services under low speed environment. In high mobility scenarios, large Doppler frequency shift, fast fading channel and fast handover issue seriously affect communication performances [3, 4].

Wireless channel play a key background role in transmission rate and quality of mobile propagation. Only after channel characteristics in a communication system are thoroughly researched, a variety of physical layer technologies are taken or adapted, such as the best modulation and coding interleaving scheme, equalizer design, or antenna configuration and subcarrier allocation in MIMO-OFDM system. Propagation prediction or channel estimation has been extensively studied in three areas: (1) to provide a theoretical performance bounds with information theory tool for a new physical technology [5]; (2) to assess various candidate schemes in the transmission system design [6]; (3) to estimate or predict channel parameters in the deployment of a new wireless system, and then optimize deployment [7].

Based on theoretical analysis method in modeling, wireless channel model can be divided into deterministic model, stochastic model, and semi-deterministic model [8]. Among them, some famous models as COST 207, COST 231, WINNER ii obtained by field measurements [9, 10] are wildly used in channel estimation. An appropriate channel model can be selected according to a particular scenario, and then its specific propagation parameters are set. Channel estimation in mobile propagation usually has two types of technologies to obtain these parameters: blind and pilot estimation [6, 11]. Pilot estimation is typically achieved by using pilot symbols strategically placed at frame heads or subcarrier. In blind estimation, channel coefficients are predicted by using statistical features of received signals. Once a channel is estimated its time-frequency characteristics, relevant parameters are used to update the pre-set model. As in any estimation application, wireless channel estimation aims to quantify the best performance of wireless systems. However, due to the unlimited number of received signal, it is a challenge to extract optimal channel coefficients.

Feedforward neural networks (FNN) is extensively used to provide models for a natural or artificial phenomena that are difficult to handle using classical parametric techniques [12]. Simsir et al. [13] demonstrated that channel estimation based on neural network ensures better performance than conventional Least Squares (LS) algorithm without any requirements for channel statistics and noise. In the meantime, [14] and [15] also proved FNN can be used in channel estimation for various wireless environments. Unfortunately, the learning speed of FNN has been a major bottleneck in many applications, and fast fading channel caused by high mobility makes this method unsuitable for channel estimation too. Unlike traditional FNN implementations, a simple learning algorithm called extreme learning machine (ELM) with good generalization performance [12, 16, 17] can learn thousands of times faster.

In this paper, we propose a channel estimation scheme based on ELM algorithm for high speed environments. Since researches in wireless channel have concentrate on large-scale and small-scale models [18], we choose path loss coefficient and fading classification as estimation objects. Compared with back-propagation (BP) algorithm, ELM shows its potential in channel estimation, especially for scenarios with high mobility.

The outline of the paper is as follows: In Sect. 2, ELM learning algorithm is present briefly. In Sect. 3, path loss estimation of wireless channel using ELM for high speed environments is proposed, and simulation results are analyzed. Section 4, fading classification estimation in COST 207 model based on ELM algorithm is provided. Conclusion is given in Sect. 5.

The performance of channel estimation based on ELM is in comparison with BP (LevenbergCMarquardt algorithm) which is a popular algorithm of FNN. All of the simulations are carried out in MATLAB 7.12.0. LevenbergCMarquardt algorithm is provided by MATLAB package, while ELM algorithm is downloaded from [19].

2 Review of ELM

Traditional FNN solution iteratively adjusts all of its parameters to minimize the cost function by using gradient-based algorithms. Although BP's gradients can be computed efficiently, an inappropriate learning rate might raise several issues, such as slow convergence, divergence, or stopping at a local minima.

ELM algorithm steps are as follow:

1. Assign a training set $\aleph = \left\{ \left(\mathbf{x}_i, \mathbf{t}_i \right) \middle| \mathbf{x}_i \in \mathbf{R}^n, \mathbf{t}_i \in \mathbf{R}^m, i = 1, 2, \ldots, N \right\}$, active function $g(x)$ and the number of hidden neurons $\tilde{N}$,
2. Randomly assign input weight vector $\mathbf{w}_i, i = 1, 2, \ldots, \tilde{N}$ and bias value $b_i, i = 1, 2, \ldots, \tilde{N}$,
3. Calculate the hidden layer output matrix $\mathbf{H}$ and its *Moore-Penrose* generalized inverse matrix $\mathbf{H}^\dagger$,
4. Calculate the output weight $\hat{\beta} = \mathbf{H}^\dagger \mathbf{T}$ with the least squares, where $\mathbf{T} = \left[\mathbf{t}_1, \mathbf{t}_2, \ldots, \mathbf{t}_N \right]^T$.

In a word, for a linear system $\mathbf{H}\beta = \mathbf{T}$, ELM algorithm finds a least-squares solution $\hat{\beta}$ rather than iterative adjustment. Seen from the steps, the learning time of ELM is mainly spent on calculating $\mathbf{H}^\dagger$. Therefore, ELM saves a lot of time in most applications. The performance evaluation in [12, 16] shows that ELM can produce good generalization performance in most cases and can learn more than hundreds of times faster than BP.

3 Large-Scale/Path Loss Channel Estimation

3.1 Large-Scale Channel Model

Large-scale/path loss channel models predict the mean signal strength for an arbitrary large transmitter-receiver distance (several hundreds or thousands of meters) in order to estimate the radio coverage area of a transmitter. Since the estimation of large-scale channel coefficients use statistical features of received signals, a blind estimation solution might work.

Both theoretical and measurement-based propagation channel models (such as free-space model, two-ray model, Okumura model, Hata model and etc.) [20] indicate that average received signal power P_r decreases logarithmically with distance [18]. Considering shadowing effects component ψ obeys a log-normal distribution, a statistical path loss model [21] is

$$
\begin{aligned}
P_r \text{dBm} &= P_t \text{dBm} + K \text{dBm} - 10\gamma \log_{10}\left[\frac{d}{d_0}\right] - \psi \\
&= P_t \text{dBm} + 20 \log_{10}\frac{\lambda}{4\pi d_0} - 10\gamma \log_{10}\left[\frac{d}{d_0}\right] - \psi
\end{aligned}
\tag{1}
$$

where P_t is the transmit power, γ is the path loss exponent indicating the rate at which path loss increases with distance, reference distance d_0 for practical systems is typically chosen to be 1 m, d is the transmitter-receiver distance, and shadowing effect exponent ψ is a zero-mean Gaussian distributed random variable with standard deviation σ_ψ (also in dBm).

γ is obtained by fitting the minimum mean square error (MMSE) of measurements

$$
F_{\text{MMSE}}(\gamma) = \min_\gamma \sum_{i=1}^{n} \left[M_{\text{measured}}(d_i) - M_{\text{model}}(d_i)\right]^2
\tag{2}
$$

where $M = P_t/P_r$, in dBm. And the variance σ_ψ^2 is given by

$$
\sigma_\psi^2 = \frac{1}{n} \sum_{i=1}^{n} \left[M_{\text{measured}}(d_i) - M_{\text{model}}(d_i)\right]^2
\tag{3}
$$

3.2 Approximation of Path Loss Exponent

In Eq. (1), path loss exponent γ and shadowing effect exponent ψ are determined by carrier frequency and propagation terrain. Typical value of γ is between 1 and 4. The smaller γ is, the less energy loss of wireless signal due to transceiver-receiver distance is. For example, in HSR environment, γ is slightly larger than 2 in rural area (within 250–3200 m) with narrow band communication system while it is near to 4 in hilly terrain (within 800–2500 m) with broadband system.

If γ calculated distance is 1500 m and the vehicle's velocity is 120 km/h, γ needs to be calculated every 45 s; if the velocity is up to 350 km/h, γ needs to be calculated every 15.4 s. According to Eqs. (2) and (3), the calculation of γ requires hundreds or thousands of receive signal measurements, the introduction of learning algorithm into γ estimation might be effective in simplifying the data processing.

We use ELM and BP algorithms to approximate the path loss exponent γ. Without loss of generality, we set velocity $v = 120$ km/h, carrier frequency $f_c = 2.35$ GHz, transmit power $P_t = 39.5$ dBm and distance d is obtained by means of GPS [22]. A training set $\left(P_{ri}, \gamma_i\right)$ and testing set $\left(P_{ri}, \gamma_i\right)$ with 1000 data, respectively are created where P_{ri} is uniformly randomly distributed on the interval $(-105, -25)$ dBm [23]. Shadowing effect exponent ψ has been added to all training samples while testing data are shadowing-free.

3.3 Simulation Results

The number of hidden neurons of ELM is initially set at 20 and active function is sigmoidal. Simulation result is shown in Fig. 1. The train accuracy measured in terms of root mean square error (RMSE) is 0.27734 due to shadowing effect, whereas the test accuracy is 0.012445. Figure 1 confirms that the estimation results of γ are accurate, and there is a visible margin of error only when $P_r > -30$ dBm.

Average 200 trails of simulation have been conducted for both ELM and BP algorithm, whose results are shown in Table 1. ELM learning algorithm spent 6.6 ms CPU time on training and 6.8 ms on testing, however, it takes 53.6 s for BP algorithm on training and 67.1 ms on testing. ELM runs 8000 times faster than BP. In high speed environment, when a vehicle's velocity is 120 km/h in cells with radius 1 km, it will carry out a handover procedure per 60 s; when a train's velocity is 350 km/h in same

Fig. 1 The estimation of path loss exponent γ by ELM learning algorithm

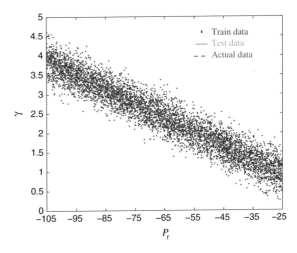

Table 1 Performance comparison for learning algoritms in large-scale channel estimation

Algorithms	Time (s)		Training (s)		Testing (s)		Hidden neurons
	Training	Testing	RMS	Dev	RMS	Dev	
ELM	0.0066	0.0068	0.2828	0.0219	0.0475	0.0500	20
BP	53.6168	0.0671	0.0745	0.0028	0.0031	0.0012	20

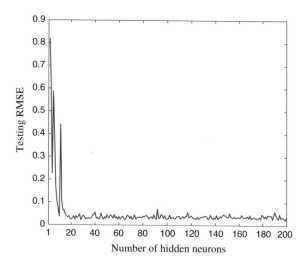

Fig. 2 The generalization performance of ELM in estimation of path loss exponent γ

cell, handover will occur per 20.6 s. Therefore, BP is too time-consuming to be used in wireless system with high mobility. Although ELM has a much higher testing error 0.0475 compared with 0.0028 in BP, this estimation error can be acceptable in our environments. In addition, assuming that network transmission rate is 1Mbps, the collection of 1000 test data takes only 1 ms, so that a packet of 125 bytes can estimate the path loss exponent γ based on ELM with accuracy rate 95 % within a time interval of less than 8 ms.

Figure 2 shows the relationship between the generalization performance of ELM and the number of hidden neurons n for γ estimation. Every n simulates 50 times. Obviously, the generalization performance of ELM is stable when $n \geq 12$. Thus, the simulation result in Fig. 1 is reasonable when n is set to 20.

Figure 3a shows the relationship between RMSE of ELM and the number of train/test data, and Fig. 3b shows the impact of this number on consuming time. Training RMSE number of train/test data is almost a constant (slightly less than 0.3) because ELM use *Moore-Penrose* inverse matrix calculation to solve the problem of finding the smallest norm least-squares output weight. Unlike training RMSE, testing RMSE decreases with increasing number of test data. The simulation confirms the conclusion in [12] that ELM has no over-trained phenomenon. Both train and

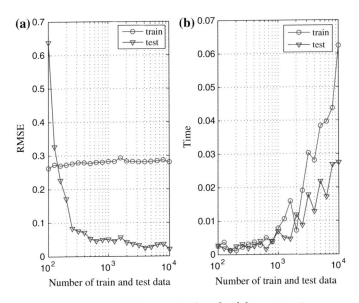

Fig. 3 Number of train/test data of ELM in estimation of path loss exponent γ

test consuming time increases with data number, however, the increase of test time is less than train time. It should also be noted that, even the number of data is up to 10^4, the time consuming of ELM is still acceptable, which is less than 70 ms.

4 Small-Scale/Fading Estimation

4.1 Small-Scale Channel Model

Small-scale/fading models characterize the rapid fluctuations of the received signal strength over very short distances (a few wavelengths) or short durations (on the order of seconds) in order to estimate the influence of multi-path propagation and the speed of a mobile terminal.

COST 207 model [9] for mobile radio specifies the power delay profiles and Doppler spread for four typical environments, i.e. rural area (RA), typical urban area (TU), bad urban area (BU) and hilly terrain (HT). The RA case consists of two distinct channel models, while the other cases each comprises four channel models. Thus, COST 207 has a total of 14 channel models: RAx4, RAx6, TUx6, TUx6alt, TUx12, TUx12alt, BUx6, BUx6alt, BUx12, BUx12alt, HTx6, HTx6alt, HTx12 and HTx12alt.

Due to radio waves' reflection and refraction, the propagation between transceiver and receiver has several paths, hence each channel model has multiple taps. For example, RAx4 is short for rural area environment with 4 taps, and HTx6alt stands

Table 2 Performance comparison for learning algorithms in small-scale channel estimation

Algorithms	Time (s)		Success rate (%)		Hidden neurons
	Training	Testing	Training	Testing	
ELM	0.0135	0.0085	86.46	72.73	20
BP	53.77	0.0694	90.60	31.08	20

for hilly terrain with 6 alternative taps. Each tap is characterized by a relative delay (with respect to the first path delay), a relative power and a Doppler spectrum category.

ELM and BP algorithms are used to estimate COST 207 channel models C_T based on modulated receive signals P_M. In order to facilitate channel estimation, each channel model is assigned an integer value [24], such as $C_T = 1$ for RAx4, $C_T = 2$ for RAx6, and etc. We still set $v = 120$ km/h, $f_c = 2.35$ GHz. Transmission rate is 1Mbps and sampling factor is 4, so that the simulation sampling rate is 4×10^6 samples per second. PSK modulation and bi-Gaussian Doppler are used in this simulation. A training set $\left(C_{Ti}, P_{Mi}\right)$ has 1000 data whereas testing set with 300 data.

4.2 Simulation Results

The hidden neurons of ELM is initially set at 20 and active function is sigmoidal. Average 50 trails of simulation have been conducted for both ELM and BP algorithm, whose results are shown in Table 2. Similarly, ELM learns up to hundreds of times faster than BP. Although BP can reach the learning rate 90.60 %, its testing rate drops to 31.08 %. On the contrary, ELM learning rate 86.46 % is slightly lower than BP, but it can achieve average testing rate 72.73 %. This is mainly because Matlab BP function *newff* doesn't support complex data. Modulated receive signals P_M must be turned into real.

5 Conclusion

In this paper, channel estimation based on ELM is proposed for high speed environments. In large-scale model, the estimation performance of path loss exponent is developed, whose experimental results show that ELM run 8000 times fast than BP learning algorithm and its testing error is acceptable. In small-scale model, fading classification estimation is provided, which shows ELM is an effective tool to classify channel type. Compared with BP, ELM still works when the elements in training set or testing set are complex. Therefore, ELM is an effective tool in channel estimation.

References

1. Shu, X., Zhu, L., Zhao, H., Tang, T.: Design and performance tests in an Integrated TD-LTE based train ground communication system. In: 17th IEEE International Conference on Intelligent Transportation Systems, pp. 747–750. IEEE Press, New York (2014)
2. Golestan K., Sattar F., Karray F., Kamel M., Seifzadeh S.: Localization in vehicular ad hoc networks using data fusion and V2V communication. Comput. Comm. (2015) (in press)
3. Zhou, Y., Ai, B.: Handover schemes and algorithms of high-speed mobile environment: a survey. Comput. Comm. **7**, 1–15 (2014)
4. Calle-Sanchez, J., Molina-Garcia, M., Alonso, J., Fernandez-Duran, A.: Long term evolution in high speed railway environments: feasibility and challenges. Bell Labs Tech. J. **18**, 237–253 (2013)
5. Singh, C., Subadar, R.: Capacity analysis of M-SC receivers over TWDP fading channels. Int. J. Electr. Comm. **68**, 166–171 (2014)
6. Vieira, M., Taylor, M., Tandon, P., Jain, M.: Mitigating multi-path fading in a mobile mesh network. Ad Hoc Netw. **11**, 1510–1521 (2013)
7. Shu, H., Simon, E., Ros, L.: On the use of tracking loops for low-complexity multi-path channel estimation in OFDM systems. Signal Process. **117**, 174–187 (2015)
8. Molisch, F., Kuchan, A., Laurila, J., Hugl, K., Schmalenberger, R.: Geometry-based directional model for mobile radio channels-principles and implementation. Eur. Trans. Telecomm. **14**, 351–359 (2003)
9. COST 207: Digital Land Mobile Radio Communications. Technical report, Office for Official Publications of the European Communities (1989)
10. 3GPP TS 45.005 V7.9.0: Radio transmission and reception (Release 7). Technical report, 3GPP GSM/EDGE Radio Access Network (2007)
11. Larsen, M., Swindlehurst, A., Svantesson, F.: Performance bounds for MIMO-OFDM channel estimation. IEEE Trans. Signal Process. **57**, 1901–1916 (2009)
12. Huang, G., Zhu, Q., Siew, C.: Extreme learning machine: theory and applications. Neuro Comput. **70**, 489–501 (2006)
13. Simsir, S., Taspinar, N.: Channel Estimation Using Neural Betwork in OFDM-IDMA System. In: 2014 Telecommunications Symposium, pp. 1–5. IEEE Press, New York (2014)
14. Cheng, C., Huang, Y., Chen, H., Yao, T.: Neural Network-Based Estimation for OFDM Channels. In: 29th IEEE International Conference Advanced Information Networking and Applications, pp. 600–604. IEEE Press, New York (2015)
15. Sotiroudis S., Siakavara K.: Mobile Radio Propagation Path Loss Prediction Using Artificial Neural Networks with Optimal Input Information for Urban Environments. In: International Journal of Electronics and Communications (2015) (in press)
16. Cao, J., Lin, Z., Fan, J.: Extreme learning machine on high dimensional and large data applications: a survey. Math. Probl. Eng. **2015**, 1–12 (2015)
17. Cao, J., Lin, Z., Fan, J.: Bayesian signal detection with compressed measurements. Inf. Sci. **289**, 241–253 (2014)
18. Rappaport, T.: Wireless communications: principles and practice. Prentice Hall Press, Cranbury (2009)
19. Huang Guang-bin. http://weibo.com/elm201x
20. Erceg, V.: An empirically based path loss model for wireless channels in suburban environments. IEEE J. Select. Areas Comm. **17**, 1205–1211 (1999)
21. Kastell, K.: Challenges And Improvements in Communication with Vehicles and Devices Moving with High-speed. In: 13th IEEE International Conference Transparent Optical Networks, pp. 26–30. IEEE Press, New York (2011)
22. Luan, F., Zhang, Y., Xiao, L., Zhou, C., Zhou, S.: Fading characteristics of wireless channel on high-speed railway in hilly terrain scenario. Int. J. Antennas Propag. **12**, 188–192 (2013)
23. Liu, J., Jin, X., Dong, F.: Wireless fading analysis in communication system in high-speed train. J. Zhejiang Univ. **46**, 1580–1584 (2012)
24. Cao J., Xiong K.: Protein sequence classification with improved extreme learning machine algorithms. BioMed Res. Int. (2014). Article ID 103054

MIMO Modeling Based on Extreme Learning Machine

Junbiao Liu, Fang Dong, Jiuwen Cao and Xinyu Jin

Abstract With multiple antennas' transmission, multiple-input multiple-output (MIMO) technique is able to utilize the space diversity to obtain high spectrum efficiency. In this study, we propose a single hidden layer feedforward network trained with extreme learning machine (ELM) to estimate channel performances in MIMO system. Bit error rate (BER) and signal-to-noise ratio (SNR) performance of back-propagation (BP) algorithm are also compared with our proposed neural network. The simulation results show that ELM has got much better time efficiency than BP in MIMO modeling. Furthermore, MIMO modeling based on ELM doesn't need to send pilot, which reduce the waste of spectrum resources.

Keywords Multiple-input multiple-output · Channel modeling · Extreme learning machine · Bit error rate · Signal-to-noise ratio

1 Introduction

The ever-expanding mobile networks are expected to deliver a wide variety of high data-rate services, such as multimedia interactive games, web browsing, broadband video, media content downloading. In order to raise transmission rate, it is

J. Liu (✉) · X. Jin
Institution of Information Science and Electrical Engineering,
Zhejiang University, Yugu Street 38, Hangzhou, China
e-mail: liujunbiao@zju.edu.cn

X. Jin
e-mail: jinxy@zju.edu.cn

F. Dong
School of Information and Electrical Engineering, Zhejiang University
City College, HuZhou Street 50, Hangzhou, China
e-mail: dongf@zucc.edu.cn

J. Cao
Institute of Information and Control, Hangzhou Dianzi University,
Erhao Street 1158, Hangzhou, China
e-mail: caojw@hdu.edu.cn

© Springer International Publishing Switzerland 2016
J. Cao et al. (eds.), *Proceedings of ELM-2015 Volume 1*,
Proceedings in Adaptation, Learning and Optimization 6,
DOI 10.1007/978-3-319-28397-5_14

imperative to improve the utilization of wireless resources, especially spectrum resources.

A key physical layer technology named multiple-input multiple-output (MIMO) is a viable option to meet the demands of mobile networks. It has been adopted in several wireless protocols, including WLAN (802.11 series) [1], WiMAX (802.16 series) [2], LTE or LTE-Advanced [3], and continues to be used in future standards like 5G [4, 5]. MIMO techniques provide multiplexing and diversity gains by using multiple antennas at both transmitter and receiver ends in a communication system [6, 7]. With the potential spatial gains, MIMO greatly improve capacity of wireless channel, and the increase is proportional to the number of antennas.

A large number of studies have reported on incorporating physical propagation characteristic in MIMO systems. In recent years, there has been increasing interest in using neural network to model MIMO channel [8–10] and estimate MIMO parameters [11, 12]. As a useful tool, neural network is able to predict an output when it recognizes a given input pattern. Thus, it can act as a reliable estimator in modeling MIMO channel without any pilot symbol bits required by other schemes. Such blind estimation would preserve bandwith and increase spectral efficiency. Sarma and Mitra [8, 9] define a complex time-delay fully recurrent neural network block and use it in MIMO channel estimation, which saves processing time than other stochastic estimation. [10] proposes a channel estimation technique based on neural network for space-time coded MIMOCOFDM systems. [11] models MIMO channels using artificial neural network with multi-layer perception (MLP), and it shows better performance than conventional MLP. In [12], neural network acts as a pre-processing block to the estimator, whose effect is measured by bit error rate (BER). Belkacem et al. [13] designed a neural network equalization for frequency selective nonlinear MIMO channels. However, all of these works focus mainly on the training-learning aspect of neural network and its capacity to deal with MIMO channel estimation, without consideration on time-varying characteristics of wireless channels. Most of these works don't evaluate the learning time and testing time of their schemes. They only concern about BER performance. Even in the algorithm [8] that claims to save processing time, the learning time of neural network is above 40s. Therefore, these MIMO channels modelled by neural network are not practical in time-varying case.

Contrary to neural network with slow learning speed, extreme learning machine (ELM) can learn very fast in most cases [14, 15]. The primary reason why neural network spend a lot of time learning is that all the parameters of neural network are iteratively adjusted base on gradient to make the learning algorithm converge in the optimal values. However, ELM approaches the best values by matrix operation rather than iteration procedure, which saves processing time.

This paper aims to provide the performance of MIMO modeling based on ELM learning and compare the simulation results with a popular neural network like back-propagation (BP) algorithm. We evaluate the possibility of using ELM in a MIMO environment with 2 transmitting antennas and 1 receiving antenna (2Tx1Rx). The performance comparison has been conducted in two MIMO channel problem: (1) estimating channel BER according to signal-to-noise ratio (SNR) at receiver end;

(2) estimating receiver's SNR according to BER at transmitter end. In both data sets, ELM seems to be a reliable tool in MIMO modeling.

The rest of the paper is organized as follows. Section 2 briefly introduces ELM algorithm. Section 3 describes our system model. Section 4 presents the BER estimation based on ELM at the MIMO receiver. Section 5 evaluates the performance of SNR estimation at the MIMO transmitter and discusses the optimization of parameters in ELM processing. Finally, Sect. 6 concludes the paper.

2   ELM Algorithms

ELM is a least-square (LS) learning algorithm developed for single hidden layer feedforward networks (SLFNs). It is tuning-free so that its learning speed is much faster than traditional gradient-based neural network algorithms. By calculating the *Moore−Penrose* generalized inverse of the hidden layer output matrix, ELM tends to reach the small norm of network output weights.

If we have a train data set $\aleph = \left\{ (\mathbf{x}_i, \mathbf{t}_i) \middle| \mathbf{x}_i \in \mathbf{R}^n, \mathbf{t}_i \in \mathbf{R}^m, i = 1, 2, \dots, N \right\}$, the basic procedure of ELM algorithm is as follows:

Step 1: for N arbitrary training data set $\aleph$, set active function $g(x)$ and the number of hidden neurons $\tilde{N}$.
Step 2: randomly assigns input weight vector $\mathbf{w}_i, i = 1, 2, \dots, \tilde{N}$ and bias value $b_i, i = 1, 2, \dots, \tilde{N}$.
Step 3: calculate the hidden layer output matrix $\mathbf{H}$ and its *Moore − Penrose* generalized inverse matrix $\mathbf{H}^{\dagger}$.
Step 4: calculate the output weight $\hat{\beta} = \mathbf{H}^{\dagger}\mathbf{T}$, where $\mathbf{T} = \left[\mathbf{t}_1, \mathbf{t}_2, \dots, \mathbf{t}_N\right]^{\mathrm{T}}$ is the target output matrix.

Obviously, the network output weights can be analytically determined by solving a linear system. Thus, The training procedure of ELM can avoid time-consuming learning iterations to achieve a good generalization performance.

3   System Description

The baseband-equivalent communication model for MIMO system [16] is depicted in Fig. 1, including transmitter and receiver structure.

In our system, the transmitter equipped with $n_T = 2$ antennas and the receiver is equipped with $n_R = 1$ antenna. Transmitter adopts BPSK data modulation. The output of the BPSK modulator generates unit power signals, i.e., signal power is 1 mw. MIMO uses OSTBC encoder to obtain potential spatial gains. The uncorrelated data stream is transmitted over Rayleigh channel. The received signal sample $y_1(n)$ of the receive antenna at time n is given by

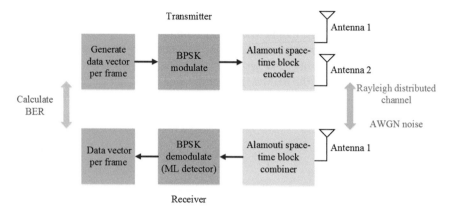

Fig. 1 MIMO model with transmitter and receiver structure

$$y_1(n) = \sum_{t=1}^{2} h_{1,t}(n) d_T(n) + v_1(n) \tag{1}$$

where d_T is the output of transmitter, $h_{1,t}(n)$ is the channel impulse response from tth transmitting antenna to the 1th receiving antenna for $t = 1, 2$, and $v_1(n)$ is AWGN samples of variance $N_0/2$ [17].

Before reaching 1th receiving antenna, the transmitted signals are affected by the propagation channel which can be modulated by an $n_R \times n_T$ propagation matrix $\mathbf{H}$. In our case, the system input-output relationship can be expressed

$$\left[y_1(n)\right] = \mathbf{H} \times \begin{bmatrix} d_1(n) \\ d_2(n) \end{bmatrix} + \mathbf{V} \tag{2}$$

where $\mathbf{H} = \left[h_{1,1}(n), h_{1,2}(n) \right]$ and $\mathbf{V} = v_1(n)$.

4 BER Estimation at Receiver

In wireless network, error occurs inevitably during data transmission due to various reasons, such as non-ideal transmission media, noise and outside interference. BER is an important factor indicating the accuracy of data transmission within a specific interval. As a basic measure of performance, BER usually shows its dependence on SNR. To obtain this characteristic, a sufficient number of BER values should be measured for various SNR at receiver end. Since terminals in wireless network are always in mobile state, the availability of an accurate and time efficient method of BER measurements is particular important [18]. We simulate BER estimation based on ELM to assess the accuracy and time efficiency of this scheme.

Table 1 The performance comparisons of BER estimation with ELM and BP

Algorithms	Time (s)		Training (s)		Testing (s)		Hidden neurons
	Training	Testing	RMS	Dev	RMS	Dev	
ELM	0.0034	0.0019	0.0111	1.0514e-17	0.0104	1.2266e-17	10
BP	42.8400	0.0659	1.2003e-4	8.2158e-20	1.1592e-4	2.7386e-20	10

4.1 BER Estimation Based on ELM

In this section, BER estimation of ELM is compared with BP algorithm. All experiments are carried in MATLAB 7.12.0 environment running an ordinary PC with 2.4GHz CPU and 2GB RAM. A training set (BER_i, SNR_i) and testing set (BER_i, SNR_i) have 1000 data, respectively, with SNR_i uniformly randomly distributed on the interval $(-10, 10)$. In Fig. 1 MIMO system, each frame in simulation consists of 100 bits, and the number of train packets and test packets are 1. If the data rate is 1 Mbps, it only takes 0.1 ms to obtain test data.

There are 10 hidden nodes for both ELM algorithm and BP algorithm. The activation function used in ELM is set as sigmoid. Average results of 50 trails of simulations are observed in Table 1. The enhancement of time efficiency is apparent. ELM spends 3.4 ms CPU time obtaining training error rate slightly above 0.01, whereas BP spends 42.84 s to reach training error rate 0.0001. Although ELM training error rate and testing error rate are much larger than BP, the learning speed of ELM runs 12600 times faster than BP and the testing time of BP is 35 times longer than ELM.

Figure 2 shows the actual BER and the estimation value of ELM algorithm. The test fitting curve are well constant with the actual data when SNR is less than 4dB. Since the number of error bits is an integer, train data sets are divided into separate segments or discrete points when SNR is large than 4dB, so that the ELM estimation result is not idealistic. However, because of the decrease in the number of errors, the standard deviation of training and testing BER estimation in both algorithms are below 1×10^{-16}.

4.2 Optimization of Parameters in ELM

There are three main parameters that affect the BER estimation performance based on ELM, i.e., the number of train packets, the number of test packet and the number of hidden neurons. We simulate the effect of them on the training root mean square error (RMSE) and the testing RMSE, and visualize the results in Fig. 3. The curves are obtained by 1-D interpolation function *interp1* with method *pchip*.

The training RMS is determined by the number of train packets, and decreases with the increase of the latter. For example, when the number of train packets changes from 1 to 5, the training RMS halves to 0.0155. When the number of train packets

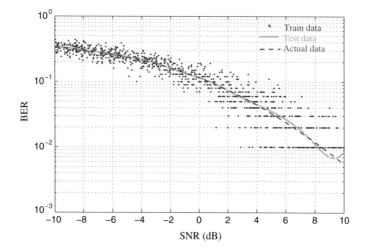

Fig. 2 BER estimation based on ELM algorithm

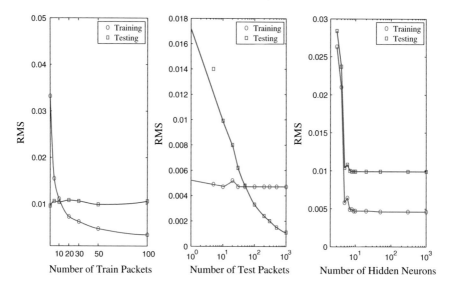

Fig. 3 Training RMSE and testing RMSE using ELM (sigmoid) with respect to the number of train packets, the number of test packet and the number of hidden neurons in BER estimation

increases to 20, the training RMS again halves to 0.0073. Similarly, the testing RMS is determined by the number of test packets. The more train and test packets are used, the higher the accuracy of estimation is. However, this led to the increase of time to form train data and test data too, so there is a tradeoff between accuracy and time efficiency.

When the number of hidden neurons is less than 10, it can also affect both the training and testing RMS. Especially when less than 5 hidden neurons are randomly generated, the BER performance takes a sharp deterioration. In contrast, the generalization performance of ELM is very stable when more than 10 hidden nodes exist.

5 SNR Estimation at Transmitter

SNR estimation of received signal plays an important role in a wireless system, because it can effectively adjust transmission scheme and relevant parameters. For instance, In addition to spectrum sensing capability required by cognitive radio (CR) at transmitter, SNR estimation of the primary signals is crucial to CR in order to adapt the transmitter's coverage area dynamically using underlay techniques [19]. The SNR estimation is mostly taken in the frequency domain using the known preamble or pilot in packets. We simulate SNR estimation based on ELM without any pilot.

5.1 SNR Estimation Based on ELM

The performance of ELM and BP are compared on the SNR estimation. The data sets of train and test consists of 1000 elements. The distribution of these data sets are unknown to both algorithms, and they are randomly generated before each trail of simulation. Each frame has 100 bits. The number of train packets is 30 while the number of test packets is 50. The performances of 50 trails of simulations are shown in Table 2. From this table, the advantage of ELM on time efficiency is quite obvious, but both algorithms can't ignore their RMSE.

The approximated function of ELM and the actual SNR data are shown in Fig. 4. Compared with BER estimation, SNR estimation obtain a worse generalization performance, especially when BER is smaller than 10^{-3}. Similar to BER estimation, the number of error bits is an integer so that it is difficult to estimate SNR value accurately by a small BER. Therefore, the RMSE of SNR estimation is above 0.8 in ELM and 0.6 in BP.

Table 2 The performance comparisons of BER estimation with ELM and BP

Algorithms	Time (s)		Training (s)		Testing (s)		Hidden neurons
	Training	Testing	RMS	Dev	RMS	Dev	
ELM	0.0037	0.0025	0.8771	1.1214e-16	0.8807	4.4859e-16	10
BP	29.0121	0.0633	0.6790	2.2429e-16	0.6962	5.6074e-16	10

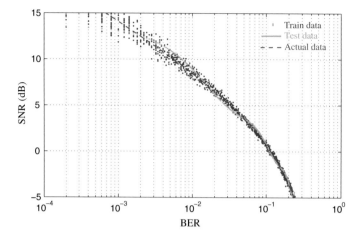

Fig. 4 SNR estimation based on ELM algorithm

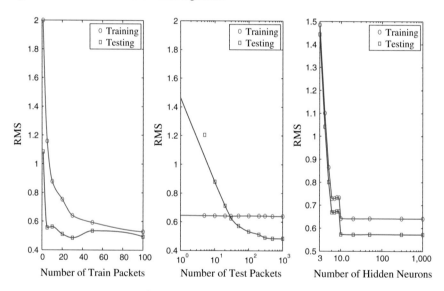

Fig. 5 Training RMSE and testing RMSE using ELM (sigmoid) with respect to the number of train packets, the number of test packet and the number of hidden neurons in SNR estimation

5.2 Optimization of Parameters in ELM

The impact of three parameters on SNR estimation performance is depicted in Fig. 5. We simulate the effect of them on generalization performance and use *interp1* function to complete interpolation. Too few train packets will lead to a large train error, and then reduce the accuracy of the SNR estimation result. But the reverse relationship is not established, i.e., the number of test packets only affect the testing RMSE. The RMSE performance is still very stable when the number of hidden neurons is larger than 10.

6 Conclusion

In this paper, the performance of MIMO modeling based on ELM learning algorithm is provided, and compared with BP algorithm. The simulation results show that ELM is a time efficient algorithm in BER estimation at receiver and SNR estimation at transmitter. ELM needs less than ten millisecond to obtain a relative accurate predicted value without sending pilot, which makes it a reliable tool in MIMO modeling.

Acknowledgments This work was supported by the Natural Science Foundation of Zhejiang province (China under Grants LY15F030017 and Q13G010016), National Natural Science Foundation of China (China under Grants 61403224).

References

1. Deek, L., Garcia-Villegas, E., Belding, E., Lee, S., Almeroth, K.: A practical framework for 802.11 MIMO rate adaptation. Comput. Netw. **83**, 332–348 (2015)
2. Li, Q., Lin, X., Zhang, J., Roh, W.: Advancement of MIMO technology in WiMAX: from IEEE 802.16d/e/j to 802.16m. IEEE Commun. Mag. **47**, 100–107 (2009)
3. Liu, L., Chen, R., Geirhofer, S., Sayana, K., Shi, Z., Zhou, Y.: Downlink MIMO in LTE-advanced: SU-MIMO vs. MU-MIMO. IEEE Commun. Mag. **50**, 140–147 (2012)
4. Zheng, K., Taleb, T., Ksentini, A., Magedanz, T., Ulema, M.: Research & standards: advanced cloud & virtualization techniques for 5G networks. IEEE Commun. Mag. **53**, 16–17 (2015)
5. Akyildiz, I., Wang, P., Lin, S.: SoftAir: a software defined networking architecture for 5G wireless systems. Comput. Netw. **85**, 1–18 (2015)
6. Foschini, J.: Layered space-time architecture for wireless communications in a fading environment when using multiple antennas. Bell Labs Tech. J. **1**, 41–59 (1996)
7. Zheng, L., Tse, D.: Diversity and multiplexing: a fundamental tradeoff in multiple-antenna channels. IEEE Trans. Inf. Theory. **49**, 1073–1096 (2003)
8. Sarma, K., Mitra, A.: Modeling MIMO channels using a class of complex recurrent neural network architectures. Int. J. Electron. Commun. **66**, 322–331 (2012)
9. Sarma, K., Mitra, A.: MIMO channel modelling using multi-layer perceptron with finite impulse response and infinite impulse response synapses. IET Commun. **7**, 1540–1549 (2013)
10. Seyman, M., Taspinar, N.: Channel estimation based on neural network in space time block coded MIMOCOFDM system. Digit. Signal Process. **23**, 275–280 (2013)
11. Sit Y., Agatonovic, M., Zwick T.: Neural network based direction of arrival estimation for a MIMO OFDM radar. In: 9th European Radar Conference, pp. 298–301. IEEE Press, New York (2012)
12. George, M., Francis, A., Mathew, B.: Neural network estimation for MIMO-OFDM receivers. In: International Conference on Circuit, Power and Computing Technologies, 2014, pp. 1329-1333. IEEE Press, New York (2014)

13. Belkacem O., Zayani R., Ammari M., Bouallegue R., Roviras, D.: Neural network equalization for frequency selective nonlinear MIMO channels. In: IEEE Symposium on Computers and Communications, 2012, pp. 18–23. IEEE Press, New York (2012)

14. Huang, G., Zhu, Q., Siew, C.: Extreme learning machine: theory and applications. Neuro comput. **70**, 489–501 (2006)

15. Cao, J., Huang, G., Liu, N.: Voting based extreme learning machine. Inf. Sci. **185**, 66–77 (2012)

16. Fang, J., Tan, Z., Tan, K.: Soft MIMO: a software radio implementation of 802.11n based on sora platform. In: 4th IET International Conference Wireless, Mobile & Multimedia Networks, pp. 165–168. IEEE Press, New York (2011)

17. Zhang, P., Chen, S., Hanzo, L.: Two-tier channel estimation aided near-capacity MIMO transceivers relying on norm-based joint transmit and receive antenna selection. IEEE Trans. Wirel. Commun. **14**, 122–137 (2015)

18. Berber, S.: An automated method for BER characteristics measurement. IEEE Trans. Instrum. Meas. **53**, 575–580 (2004)

19. Sharma S., Chatzinotas S., Ottersten B.: Eigenvalue based SNR estimation for cognitive radio in presence of channel correlation. In: 56th IEEE Global Communications Conference, pp. 1107–1121. IEEE Press, New York (2013)

Graph Classification Based on Sparse Graph Feature Selection and Extreme Learning Machine

Yajun Yu, Zhisong Pan and Guyu Hu

Abstract Identification and classification of graph data is a hot research issue in pattern recognition. The conventional methods of graph classification usually convert the graph data to vector representation which ignore the sparsity of graph data. In this paper, we propose a new graph classification algorithm called graph classification based on sparse graph feature selection and extreme learning machine. The key of our method is using lasso to select sparse feature because of the sparsity of the corresponding feature space of the graph data, and extreme learning machine (ELM) is introduced to the following classification task due to its good performance. Extensive experimental results on a series of benchmark graph datasets validate the effectiveness of the proposed methods.

Keywords Graph kernel · Graph classification · Extreme learning machine · Lasso

1 Introduction

Most existing machine learning algorithms such as support vector machine (SVM) are only apply to deal with data with vector representation. But in many practical applications such as bioinformatics, drug discovery, web data mining and social networks involves the study of relationships between structured objects [1], which can not be represented in vector forms. In recent years, the research about structured data has become a hot research issue in machine learning and data mining. Graphs are usually employed to represent the structured objects, and the nodes of graphs represent objects while edges indicate the relationships between objects. To analyse graph data, the most important thing is the similarity measurement of two graphs. Kernel method is a good way to study graph data and kernel function can be used to measure the similarity between two graphs.

Y. Yu (✉) · Z. Pan · G. Hu
College of Command Information System, PLA University of Science
and Technology, Nanjing 210007, China
e-mail: 492675818@qq.com

© Springer International Publishing Switzerland 2016
J. Cao et al. (eds.), *Proceedings of ELM-2015 Volume 1*,
Proceedings in Adaptation, Learning and Optimization 6,
DOI 10.1007/978-3-319-28397-5_15

179

The general method of graph classification is using graph kernel method to map the graph data into higher dimensional vector described features space and computing the kernel function K consisting of each similarity of two graphs, then the original graph dataset can be classified by SVM [2]. However, on the one hand, this method neglects the sparsity of the graph data. Intuitively, because of the structure of the graph changes, a specific structure contained in a graph doesn't exists in others, and with the number of nodes and edges of the graph increases, the sparsity of the graph data will be more evident. So in this paper, we will show the sparsity of the graph feature space through experiment, and utilize lasso to select sparse feature.

On the other hand, SVM is suitable to solve small-scale samples, with the increase of the sample, the storage and computation of the matrix will cost a lot of memory and computing time. Since ELM has better identification accuracy and faster speed, we use ELM to classify the graph dataset.

The main contributions of this paper include:

- Using Weisfeiler-Lehman graph kernel feature mapping method [3] to test the sparsity of graph feature;
- Proposing graph feature selection via lasso;
- Using ELM to classify graph.

The rest of this paper is organized as follows. In Sect. 2, we introduce the concepts of graph kernel. And we introduce the idea of lasso for sparse graph feature selection, and verify the sparsity of the graph data feature space in Sect. 3. In Sect. 4, we describe the extreme learning machine. And we propose graph classification based on sparse graph feature selection and ELM in Sect. 5. Experimental results are presented in Sect. 6, and finally, we conclude this paper in Sect. 7.

2 Graph Kernel

We define a graph G as a four tuple (V, E, l_v, l_e). Here, V is the set of vertices, $V = \{v_1, v_2, \ldots, v_n\}$; E is the set of undirected edges, $E = \{e_1, e_2, \ldots, e_m\}$; $l_v(l_e) : V, E \rightarrow \Sigma$ is a function mapping from nodes or edges in the graph to an alphabet Σ.

Kernel method has been applied widely in the analysis of vector data. Informally, a kernel is a function of two objects that quantifies their similarity. Define the kernel function between two graphs called graph kernel. Given a feature mapping φ, the graph in the original space can be mapped to the high-dimensional and even infinite dimensional vector space through it, and formula (1) is workable.

$$K(G_1, G_2) = \langle \varphi(G_1), \varphi(G_2) \rangle \tag{1}$$

The similarity between two graphs can be measured by the kernel. In recent years, several different graph kernels were proposed. They can be categorized into three classes: (1) Graph kernels based on walks and paths, such as shortest-path kernels [4], Joint kernels [5] and kernels based on label pairs [6]; (2) Graph kernels based on limited-size subgraphs, such as cyclic pattern kernels [7]; (3) Graph kernels based on subtree patterns such as graph kernels based on tree patterns [8] and fast subtree kernels [9].

This paper uses the Weisfeiler-Lehman subtree kernel. The key idea of the kernel is [3]: Let Σ_0 be the set of original node labels of G and G'. Assume all Σ_i are pairwise disjoint. Assume that every $\Sigma_i = \{\sigma_{i1}, \ldots, \sigma_{i|\Sigma_i|}\}$ is ordered. Define a map $c_i : \{G, G'\} \times \Sigma_i \to N$, so that $c_i(G, \sigma_{ij})$ is the number of occurrences of the letter σ_{ij} in the graph G. The Weisfeiler-Lehman subtree kernel on two graphs G and G' with h iterations is defined as:

$$k^{(h)}_{WLsubtree}(G, G') = \langle \phi^{(h)}_{WLsubtree}(G), \phi^{(h)}_{WLsubtree}(G') \rangle \qquad (2)$$

where $\phi^{(h)}_{WLsubtree}(G) = (c_0(G, \sigma_{01}), \ldots, c_0(G, \sigma_{0|\Sigma_0|}), \ldots, c_h(G, \sigma_{h1}), \ldots, c_h(G, \sigma_{h|\Sigma_h|}))$, and $\phi^{(h)}_{WLsubtree}(G') = (c_0(G', \sigma_{01}), \ldots, c_0(G', \sigma_{0|\Sigma_0|}), \ldots, c_h(G', \sigma_{h1}), \ldots, c_h(G', \sigma_{h|\Sigma_h|}))$.

See Fig. 1, 1–5 for an illustration of one iteration of the Weisfeiler-Lehman subtree kernel.

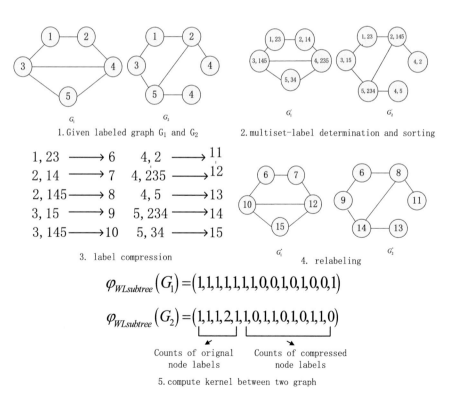

Fig. 1 Illustration of one iteration of the Weisfeiler-Lehman subtree kernel

3   Graph Feature Selection Based on Lasso

Intuitively, if two graphs are not isomorphic, then the set of node labels between them obtained by Weisfeiler-Lehman test of graph isomorphism algorithm [3] is different, and it will result in a large part of a corresponding feature vector of the graph data are 0. All corresponding feature vectors of graph data form a feature matrix. Figure 2 shows the sparsity of graph datasets. We use the graph datasets used in the literature 3. They are MUTAG, NCI1, NCI109, ENZYMES and DD. We use the Weisfeiler-Lehman graph kernel feature mapping method to get a feature matrix, and then calculate the number of zero element accounted for the proportion of all elements of each feature matrix. It can be seen from Fig. 2 that when $h = 4$, the sparse rate of these five graph data sets nearly come close to 99 %. It means only a small number of features contribute to its classification, the representation of features are sparse.

Because of the sparsity of graph data, when classifying graph data we should select key features of the graph data, so that we can get common features of the graph data that most able to distinguish graph data. And it can not only speed up the whole learning process, but also improve the distinction accuracy rate.

In order to find sparse representation of features, we utilize lasso to select features. It minimizes the residual sum of squares subject to the sum of absolute value of the coefficients being less than a constant, it shrinks some coefficients and sets others to 0, and then achieve the purpose of feature selection [10]. Suppose the dataset processed by Weisfeiler-Lehman subtree graph kernel are $(x_i, y_i), i = 1, 2, 3 \ldots, N$, Here, $x_i = (x_{i1}, x_{i2}, \ldots, x_{ip})$ are input variables, and y_i are responses. Assume that the x_{ij} are standardized, and $\widehat{\omega} = (\widehat{\omega}_0, \widehat{\omega}_1, \ldots \widehat{\omega}_p)^T$, then the lasso estimate $\widehat{\omega}$ is defined by

$$\widehat{\omega} = argmin\{\sum_{i=1}^{N}(y_i - \sum_j \omega_j x_{ij})^2\} \ subject \ to \ \sum_j |\beta_j| \le t \qquad (3)$$

Fig. 2 Sparse rate of each data sets

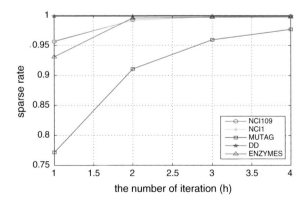

where $t \geq 0$ is a turning parameter, it controls the degree of sparsity. Appropriate parameter t will cause shrinkage of the solutions towards 0, and some coefficients may be exactly to 0. Then the solution to this problem causes many elements in x will be set to zero. Formula (3) is difficult to solve since it's a nonsmooth convex problem. This problem can be solved by many existing software packages, such as SLEP [11].

4 Extreme Learning Machine

ELM is a fast training algorithm for single hidden feedforward layer neural network (SLFN), it can randomly chooses the input weights and analytically determines the output weights of SLFNs [12]. So, the network parameters can be determined without any iteration step, and the adjustment time of the network parameters is greatly reduced. Compared with the traditional artificial neural networks, this method has not only the advantages of fast learning speed, but also good generalization performance. Figure 3 shows the structure of ELM network. The ELM network has received wide attention in recent years.

Consider N arbitrary distinct samples (x_i, t_i), where $x_i = (x_{i1}, x_{i2}, \dots, x_{in})^T \in \mathbb{R}^n$ and $t_i = (t_{i1}, t_{i2}, \dots, t_{im}) \in \mathbb{R}^m$, an ELM with L hidden nodes and an activation function $g(x)$ is mathematically modeled as:

$$\sum_{i=1}^{L} \beta_i g(w_i \cdot x_k + b_i) = o_k, k = 1, \dots, N \tag{4}$$

where $w_i = (w_{i1}, w_{i2}, \dots, w_{in})^T$ is the weight vector connecting the ith hidden neuron and the input neurons, $\beta_i = (\beta_{i1}, \beta_{i2}, \dots, \beta_{im})^T$ is the weight vector connecting the ith

Fig. 3 The structure of ELM. It contains three layers: input layer, hidden layer and output layer, where the hidden layer includes L hidden neurons, and the L is much less than N in general, the output of the output layer is a m dimensional vector

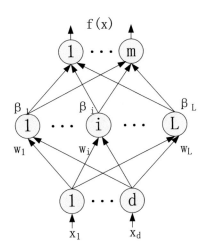

hidden neuron and the output neurons, b_i is the threshold of the ith hidden neuron and $o_k = (\beta_{k1}, \beta_{k2}, \ldots, \beta_{km})^T$ is the output vector of the SLFN. $w_i \cdot x_k$ denotes the inner product of w_i and x_k. The ELM with L hidden nodes and activation function g(x) reliably approximates N samples with minimum error:

$$\sum_{i=1}^{L} \beta_i g(w_i \cdot x_k + b_i) = t_k, k = 1, \ldots, N \tag{5}$$

The above N equations can be written compactly as:

$$H\beta = T \tag{6}$$

where

$$H = \begin{bmatrix} g(w_1 \cdot x_1 + b_1) & \cdots & g(w_L \cdot x_1 + b_L) \\ \vdots & \ddots & \vdots \\ g(w_1 \cdot x_N + b_1) & \cdots & g(w_L \cdot x_N + b_L) \end{bmatrix}_{N \times m} \tag{7}$$

$$\beta = \begin{bmatrix} \beta_1^T \\ \vdots \\ \beta_L^T \end{bmatrix}_{L \times m} \quad and \quad T = \begin{bmatrix} t_1^T \\ \vdots \\ t_N^T \end{bmatrix}_{N \times m} \tag{8}$$

where H is called the hidden layer output matrix of the neural network. If the number of neurons in the hidden layer is equal to the number of samples, then H is square and invertible. Otherwise, the system of equations needs to be solved by numerical methods, concretely by solving

$$\min_{\beta} \|H\beta - T\| \tag{9}$$

The result that minimizes the norm of this least squares equation is

$$\hat{\beta} = H^\dagger T \tag{10}$$

where $H^\dagger$ is called Moore-Penrose generalized inverse [13].

5 Proposed Graph Classification Algorithm

Assume we have graph dataset $(G_i)_{i=1}^{N}$, after the Weisfeiler-Lehman subtree kernel mapping, we can get a kernel matrix $K = (k(G_i, G_j))_{N \times N}$. Then we use lasso to select sparse features that key to classification. After that, these selected features are used to train and test an ELM classifier. Figure 4 shows the approximate process of graph classification based on sparse graph feature selection and extreme learning machine (GC-LASSO-ELM).

Fig. 4 The approximate process of GC-LASSO-ELM

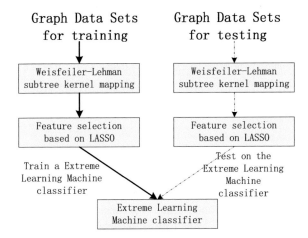

As we can see in Fig. 4, our method has three parts: Weisfeiler-Lehman subtree kernel mapping, sparse feature selection and classification using ELM. We get the optimal parameters through experiments, and parameters gained based on the training set. The detailed procedure of GC-LASSO-ELM is listed in Algorithm 1.

Algorithm 1 GC-LASSO-ELM Algorithm

Require:
    A graph dataset of N graphs $G = \{G_1, G_2, \ldots, G_N\}$.
Ensure:
    The label of graphs.
    1. Compute Kernel matrix $K = (k_{ij})_{i,j=1}^{N}$ using Weisfeiler-Lehman subtree graph kernel $k_{WLsubtree}^{(h)}(G, G') = \langle \phi_{WLsubtree}^{(h)}(G), \phi_{WLsubtree}^{(h)}(G') \rangle$.
    2. Use lasso to select sparse features, and represent graph data with these new features.
    3. Classify graph data processed through 1 and 2 based on ELM.

---

6 Experimental Results

6.1 Datasets

In this section, we validate our method on the following datasets: MUTAG [14], PTC_MM, PTC_FM, PTC_MR and PTC_FR. MUTAG is the dataset of mutable molecules, it contains 188 chemical compounds, and it can be divided into two classes according to whether they are mutagenic or not, where 125 of them are positive and 63 are negative. The PTC [15] is the datasets of carcinogenic molecules. It contain 417 chemical compounds, and has four types dataset: Male

Table 1 Graph data sets

	MUTAG	PTC_MM	PTC_FM	PTC_MR	PTC_FR
Number of positive	125(66.5%)	66(37.7%)	77(42.4%)	62(35.6%)	61(32.6%)
Number of negative	63(33.5%)	109(62.3%)	109(58.6%)	112(64.4%)	126(67.4%)
Total number	188	175	186	174	187
Average node	17.93	25.05	25.25	25.56	26.08

Mouse(PTC_MM), Female Mouse(PTC_FM), Male Rat(PTC_MR) and Female Rat
[] (PTC_FR). Each molecule is assigned one of the eight labels: EE, IS, E, CE, SE, P,
NE, N according to its carcinogenicity. EE, IS and E denote negative, CE, SE and P
denote positive, and NE and N is considered as can not discriminant, so not involved
in classification. Table 1 gives the detail information of the five graph datasets used
in this paper.

6.2 Experiments Settings

In our method, we use the Weisfeiler-Lehman subtree kernel proposed by Nino Shervashidze because of its significant computer speed on large graph datasets, and we
choose h = 5 in the Weisfeiler-Lehman subtree kernel. For lasso, the parameter λ
is a tunable parameter that controls the degree of the sparsity. For ELM, the parameter involves the number of hidden neurons. We select the optimal λ and the number of hidden neurons through contrast experiment, parameters is set as follows:
$\lambda = [0.3, 0.5, 0.7, 0.9]$, the number of hidden neurons between 1 and 100. Figures 5,
6, 7, 8 and 9 show the classify accuracy of different λ and the number of hidden
neurons on five graph datasets.

Fig. 5 Classification
accuracy at varying λ and
number of neurons on
MUTAG

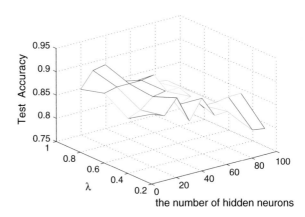

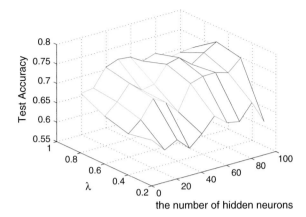

Fig. 6 Classification accuracy at varying λ and number of neurons on *PTC_MR*

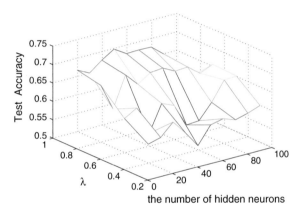

Fig. 7 Classification accuracy at varying λ and number of neurons on *PTC_FM*

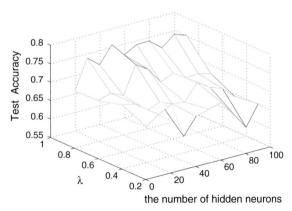

Fig. 8 Classification accuracy at varying λ and number of neurons on *PTC_FR*

As we can see in Fig. 5, we choose $\lambda = 0.9$ and 30 neurons when apply our method on MUTAG because of its higher accuracy. Figures 6, 7, 8 and 9 are test on PTC datasets, we choose $\lambda = 0.9$ and 40 neurons when apply our method on PTC, because it has higher accuracy on all the four type datasets.

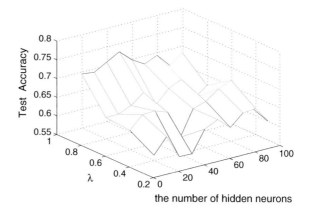

Fig. 9 Classification accuracy at varying λ and number of neurons on *PTC_MM*

6.3   Evaluation of Classification Performance

After choose the optimal λ and the number of neurons, we use these choosing parameters to perform 10-fold cross-validation using 9 folds for training and 1 for testing on every datasets and repeat each experiment 10 times. We compare our method with Graph Kernel based Dimensionality Reduction(GK-DR) [2] which using PCA for dimensionality reduction and SVM for classification, and with Graph Classification using lasso for feature selection and SVM for classification (GC-LASSO-SVM) on five datasets. See Table 2, it lists the average classification accuracy on each dataset. It shows that in all of these five datasets, the classification accuracy of GC-LASSO-SVM and GC-LASSO-ELM which use lasso to select graph feature is better then GK-DR which use PCA to select graph feature. And it shows that with the optimal parameter, the classification accuracy of GC-LASSO-ELM is better than GC-LASSO-SVM. More notable is that the classification accuracy on PTC datasets of graph classification using SVM directly after graph kernel mapping are: 61.0 %(*PTC_MM*), 61.0 %(*PTC_FM*), 62.8 %(*PTC_MR*), 66.7 %(*PTC_MM*) [5].

Figure 10 shows the classification accuracies of GK-DR, GC-LASSO-SVM and GC-LASSO-ELM on MUTAG dataset. It can be seen in the figure that GC-LASSO-ELM has high classification accuracy than GK-DR and GK-LASSO. Figures 11, 12, 13 and 14 show the classification accuracies of GK-DR, GC-LASSO-SVM and

Table 2 Mean classification accuracy

	GK-DR (%)	GC-LASSO-SVM (%)	GC-LASSO-ELM (%)
MUTAG	83.41	84.48	**85.62**
PTC_MM	62.28	65.03	**67.67**
PTC_FM	61.42	62.37	**69.4**
PTC_MR	64.28	64.05	**72.80**
PTC_FR	67.28	68.52	**71.01**

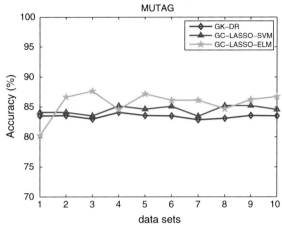

Fig. 10 Average classification accuracy on MUTAG

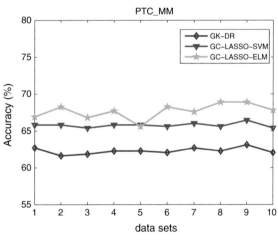

Fig. 11 Average accuracy on *PTC_MM*

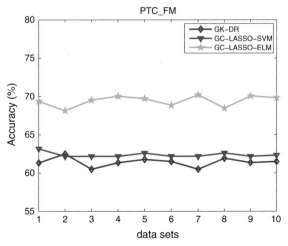

Fig. 12 Average accuracy on *PTC_FM*

Fig. 13 Average accuracy
on *PTC_MR*

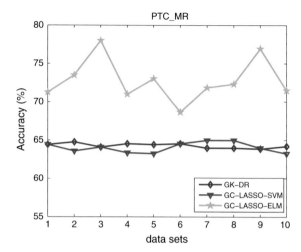

Fig. 14 Average accuracy
on *PTC_FR*

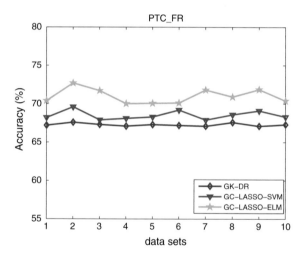

GC-LASSO-ELM on PTC datasets. It's obviously that the classification accuracy of
GC-LASSO-ELM is much better then GK-DR and GC-LASSO-SVM.

7 Conclusion

In this paper, an efficient graph classification method is proposed. We utilize lasso for
graph feature selection, and ELM for classification. Experimental results on MUTAG
and PTC show that when use lasso to select graph feature, the classification accuracy is better then use PCA, and both of them are better then the method of graph
classification using SVM directly after graph kernel mapping which have not select

graph feature. It's much better when we use ELM for classification then we use SVM. In the future work, we will use different kernel to study graph and compare there accuracy on ELM.

References

1. Vishwanathan, S.V.N., Schraudilph, N.N., Kondor, R.: Graph kernels. J. Mach. Learn. Res. **11**, 1201–1242 (2010)
2. Wu X., Zhang, D.: Graph kernel based semi-supervised dimensionality reduction method. J. Front. Comput. Sci. Technol. (2010)
3. Shervashidze, N., Schweitzer, P., van Leeuwen, E.J., Mehlhorn, K., Borgwardt, K.M.: Weisfeiler-lehman graph kernels. J. Mach. Learn. Res. **12**, 2536–2561 (2011)
4. Borgwardt, K.M., Kriegel, H.P. M.: Shortest-path kernels on graphs. In: The 5th IEEE International Conference on Data Mining (2005)
5. Kashima, H., Tsuda, K., Inokuchi, A.: Marginalized kernels between labeled graphs. In: The 20th International Conference on Machine Learning (ICML) (2003)
6. Gartner, T., Flach, P.A., Wrobel, S.: On graph kernels: Hardness results and efficient alternatives. In: The 16th Annual Conference on Computational Learning Theory and 7th Kernel Workshop (COLT) (2003)
7. Horvath, T., Gartner, T., Wrobel, S.: Cyclic pattern kernels for predictive graph mining. In: The International Conference On Knowledge Discovery and Data Mining (2004)
8. Mahe, P., Vert, J.P.: Graph kernels based on tree patterns for molecules. Mach. Learn. 205–214 (2009)
9. Shervashidze, N., Borgwardt, K.: Fast subtree kernels on graphs. Int. Conf. Neural Inf. Process. Syst. (2009)
10. Tishirani, R.: Regression Shrinkage and Selection via the Lasso. J. R. Stat. Soc. Seris B, 267–288 (1996)
11. Liu, J., Ji, S., Ye, J.: SLEP: Sparse learning with efficient projections. Arizona State University, 6 (2009)
12. Huang, G.B., Chen, L., Siew, C.K.: Universal approximation using incremental constructive feedforward networks with random hidden nodes. IEEE Trans. Neural Netw. **17**(4), 879–892 (2006)
13. Serre, D.: Matrices: Theory and Applications. Springer, New York (2002)
14. Debnath, A.K., Compadre, R.D., Debnath, G., et al.: Structure-activity relationship of mutagenic aromatic and heteroaromatic nitro compounds: correlation with molecular orbital energies and hydrophobicity. J. Med. Chem. **34**, 786–797 (1991)
15. Helma, C., King, R.D., Kramer, S., et al.: The predictive toxicology challenge. Bioinformatics, 107–108 (2001)

Time Series Prediction Based on Online Sequential Improved Error Minimized Extreme Learning Machine

Jiao Xue, Zeshen Liu, Yong Gong and Zhisong Pan

Abstract Nowadays, time series prediction is a hot issue in machine learning, however, how to predict time series fast and accurately remains extremely challengeable. In this paper, we proposed an Improved Error Minimized Extreme Learning Machine (IEM-ELM) algorithm which has better accuracy and prediction on change in direction (POCID) compared with Error Minimized Extreme Learning Machine (EM-ELM) for stock price prediction, meanwhile we implement the Online Sequential algorithm based on IEM-ELM (OSIEM-ELM) which fully inherits the merits of IEM-ELM. The performance of IEM-ELM and OSIEM-ELM are evaluated and compared with EM-ELM and OSEM-ELM respectively, and the experiments are carried out on three stock datasets, experimental results show that IEM-ELM and OSIEM-ELM produces better POCID performance than EM-ELM and OSEM-ELM at fast learning speed.

Keywords Extreme learning machine (ELM) · Error minimized extreme learning machine (EM-ELM) · Online sequential learning algorithm · Time series prediction · Stock price prediction

1 Introduction

A lots of methods have been applied in time sires prediction nowadays, namely polynomial method, radial basis function (RBF) [1, 2], support vector machine (SVM) [3–5], neural networks [6–9] and so on. However, all the existing methods and its variants for time sires prediction have some drawbacks for non-linear time series or

J. Xue (✉) · Z. Liu · Y. Gong · Z. Pan
College of Command Information System, PLA University of Science and Technology,
Nanjing 210007, China
e-mail: 1002218160@qq.com

Z. Pan
e-mail: hotpzs@hotmail.com

© Springer International Publishing Switzerland 2016
J. Cao et al. (eds.), *Proceedings of ELM-2015 Volume 1*,
Proceedings in Adaptation, Learning and Optimization 6,
DOI 10.1007/978-3-319-28397-5_16

complex data structure, such as over fit, low generalization and time consuming, etc. For example, the applications of BP neural network will easily lead to over fit or local optimum while SVM or neural network is time consuming.

Affected by many external factors, such as social, political, highly interrelated economic, stock time series is non-linear and vary complex and it is very difficult to be predicted accurately with conventional learning algorithms mentioned above. ELM [10, 11] has been discussed thoroughly due to its fast learning speed and good generalization recent years, and many variants of ELM have been proposed [12–16]. Huang et al. proposed EM-ELM algorithm in [16] and proved that EM-ELM had better performance on both regression and classification than ELM, incremental extreme learning machine (I-ELM) and resource allocation network (RAN). However, when we applied it on stock time series prediction, although EM-ELM can still obtain good root mean square error (RMSE) and good mean absolute percentage (MAPE), but get poor POCID which is a vary important index for stock prediction. Thus, in this paper, we proposed the improved EM-ELM algorithm which called IEM-ELM and proved its better performance of POCID against the comparative algorithm. At the same time, we implement the online version of IEM-ELM algorithm, named OSIEM-ELM (Online Sequential Improved Error Minimized Extreme Learning Machine) and proved its validity in the experiments.

The main contributions of this paper include:

- Proposing an Improved Error Minimized Extreme Learning Machine (IEM-ELM) algorithm.
- Implementing the online version of IEM-ELM algorithm, named OSIEM-ELM (Online Sequential Improved Error Minimized Extreme Learning Machine).
- Using IEM-ELM and OSIEM-ELM to predict stock price and verifying its better performance of POCID against EM-ELM and OSEM-ELM.

The remainder of this paper is organized as follows. After reviewing the EM-ELM algorithm in Sect. 2, in Sect. 3 we describe the algorithm IEM-ELM and introduces its online algorithm in Sect. 4. In Sect. 5, we evaluate out methods on some stock time series datasets and we conclude this paper in the end.

2 Review of Error Minimized Extreme Learning Machine (EM-ELM)

This section we will briefly review the essential parts of EM-ELM proposed by Huang et al. [16].

EM-ELM was proposed to overcome the drawback of ELM [10, 11] which determined the network architectures randomly. With growing hidden nodes and incrementally updating output weights, EM-ELM could not only choose the optimal number of hidden nodes but also reduces the computation complexity by only updating the output weights incrementally whenever the network architecture is changed when ELM recalculates the output weights based on the entire new hidden layer output matrix.

The algorithm of EM-ELM is represented as follows:

EM-ELM Algorithm: Given N training data (x_i, t_i) where $i = 1, \dots, N$, the maximum number of hidden nodes L_{max}, and the expected learning accuracy ε, the EM-ELM algorithm can be shown in two phases.

Phase 1-Initialization Phase:

(1) Initialize the networks with L_0 hidden nodes and randomly generated parameters $(a_i, b_i)_{i=1}^{L_0}$, here L_0 is a small positive integer given by users.
(2) Calculate the output matrix H_1 of the hidden layer

$$
H_1 = \begin{bmatrix} G(a_1, b_1, X_1) & \cdots & G(a_L, b_L, X_1) \\ \vdots & \ddots & \vdots \\ G(a_1, b_1, X_N) & \cdots & G(a_L, b_L, X_N) \end{bmatrix}_{N \times L_0} \tag{1}
$$

(3) Calculate the corresponding output error $E(H_1) = \| H_1 H_1^{\dagger} T - T \|$.

Phase 2-Hidden layer nodes Growing Phase:

Let $k = 0$. While $L_k < L_{max}$ and $E(H_k) > \varepsilon$:
(1) $k = k + 1$.
(2) Randomly add γL_{k-1} hidden nodes to the existing networks. The total number of hidden layer nodes becomes $L_k = L_{k-1} + \gamma L_{k-1}$ and the corresponding hidden layer output matrix $H_{k+1} = [H_k, \gamma H_k]$, where

$$
\gamma H_k = \begin{bmatrix} G(a_{L_{k-1}} + 1, b_{L_{k-1}} + 1, X_1) & \cdots & G(a_{L_k}, b_{L_k}, X_1) \\ \vdots & \ddots & \vdots \\ G(a_{L_{k-1}} + 1, b_{L_{k-1}} + 1, X_N) & \cdots & G(a_{L_k}, b_{L_k}, X_N) \end{bmatrix}_{N \times \gamma L_{k-1}} \tag{2}
$$

(3) The output weights β are recursively updated as

$$
\begin{aligned}
D_k &= ((I - H_k H_k^{\dagger})\gamma H_k)^{\dagger} \\
U_k &= H_k^{\dagger}(I - \gamma H_k^T D_k) \\
\beta_{k+1} &= H_{k+1}^{\dagger} T = \begin{bmatrix} U_k \\ D_k \end{bmatrix} T.
\end{aligned} \tag{3}
$$

End While.

3   Improved Error Minimized Extreme Learning Machine

We proposed the Improved Error Minimized Extreme Learning Machine (IEM-ELM) algorithm used for stock time series prediction in this section.

The caring about RMSE and POCID makes stock time series prediction different from original time series prediction which only concern for RMSE of an algorithm. POCID plays an important role in stock prediction, as it directly related to the interests of people. In the EM-ELM algorithm, the ending condition of the Hidden Layer Nodes Growing Phase is the output error $E(H_k) < \varepsilon$ before the number of hidden nodes L_k ups to L_{max}. It gives no consideration to POCID, meanwhile, the number of new hidden nodes to be added every time is indeterminate. Comparing with EM-ELM, we add the ending condition $P(H_k) < \xi$ in the Hidden Layer Nodes Growing Phase of the IEM-ELM algorithm and determine the number of new hidden nodes to be added according to the subtraction of $P(H_k)$ and ξ, that is the absolute of 0.5 $(P(H_k) - \xi)$. At the same time, considering when ε is too small and L_{max} is very large, it will never happen that the output error $E(H_k)$ ended with ε, that's means the circulation will end with L_k equals L_{max} which is we don't expected. So we add an additional ending condition in the Hidden Layer Nodes Growing Phase which is $| E(H_{k-1}) - E(H_k) |$ approximately equals 0 and do the same processing with POCID.

The whole algorithm of IEM-ELM is given as follows:

IEM-ELM Algorithm: Given a set of training data $(x_i, t_i)_{i=1}^{N}$, the maximum number of hidden nodes L_{max}, and the expected learning accuracy ε and the expected POCID ξ, the IEM-ELM algorithm can be shown in two phases.

Phase 1-Initialization Phase:

(1) Initialize the networks with L_0 hidden nodes and randomly generated parameters $(a_i, b_i)_{i=1}^{L_0}$, here L_0 is a small positive integer given by users.
(2) Calculate output matrix H_1 of the hidden layer

$$H_1 = \begin{bmatrix} G(a_1, b_1, X_1) & \cdots & G(a_L, b_L, X_1) \\ \vdots & \ddots & \vdots \\ G(a_1, b_1, X_N) & \cdots & G(a_L, b_L, X_N) \end{bmatrix}_{N \times L_0} \qquad (4)$$

(3) Calculate the corresponding output error $E(H_1) = \| H_1 H_1^\dagger T - T \|$.

Phase 2-Hidden Layer Nodes Growing Phase: Let $k = 0$.
While $L_k < L_{max}$ and $E(H_k) > \varepsilon$ and $P(H_k) < \xi$:

(1) $k = k + 1$.
(2) Add $| 0.5(P(H_k) - \xi)|$ hidden nodes to the existing networks. The total number of hidden nodes becomes $L_k = L_{k-1} + | 0.5(P(H_k) - \xi)|$ and the corresponding hidden layer output matrix $H_{k+1} = [H_k, \delta H_k]$, where

$$\delta H_k = \begin{bmatrix} G(a_{L_{k-1}} + 1, b_{L_{k-1}} + 1, X_1) & \cdots & G(a_{L_k}, b_{L_k}, X_1) \\ \vdots & \ddots & \vdots \\ G(a_{L_{k-1}} + 1, b_{L_{k-1}} + 1, X_N) & \cdots & G(a_{L_k}, b_{L_k}, X_N) \end{bmatrix}_{N \times |0.5(P(H_k) - \xi)|} \tag{5}$$

(3) The output weights β are recursively updated as

$$D_k = ((I - H_k H_k^\dagger) \delta H_k)^\dagger$$
$$U_k = H_k^\dagger (I - \delta H_k^T D_k)$$
$$\beta_{k+1} = H_{k+1}^\dagger T = \begin{bmatrix} U_k \\ D_k \end{bmatrix} T. \tag{6}$$

(4) if $(| E(H_{k-1}) - E(H_k) | < 1e^{-5}$ && $| P(H_{k-1}) - P(H_k) | < 1e^{-5})$ break;

End While.

4 Online Sequential Improved Error Minimized Extreme Learning Machine (OSIEM-ELM)

The IEM-ELM algorithm assuming that all the training data is available for training makes it improper for stock prediction as stock data can only arrived sequentially. So we implement the online sequential learning algorithm of IEM-ELM which been called OSIEM-ELM. OSIEM-ELM consists of two parts, namely IEM-ELM training part and online sequential learning part. The IEM-ELM training part is the same as the IEM-ELM algorithm, refers to Sect. 3. The online sequential learning part also consists of two phase which is the boosting phase and the sequential learning phase just as OS-ELM algorithm [17, 18].

Now, the OSIEM-ELM algorithm is presented as follows:

Proposed OSIEM-ELM Algorithm: Given the activation function g, the sequentially arrived training data $\aleph = \{(x_i, t_i) x_i \in R^n, t_i \in R^m, i = 1, \dots\}$.

Part (1)   **IEM-ELM Training Part**:

Given the maximum number of hidden nodes L_{max}, and the expected learning accuracy ε and the expected POCID ξ, the all available training data $(x_i, t_i)_{i=1}^{N_{now}}$ at now.

Phase 1-Initialization Phase:

(1) Initialize the SLFN with a small group of randomly generated hidden nodes $(a_i, b_i)_{i=1}^{L_0}$ where L_0 is a small positive integer given by users.
(2) Calculate the hidden layer output matrix H_1

$$H_1 = \begin{bmatrix} G(a_1, b_1, X_1) & \cdots & G(a_L, b_L, X_1) \\ \vdots & \ddots & \vdots \\ G(a_1, b_1, X_N) & \cdots & G(a_L, b_L, X_N) \end{bmatrix}_{N \times L_0} \tag{7}$$

(3) Calculate the corresponding output error $E(H_1) = \| H_1 H_1^\dagger T - T \|$.

Phase 2-Recursively Growing Phase: Let $k = 0$.
While $L_k < L_{max}$ and $E(H_k) > \varepsilon$ and $P(H_k) < \xi$:

(1) $k = k + 1$.
(2) Add $| 0.5(P(H_k) - \xi)|$ hidden nodes to the existing SLFN. The total number of hidden nodes becomes $L_k = L_{k-1} + | 0.5(P(H_k) - \xi)|$ and the corresponding hidden layer output matrix $H_{k+1} = [H_k, \delta H_k]$, where

$$\delta H_k = \begin{bmatrix} G(a_{L_{k-1}} + 1, b_{L_{k-1}} + 1, X_1) & \cdots & G(a_{L_k}, b_{L_k}, X_1) \\ \vdots & \ddots & \vdots \\ G(a_{L_{k-1}} + 1, b_{L_{k-1}} + 1, X_N) & \cdots & G(a_{L_k}, b_{L_k}, X_N) \end{bmatrix}_{N \times |0.5(P(H_k) - \xi)|} \tag{8}$$

(3) The output weights β are updated in a fast recursive way as

$$D_k = ((I - H_k H_k^\dagger) \delta H_k)^\dagger$$
$$U_k = H_k^\dagger (I - \delta H_k^T D_k)$$
$$\beta_{k+1} = H_{k+1}^\dagger T = \begin{bmatrix} U_k \\ D_k \end{bmatrix} T. \tag{9}$$

(4) if $(| E(H_{k-1}) - E(H_k) | < 1e^{-5}$ && $| P(H_{k-1}) - P(H_k) | < 1e^{-5})$ break;
(5) End While.
(6) set the final number of hidden nodes $L = L_k$.

Part (2)   **Online Sequential Learning Part**:

Phase 1-Boosting Phase: Initialize the learning using a small chunk of initial training data $\aleph_0 = (x_i, t_i)_{i=1}^{N_0}$ from the given training set $\aleph = (x_i, t_i) x_i \in R^n, t_i \in R^m$, $i = 1, \ldots, N_0 \geq L$.

(1) Assign arbitrary input weight a_i and bias b_i or center u_i and impact factor c_i, $i = 1, \ldots, L$.
(2) Calculate the initial hidden layer output matrix H_0

$$H_0 = \begin{bmatrix} g(a_1, b_1, X_1) & \cdots & g(a_L, b_L, X_1) \\ \vdots & \ddots & \vdots \\ g(a_1, b_1, X_{N_0}) & \cdots & g(a_L, b_L, X_{N_0}) \end{bmatrix}_{N_0 \times L} \tag{10}$$

(3) Estimate the initial output weight $\beta^0 = Q_0 H_0^T T_0$, where $Q_0 = (H_0^T H_0)^{-1}$ and $T_0 = [t_1, \ldots, t_{N_0}]^T$.
(4) Set $k = 0$.

Phase 2-Sequential Learning Phase: For each further coming observations $\aleph_{k+1} = (x_i, t_i)_{i=(\sum_{j=0}^{k} N_j)+1}^{\sum_{j=0}^{k=1} N_j}$ where N_{k+1} denotes the number of observations in the $(k+1)$th chunk.

(1) Calculate the hidden layer output matrix H_{k+1} for the $(k+1)$th chunk of data $\aleph_{k+1}$. Here H_{k+1} is

$$
H_{k+1} = \begin{bmatrix} g(a_1, b_1, X_{(\sum_{j=0}^{k} N_j)+1}) & \cdots & g(a_L, b_L, X_{(\sum_{j=0}^{k} N_j)+1}) \\ \vdots & \ddots & \vdots \\ g(a_1, b_1, X_{\sum_{j=0}^{k+1} N_j}) & \cdots & g(a_L, b_L, X_{\sum_{j=0}^{k+1} N_j}) \end{bmatrix}_{N_0 \times L} \tag{11}
$$

(2) Set $T_{k+1} = [T_{(\sum_{j=0}^{k} N_j)+1)}, \ldots, T_{\sum_{j=0}^{k+1} N_j}]^T$.

(3) Calculate the output weight β^{k+1}

$$
Q_{k+1} = Q_k - Q_k H_{k+1}^T (I + H_{k+1} Q_k H_{k+1}^T)^{-1} H_{k+1} Q_k \tag{12}
$$

$$
\beta^{k+1} = \beta^k + Q_{k+1} H_{k+1}^T (T_{k+1} - H_{k+1}\beta^k) \tag{13}
$$

(4) Set $k = k + 1$. Go to phase 2.

Remark If $N_0 = N$ and $N_{now} = N$, then OSIEM-ELM becomes IEM-ELM. Thus, batch IEM-ELM can be considered as a special case of OSIEM-ELM when all the training data are present in one learning iteration.

5   Experiments

In this section, we will investigate the performances of the proposed algorithms IEM-ELM and OSIEM-ELM in three stock time series prediction applications by comparing them with EM-ELM and OSEM-ELM respectively. We first introduce the datasets and environment settings of the experiments. After that, we give the evaluation indexes and the results of the experiments. Finally, the evaluations are done.

5.1   Datasets Description

Due to the differences of the stock market in China and America, we select one American stock time series datasets and two Chinese stock time series datasets in order to show the effectiveness of IEM-ELM and OSIEM-ELM. The details of these three stock datasets are shown in Table 1. Four attributes of the stock have been selected in the experiment, they are open price, highest price, lowest price and close

Table 1 Details of the stock datasets used in the experiments

Stock names	Begin time	End time	Training data	Testing data	Country
Dow Jones Industrial Average	1996-01-01	2014-12-31	4061	716	America
Shanghai Composite Index	2007-01-01	2014-12-31	1652	291	China
Shenzhen Composite Index	2007-01-01	2014-12-31	1652	291	China

price. Open price, highest price and lowest price are selected as the inputs of the network and close price is selected as the output of the network.

5.2 Experiment Settings

All the experiments are been conducted in Matlab R2013a running on a desktop PC with 3.10 GHZ CPU. The sigmoid function $g(x) = 1/(1 + exp(-x))$ is selected as the activation function for all the algorithms. The hidden nodes are setted as follows. For EM-ELM, the initial hidden nodes L_0 is setted as 5 in the initialization phase, in the recursively growing phase, the minimum error ε is setted as 0.005 and the algorithm increases the hidden noses one by one. In the implementation of IEM-ELM, the initial hidden nodes L_0 is also given as 5, minimum RMSE error ε and POCID ξ are given as 0.005 and 75. The new hidden nodes to be added is $| 0.5(P(H_k) - \xi)|$. For OSEM-ELM and OSIEM-ELM, L is setted the same way as EM-ELM and IEM-ELM respectively. The number of the initial training set is $(L + 150)$ in all the two algorithms, and add 100 samples each time in the sequential learning phase.

In the comparisons of EM-ELM and IEM-ELM, time windows is setted from 1 to 10. When time windows is 2 which means the open price, highest price, lowest price of today and the close price of the prior 2 days before today is used as the input vector of a sample, the close price of today is used as the target of a sample. In the comparisons of OSEM-ELM and OSIEM-ELM, time windows is setted as 3.

5.3 Evaluation Indexes

Five metrics have been used in the experiments to measure the performance of IEM-ELM and OSIEM-ELM: RMSE (Root Mean Square Error), MAPE (Mean Absolute Percentage Error), POCID (Prediction on Change In Direction), training times and testing times. Now we will give the definitions of RMSE, MAPE and POCID.

RMSE is defined as follows:

$$RMSE = \sqrt{\frac{1}{N}\sum_{i=1}^{N}(target_i - output_i)^2} \tag{14}$$

RMSE reflects the differences between the actual outputs and the expected outputs on the whole testing samples, the smaller of RMSE the better.

MAPE is defined as follows:

$$MAPE = \frac{1}{N}\sum_{i=1}^{N}|\frac{target_i - output_i}{output_i}| \tag{15}$$

MAPE is the percentage of the output error, the smaller of MAPE the better.

POCID is defined as follows:

$$POCID = 100\frac{\sum_{i=1}^{L}D_t}{N} \tag{16}$$

$$D_t = \begin{cases} 1, & if(target_t - target_{t-1})(output_t - outputt - 1) > 0 \\ 0, & otherwise \end{cases} \tag{17}$$

POCID is the percentage of the stock time series directions predicted correctly by the algorithm, the bigger of POCID the better.

5.4 Results and Evaluations

Tables 2, 3 and 4 show the comparison results of EM-ELM and IEM-ELM and Figs. 1, 2 and 3 display the outputs of EM-ELM and IEM-ELM when time windows is 3.

As we can see from Tables 2, 3 and 4, the RMSE, MAPE and POCID of IEM-ELM are better than EM-ELM in almost all the datasets no matter what the time windows is. Meanwhile, IEM-ELM has faster training speed than EM-ELM on Dow Jones Industrial Average, and Shanghai Composite Index with all the time windows. At the same time IEM-ELM has the same or better testing speed with EM-ELM on average. Figures 1, 2 and 3 show that the outputs of IEM-ELM are more closer with the raw data than EM-ELM. So We can tell that the proposed IEM-ELM has better POCID and RMSE than EM-ELM on stock time series at faster speed.

Table 2 Comparison of EM-ELM and IEM-ELM on Dow Jones Industrial Average dataset

Time windows	Algorithms	RMSE	MAPE	POCID	Training times (s)	Testing times (s)
1	EM-ELM	0.0026	0.0107	78.8827	6.1394	7.80e-04
	IEM-ELM	9.20e-04	0.0035	84.4693	1.5304	7.80e-04
2	EM-ELM	0.0055	0.0262	70.6215	5.9928	0.0016
	IEM-ELM	0.0015	0.0058	82.5978	2.3938	7.80e-04
3	EM-ELM	0.0035	0.0157	70.7682	7.6121	7.80e-04
	IEM-ELM	0.0014	0.0056	81.9902	3.3953	0.0012
4	EM-ELM	0.0055	0.0248	68.1983	6.7892	7.80e-04
	IEM-ELM	0.0012	0.0048	82.3593	4.3547	0.0002
5	EM-ELM	0.0055	0.0240	70.6983	7.3203	0.0031
	IEM-ELM	0.0015	0.0059	82.1788	5.4257	0.0016
6	EM-ELM	0.0042	0.0182	70.1047	9.7150	0.0016
	IEM-ELM	0.0013	0.0052	80.3422	6.9202	7.80e-04
7	EM-ELM	0.0050	0.0223	65.6774	7.3929	7.80e-04
	IEM-ELM	0.0020	0.0086	81.6201	6.6636	0.0023
8	EM-ELM	0.0065	0.0298	66.7388	7.8211	7.80e-04
	IEM-ELM	0.0014	0.0056	82.0182	7.4459	7.80e-04
9	EM-ELM	0.0061	0.0274	65.7682	8.8304	0.0039
	IEM-ELM	0.0014	0.0055	81.7738	8.3617	7.80e-04
10	EM-ELM	0.0061	0.0273	64.5810	7.4249	0.0036
	IEM-ELM	0.0013	0.0051	81.6899	9.3335	7.80e-04

Table 3 Comparison of EM-ELM and IEM-ELM on Shanghai Composite Index dataset

Time windows	Algorithms	RMSE	MAPE	POCID	Training times (s)	Testing times (s)
1	EM-ELM	**0.0035**	0.0121	**76.7010**	1.6427	0.0051
	IEM-ELM	**0.0035**	**0.0119**	76.5636	**0.6958**	**0.0011**
2	EM-ELM	0.0039	0.0144	74.5017	1.7381	4.70e-04
	IEM-ELM	**0.0034**	**0.0122**	**76.6323**	**0.6708**	**0**
3	EM-ELM	0.0047	0.0172	73.3334	1.7472	**4.70e-04**
	IEM-ELM	**0.0035**	**0.0122**	**77.4570**	**0.5959**	0.0062
4	EM-ELM	0.0039	0.0147	76.0825	2.3821	0.0016
	IEM-ELM	**0.0034**	**0.0122**	**76.9072**	**1.1825**	**0.0005**
5	EM-ELM	0.0052	0.0202	73.4708	2.3197	**0.0012**
	IEM-ELM	**0.0035**	**0.0125**	**77.2508**	**0.9953**	0.0031
6	EM-ELM	0.0038	0.0137	76.3917	3.2448	0.0062
	IEM-ELM	**0.0034**	**0.0123**	**76.5636**	**1.3822**	**0.0015**
7	EM-ELM	0.0072	0.0328	68.4536	2.0561	0.0003
	IEM-ELM	**0.0035**	**0.0124**	**76.9072**	**1.4009**	**3.20e-04**
8	EM-ELM	0.0067	0.0267	66.9072	2.1715	0.0014
	IEM-ELM	**0.0035**	**0.0127**	**76.8385**	**1.8283**	**2.40e-04**
9	EM-ELM	0.0063	0.0270	70.9278	2.5210	0.0031
	IEM-ELM	**0.0035**	**0.0123**	**76.9072**	**1.8502**	**0.0016**
10	EM-ELM	0.0053	0.0342	65.7388	2.2464	0.0016
	IEM-ELM	**0.0034**	**0.0122**	**76.8382**	2.6988	0.0094

J. Xue et al.

Table 4 Comparison of EM-ELM and IEM-ELM on Shenzhen Composite Index dataset

Time windows	Algorithms	RMSE	MAPE	POCID	Training times (s)	Testing times (s)
1	EM-ELM	0.0036	0.0035	**84.0893**	**0.6271**	0.0031
	IEM-ELM	**0.0034**	**0.0034**	83.8144	1.1293	**0.0016**
2	EM-ELM	**0.0035**	**0.0034**	**84.3643**	**0.7438**	**0.0013**
	IEM-ELM	**0.0035**	0.0035	84.3299	1.1622	0.0016
3	EM-ELM	0.0036	0.0036	**84.3299**	**1.1528**	**0.0008**
	IEM-ELM	**0.0035**	**0.0035**	83.9519	1.9984	0.0016
4	EM-ELM	0.0037	0.0037	83.9519	**1.3322**	0.0031
	IEM-ELM	**0.0035**	**0.0035**	**84.2612**	1.8751	**0.0012**
5	EM-ELM	0.0037	0.0036	84.2268	**1.7363**	0.0047
	IEM-ELM	**0.0035**	**0.0035**	**84.4330**	1.8954	**0.0016**
6	EM-ELM	0.0037	0.0037	83.3333	**1.8190**	0.0015
	IEM-ELM	**0.0035**	**0.0035**	**84.3643**	2.2277	**0.0012**
7	EM-ELM	0.0050	0.0051	79.9313	**1.6583**	0.0042
	IEM-ELM	**0.0035**	**0.0035**	**84.4673**	2.4040	**0.0016**
8	EM-ELM	0.0043	0.0043	81.7526	2.0966	**3.20e-04**
	IEM-ELM	**0.0035**	**0.0035**	**84.3299**	**2.0208**	0.0016
9	EM-ELM	0.0047	0.0047	81.1306	2.6255	**0.0047**
	IEM-ELM	**0.0035**	**0.0035**	**84.8454**	**2.1528**	0.0062
10	EM-ELM	0.0067	0.0068	75.1890	**1.8970**	0.0014
	IEM-ELM	**0.0035**	**0.0035**	**84.7766**	2.8220	**2.70e-04**

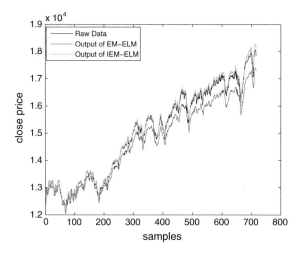

Fig. 1 Outputs of EM-ELM and IEM-ELM on Dow Jones Industrial Average dataset

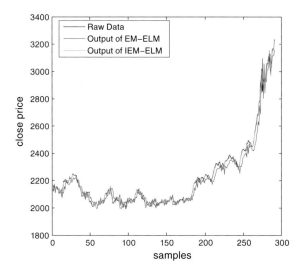

Fig. 2 Outputs of EM-ELM and IEM-ELM on Shanghai Composite Index dataset

From Tables 2, 3 and 4, we can also observed that when the time windows is setted as 3, the algorithms can get better results on almost all the datasets.

Table 5 shows the comparison results of OSEM-ELM and OSIEM-ELM on the three datasets and the outputs of OSEM-ELM and OSIEM-ELM are shown in Figs. 4, 5 and 6. We can see that the proposed algorithm OSIEM-ELM has better

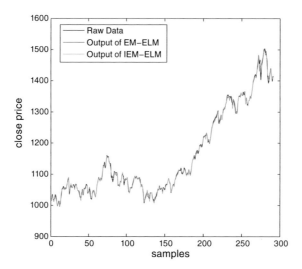

Fig. 3 Outputs of EM-ELM and IEM-ELM on Shenzhen Composite Index dataset

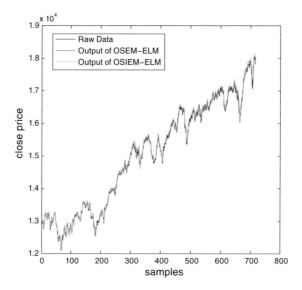

Fig. 4 Outputs of OSEM-ELM and OSIEM-ELM on Dow Jones Industrial Average dataset

Table 5 Comparison of OSEM-ELM and OSIEM-ELM

DataSets	Algorithms	RMSE	MAPE	POCID	Training times (seconds)
Dow Jones Industrial Average	OSEM-ELM	0.0040	0.0179	81.1453	**0.0468**
	OSIEM-ELM	**0.0016**	**0.0061**	**84.3575**	0.0499
Shanghai Composite Index	OSEM-ELM	0.0065	0.0280	68.7113	**0.0328**
	OSIEM-ELM	**0.0048**	**0.0197**	**73.3162**	0.0343
Shenzhen Composite Index	OSEM-ELM	**0.0053**	**0.0054**	79.9485	0.0413
	OSIEM-ELM	0.0054	**0.0054**	**82.1306**	**0.0211**

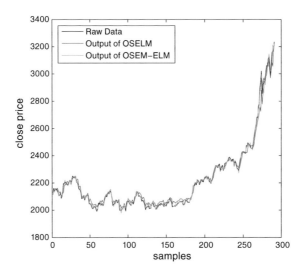

Fig. 5 Outputs of OSEM-ELM and OSIEM-ELM on Shanghai Composite Index dataset

POCID performance on all the datasets and has litter RMSE, MAPE, training times and testing times on two datasets than OSEM-ELM. Thus, we can conclude that OSIEM-ELM not only improved POCID of the stock time series but also is more accurate and faster than OSEM-ELM.

6 Conclusion

Stock price prediction is concerned by more and more people nowadays. In this paper, we proposed the Improved Error Minimized Extreme Learning Machine (EM-ELM) and its online sequential algorithm OSIEM-ELM for stock time series prediction. Two experiments on four stock time series datasets have been done to prove

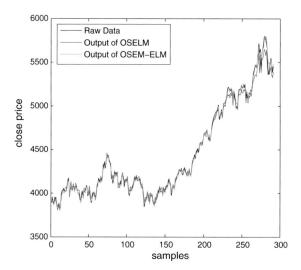

Fig. 6 Outputs of OSEM-ELM and OSIEM-ELM on Shenzhen Composite Index dataset

the better performance of IEM-ELM and OSIEM-ELM than EM-ELM and OSEM-ELM respectively. The results of the experiments showed that the proposed algorithms IEM-ELM and OSIEM-ELM can not only predict the stock tendency more exactly but also can predict the stock price more accurately.

However, only one-day prediction of the stock closing price is done in this paper, which will not offer enough information for the decision makers, so multi-steps prediction of stock price will be done in the future work.

References

1. Yee, P., Haykin, S.: A dynamic regularized radial basis function network for nonlinear, non-stationary time series prediction [J]. IEEE Trans. Signal Process. **47**(9), 2503–2521 (1999)
2. Iranmanesh, H., Abdollahzade, M., Samiee, S.: RBF neural networks and PSO algorithm for time series prediction application to energy demand prediction. In: Proceedings of IEEE International Conference on Intelligent Computing and Intelligent Systems (ICIS'2011), vol. 03 (2011)
3. Liu, P.: Application of least square support vector machine based on particle swarm optimization to chaotic time series prediction. In: IEEE International Conference on Intelligent Computing and Intelligent Systems (ICIS2009), vol. 4 (2009)
4. Xie, J.: Time series prediction based on recurrent $LS_S VM$ with mixed kernel [A]. Proceedings of 2009 Asia-Pacific Conference on Information Processing, vol 1 [C] (2009)
5. Weiwei, W.: Time series prediction based on SVM and GA. In: Proceedings of 8th International Conference on Electronic Measurement and Instruments (ICEMI'2007), vol. II (2007)
6. Fishman, M., Barr, D.S., Loick, W.J.: Using neural nets in market analysis [J]. Tech. Anal. Stocks Commod. **9**(4), 18–20 (1991)

7. Kai, F., Wenhua, X.: Training neural network with genetic algorithms for forecasting the stock price index. In: IEEE International Conference Intelligent Processing Systems, 1997, ICIPS '97, pp. 401–403. Beijing, China (1997)

8. Chi, S.-C., Chen, H.-P., Cheng, C.-H.: A forecasting approach for stock index future using grey theory and neural networks. In: International Joint Conference on Neural Networks, IJCNN '99, pp. 3850–3855. Washington DC, USA (1999)

9. Pan, W.-M., Shen, L.: Design of neural networks based time series online predictor. In: Proceedings of International Conference on Neural Networks and Brain (ICNN& B'98) (1998)

10. Huang, G.B., Zhu, Q.Y., Siew, C.K.: Extreme learning machine: a new learning scheme of feedforward neural networks. In: Proceedings of the IEEE International Joint Conference on Neural Networks, 2004, vol. 2, pp. 985–990. IEEE (2004, July)

11. Huang, G.B., Zhu, Q.Y., Siew, C.K.: Extreme learning machine: theory and applications [J]. Neural Comput. **70**, 489–501 (2006)

12. Huang, G.-B., Li, M.-B., Chen, L., Siew, C.-K.: Incremental extreme learning machine with fully complex hidden nodes. Neurocomputing **71**, 576–583 (2008)

13. Bai, Z., Huang, G.-B., Wang, D., Wang, H., Westover, M.B.: Sparse extreme learning machine for classification. In: IEEE Transactions on Cybernetics (2014) (in press)

14. Huang, G., Song, S., Gupta, J.N.D., Wu, C.: Semi-supervised and unsupervised extreme learning machines. In: IEEE Transactions on Cybernetics (2014) (in press)

15. Huang, G.-B., Chen, L.: Convex incremental extreme learning machine. Neurocomputing **70**, 3056–3062 (2007)

16. Feng, G., Huang, G.-B., Lin, Q., Gay, R.: Error minimized extreme learning machine with growth of hidden nodes and incremental learning. IEEE Trans. Neural Netw. **20**(8), 1352–1357 (2009)

17. Liang, N.-Y., Huang, G.-B.: A fast and accurate online sequential learning algorithm for feedforward networks. In: IEEE Transactions on Neural Networks, vol. 17, no. 6. November 2006

18. Huang, G.-B., Liang, N.-Y., Rong, H.-J., Saratchandran, P., Sundararajan, N.: On-line sequential extreme learning machine. In: Presented at the IASTED International Conference on Intelligent Computing (CI 2005), p. 46. Calgary, AB, Canada. July 2005

Adaptive Input Shaping for Flexible Systems Using an Extreme Learning Machine Algorithm Identification

Jun Hu and Zhongyi Chu

Abstract In this paper, a promoted adaptive input-shaping (AIS) with extreme learning machine (ELM) is presented to get zero residual vibration (ZRV) of severely time-varying flexible systems. Firstly, the ZRV condition and the traditional adaptive input-shaper is reviewed, together with its disadvantages of insufficient adaptability caused by giant amount of data and low-accuracy calculation caused by noise. After that, online sequential-ELM (OS-ELM) algorithm is introduced to identify the impulse response sequences of the flexible system, its fitting impulse response sequences are gotten to update the shaper parameters with fixed length and less noise; therefore, the above-mentioned problems of traditional AIS could be significantly avoided; that is to say, AIS's adaptability and identification-accuracy could be improved apparently, which means better performance to suppress the residual vibration of the flexible system. Finally, the verification experiments of presented AIS are implemented on a two-links flexible manipulator, which is a classical flexible system with severely time-varying dynamics; the results proves the effectiveness of the presented AIS method for the vibration control of severely time-varying flexible systems.

Keywords Vibration control · Adaptive input shaping · Extreme learning machine · Flexible system

1 Introduction

The solutions to reduce the residual vibration of flexible systems could be roughly divided into passive approaches and proactive approaches. The former include adding damping materials and modifying the design of mechanics [1], and the latter

J. Hu (✉) · Z. Chu
School of Instrumental Science and Opto-electronics Engineering,
Beihang University, Beijing, China
e-mail: flankerhu@163.com

© Springer International Publishing Switzerland 2016
J. Cao et al. (eds.), *Proceedings of ELM-2015 Volume 1*,
Proceedings in Adaptation, Learning and Optimization 6,
DOI 10.1007/978-3-319-28397-5_17

include feedback control and feedforward control. Among them, the input-shaping, because of its low costs and low difficulty, has gotten a lot of attention and become the hotspot in the field [2].

Input-shaping puts a bandstop filter before the flexible systems, banning the modal frequencies of the flexible systems; thus, the vibration of flexible systems wouldn't be activated. However, the original input-shaper is robust less, which makes input-shaping could hardly get good performance in applications. To improve its robustness, robust input-shaper are developed. Robust input-shaper could get better performance [3, 4], but it's in expense of additional shaping impulses [5], which means it sacrifices some response velocity. Consequently, adaptive-input-shaping (AIS) got more attention [6], adapting its impulse amplitudes and each impulse lag times to the changing system dynamic properties. Early AIS adapts its coefficients of input-shaper by empirical transfer function estimate (ETFE), which gets the modals by doing Fourier transform of I/O data from the flexible system, called indirect-AIS [2, 7]. It brings heavy burden of computation, and no high accuracy of the flexible system dynamics. Thus, direct-AIS was presented, adapting its coefficients of input-shaper by the algorithms such as recursive least square (RLS) [8], algebraic identification (AI) [9], and neural network (NN) [10]. Direct-AIS calculates the impulse response sequences of the flexible system using I/O data directly, and the corresponding updated parameters in input-shaper could be obtained.

Because of little computation and easy operation, RLS algorithm got a lot of application in direct-AIS [8, 11, 12]; however, it could hardly achieve high-accuracy identification of severely time-varying flexible system because of the insufficient adaptability and noise effect. To get high-accuracy identification in adaptive control, a lot of approaches are presented, including fuzzy adaptive control [13], NN adaptive control [14, 15], and so on. However, the traditional NN algorithms have giant computation quantity, which causes their high demand on control systems. Extreme learning machine (ELM) algorithm, as a newly presented neural network algorithm in 2004 [16], could achieve much less computation but keep the advantages of high-accuracy [17]. After that, ELM algorithm attracts much attention, and many ELM based algorithms are promoted, including online sequential ELM (OS-ELM) [18], incremental ELM (I-ELM) [19], and so on. However, to the best knowledge of authors, it is still a blank in how to achieve real-time identification of online sequences for flexible system using the ELM algorithm.

To achieve satisfactory performance of zero residual vibration (ZRV) for severely time-varying flexible systems, the promoted AIS with OS-ELM is proposed in this paper. To improve the adaptability and identification accuracy of AIS, the OS-ELM algorithm is introduced, identifying the impulse response sequences of the flexible system. The effect of noise would be suppressed obviously in ELM's fitting impulse response sequences, which means the accuracy of calculation would be improved. Furthermore, with the fixed length of identified impulse response sequences, the adaptability of the recursive calculation could be strengthened. Finally, to prove the correctness of the presented AIS method, verification experiments are conducted on a two-link flexible manipulator, which is belonging to a classical severely time-varying system. Compared with the traditional AIS, the adaptability of the proposed method is obviously improved; thus, it could satisfy the demand of real-time vibration suppression of severely time-varying flexible systems.

This paper is organized as follows. In Sect. 2, there is a review of ZRV of flexible systems and the design of traditional direct-adaptive input-shaper. Subsequently, the promoted AIS based on OS-ELM identification is presented. In Sect. 3, there are verification experiments, in which a two-link flexible manipulator is introduced as a classical severely time-varying flexible system, the results of the experiments prove the expectant improvement of AIS with ELM on adaptation ability and high-accuracy identification of the input-shaper coefficients corresponding to the flexible system. Finally, conclusions are summarized in Sect. 4.

2 AIS Based on ELM Identification for Flexible System

2.1 Review of ZRV Condition and Traditional AIS

A flexible system could be commonly presented by its flexible segment G and overall motion segment P, as Fig. 1 shows. The shaped input sequences u could get corresponding vibration output sequences y assuming that the impulse response sequences of G is g. The coefficients of input-shaper are $h = \{h_0, h_1, h_2, \ldots, h_Q\}^T$, where the number of impulse should satisfy $Q + 1 \geq 2M + 1$, M is the number of modals that need to be reduced. The delay time of two impulse in input-shaper is $\Delta_x = n_x t_s$, $x = 1, 2, \ldots, Q$, and n_x belong to positive integer. The total transfer segment from unshaped input i to vibration output y is f = h * g.

According to the ZRV condition [11], the recursive calculation from flexible system's input u and output y to the values of coefficients in input-shaper h could be as

$$e_N = s_N^T \begin{bmatrix} \omega \\ x \end{bmatrix}_N + \psi_N^T A b \tag{1}$$

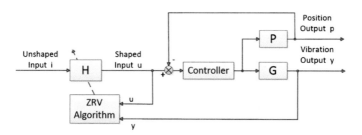

Fig. 1 Structure of traditional AIS with a flexible system

$$d_N = -\left(\frac{1}{1 + s_N^T R_N s_N}\right) R_N s_N \tag{2}$$

$$\begin{bmatrix} \omega \\ x \end{bmatrix}_{N+1} = \begin{bmatrix} \omega \\ x \end{bmatrix}_N + d_N e_N \tag{3}$$

$$R_{N+1} = \left(I + d_N s_N^T\right) R_N \tag{4}$$

$$h = T\omega + b \tag{5}$$

$$b = \frac{1}{Q+1} \begin{bmatrix} 1 \\ \cdots \\ 1 \end{bmatrix}_{[Q+1] \times 1} \tag{6}$$

$$s_N = \begin{bmatrix} T^T A^T \psi_N \\ \nu_N \end{bmatrix} \tag{7}$$

$$\psi_N = [y_N \, y_{N-1} \ldots y_{N-K}]^T \tag{8}$$

$$\nu_N = [u_N u_{N-1} \ldots u_{N-K}]^T \tag{9}$$

and $T \in R^{[K+1] \times K}$ is an orthogonal complement of b to ensure the sum of input-shaper coefficients equal 1, $A \in R^{[K+1] \times [Q+1]}$ containing 0 and 1 is used to assign the delay time of every two impulses in input-shaper and its corresponding sample time t_s in the system. ω and x are two intermediate vectors in calculation.

ZRV algorithm of Eqs. (1)–(4) is to calculate the corresponding values of input shaper by making the error's quadratic sum minimum. However, the traditional AIS would calculate the least-square solution of I/O data from time *zero* to time N. Thus, with the increase of N in long-running and the larger total amount of I/O data, the problem of insufficient adaptability is produced. Furthermore, the least-square solution of all the I/O data from time *zero* to time N would bring another problem of noise-caused low-accuracy calculation.

$$\hat{y} = y + d \tag{10}$$

where d is the noise with zero mean value and variance $E(d^2) = \sigma^2$. So the error prediction in terms of h and f is given as

$$\hat{e} = H\hat{y} + Fu \tag{11}$$

where H and F is the matrix form of h and f. Thus, the noise-perturbed quadratic cost with the length N data is

$$\hat{J} = \hat{e}^T \hat{e} = h^T \hat{Y}_N^T \hat{Y}_N h + 2h^T \hat{Y}_N^T U_N f + f^T U_N^T U_N f \qquad (12)$$

where $\hat{Y}_N = Y_N + D_N$. Then, the noise-perturbed cost's expectation is

$$E(\hat{J}) = J + h^T E(D_N^T D_N) h = H + \sigma^2 N ||h||^2 \qquad (13)$$

Equation (13) shows the effects of noise to ZRV calculation. With the recursive algorithm running, the value of N would become bigger and bigger, and so does the $\sigma^2 N ||h||^2$ in Eq. (13). However, the goal of ZRV is to get minimum value of $E(\hat{J})$. Consequently, the $||h||^2$ in Eq. (13) has to be less and less. When the parameters of input-shaper h are all equal, $||h||^2$ get minimum. Consequently, noise accumulation will make the coefficients of input-shaper closer and closer, which could cause the deterioration of vibration suppression's performance.

From the mathematical analysis, it could be known that traditional AIS faces the problems of low-accuracy calculation caused by insufficient adaptability and noise. This means that it could hardly get good performance of vibration suppression towards severely time-varying flexible system.

2.2 AIS with ELM Identification

To solve the problems that traditional AIS faces, an OS-ELM algorithm is introduced to achieve real-time identification of the flexible system, as shown in Fig. 2. Transport the flexible system's input sequences u and vibration output y to the OS-ELM, the OS-ELM would fitting the flexible segment G in real-time recursive calculation. After every time of training, given the OS-ELM network a unit impulse, the fitting impulse response sequences g of the flexible system could be gotten. Then the coefficients of input-shaper are calculated using g. With the adaptation of ELM algorithm, the coefficients of input-shaper will adjust in real-time. There will be the basic introduction of OS-ELM that applicated in AIS thereafter.

For a classical single hidden layer feedforward neural network, its number of input nodes, hidden layer nodes, and output nodes is D, L, and M. Given N groups of I/O data $\{u_j, y_j\}$. There is

$$H\beta = T \qquad (14)$$

where $H = \begin{bmatrix} h_1(u_1) & \cdots & h_L(u_1) \\ \cdots & \cdots & \cdots \\ h_1(u_N) & \cdots & h_L(u_N) \end{bmatrix}_{N \times L}$, $u_i = \begin{bmatrix} u_i \\ \cdots \\ u_{i+D-1} \end{bmatrix}_{D \times 1}$, $h_i(x) = G_i(a_i \cdot u_i + b)$

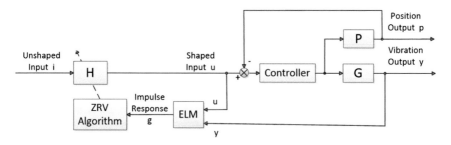

Fig. 2 Structure of AIS introduced OS-ELM

is the mapping function, $\beta = \begin{bmatrix} \beta_1 \\ \cdots \\ \beta_L \end{bmatrix}_{L \times m}$ is the matrix of output weight,

$T = \begin{bmatrix} y_1 & \cdots & y_M \\ \cdots & \cdots & \cdots \\ y_N & \cdots & y_{N+M-1} \end{bmatrix}_{N \times M}$ is the matrix consist of output data. To align the

time of input and output, the number of input nodes and output nodes are set equal, that is to say, $D = L$.

In ELM, the mapping function $h_i(x) = G_i(a_i \cdot u + b)$ is given by oneself, and its input weights and input biases are randomly generalized, the neural network could fitting any function by different β. Thus, the T matrix is known. The H matrix is also known after the mapping function, input weights, input biases generalized and input data substituted. Consequently, calculate the matrix β now is obvious.

$$\beta = H^+ T = \left(H^T H \right)^{-1} H^T T \tag{15}$$

With the online training using historical I/O data, the output weight matrix β is calculated in real-time. And the neural network could simulate the goal function $G(u)$, and get fitting output from given input $\tilde{u}$

$$G(\tilde{u}) = H(\tilde{u})\beta \tag{16}$$

To make ELM solve the real-time identification from online sequences, OS-ELM is promoted [19], reforming the calculation in (15) into recursive calculation. Given historical matrix H_0 and T_0, there is

$$\beta_0 = \left(H_0^T H_0 \right)^{-1} H_0^T T_0 \tag{17}$$

After new I/O data coming, the new added matrix H_1 and T_1 is generalized. So the new equality is

$$\beta_1 = \left(\begin{bmatrix} H_0 \\ H_1 \end{bmatrix}^T \begin{bmatrix} H_0 \\ H_1 \end{bmatrix} \right)^{-1} \begin{bmatrix} H_0 \\ H_1 \end{bmatrix}^T \begin{bmatrix} T_0 \\ T_1 \end{bmatrix} \tag{18}$$

Given $K_1 = \begin{bmatrix} H_0 \\ H_1 \end{bmatrix}^T \begin{bmatrix} H_0 \\ H_1 \end{bmatrix}$, and which could be derived is

$$K_1 = K_0 + H_1^T H_1 \tag{19}$$

On the other hand, the $\begin{bmatrix} H_0 \\ H_1 \end{bmatrix}^T \begin{bmatrix} T_0 \\ T_1 \end{bmatrix}$ in Eq. (18) satisfy

$$\begin{bmatrix} H_0 \\ H_1 \end{bmatrix}^T \begin{bmatrix} T_0 \\ T_1 \end{bmatrix} = H_0^T T_0 + H_1^T T_1 = K_1 \beta_0 - H_1^T H_1 \beta_0 + H_1^T T_1 \tag{20}$$

Given $K^{-1} = P$; thus, the recursive calculation of OS-ELM in AIS could be described as follows.

When the $N + 1$ I/O data come, there are

$$P_{N+1} = P_N - P_N H_{N+1}^T \left(I + H_{N+1} P_N H_{N+1}^T \right)^{-1} H_{N+1} P_N \tag{21}$$

$$\beta_{N+1} = \beta_N + P_{N+1} H_{N+1}^T \left(T_{N+1} - H_{N+1} \beta_N \right) \tag{22}$$

Then make $N = N + 1$, and prepare for the next step of recursive calculation. The initial matrix H_0 and T_0 are calculated according to the first I/O data, and the initial matrix β_0 and P_0 are given approximately. The OS-ELM algorithm would fit the flexible system in online training, and the I/O sequences would update when there is new data coming. To reduce the computation quantity, the online training could be executed in bigger interval time. That is to say, the updation could be executed until u_{N+a} and y_{N+a} come, where a belongs to a given positive integer.

After the addition of OS-ELM, the variance of noise σ^2 in g could be smaller than before [17], so the error quantic in Eq. (13) could be reduced. Secondly, the length of fitting impulse response sequences is fixed, which means the N in Eq. (13) won't become bigger and bigger when in long-running. Consequently, the adaptability and calculation accuracy would be obviously improved.

3 Experiment on a Two-Link Flexible Manipulator

3.1 Experimental Setup

To prove the effectiveness of promoted AIS, test experiments will be executed on a lab-scale two-link flexible manipulator, as Fig. 3 show. The materiel of two links is aluminum with modulus of elasticity 69 GPa. Specific information of the two links and tip mass are listed in Table 1. The two joints' actuator is DC brushless motors,

Fig. 3 Two-link flexible manipulator

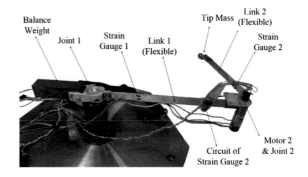

Table 1 Specification of two-links and tip mass

	Length (mm)	Width (mm)	Thickness (mm)	Weight (g)
Link 1	300	20	2	65
Link 2	200	15	1.5	16
Tip mass	–	–	–	8

Table 2 Specification of two motors, strain gauges and amplifier circuits

	Manufacture	Model numbers	Weight (g)
Motor 1	FAULHABER	2232024CSD	239
Motor 2	MOTEC	DBM 22.33.03.52.01	94
Strain gauges	DJET	BF350-3AA	–
Amplifier circuits	DJET	RC-A3N	3

and their specific information is listed in Table 2. The two joints' angles θ_1 and θ_2 are measured by incremental encoders. The angle θ_1 of link 1 is presented in the intersection angle with a given datum, and angle θ_2 of link 2 is presented in the intersection angle with the extended line from link 1, as shown in Fig. 4. Both θ_1 and θ_2 are positive when clockwise and negative when counterclockwise. Two strain gauges are placed close to the back end of the links, measuring the bending of

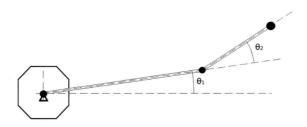

Fig. 4 Representation of two angles

two joints. The original signals from strain gauges would be amplified and filtered by a circuit, the specific information of strain gauges and amplifying circuits are listed in Table 2. Thus, link deflection and residual vibration could be evaluated using the signals from amplifying circuits.

To achieve the high angle accuracy together with rapid response, PID feedback controllers of joint angles are used in rotation control. The servo-control loop and adaptive input shaper's algorithm are achieved by a DSP 2812 control circuit, it could control the two motors and transport the result of experiment to PC in real-time. The structure of experiment manipulator is shown as Fig. 5. It should be noted that the presentation θ in Fig. 5 stands for both angles. Specific information of the two DC brushless motors and their servo-control loops is given in Table 3.

The coefficients of the input shapers are chosen as shown in Table 4, under the sampling time 10-ms-t_s. The initial values of the R_N is set at

$$R_0 = \begin{bmatrix} 10^7 & \cdots & 0 \\ \vdots & \ddots & \vdots \\ 0 & \cdots & 10^7 \end{bmatrix}_{[K+Q+1] \times [K+Q+1]} \tag{23}$$

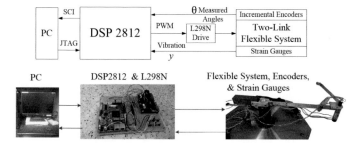

Fig. 5 Experimental settlement of two-link flexible manipulator with AIS

Table 3 Servo-control loop parameters of two joints

Joint	Nominal voltage (V)	Reduction ratio	Torque constant (mNm/A)	Peak current (A)	P gain	D gain	I gain
1	24	246	31.4	1.5	72	8	148
2	3.6	162.7	6.68	0.85	43	1.5	21.5

Table 4 Set values of input shaper

	Duration time (K)	Nonzero coefficient number (Q + 1)	Delay time Δ_x	Initial value c_n of input shaper
Values	60 t_s	5	15 t_s	$1/(Q+1)$

There are two AIS methods tested in the experiment. Traditional AIS and the presented AIS with ELM. The number of nodes in ELM is 25, and its numbers of input-nodes and output-nodes are both 200. The mapping function used in ELM is sigmoid function, the values of input-weight and input-bias are given randomly.

3.2 Experimental Results

Firstly, the command track of the two joints' angles in the experiment as shown in Figs. 6a and 7a. One motion period's duration is 8 s. In each period, the two joints will rotate a predetermined angle in rest-to-rest and stop for a moment, after that they will rotate back to the original angle in the same method. In the first four periods, the rotation angles of two joints in each period are decreased from the last period, because it could make the dynamics of flexible system varying severely in different periods. However, in the second four-periods, the rotation angles will repeat the first four-periods, so that the dynamics of the flexible system will repeat similar variation in two four-periods. Consequently, there are totally 8 periods in the verification experiments. Besides, to show the effects of input shaping and give initialization time of AIS, there is no input shaping of two joints in the first period. Different input shaping will be started from the second period, including traditional AIS, and the presented AIS with ELM.

The vibration could be linearly represented by the outputs of strain gauges; consequently, the curve of the strain gauges' output could reflect the strain's variety of the flexible system in linearity. Joint 1's output contrast from strain gauges with two different methods is shown in Fig. 6b, and joint 2's output contrast is shown in Fig. 7b. All the unit of these curves are V.

To give clear contrast of the two different AIS approaches, the arbitrarily chosen period from 18 to 26 s is magnified. Correspondingly, the maximum residual-vibration and the mean square error of residual-vibration are given in Table 5.

From Table 5, it could be seen that the proposed method of AIS with ELM achieve better performance in vibration suppression. To analyze the source of better performance, the corresponding filter coefficients of adaptive input-shaper are shown in Figs. 8 and 9. It could be seen from Fig. 8 that the filter coefficients tend to close and could hardly achieve real-time adjustment using traditional AIS. After the addition of ELM, the filter coefficients of AIS with ELM are significantly different from traditional AIS, and similar change of coefficients could be observed from 4 to 34 s and 36 to 66 s, corresponding to the same angle tracks in these times. This means that the AIS with ELM could adjust to the time-varying dynamics of the flexible system; thus, it could achieve better performance in vibration suppression.

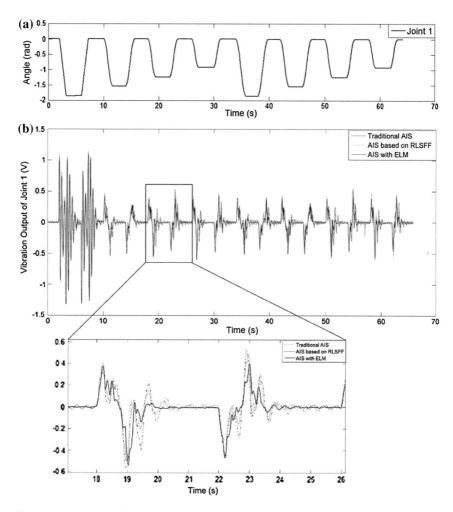

Fig. 6 a Angle track of joint 1. **b** Joint 1's result of vibration suppression with two AIS methods

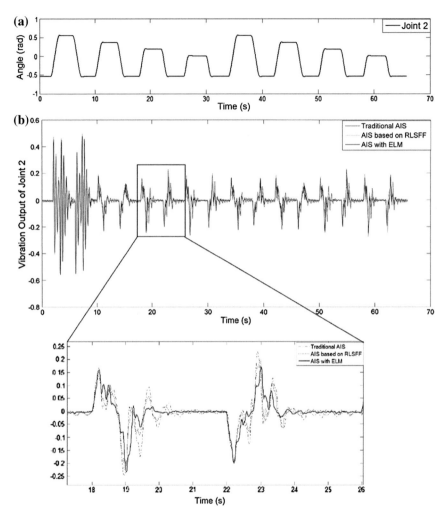

Fig. 7 a Angle track of joint 2. **b** Joint 2's result of vibration suppression with two AIS methods

Table 5 Results contrast of the vibration with two AIS methods

	Maximum deflection (V)	Maximum residual vibration (V)	Mean square error (V)
(a) Contrast of joint 1 with two AIS methods			
Traditional AIS	0.6716	0.4502	0.1089
AIS with ELM	0.6331	0.1830	0.0689
(b) Contrast of joint 2 with two AIS methods			
Traditional AIS	0.2802	0.1840	0.0405
AIS with ELM	0.2620	0.0711	0.0216

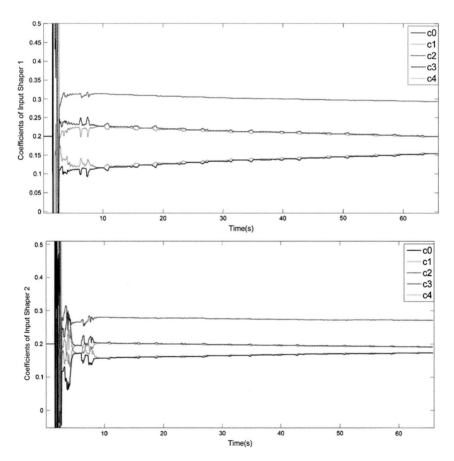

Fig. 8 Filter coefficients in traditional AIS

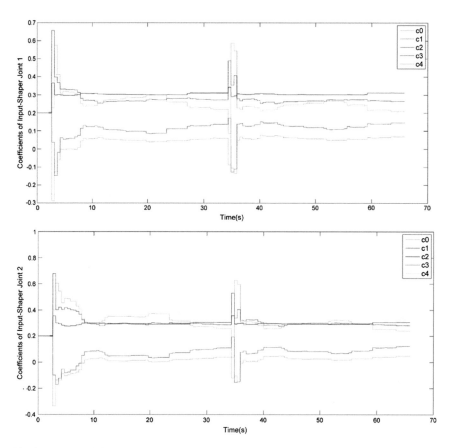

Fig. 9 Filter coefficients in AIS with ELM

4 Conclusions

In this paper, a promoted approach of AIS with ELM identification is proposed. To improve the adaptability and calculation accuracy of traditional AIS. Firstly, using the characteristic that ELM could fitting any non-linear function, the impulse response sequences of flexible system is being fitted, with better suppression of noise. Thus, the noise-caused low accuracy in calculation could be reduced significantly. Secondly, with the fixed length of the fitting impulse response sequences, the adaptivity of AIS could be improved obviously, as well as better calculation accuracy. After that, verification experiments are executed on a two-link flexible manipulator. The results of the verification experiment certified the promotion of AIS with ELM in calculation accuracy and its performance of residual vibration reduction.

Acknowledgements The authors acknowledge the financial support from the Natural Science Foundation of China (51375034, 61327809).

References

1. Chasalevris, A., Dohnal, F.: A journal bearing with variable geometry for the suppression of vibrations in rotating shafts: simulation, design, construction and experiment. Mech. Syst. Signal Process. **52–53**, 506–528 (2015)
2. Singhose, W.: Command shaping for flexible systems: a review of the first 50 years. Int. J. Precis. Eng. Manufact. **10**(4), 153–168 (2009)
3. Pridgen, B., Bai, K., Singhose, W.: Slosh suppression by robust input shaping. In: The 49th IEEE Conference on Decision and Control, pp. 2316–2321. Atlanta, GA, USA (2010)
4. Singhose, W., Eloundou, R., Lawrence, J.: Command generation for flexible systems by input shaping and command smoothing. J. Guid. Control Dyn. **33**(6), 1697–1707 (2010)
5. Xie, X., Huang, J., Liang, Z.: Vibration reduction for flexible systems by command smoothing. Mech. Syst. Signal Process. **39**, 461–470 (2013)
6. Van den Broeck, L., Diehl, M., Swevers, J.: Embedded optimization for input shaping. Trans. Control Syst. Technol. **18**(5), 1146–1154 (2010)
7. Chu, Z., Cui, J., Ren, S., Ge, S.S.: Semi-physical experimental study of adaptive disturbance rejection filter approach for vibration control of a flexible spacecraft. J. Aerosp. Eng. doi:10. 1177/0954410014565677 (2015)
8. Rhim, S., Book, W.J.: Noise effect on adaptive command shaping methods for flexible manipulator control. Trans. Control Syst. Technol. **9**(1), 84–92 (2001)
9. Pereira, E., Trapero, J.R., Díaz, I.M., Feliu, V.: Adaptive input shaping for single-link flexible manipulators using an algebraic identification. Control Eng. Pract. **20**, 138–145 (2012)
10. MacKunis, W., Dupree, K., Bhasin, S., et al.: Adaptive neural network satellite attitude control in the presence of inertia and CMG actuator uncertainties. In: American Control Conference, Seattle, Washington, USA (2008)
11. Park, J.H., Rhim, S.: Estimation of optimal time-delay in adaptive command shaping filter for residual vibration suppression. In: The 16th IEEE International Conference on Control Applications, Singapore (2007)
12. Cole, M.O.T., Wongratanaphisan, T.: Optimal FIR input shaper designs for motion control with zero residual vibration. J. Dyn. Syst. Measur. Control **133**(021008), 1–9 (2011)
13. Ahmad, M.A., Raja-Ismais, R.M.T., et al.: Vibration control of flexible joint manipulator using input shaping with PD-type fuzzy logic control. In: IEEE International Symposium on Industrial Electronics, Seoul, Korea (2009)
14. Ye, X., Zhang, W., et al.: Neural network adaptive control for X-Y position platform with uncertainty. TELKOMNIKA **12**(1), 79–86 (2014)
15. Zhen, H., Qi, X., et al.: Neural network L_1 adaptive control of MIMO systems with nonlinear uncertainty. Sci. World J. doi:10.1155/2014/942094 (2014)
16. Huangm, G.B., Zhu, Q.Y., Siew, C.K.: Extreme learning machine: A new learning scheme of feedforward neural networks. In: Proceedings of International Joint Conference Neural Network, vol. 2, pp. 985–990 (2006)
17. Huang, G.B., Zhu, Q.Y., Siew, C.K.: Extreme learning machine: theory and applications. Neurocomputing **70**, 489–501 (2006)
18. Liang, N.Y., Huang, G.B., Sundararajan, N.: A fast and accurate online sequential learning algorithm for feedforward networks. Trans. Neural Network **17**(6), 1411–1423 (2006)
19. Huang, G., Huang, G.B., Song, S., et al.: Trends in extreme learning machines: a review. Neural Networks **61**, 32–48 (2015)

Kernel Based Semi-supervised Extreme Learning Machine and the Application in Traffic Congestion Evaluation

Qing Shen, Xiaojuan Ban, Chong Guo and Cong Wang

Abstract Extreme learning machine (ELM) has proven to be an efficient and effective learning paradigm for a wide field. With the method of kernel function instead of the hidden layer, Kernel-ELM overcame the problem of variation caused by randomly assigned weights. In this paper, Kernel based optimization is introduced in semi-supervised extreme learning machine (SSELM) and the improvements of performance are evaluated by the experiment. The result shows that optimized by kernel function, Kernel-SSELM can achieve higher classification accuracy and robustness. In addition, The Kernel-SSELM is used to train the traffic congestion evaluation framework in Urban Transportation Assessment and Forecast System.

Keywords Semi-supervised ELM · Kernel function · Traffic congestion evaluation

Qing Shen, male, 1988, doctoral candidate, main research direction: action analysis, machine learning, E-mail: shenqingcc222333@gmail.com. Xiaojuan Ban, female, 1970, doctor, doctoral student supervisor, main research direction: artificial life, intelligent control, E-mail: banxj@ustb.edu.cn.

Q. Shen (✉) · X. Ban · C. Guo · C. Wang
School of Computer and Communication Engineering, University of Science and Technology Beijing, Beijing 100083, China
e-mail: shenqingcc222333@gmail.com

X. Ban
e-mail: banxj@ustb.edu.cn

C. Guo
e-mail: kotrue2015@gmail.com

C. Wang
e-mail: 1303003723@qq.com

© Springer International Publishing Switzerland 2016
J. Cao et al. (eds.), *Proceedings of ELM-2015 Volume 1*,
Proceedings in Adaptation, Learning and Optimization 6,
DOI 10.1007/978-3-319-28397-5_18

1 Introduction

Primarily, ELM was applied to supervised learning problems in full labeled data. Gao Huang et al. [1] proposed the semi-supervised framework of ELM to extend the capacity to deal with unlabeled data. SSELM greatly extend the application of ELM, for instance, in the field of text classification, information retrieval and fault evaluation as the collection of labeled data is bound to cost a lot of money and time while the unlabeled data is easy to collect and its number is large.

Although ELM improves the training efficiency to a high extent, the random distribution of input layer and the hidden layer parameters cause great variation of classification accuracy under the circumstance of same training data and model parameters which significantly influences the stability of ELM [2]. On the other hand, the number of hidden layer nodes also has a huge impact on the accuracy. In many studies, the number of hidden layer nodes is set to a large number that is usually greater than the number of training samples. However, the experiments show that the more hidden layer nodes is not better. The relationship between optimal accuracy of different datasets and the number of hidden layer nodes is complicated.

The approach replacing ELM hidden layer with kernel function make ELM does not need random hidden layer and input layer because the calculation of hidden input is carried out by kernel function. Kernel-ELM solves the problem resulted from random distribution of input layer and hidden layer parameters in ELM and gain higher relevance to corresponding datasets as well as higher stability [3] with the sacrifice of training speed.

SSELM and ELM have a unified framework. As a result of randomly generated feature mapping, stability problem is existed in the SSELM. This paper introduces the kernel function into the SSELM of Gao Huang et al. [1] and evaluates the improvements in stability and accuracy of SSELM optimized by kernel function.

The rest of the paper is organized as follows. Section 2 reviews the current research progress in the field of semi-supervised learning and kernel function at present in. Section 3 presents the algorithms framework of Kernel-SSELM. The evaluation experiment of efficiency is conducted in Sect. 4. Section 5 elaborates the application of Kernel-SSELM in the Traffic congestion evaluation system based on floating car data. Finally, Sect. 6 draws the conclusion and our future plan.

2 Related Research

Only a few existing research studies ELMs have dealt with the problem of semi-supervised learning. In the earlier days the manifold regularization framework was introduced into the ELMs model to leverage unlabeled data extending ELMs for semi-supervised learning [4, 5]. Li et al. [6] propose a training algorithm that assigns the most reliable predicted value to unlabeled sample in the repeated trainings of ELM for purpose of expanding the labeled sample sets continuously.

The proposed SSELM of Gao Huang et al. [1] takes example by the state-of-the-art semi-supervised learning framework to optimize the cost equation of ELM's processing unlabeled samples. Related to Laplacian support vector machines (LapSVM) and Laplacian regularized least squares (LapRLS), it is involved with the manifold assumption and simplifies the problem into the regularized least square problem.

Ever since the optimization based on kernel function was introduced into the ELM [2], many researchers have made advances in the practical application of theories. The significant solved problems are from two aspects. One aspect aims to choices of specific application's kernel function and optimization [7, 8]. The other aspect aims to the information fusion of ELM [9].

3 Kernel-Based SSELM

Gao Huang et al. [1] introduced manifold assumption into ELM, and proposed the solution of β in SSELM. For a training data set having 1 number of labeled samples and u number of unlabeled samples, the output weights β of a SSELM is:

$$\beta = H^T \left(I + \tilde{C}HH^T + \lambda LHH^T \right)^{-1} \tilde{C}\tilde{Y} \tag{1}$$

The formulate is valid when the number of hidden nodes is more than the number of labeled samples 1. The $\tilde{Y}$ is the training target including the first 1 rows of labeled data equal to Y and the rest equal to 0. λ is user-defined semi-supervised learning rate. $\tilde{C}$ is a $(1+u) \times (1+u)$ diagonal matrix with the first 1 diagonal elements of cost coefficient and the rest equal to 0. $\tilde{C}$ can be calculated as:

$$C_i = \frac{C_0}{N_{P_i}} \quad i = 1, \ldots, l \tag{2}$$

where C_0 is user-defined cost coefficient, and N_{P_i} represents the sample quantity of the pattern of ith sample. L is Laplacian matrix, which can be calculated as $L = D - W$. $W = \left[w_{i,j} \right]$ is the similarity matrix of all the labeled and unlabeled samples. D is a diagonal matrix with its diagonal elements $D_{ii} = \sum_{j=1}^{n} w_{ij}$.

Huang et al. [2] suggested using a kernel function if the hidden layer feature mapping h(x) is unknown. The kernel matrix χ for ELM can be written as follows, where $K(x_i, y_i)$ is kernel function:

$$\chi_{ELM} = HH^T \quad \chi_{ELM_{i,j}} = h(x_i) \cdot h(y_i) = K(x_i, y_i) \tag{3}$$

Then the output function of Kernel-SSELM can be written as:

$$y = F_{SSELM}(x) = h(x)\beta = \begin{bmatrix} K(x, x_1) \\ \vdots \\ K(x, x_n) \end{bmatrix} \left(I + \tilde{C}\chi_{ELM} + \lambda L \chi_{ELM} \right)^{-1} \tilde{C}\tilde{Y} \tag{4}$$

4 Experiment Result

4.1 Experimental Setup

We evaluated the performance of Kernel-SSELM on various semi-supervised tasks. All experiments were implemented using Matlab R2013b on a 3.40 GHz machine with 4 GB of memory.

The experiment was implemented on 4 popular data sets, which have been widely used for evaluating semi-supervised algorithms. In particular, USPST data set is the testing set of USPS, which is a classical handwritten digit recognition data set.

Each data set was randomly divided into 4 equal folds. Each of the folds was used as the testing set once and the rest were used for training (4-fold cross-validation). The random generation process was repeated 3 times, so that there were 12 different experiment groups for each data set. For each group, the training set was split into 3 different folds again as Table 1. In Table 1, L is the labeled data set for training, U is the unlabeled data set, and V represents the validation set.

4.2 Comparisons with Related Algorithms

In the experiment, we compared the Kernel-SSELM and SSELM with the other state-of-the-art semi-supervised learning algorithms such as TSVM, LDS, LapRLS, and LapSVM. The validation set V was used to select the optimal model parameter for every algorithm. In particular, for Kernel-SSELM and SSELM, the cost coefficient C_0 and the semi-supervised rate λ were selected from the exponential

Table 1 Details of the division of the data sets

Dataset	Classes	Dims	L	U	V	T
G50C	2	50	50	314	50	136
G10N	2	10	50	314	50	136
COIL20	20	1024	40	1000	40	360
USPST	10	256	50	1409	50	498

Table 2 Performance comparison between different semi-supervised algorithms

Dataset	Subset	TSVM	LDS	LapRLS	LapSVM	SSELM	Kernel-SSELM
G50C	U	6.43 (2.11)	5.61 (1.46)	6.23 (1.52)	**5.16** **(1.45)**	5.92 (2.34)	5.41(1.49)
	T	6.93 (2.37)	5.83 (2.03)	6.84 (2.41)	5.37 (1.56)	6.16 (2.87)	**5.23(1.91)**
G10N	U	13.91 (3.09)	9.79 (2.05)	**9.04** **(2.31)**	9.27 (2.63)	9.96 (3.65)	9.17(1.86)
	T	14.36 (3.68)	9.72 (1.9)	**9.48** **(2.63)**	9.82 (2.03)	10.44 (3.8)	9.83(2.15)
COIL20	U	26.35 (4.63)	14.68 (4.81)	**10.22** **(4.17)**	10.53 (2.47)	11.41 (3.35)	10.62(2.04)
	T	25.87 (4.52)	15.09 (3.79)	11.3 (3.3)	11.59 (2.82)	12.05 (3.57)	**11.2(2.16)**
USPST	U	24.98 (4.89)	15.53 (3.35)	15.38 (4.17)	15.93 (3.56)	14.61 (3.89)	**13.81(2.47)**
	T	26.5 (4.69)	16.8 (3.54)	16.81 (3.28)	16.76 (3.98)	14.76 (3.64)	**13.43(1.95)**

Bold values indicate the best result in the dataset

sequence $\{10^{-6}, 10^{-5}, \ldots, 10^{6}\}$. The number of hidden layer nodes of SSELM was fixed to 1000 for G50C and G10 N, and 2000 for COIL20 and USPST. The Kernel function of the Kernel-SSELM was radial basis function (RBF), and its parameter γ was selected in $\{2^{0}, 2^{1}, \ldots, 2^{10}\}$.

Table 2 shows the error rate (with the standard deviation) of each algorithm. Kernel-SSELM and SSELM can achieve comparable result with the other 4 algorithms. Particularly, for the multi-class problems on the high dimension data such as COIL20 and USPST, Kernel-SSELM gave better performances than the others. Compared with SSELM, Kernel-SSELM yielded higher accuracy and lower deviation on all dataset. It is obvious to find that the algorithm with kernel function could build more stable model in classification task.

Table 3 displays the training efficiency of each algorithm on the 4 experiment datasets. SSELM was the fastest, while Kernel-SSELM was a bit slower but still stayed on the same level. On the binary-problem dataset, Kernel-SSELM and SSELM did not show much advantage to LapRLS, and LapSVM. This result in the they all need to calculate the Laplacian matrix which is a time consuming process

Table 3 Training time of different semi-supervised algorithms

Dataset	TSVM	LDS	LapRLS	LapSVM	SSELM	Kernel-SSELM
G50C	0.539	0.651	0.083	0.089	0.047	0.053
G10N	0.386	0.427	0.046	0.048	0.032	0.036
COIL20	34.32	39.18	11.98	8.367	1.201	1.634
USPST	188.7	205.3	15.27	13.84	2.932	3.524

and dominates the computation cost. However, for multi-class problem, the extreme learning methods showed significant advantage in training efficiency.

In all, from the two tables, we could found that Kernel-SSELM can give higher accuracy and stability in the cost of a little training speed.

5 Application in Traffic Congestion Evaluation

5.1 Traffic Congestion Evaluation

Urban Transportation Assessment and Forecast System analyzes the traffic congestion of transportation network in a city of southwest China and shows the evaluation results of the real-time traffic states on the GIS map using different colors on the foundation of the floating cars' GPS information (Fig. 1).

Seen from Fig. 2, traffic congestion evaluation system based on floating car data is the fundamental part of core function. In previous work, traditional method evaluating the present road congestion through fixed empirical evaluation standard is easy to implement and consumes a little system resources. But it does have the following drawbacks: First, the empirical evaluation frameworks do not take full consideration of the road information and network conditions. Second, it causes a significant gap between the congestion information on the map and users' experience.

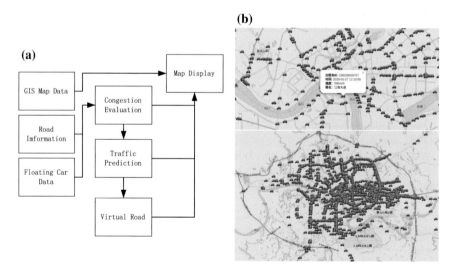

Fig. 1 Urban transportation assessment and forecast system. **a** Structure of urban transportation assessment and forecast system. **b** Floating car distribution on the map

Fig. 2 Real-time traffic evaluation

To overcome the shortcomings above, machine learning methods are introduced into the traffic congestion evaluation system. However, many approaches such as SVM have deficiencies when applied to huge data and semi-supervised task in this traffic congestion evaluation system.

Applying Kernel-SSELM to traffic congestion evaluation system based on floating car data, this paper has the following strengths:

1. Though the congestion value of unlabeled data is uncertain, it represents the different traffic conditions which reflect the distribution information of traffic data. Kernel-SSELM improves the recognition accuracy of evaluation models by involving unlabeled data in the training.
2. Extreme learning machine has high training efficiency and is easy to implement. In the case of large data scales, high training speed ensures that despite traffic conditions changes it can still renew training for several times to choose a better model. At the same time, extreme learning machine is able to be modified into incremental learning easily so that we can make use of the latest information to update the evaluation network in real time.
3. With the neglecting the number of hidden layer nodes, the optimization of kernel function improves the stability of SSELM.

The evaluation system optimized by Kernel- SSELM improves the evaluation accuracy and is more in accord with the evaluation of congestions from local residents. As for the urban administration, the traffic congestion evaluation system plays an assistant role in management and supplies solutions for alleviating urban traffic. As for citizens, they may choose the right way to get around or the optimal driving route via the precise congestion evaluation.

5.2 Congestion Eigenvalue and Congestion Value

Traffic congestion evaluation system takes the road sections as the individual samples. Be specific, a road section demonstrates a portion of a road in a single direction. Its traffic congestion evaluation originates from two sources. The first part of source is the essential information of the road section from the Transportation Department, including Number of lanes, numbers of lanes of the entrance and exit, number of traffic lights and road grades. The second part of source is the real-time speed information of the road section from the floating car data, including average speed, speed distribution, and average stopping time.

The work of labeling training samples is completed by 5 experts from the Transportation Department of the city. Through surveillance cameras experts recorded information and gave evaluation of the traffic congestion at that time. Congestion evaluation is divided into three grades: Smooth, Average and Congested. The final label is in the grade which receives the most votes in 5 experts.

5.3 Evaluation Experiment

The environment of the experiment is the same as Sect. 4. In the experiment, we collect the floating car data from June 15th to June 16th 2015, and the quantity is more than 30,000,000. The data is grouped in interval for 5 min and matched to the corresponding road section. Finally we collect 13,681 samples. The evaluation of experts is based on the video from surveillance cameras about 30 typical road sections in the city. 537 valid samples were finally collected, and the rest 13,144 samples were unlabeled.

For comparison, we tested the SSELM, Kernel-SSELM and the empirical rule in Table 1. The test set had 100 samples randomly selected from the labeled sample, and the random generation process was repeated in 10 times. The cost coefficient C_0 was fixed to 100 and the semi-supervised rate λ was fixed to 0.001. The kernel function of Kernel-SSELM is RBF with the parameter γ fixed to 100. The number of hidden layer nodes of SSELM was set to 5000.

Table 4 shows that the evaluation model trained by Kernel-SSELM had the highest average accuracy at 86.2 %. In addition, Kernel-SSELM only takes 48.2 s for training, which keep the high training efficiency of SSELM.

Table 4 The result of evaluation experiment

	Empirical rule	SSELM	Kernel-SSELM
Average accuracy	68.9 %	82.6 %	86.2 %
Best accuracy	73 %	87.5 %	88 %
Training time	–	41.6	48.2

The trained model was used in the Urban Transportation Assessment and Forecast System. Figure 2 displays the real-time traffic condition. In the map, Green represents smooth traffic, yellow shows average condition, and red means the road is congested. Seen from the image taken by surveillance cameras, the traffic evaluation accurately reflects the road traffic congestion at that time.

6 Conclusion and Future Work

In this paper, a kernel based optimization is proposed to promote the SSELM. Experiments show that Kernel-SSELM can achieve higher accuracy and model stability, because kernel function avoids the problem of setting hidden layer. Compared with the other state-of-the-art semi-supervised learning algorithms, Kernel-SSELM shows significant advantages in training efficiency and multi-classification ability. In the application of traffic congestion evaluation, Kernel-SSELM was used to train the evaluation model on the large-scale data set. Both the experiment and the real-time application show the evaluation system can precisely reflect the traffic condition.

Since the type of kernel function and its parameter also have much influence on the training model, how to choose an optimized kernel function is still an important problem in the particular application of Kernel-SSELM. There is a general that assume a linear combination of a group of base kernels could be the optimal choice. In the future, we plan to research the multi-kernel framework for promoting the Kernel-SSELM in the traffic application.

Acknowledgment This work was supported by National Nature Science Foundation of P. R. China (No. 61272357, 61300074, 61572075).

References

1. Huang, G., Song, S., Gupta, J.N.: Semi-supervised and unsupervised extreme learning machines. IEEE Trans. Syst. Man. Cybern. Part B **44**(12), 2405–2417 (2014)
2. Huang, G.-B., Zhou, H., Ding, X., Zhang, R.: Extreme learning machine for regression and multiclass classification. IEEE Trans. Syst. Man Cybern. Part B **42**(2), 513–529 (2012)
3. Pal, M., Maxwell, A.E., Warner, T.A.: Kernel-based extreme learning machine for remote-sensing image classification. Remote Sens. Lett. **4**(9), 853–862 (2013)
4. Liu, J., Chen, Y., Liu, M., Zhao, Z.: SELM: semi-supervised ELM with application in sparse calibrated location estimation. Neurocomputing **74**(16), 2566–2572 (2011)
5. Li, L., Liu, D., Ouyang, J.: A new regularization classification method based on extreme learning machine in network data. J. Inf. Comput. Sci. **9**(12), 3351–3363 (2012)
6. Li, K., Zhang, J., Xu, H., Luo, S., Li, H.: A semi-supervised extreme learning machine method based on co-training. J. Comput. Inf. Syst. **9**(1), 207–214 (2013)
7. Gu, J., Zhou, S., Yan, X., et al.: Formation mechanism of traffic congestion in view of spatio-temporal agglomeration of residents' daily activities: a case study of Guangzhou. Sci. Geogr. Sinica **32**(8), 921–927 (2012)

8. Liu, X., Yin, J., Wang, L., Liu, L., Liu, J., Hou, C., Zhang, J.: An adaptive approach to learning optimal neighborhood kernels. IEEE Trans. Cybern. **43**(1), 371–384 (2013)
9. Cao, L.L., Huang, W.B., Sun, F.C.: Optimization-based multi kernel extreme learning for multimodal object image classification. International Conference on Multisensor Fusion and Information Integration for Intelligent Systems (MFI) pp. 1–9 (2014)

Improvement of ELM Algorithm for Multi-object Identification in Gesture Interaction

Liang Diao, Liguo Shuai, Huiling Chen and Weihang Zhu

Abstract ELM algorithm has been widely applied in gesture recognition. When the dataset is multi-objective, however, using classical ELM algorithm directly may produce a big recognition error. To address this problem, an improved ELM recognition algorithm is proposed. The presented ELM algorithm is characterized as building separated ELM network for each gesture instead of constructing a unified ELM network for all gestures. A simplified and optimized feature is proposed. Comparison experiment between the classical ELM algorithm and the optimized ELM algorithm aiming to four classical gestures are conducted. The result shows that the training accuracy of the improved algorithm is about 5.25 times of the classical algorithm, and the right recognition ability of the improved algorithm is more than 1.8 times than the classical algorithm. The training time of the optimized algorithm is less than that of the classical algorithm.

Keywords Gesture recognition · Multi-objects · Simplified-gesture features · Distributed network

1 Introduction

In human-computer interaction (HCI) fields, Gesture recognition has gained more and more attention from the world. At present, it is commonly used two types of approaches in gesture recognition: data glove-based recognition and vision-based recognition [1]. (1) The first manner data glove-based recognition, employing data

L. Diao (✉) · L. Shuai · H. Chen
School of Mechanical Engineering, Southeast University, Nanjing 211189,
People's Republic of China
e-mail: zhongtianfang@163.com

W. Zhu
Department of Industrial Engineering, Lamar University, Beaumont, TX 77710, USA
e-mail: humorstar@yahoo.com

© Springer International Publishing Switzerland 2016
J. Cao et al. (eds.), *Proceedings of ELM-2015 Volume 1*,
Proceedings in Adaptation, Learning and Optimization 6,
DOI 10.1007/978-3-319-28397-5_19

glove, installs acceleration sensor and geomagnetic sensor, to calculate the gesture from the fingers' position. Having strong adaptability, this manner can recognize not only statistic gesture but the dynamic gesture. However, because of high-cost, it has not been widely used [2]. (2) The second manner, vision-based recognition, involves the use of one or more cameras to collect gesture picture, from which gesture characteristics would be extracted through image processing techniques, then pattern recognition would be used to perform gesture classifications. The second manner, with no restriction of data glove, is low-cost, and makes the HCI of higher efficiency and more convenient. A complete recognition based on camera involves three steps [1, 3]. The first step is collecting gesture pictures with camera, the second one is detecting and tracking hand feature and the third one is recognition of hand feature. Recently, the former two steps have become full-fledged, while the recognition is various due to different algorithms selected such as Hidden Markov Models (HMMs), Dynamic Time Warping (DTW), Template Matching (TM), (Support Vector Machine) SVR, Artificial Neural Networks (ANN), etc. These model recognition algorithms mentioned above, HMM and DTW are available for dynamic gesture recognition, especially in the action recognition with time span [4–6]. TM is a simple gesture recognition algorithm used in the early stage, but because of its rigid matching process, the recognition may failed if the matching templates are twisted or beyond a certain limit [7, 8]. SVM is also a popular static gesture recognition algorithm, for having good real-time performance, however, it is not suitable for processing large sample data [9]. In contrast, the AAN has inherent advantages in dealing with large sample data. Extreme Learning Machine (ELM) is a kind of AAN with ability of extremely fast learning, which has raised a great concern in the field of HCI after it was first proposed in 2004. Then a large number of pattern recognition algorithms based on ELM are developed by researchers all around the world [10]. Many improvements focused on network structure and hidden nodes number has been proposed. Among these algorithms, a land cover classifier, Proposed by Chen et al. [11], trains multiple ELM networks parallelly, which not only expands the selectable range of the hidden nodes number but improves the recognition accuracy. Arif introduced an Online Sequential Extreme Machine Learning to recognize human action [12], which has the capability to handle both additive and RBF nodes in a unified framework and can learn the training data chunk-by-chunk. For the purpose of avoiding big error caused by the multi-objective recognition, in this paper, an effective feature tracking algorithm is developed, and multiple ELM networks are trained respectively for each kinds of hand gesture, in order to decrease the training error and recognition error.

2 Methodology

ELM algorithm is a kind of feed-forward neural network algorithm with single-hidden layer [13]. Once the number of input layer nodes, hidden layer nodes and output layer nodes are determined, the connection weights between hidden layer nodes and output nodes would be calculated by least square method immediately. Therefore ELM algorithm has strong ability to cut training time. But when the training sample is relatively complex and the number of hidden layer nodes is too small, the model trained by ELM may produce large training error. The value of training error is related to the number of hidden nodes and the training sample complexity, where the sample complexity depends on the sample size [14] and the differences of samples [15]. The smaller number of hidden nodes and the more complexity of sample will lead to a bigger recognition error. But the sample size is mainly determined by the objective conditions. The factors such as uneven lighting and different shooting angle of gestures will lead to large number of variations in one type of gesture image, which is the reason why the gesture sample size is general extremely large. In order to reduce recognition error and reduce training time when the training objects are multi-gestures, a method for feature optimization has been proposed in this paper. Multi-ELM models are trained with these matrices parallelly, each of which can recognize one type of gesture. As hand gesture is recognized with this improved algorithm, optimized characteristic vectors should be extracted first, the number of these vectors is same as the type of gestures, then the optimized vectors would be inputted into multi-models to get the recognition errors, finally when the minimum is under the permissible error, the test gesture would be recognized by the model whose recognition error is the minimum.

2.1 Model Training

The topology structure of improved algorithm is shown in Fig. 1. Multi-models are trained parallely with X_i. The output nodes number of each model is 1. And each

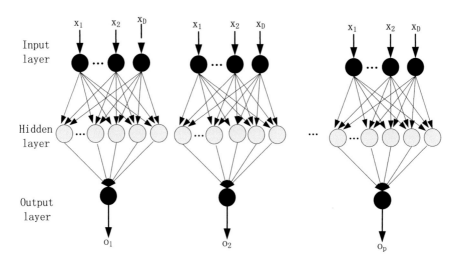

Fig. 1 The topology of the proposed model

model is independent with other models. The training process can be divided into 4 steps.

1. Firstly, the ideal output (O_i) should be set for each feature optimized matrix (X_i). As the output node number is 1, O_i is a column vector shown in Eq. (1). Only two values can be found in O_i, one is 0 and another is 1, where 1 is the ideal output of the identified gesture while 0 is the output of other gestures.

$$O_i = [o_1 \quad o_2 \quad \cdots \quad o_N]^T \quad (i = 1, 2, 3, \ldots, p) \tag{1}$$

2. Secondly, the threshold (B_i) of the hidden node and the weight matrix ($W1_i$) that connect the hidden nodes and the input nodes should be initialized for each models. In Eqs. (2) and (3), $w1$ and b is decimal numbers assigned between 0 and 1 in random. The row number of weight matrix is same as feature extraction number (D), and the column number is same as hidden nodes number (L).

$$W1_i = [w1_{j1} \quad w1_{j2} \quad \cdots \quad w1_{jL}] \quad (i = 1, 2, 3, \ldots, p; j = 1, 2, \ldots, D) \tag{2}$$

$$B_i = \begin{bmatrix} b_1 & b_2 & \cdots & b_L \\ b_1 & b_2 & \cdots & b_L \\ \vdots & \vdots & \vdots & \vdots \\ b_1 & b_2 & \cdots & b_L \end{bmatrix}_{N*L} \quad (i = 1, 2, 3, \ldots, p) \tag{3}$$

3. Thirdly, calculating Hp_i by Eq. (4), where g is the inactivate value, which should be activated with activation function, where the most common one is sigmoid, whose equation is shown in (5). H_i is the activation matrix, in which G is the activation value with g. From (4) (5) (6), there is p H_i obtained.

$$\begin{aligned} Hp_i = [g_{j1} \quad g_{j2} \quad \cdots \quad g_{jL}] = X_i * W1_i + B_i \\ (i = 1, 2, 3, \ldots, p; j = 1, 2, \ldots, N) \end{aligned} \tag{4}$$

$$G_{ij} = \frac{1}{1 + \exp(-g_{ij})} \quad (i = 1, 2, 3, \ldots, N; j = 1, 2, 3, \ldots, L) \tag{5}$$

$$H_i = [G_{j1} \quad G_{j2} \quad \cdots \quad G_{jL}] \quad (i = 1, 2, 3, \ldots, p; j = 1, 2, \ldots, N) \tag{6}$$

4. Finally, calculating the generalized inverse matrix (H_i^{+-1}) for each H_i, where H_i^+ is obtained by Eq. (7). Then substituting H_i^{+-1} and O_i into Eq. (8) finally gain weight matrix ($W2_i$) that connect the hidden nodes and the input nodes, where $W2_i$ is a dimensional vector same as O_i. Once $W2_i$ of each model is obtained, the whole training process of improved model is over.

$$H_i^+ = H_i^T H_i (i = 1, 2, 3, \ldots, p) \tag{7}$$

$$W2_i = H_i^{+-1} O_i^T (i = 1, 2, 3, \ldots, p) \tag{8}$$

$$W2_i = [w2_1 \quad w2_2 \quad \ldots \quad w2_L]^T (i = 1, 2, 3, \ldots, p) \tag{9}$$

For the purpose of evaluating the quality of training model, we use error Eq. (10) to calculate training error and recognition error, where E means the output of the recognized object. When the recognized object is training matrix, J is the training error. When the recognized object is gesture that is not contained in training matrix is recognition error.

$$J = \left\| (E - O) * (E - O)^T \right\|_2 \tag{10}$$

3 Experiments

In this part, we planned to employ the classical algorithm and improved algorithm respectively to train 4 typical hand gestures of FIVE, STONE, OK, and VICTORY. Then the data of these two kinds of models, including training error, recognition error, training time and recognition time, are calculated and compared respectively, which are derived by simulation implemented in MATLAB. There are 100 available images collected at natural state for each gesture type. Thus, there are 400 gesture images in all. When selecting the samples, hand should be completely relaxed, put in the focus area of camera and parallel to desk.

3.1 Training Error

1. For the classical model, using classical ELM algorithm to train the classic characteristic matrix, the ideal output array of each gesture is shown in Table 1. After model training, calculating the outputs of these $400 \vec{y}$. According to the outputs, the train error is obtained with Eq. (10). In this experiment, $L = 200$–500 is set, the training error would be added up as the model is repetitively trained for each L, Finally the average training error is calculated

Table 1 Ideal outputs of classical model

No.	1	2	3	4
FIVE	1	0	0	0
OK	0	1	0	0
STONE	0	0	1	0
VICTORY	0	0	0	1

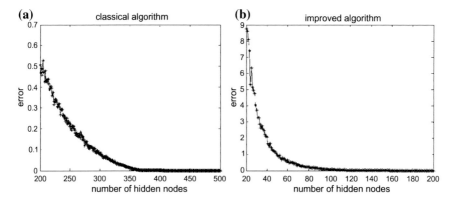

Fig. 2 Training error. **a** Training error of classical model. **b** Training error of improved model

after the model is trained 10 times, by which the curve of error and hidden nodes number is derived in Fig. 2a. In the figure, as hidden nodes number increase, the training error decreases gradually. When L = 360, the train error is 0.0046, and then the training error gradually converge to 0.

2. For the improved model, according to the characteristic extraction (C_i), optimized characteristic matrix is extracted and t1 = t2 = 25 are set, when the value of feature extraction is D = 50, and four groups of optimized characteristic matrix ($X_i = \begin{bmatrix} \overrightarrow{x_{1i}} & \overrightarrow{x_{2i}} & \dots & \overrightarrow{x_{400i}} \end{bmatrix}^T$ ($i = 1, 2, 3, 4$)) are obtained. Then the independent ELM models are trained with above 4 Xi, each model is corresponding to 50 input nodes and one output node. After the development of networks, these 4 Xi are substituted into each model respectively to calculate output matrix, then we obtain training error with output matrix and substitute it into formula (10). We define the sum error of these 4 models as the training error of improved structure. In the experiment, L = 20–200 is set and model is repeatedly trained for 10 times at each hidden node. Average training error is calculated, and the curve of training error is shown in Fig. 2b. In the figure, When L = 120, the error is 0.024, and then gradually converge to 0.

3.2 Recognition Error

To further test the recognition effect of classical and improved models. We chose 80 test images that do not belong to the training samples, and each one type of gesture has 20 test images.

Fig. 3 Recognition error

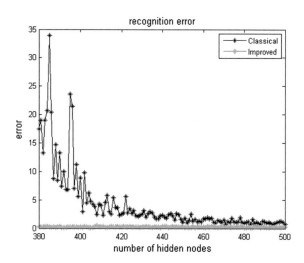

1. For the classical model, calculating the outputs with the 80 test inputs, then we will get the recognition error by Eq. (10). And when L = 360, the training error is nearly 0, which is verified in experiment (Sect. 3.1).

   Therefore L = 380–500 is set to calculate the whole recognition error of these 80 test inputs. Average training error is calculated, as the model is repeatedly trained for 10 times at each hidden node.

2. For the improved model. The optimized feature vector should be extracted according to C_i. Then we will obtain 4 groups optimized inputs, each group has 80 optimized feature vectors. Calculating the outputs with these optimized inputs, then we will get the recognition error by Eq. (10). And when L = 130, the training error is nearly 0. Therefore L = 380–500 is set to calculate the whole recognition error of these 360 optimized inputs. Average training error is calculated, as the improved model is repeatedly trained for 10 times at each hidden node. Finally, the test error curves of classical and improved models are shown in Fig. 3.

   As shown in Fig. 3, the recognition errors of classical structure decrease progressively as the hidden nodes increase. When the L = 500, the recognition error is at its minimum of 0.7386, while the recognition error of the improved model is only 0.2628 with the decreasing rate of 1.81. When L = 385, the recognition error of classical model is at its maximum of 33.9, while the recognition error of improved model is only 0.2227, which is far lower than the classical one. As shown in the curves, the recognition error of improved model has been already coverage, while the classical one is decreasing progressively, and the former one is always less than the latter one with the same hidden nodes number.

3.3   The Training and Recognition Time

In the improved model, the input node and output node have been simplified, while
due to the model has multi-ELM pattern, the training time would be affected. To
exam the discrimination of training efficiency between the improved and classical
models, on the condition of the same recognition error level, the experiment is
designed to calculate the training time on both models with the former training
samples. According to the pre-experiment, when L = 750–850, the average
recognition error of the classical one is 0.3201, which is basically similar with the
average error (0.3155) of improved one as its L = 200–300. Therefore the experiment would be conducted in above range of L. The computer is configured with
Intel® Core™ i3-T6600 CPU @ 2.2 GHz; RAM 3.0 GB, and 32 bit Windows
Operation System.

3.3.1   Training Time

For the classical model, L = 750–850 is set, the curve of classical training time is
shown in Fig. 4a. For the improved model, the training time is the sum of the time
to training these 4 models. The curve of improved training time is shown in Fig. 4b,
as its L = 200–300.

In Fig. 4, the training time of the both algorithms increase gradually, as the
hidden nodes number increases. It indicates that the training time of the improved
one is much less than the classical one. When the L of classical model is 847, the
training time is at its maximum of 2.053 s. However, the training time of the
improved model is at its peak value of 0.391 s, when the L = 297, which is 4.25
times less than the classical training time.

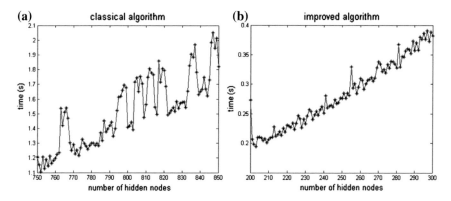

Fig. 4   Training time. **a** Training time of classical model. **b** Training time of improved model

3.3.2 Recognition Time

The process of improved model, compared with the classical one, is more complex, because the optimized input should be extracted in advance and the inputs should be taken into multi-networks. To exam the recognition speed of improved model, the experiment of recognition time is conducted with the former 80 test inputs.

1. For the classical model, L = 750–850 is set, calculating the average recognition time of the test inputs, then the curve of classical training time is shown in Fig. 5a.
2. For the improved model, the recognition time is from the feature optimizing to the calculation of the recognition error. The curve of improved training time is shown in Fig. 5b, as its L = 200–300.

The average recognition time of the improved model is 1.7762e-004 s, compared with the average recognition time of classical model which is 1e-004 s, increases slightly. Although the recognition speed of the improved model reduces slightly, it will not affect the efficiency of HCI, and the recognition time of the improved model is much less than the hand response time, which is about 200 ms [16].

3.4 Effect of Optimized Feature Extraction Number

In these experiments above, D = 50, while the recognition error may be affected by the feature extraction number. Therefore, L = 200 is limited, D changes from 20 to 200 on the condition of t1 = t2 to calculate the recognition error of the 80 test inputs. Then the recognition curve (D = 20–200; L = 200) is shown in Fig. 6. The recognition error firstly decreases to the minimum (0.088), when D = 112, then

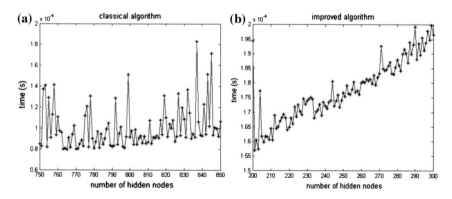

Fig. 5 Recognition time. **a** Recognition time of classical model. **b** Recognition time of improved model

Fig. 6 Recognition error
(D = 20–200; L = 800)

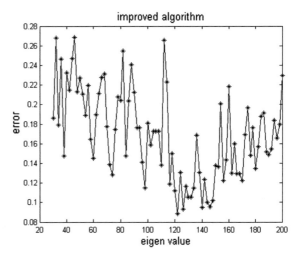

Fig. 7 Recognition error
(D = 200; L = 1000–1100)

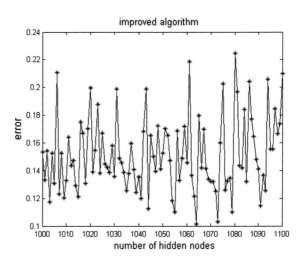

increase to 0.2269 whose D = 200. The reason why the error increases after D = 112 may be caused by less L. Therefore further experiment is conducted, where D = 200 and L = 1000–1100. The curve of the error is shown in Fig. 7.

In Fig. 6, the recognition error is oscillating slightly. While the minimum value is 0.1014 at L = 1064, which is still bigger than the value at D = 112 and L = 800. This experiment verified that the valid features are almost extracted as D = 112. When D > 112, the redundant pixels may lead to big error. Therefore D should be selected in order to avoid introducing unnecessary noise.

4 Conclusions

In this paper, it is proposed an improved feature-extraction algorithm and proposed a constructing distributed-network for each kind of processed gesture features on the basis of ELM. Experiments of training error, recognition error, training time and recognition time are conducted in our study. Results show that the training error of the improved algorithm can converge to 0 faster. The recognition error of the improved model is 1.81 times less than the classical ELM on the condition of 500 hidden nodes. And when these two kinds of model have the same recognition error, the training time of the improved one has decreased more than 4.25 times. Although the recognition time of the improved algorithm is more than the classical one, the increments are so small that it can be neglected.

The improved algorithm also has some shortcomings, when the training samples are extremely complicated, the recognition time may further increase for high accuracy. Therefore, as our future works, we aim to balance the relation of the efficiency and accuracy. In addition, the improved algorithm is not only suitable for gesture recognition applications, but also for face recognition fields, as the features of human face are relatively stable. Some features, in one particular face, extracted by improved algorithm are typical to all other faces, which may make the face recognition more efficiency and accuracy. Thus in the future, we will also set about to do the research of our improve algorithm in face recognition applications.

Acknowledgements This work is supported by the National Natural Science Foundation of China (61175069), the National Science Foundation's Course, Curriculum, and Laboratory Improvement (CCLI) program (0737173), the Natural Science Foundation of Jiangsu Province of China (BK20131205) and the Prospective Project of Jiangsu Province for Joint Research (SBY201320601).

References

1. Badi, H., Hussein, S.H., Kareem, S.A.: Feature extraction and ML techniques for static gesture recognition. Neural Comput. Appl. **25**(3), 733–741 (2014)
2. Lin, B.S., Lee, I.-J., Hsiao, P.-C., Yang, S.-Y., Chou, W.: Data glove embedded with 6-DOF inertial sensors for hand rehabilitation. In: International Conference on Intelligent Information Hiding and Multimedia Signal Processing, pp. 25–28 (2014)
3. Kumar, V., Nandi, G.C., Kala, R.: Static hand gesture recognition using stacked denoising sparse autoencoders. In: 2014 Seventh International Conference on Contemporary Computing (IC3), pp. 99–104 (2014)
4. Badi, H.S., Hussein, S.: Neural Comput. Appl. **25**(3), 871–878 (2014)
5. Zabulis, X.: Vision-based hand gesture recognition for human–computer interaction. The Universal Access Handbook, LEA (2009)
6. Vieriu, R.L., Goras, B., Goras, L.: On HMM static hand gesture recognition. In: 10th International Symposium on Signals, Circuits and Systems (ISSCS), pp. 1–4 (2011)
7. Jain Anil, K.: Statistical pattern recognition: a review, pattern analysis and machine intelligence. IEEE Trans. **22**, 4–37 (2000)

8. Watanabe, T., Lee, C.-W., Yachida, M.: Recognition of complicated gesture in real-time interactive system. In: 5th IEEE International Workshop on Robot and Human Communication, ROMAN.1996.568842, pp. 268–273 (1996)
9. Zhang, J.-P., Li, Z.-W., Yang, J.: A parallel SVM training algorithm on large-scale classification problems. In: 2005 International Conference on Machine Learning and Cybernetics, ICMLC 2005, pp. 1637–1641 (2005)
10. Yu, H., Chen, Y., Liu, J., Huang, G.-B.: An adaptive and iterative online sequential ELM-based multi-degree-of-freedom gesture recognition system. IEEE Intell. Syst. **28**(6), 55–59 (2013)
11. Chen, J., Zheng, G., Chen, H.: ELM-MapReduce: MapReduce accelerated extreme learning machine for big spatial data analysis. In: 10th IEEE International Conference on Control and Automation (ICCA), pp. 400–405 (2013)
12. Budiman, A., Fanany, M.I., Basaruddin, C.: Constructive, robust and adaptive OS-ELM in human action recognition. In: International Conference on Industrial Automation, Information and Communications Technology (IAICT), pp. 39–45 (2014)
13. Huang, G.B., Zhu, Q.Y., Siew, C.K.: Extreme learning machine: theory and applications. Neurocomputing **70**(1–3), 489–501 (2006)
14. Cui, Y.-J., Davis, S., Cheng, C.-K., Bai, X.: A study of sample size with neural network. In: International Conference on Machine Learning and Cybernetics, vol. 6(26–29), pp. 3444–3448 (2004)
15. Pestov, V.: A note on sample complexity of learning binary output neural networks under fixed input distributions. In: 2010 Eleventh Brazilian Symposium on Neural Networks (SBRN), pp. 7–12 (2010)
16. Rosenbaum, D.A.: Human factors and behavioral science: central control of movement timing. Bell Syst. Tech. J. **162**(6), 1647–1657 (1983)

SVM and ELM: Who Wins? Object Recognition with Deep Convolutional Features from ImageNet

Lei Zhang, David Zhang and Fengchun Tian

Abstract Deep learning with a convolutional neural network (CNN) has been proved to be very effective in feature extraction and representation of images. For image classification problems, this work aim at finding which classifier is more competitive based on high-level deep features of images. In this paper, we have discussed the nearest neighbor, support vector machines and extreme learning machines for image classification under deep convolutional activation feature representation. Specifically, we adopt the benchmark object recognition dataset from multiple sources with domain bias for evaluating different classifiers. The deep features of the object dataset are obtained by a well-trained CNN with five convolutional layers and three fully-connected layers on the challenging ImageNet. Experiments demonstrate that the ELMs outperform SVMs in cross-domain recognition tasks. In particular, state-of-the-art results are obtained by kernel ELM which outperforms SVMs with about 4 % of the average accuracy. The Features and MATLAB codes in this paper are available in http://www.escience.cn/people/lei/index.html.

Keywords Extreme learning machine · Deep learning · Image classification · Support vector machine · Object recognition

L. Zhang (✉) · F. Tian
College of Communication Engineering, Chongqing University, 174 Shazheng Street, ShapingBa District, Chongqing 400044, China
e-mail: leizhang@cqu.edu.cn

F. Tian
e-mail: fengchuntian@cqu.edu.cn

L. Zhang · D. Zhang
Department of Computing, The Hong Kong Polytechnic Unviersity, Hung Hom, Hong Kong, China
e-mail: csdzhang@comp.polyu.edu.hk

© Springer International Publishing Switzerland 2016
J. Cao et al. (eds.), *Proceedings of ELM-2015 Volume 1*,
Proceedings in Adaptation, Learning and Optimization 6,
DOI 10.1007/978-3-319-28397-5_20

1 Introduction

Recently, deep learning as the hottest learning technique has been widely explored in machine learning, computer vision, natural language processing and data mining. In the early, convolutional neural network (CNN), as the most important deep net in deep learning, has been applied to document recognition and face recognition [1, 2]. Moreover, some deep learning algorithms with multi-layer fully connected networks (e.g. multi-layer perceptrons, MLP) for auto-encoder have been proposed, for examples, stacked auto encoders (SAE) [3], deep belief networks (DBN) [4] and deep Boltzmann machines (DBM) [5]. However, in large-scale learning problems, e.g. image classification in computer vision, CNNs with convolutioanl layers, pooling layers and fully-connected layers are widely investigated for its strong deep feature representation ability and state-of-the-art performance in challenged big datasets like ImageNet, Pascal VOC, etc. In the latest progress of deep learning, researchers have broken the new record in face verification by using CNNs with different structures [6–9]. The latest verification accuracy on LFW data is 99.7 % by Face++ team. Besides the faces, CNN has also achieved very competitive results on ImageNet for image classification and Pascal VOC data [10–17]. From these works, CNNs have been proved to be highly effective for deep feature representation with large-scale parameters. The main advantages of deep learning can be shown in three facets. (1) Feature representation. CNN integrates feature extraction (raw pixels) and model learning together, without using any other advanced low-level feature descriptors. (2) Large-scale learning. With the adjustable network structures, big data in millions can be learned by a CNN at one time. (3) Parameter learning. Due to the scalable network structures, millions of parameters can be trained. Therefore, CNN based deep method can be state-of-the-art parameter learning technique.

In this paper, we would like to discuss about the deep feature representation capability of CNN by using traditional classification method with high-level deep features of images, and find which classifier is the best under the deep representation. Therefore, we mainly exploit the nearest neighbor (NN) [18], support vector machine (SVM) [19], least-square support vector machine (LSSVM) [20], extreme learning machine (ELM) [21] and kernel extreme learning machine (KELM) [22]. These classifiers are well-known in many different applications. Specially, ELM was initially proposed for generalized single-hidden-layer feed-forward neural networks and overcome the local minima, learning rate, stopping criteria and learning epochs that exist in gradient-based methods such as back-propagation (BP) algorithm. In recent years, ELMs are widely used due to some significant advantages such as learning speed, ease of implementation and minimal human intervention. The potential for large scale learning and artificial intelligence is preserved. The main steps of ELM include the random projection of hidden layer with random input weights and analytically determined solution by using Moore-Penrose generalized inverse. With similar impact with SVM, it has been proved to be efficient and effective for regression and classification tasks [23, 24]. The latest work about the principles and brain-alike learning of ELM has been presented [25]. Many improvement and new applications of ELMs have been proposed by researchers. The newest work about ELM for deep auto-encoder, local receptive fields for deep learning, transfer learning, and semi-supervised learning have also been proposed [26–30]. With the Mercer condition applied, a kernel ELM (KELM) that computes a kernel matrix of hidden layers has also been proposed [22]. A salient feature of KELM is that the random input weights and bias can be avoided.

In this paper, we will present a study of NN, SVM, LSSVM, ELM and KELM for object recognition on the deep convolutional activation features trained by CNN on ImageNet, and have an insight of which one is the best for classification on deep representation.

The rest of this paper is organized as follows. Section 2 presents a method review of support vector machines and extreme learning machines. Section 3 shows the training and testing protocol of CNN for deep representation of images. Section 4 presents the experiments and results. Finally, Sect. 5 concludes this paper.

2 Overview of SVMs and ELMs

2.1 Support Vector Machine (SVM)

In this section, the principle of SVM for classification problems is briefly reviewed. More details can be referred to [19].

Given a training set of N data points $\{\mathbf{x}_i, y_i\}_{i=1}^{N}$, where the label $y_i \in \{-1, 1\}$, $i = 1, \ldots, N$. According to the structural risk minimization principle, SVM aims at solving the following risk bound minimization problem with inequality constraint.

$$\min_{\mathbf{w}, \xi_i} \frac{1}{2} \|\mathbf{w}\|^2 + C \cdot \sum_{i=1}^{N} \xi_i,$$
$$s.t. \ \xi_i \geq 0, \ y_i \left[\mathbf{w}^\mathrm{T} \varphi(\mathbf{x}_i) + b \right] \geq 1 - \xi_i \tag{1}$$

where $\varphi(\cdot)$ is a linear/nonlinear mapping function, $\mathbf{w}$ and b are the parameters of classifier hyper-plane.

Generally, for optimization, the original problem (1) of SVM can be transformed into its dual formulation with equality constraint by using Lagrange multiplier method. One can construct the Lagrange function,

$$L(\mathbf{w}, b, \xi_i; \alpha_i, \lambda_i) = \frac{1}{2} \|\mathbf{w}\|^2 + C \cdot \sum_{i=1}^{N} \xi_i - \sum_{i=1}^{N} \alpha_i \left(y_i \left[\mathbf{w}^\mathrm{T} \varphi(\mathbf{x}_i) + b \right] - 1 + \xi_i \right)$$
$$- \sum_{i=1}^{N} \lambda_i \xi_i \tag{2}$$

where $\alpha_i \geq 0$ and $\lambda_i \geq 0$ are Lagrange multipliers. The solution can be given by the saddle point of Lagrange function (2) by solving

$$\max_{\alpha_i, \lambda_i} \min_{\mathbf{w}, b, \xi_i} L(\mathbf{w}, b, \xi_i; \alpha_i, \lambda_i) \tag{3}$$

By calculating the partial derivatives of Lagrange function (2) with respect to $\mathbf{w}$, b and ξ_i, one can obtain

$$\begin{cases} \frac{\partial L(\mathbf{w}, b, \xi_i; \alpha_i, \lambda_i)}{\partial \mathbf{w}} = 0 & \rightarrow \quad \mathbf{w} = \sum_{i=1}^{N} \alpha_i y_i \varphi(\mathbf{x}_i) \\ \frac{\partial L(\mathbf{w}, b, \xi_i; \alpha_i, \lambda_i)}{\partial b} = 0 & \rightarrow \quad \sum_{i=1}^{N} \alpha_i y_i = 0 \\ \frac{\partial L(\mathbf{w}, b, \xi_i; \alpha_i, \lambda_i)}{\partial \xi_i} = 0 & \rightarrow \quad 0 \leq \alpha_i \leq C \end{cases} \tag{4}$$

Then one can rewrite (3) as

$$\max_{\boldsymbol{\alpha}} \sum_i \boldsymbol{\alpha}_i - \frac{1}{2} \sum_{i,j} y_i y_j \boldsymbol{\alpha}_i \boldsymbol{\alpha}_j \varphi(\mathbf{x}_i)^{\mathrm{T}} \varphi(\mathbf{x}_j)$$

$$s.t. \ \sum_{i=1}^{N} \alpha_i y_i = 0, \ 0 \leq \alpha_i \leq C \tag{5}$$

By solving α of the dual problem (5) with a quadratic programming, the goal of SVM is to construct the following decision function (classifier),

$$f(\mathbf{x}) = \mathrm{sgn}\left(\sum_{i=1}^{M} \boldsymbol{\alpha}_i y_i \kappa(\mathbf{x}_i, \mathbf{x}) + b \right) \tag{6}$$

where $\kappa(\cdot)$ is a kernel function. $\kappa(\mathbf{x}_i, \mathbf{x}) = \varphi(\mathbf{x}_i)^{\mathrm{T}} \varphi(\mathbf{x}) = \mathbf{x}_i^{\mathrm{T}} \mathbf{x}$ for linear SVM and $\kappa(\mathbf{x}_i, \mathbf{x}) = \exp(-\mathbf{x}_i - \mathbf{x}^2 / \sigma^2)$ for RBF-SVM.

2.2 Least Square Support Vector Machine (LSSVM)

LSSVM is an improved and simplified version of SVM. The details can be referred to [20]. We briefly introduce the basic principle of LSSVM for classification problems. By introducing the square error and equality constraint, LSSVM can be formulated as

$$\min_{\mathbf{w}, \xi_i} \frac{1}{2} \|\mathbf{w}\|^2 + C \cdot \frac{1}{2} \sum_{i=1}^{N} \xi_i^2,$$

$$s.t. \ y_i \left[\mathbf{w}^{\mathrm{T}} \varphi(\mathbf{x}_i) + b \right] = 1 - \xi_i, \quad i = 1, \ldots, N \tag{7}$$

The Lagrange function of (7) can be defined as

$$L(\mathbf{w}, b, \xi_i; \alpha_i) = \frac{1}{2}\|\mathbf{w}\|^2 + C \cdot \frac{1}{2}\sum_{i=1}^{N}\xi_i^2 - \sum_{i=1}^{N}\alpha_i\left(y_i\left[\mathbf{w}^T\varphi(\mathbf{x}_i) + b\right] - 1 + \xi_i\right) \quad (8)$$

where α_i is the Lagrange multiplier.

The optimality conditions can be obtained by computing the partial derivatives of (8) with respect to the four variables as

$$\begin{cases} \frac{\partial L(\mathbf{w}, b, \xi_i; \alpha_i)}{\partial \mathbf{w}} = 0 & \rightarrow \quad \mathbf{w} = \sum_{i=1}^{N}\alpha_i y_i \varphi(\mathbf{x}_i) \\[2mm] \frac{\partial L(\mathbf{w}, b, \xi_i; \alpha_i)}{\partial b} = 0 & \rightarrow \quad \sum_{i=1}^{N}\alpha_i y_i = 0 \\[2mm] \frac{\partial L(\mathbf{w}, b, \xi_i; \alpha_i)}{\partial \xi_i} = 0 & \rightarrow \quad \alpha_i = C\xi_i \\[2mm] \frac{\partial L(\mathbf{w}, b, \xi_i; \alpha_i)}{\partial \alpha_i} = 0 & \rightarrow \quad y_i\left[\mathbf{w}^T\varphi(\mathbf{x}_i) + b\right] - 1 + \xi_i = 0 \end{cases} \quad (9)$$

The equation group (9) can be written in linear equation as

$$\begin{bmatrix} \mathbf{I} & 0 & 0 & -\mathbf{Z}^T \\ 0 & 0 & 0 & -\mathbf{Y}^T \\ 0 & 0 & C\mathbf{I} & -\mathbf{I} \\ \mathbf{Z} & \mathbf{Y} & \mathbf{I} & 0 \end{bmatrix} \begin{bmatrix} \mathbf{w} \\ b \\ \xi \\ \alpha \end{bmatrix} = \begin{bmatrix} \mathbf{0} \\ 0 \\ 0 \\ \vec{1} \end{bmatrix} \quad (10)$$

where $\mathbf{Z} = [\varphi(\mathbf{x}_1)y_1, \ldots, \varphi(\mathbf{x}_N)y_N]^T$, $\mathbf{Y} = [y_1, \ldots, y_N]^T$, $\vec{1} = [1, \ldots, 1]^T$, $\xi = [\xi_1, \ldots, \xi_N]^T$, $\alpha = [\alpha_1, \ldots, \alpha_N]^T$. The solution of α and b can also be given by

$$\begin{bmatrix} 0 & -\mathbf{Y}^T \\ \mathbf{Y} & \mathbf{Z}\mathbf{Z}^T + C^{-1}\mathbf{I} \end{bmatrix} \begin{bmatrix} b \\ \alpha \end{bmatrix} = \begin{bmatrix} \mathbf{0} \\ \vec{1} \end{bmatrix} \quad (11)$$

Let $\Omega = \mathbf{Z}\mathbf{Z}^T$, with the Mercer condition, there is

$$\Omega_{k,l} = y_k y_l \varphi(\mathbf{x}_k)^T \varphi(\mathbf{x}_l) = y_k y_l \kappa(\mathbf{x}_k, \mathbf{x}_l), \quad k, l = 1, \ldots, N \quad (12)$$

By substituting (12) into (11), the solution can be obtained by solving a linear equation instead of a quadratic programming problem in SVM. The final decision function of LSSVM is the same as SVM shown as (6).

2.3 Extreme Learning Machine (ELM)

ELM aims to solve the output weights of a single layer feed-forward neural network (SLFN) by minimizing the squared loss of predicted errors and the norm of the output weights in both classification and regression problems. We briefly introduce the principle of ELM for classification problems. Given a dataset $\mathbf{X} = [\mathbf{x}_1, \mathbf{x}_2, \ldots, \mathbf{x}_N] \in \mathfrak{R}^{d \times N}$ of N samples with label $\mathbf{T} = [\mathbf{t}_1, \mathbf{t}_2, \ldots, \mathbf{t}_N] \in \mathfrak{R}^{c \times N}$, where d is the dimension of sample and c is the number of classes. Note that if $\mathbf{x}_i (i = 1, \ldots, N)$ belongs to the k-th class, the k-th position of $\mathbf{t}_i (i = 1, \ldots, N)$ is set as 1, and -1 otherwise. The hidden layer output matrix $\mathbf{H}$ with L hidden neurons can be computed as

$$\mathbf{H} = \begin{bmatrix} h\left(\mathbf{w}_1^T \mathbf{x}_1 + b_1\right) & h\left(\mathbf{w}_2^T \mathbf{x}_1 + b_2\right) & \cdots & h\left(\mathbf{w}_L^T \mathbf{x}_1 + b_L\right) \\ \vdots & \vdots & \vdots & \vdots \\ h\left(\mathbf{w}_1^T \mathbf{x}_N + b_1\right) & h\left(\mathbf{w}_2^T \mathbf{x}_N + b_2\right) & \cdots & h\left(\mathbf{w}_L^T \mathbf{x}_N + b_L\right) \end{bmatrix} \tag{13}$$

where $h(\cdot)$ is the activation function of hidden layer, $\mathbf{W} = [\mathbf{w}_1, \ldots, \mathbf{w}_L] \in \mathfrak{R}^{d \times L}$ and $\mathbf{B} = [b_1, \ldots, b_L]^T \in \mathfrak{R}^L$ are randomly generated input weights and bias between the input layer and hidden layer. With such a hidden layer output matrix $\mathbf{H}$, ELM can be formulated as follows

$$\min_{\boldsymbol{\beta} \in \mathfrak{R}^{L \times c}} \frac{1}{2} \|\boldsymbol{\beta}\|^2 + C \cdot \frac{1}{2} \sum_{i=1}^{N} \|\boldsymbol{\xi}_i\|^2 \tag{14}$$
$$s.t. \ h(\mathbf{x}_i)\boldsymbol{\beta} = \mathbf{t}_i^T - \boldsymbol{\xi}_i^T, \ i = 1, \ldots, N \Leftrightarrow \mathbf{H}\beta = \mathbf{T}^T - \boldsymbol{\xi}^T$$

where $\boldsymbol{\beta} \in \mathfrak{R}^{L \times c}$ denotes the output weights between hidden layer and output layer, $\boldsymbol{\xi} = [\boldsymbol{\xi}_1, \ldots, \boldsymbol{\xi}_N]$ denotes the prediction error matrix with respect to the training data, and C is a penalty constant on the training errors.

The closed form solution $\boldsymbol{\beta}$ of (14) can be easily solved. First, if the number N of training patterns is larger than L, the gradient equation is over-determined, and the closed form solution of (14) can be obtained as

$$\boldsymbol{\beta}^* = \mathbf{H}^+ \mathbf{T} = \left(\mathbf{H}^T \mathbf{H} + \frac{\mathbf{I}_{L \times L}}{C}\right)^{-1} \mathbf{H}^T \mathbf{T} \tag{15}$$

where $\mathbf{I}_{L \times L}$ denotes the identity matrix with size of L, and $\mathbf{H}^+$ is the Moore-Penrose generalized inverse of $\mathbf{H}$.

If the number N of training patterns is smaller than L, an under-determined least square problem would be handled. In this case, the solution of (14) can be obtained as

$$\boldsymbol{\beta}^* = \mathbf{H}^+ \mathbf{T} = \mathbf{H}^T \left(\mathbf{H}\mathbf{H}^T + \frac{\mathbf{I}_{N \times N}}{C} \right)^{-1} \mathbf{T} \tag{16}$$

where $\mathbf{I}_{N \times N}$ denotes the identity matrix.

Then the predicted output of a new observation $\mathbf{z}$ can be computed as

$$\mathbf{y} = h(\mathbf{z})\boldsymbol{\beta}^* = \begin{cases} h(\mathbf{z}) \cdot \left(\mathbf{H}^T\mathbf{H} + \frac{\mathbf{I}_{L \times L}}{C} \right)^{-1} \mathbf{H}^T\mathbf{T}, & \text{if } N \geq L \\ h(\mathbf{z}) \cdot \mathbf{H}^T \left(\mathbf{H}\mathbf{H}^T + \frac{\mathbf{I}_{N \times N}}{C} \right)^{-1} \mathbf{T}, & \text{if } N < L \end{cases} \tag{17}$$

2.4 Kernelized Extreme Learning Machine (KELM)

One can also apply Mercer condition to ELM and thus a KELM is formulated. The KELM can be described as follows. Let $\boldsymbol{\Omega} = \mathbf{H}\mathbf{H}^T \in \mathfrak{R}^{N \times N}$, where $\Omega_{i,j} = h(\mathbf{x}_i)h(\mathbf{x}_j)^T = \kappa(\mathbf{x}_i, \mathbf{x}_j)$ and $\kappa(\cdot)$ is the kernel function. With the expression of solution $\boldsymbol{\beta}$ (16), the predicted output of a new observation $\mathbf{z}$ can be computed as

$$\begin{aligned} \mathbf{y} &= h(\mathbf{z})\boldsymbol{\beta}^* \\ &= h(\mathbf{z}) \cdot \mathbf{H}^T \left(\mathbf{H}\mathbf{H}^T + \frac{\mathbf{I}_{N \times N}}{C} \right)^{-1} \mathbf{T} \\ &= \begin{bmatrix} \kappa(\mathbf{z}, \mathbf{x}_1) \\ \vdots \\ \kappa(\mathbf{z}, \mathbf{x}_1) \end{bmatrix}^T \left(\boldsymbol{\Omega} + \frac{\mathbf{I}_{N \times N}}{C} \right)^{-1} \mathbf{T} \end{aligned} \tag{18}$$

Note that due to the kernel matrix of training data is $\boldsymbol{\Omega} \in \mathfrak{R}^{N \times N}$, therefore, the number L of hidden neurons is not explicit and the decision function of KELM can be expressed uniquely in (18).

3 Training and Testing Protocol

3.1 CNN Training on ImageNet

In this paper, we aim at proposing a comparative investigation on SVMs and ELMs for classification based on deep convolutional features. Therefore, we adopt the deep convolutional activated features (DeCAF) from [17] for experiments. The structures of CNN for training on the ImageNet with 1000 categories are the same as the proposed CNN in [10]. The basic structure of the adopted is illustrated in Fig. 1, which includes 5 convolutional layers and 3 fully-connected layers. Further details of the CNN training architecture and features can be referred to [10, 17].

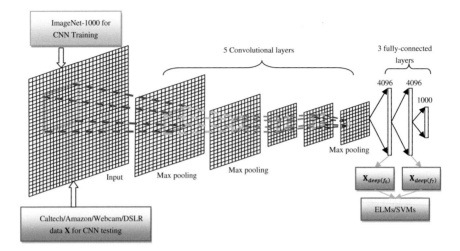

Fig. 1 Diagram of the training and testing protocol in this paper

3.2   CNN Testing

The well-trained network parameters shown in Fig. 1 are used for deep representation of the 4DA (domain adaptation) dataset [31, 32]. The CNN outputs of the 6-th (f_6) and 7-th (f_7) fully-connected layers are used as inputs of SVMs and ELMs for classification, respectively. The 4DA dataset includes four domains such as Caltech 256 (C), Amazon (A), Webcam (W) and Dslr (D) sampled from different sources, in which 10 object classes are selected. As can be seen from Fig. 1, the dimension of features from f_6 and f_7 is 4096. The detail of 4DA dataset with deep features is summarized in Table 1.

3.3   Classification

The 4DA dataset is commonly used for evaluating domain adaptation and transfer learning tasks. So, in this paper, we investigate the classification ability of deep

Table 1 Details of 4DA-CNN datasets

Dataset	#Class	#Dimension	#Samples	n_s/c	n_t/c
Amazon	10	4096	958	20	3
DSLR	10	4096	157	8	3
Webcam	10	4096	295	8	3
Caltech	10	4096	1123	8	3

representation on domain shifted data. We adopt the deep features for SVMs/ELMs training, and compare the classification accuracy. The specific experimental setup is described in Experiments section.

4 Experiments

4.1 Experimental Setup

In the experiment, three settings are investigated respectively, as follows.

1. **Setting 1**: *single-domain* recognition task.
   For example, we train a model on the training data of Amazon, and report the test accuracy on the remaining data of Amazon. As shown in Table 1 (n_s/c), 20, 8, 8, and 8 samples per class are randomly selected for training from Amazon, DSLR, Webcam and Caltech domains, respectively, and the remaining are used as test samples for each domain. 20 random train/test splits are run, and the average recognition accuracy for each method is reported.

2. **Setting 2**: *cross-domain* recognition tasks–source only.
   We perform a cross-domain recognition task. For example, we train a SVM/ELM on the Amazon and test on DSLR, i.e. A → D. Totally, 12 cross-domain tasks among the four domains are conducted. Note that the training data is source data only (source only) without leveraging the data from target domain. The number of training data is 20, 8, 8 and 8 per class for Amazon, DSLR, Webcam and Caltech domains, respectively, when used as source domain. 20 random train/test splits are run, and the average recognition accuracy for each method is reported.

3. **Setting 3**: *cross-domain* recognition tasks—source and target.
   Similar to Setting 2, we perform a cross-domain recognition task. For example, we train a SVM/ELM on the Amazon and test on DSLR, i.e. A → D. Totally, 12 cross-domain tasks among the four domains are conducted. However, the difference from Setting 2 lies in that the training data includes the labeled source data and few labeled target data. The number of training data is 20, 8, 8 and 8 per class for Amazon, DSLR, Webcam and Caltech domains, respectively, when used as source domain. The number of few labeled target data is 3 per class for each domain when they are used as target domain, as shown in Table 1 (n_t/c). 20 random train/test splits are run, and the average recognition accuracy for each method is reported.

4.2   Parameter Setting

To make sure that the best result of each method can be obtained, we have adjusted the parameters. For SVM the penalty coefficient C and kernel parameter σ are set as 1000 and 1, respectively, by using *Libsvm-3.12* toolbox. For LSSVM, the two coefficients are automatically optimized with a grid search by using *LSSVM-1.7* toolbox. For ELM, the penalty coefficient C and the number L of hidden neurons are set as 100 and 5000, respectively. For KELM, the penalty coefficient C and kernel parameter σ are set as 100 and 0.01, respectively. Note that the penalty coefficient C and kernel parameter σ for SVM, ELM, and KELM are adjusted from the set $C = \{1, 100, 10000\}$ and $\sigma = \{0.0001, 0.01, 1, 100\}$.

4.3   Experimental Results

1. Results of **Setting 1**.
   For experimental **Setting 1**, the average accuracy of 20 randomly generated train/test splits for five methods including NN, SVM, LSSVM, ELM and KELM are reported in Table 2. We can observe that the recognition performance based on the deep features from the 6-th layer (f_6) and 7-th layer (f_7) is slightly different. The best two methods are highlighted with bold face. From the comparisons, we can find that ELMs outperforms SVMs and NN methods for all domains, and KELM shows a more competitive performance. Specifically, by comparing KELM and SVM, the improvement in accuracy for the deep features f_6 is 0.8, 0.2, 1.1 and 2.1 % for Amazon, DSLR, Webcam, and Caltech, respectively. For the deep features f_7, the improvement is 1.0, 0.6, 0.8, and 2.5 %, respectively.
2. Results of **Setting 2**.
   Table 3 presents the average recognition accuracy of 20 randomly generated train/test splits based on the experimental setting 2. Totally, 12 cross-domain recognition tasks are conducted. The first two highest accuracies are highlighted in bold face. We can observe that (1) the recognition performance with deep feature f_7 clearly outperforms that of f_6, which demonstrates the effectiveness of "deep"; (2) the performance of ELM and KELM is significantly better than SVM and LSSVM, the average improvement of 12 tasks of KELM is 4 % better than that of SVM. The results demonstrate that for more difficult problems (i.e. cross-domain tasks), the ELM based methods show a more competitive and robust advantage for classification.
3. Results of **Setting 3**.
   The results under experimental Setting 3 are reported in Table 4, from which we can find that ELMs especially KELM outperform other methods. Note that the digits in bold denote the first two highest accuracies for each task. Due to that

Table 2 Recognition accuracy of each method for different domains in **Setting 1**

Method	CNN_layer	Amazon	DSLR	Webcam	Caltech	CNN_layer	Amazon	DSLR	Webcam	Caltech
NN	f_6	91.0 ± 0.3	97.3 ± 0.6	95.0 ± 0.4	75.0 ± 0.4	f_7	92.4 ± 0.2	96.8 ± 0.5	95.3 ± 0.5	76.2 ± 0.5
SVM	f_6	**92.9 ± 0.1**	97.6 ± 0.6	96.7 ± 0.3	83.9 ± 0.4	f_7	93.2 ± 0.1	96.9 ± 0.5	96.5 ± 0.4	83.2 ± 0.5
LSSVM	f_6	**92.9 ± 0.2**	97.5 ± 0.4	96.4 ± 0.4	84.6 ± 0.3	f_7	93.5 ± 0.1	96.3 ± 0.6	95.4 ± 0.4	83.9 ± 0.4
ELM	f_6	**92.9 ± 0.1**	**98.0 ± 0.3**	**97.7 ± 0.2**	**84.8 ± 0.3**	f_7	**93.6 ± 0.1**	**97.2 ± 0.4**	**97.4 ± 0.3**	85.0 ± 0.3
KELM	f_6	**93.7 ± 0.1**	**97.8 ± 0.3**	**97.8 ± 0.2**	**86.0 ± 0.3**	f_7	**94.2 ± 0.1**	**97.5 ± 0.4**	**97.3 ± 0.4**	**85.7 ± 0.3**

Table 3 Recognition accuracy of each method with **Setting 2**, where the training data is from source domain only

Method	CNN_layer	A → D	C → D	W → D	A → C	W → C	D → C	D → A	W → A	C → A	C → W	D → W	A → W
NN	f_6	71.9 ± 0.9	72.0 ± 1.7	92.7 ± 0.5	76.8 ± 0.3	56.6 ± 0.9	64.4 ± 0.4	75.1 ± 0.7	64.0 ± 0.6	78.1 ± 0.8	61.5 ± 1.1	95.8 ± 0.4	65.1 ± 1.0
	f_7	78.7 ± 0.5	75.6 ± 1.3	96.9 ± 0.4	77.2 ± 0.4	66.2 ± 0.5	70.7 ± 0.4	75.0 ± 0.7	66.3 ± 0.8	83.6 ± 0.4	60.7 ± 1.2	95.2 ± 0.4	68.5 ± 0.8
SVM	f_6	79.6 ± 0.7	75.1 ± 1.8	96.7 ± 0.4	79.5 ± 0.4	59.5 ± 0.9	67.3 ± 1.2	77.0 ± 1.0	66.8 ± 1.0	85.8 ± 0.4	67.1 ± 1.1	95.4 ± 0.4	70.6 ± 0.8
	f_7	80.6 ± 0.8	76.4 ± 1.4	96.7 ± 0.4	79.6 ± 0.4	68.1 ± 0.6	74.3 ± 0.6	81.8 ± 0.5	73.4 ± 0.7	86.5 ± 0.5	67.8 ± 1.1	95.3 ± 0.5	71.0 ± 0.8
LSSVM	f_6	77.1 ± 0.9	76.8 ± 1.2	96.1 ± 0.3	77.5 ± 0.6	61.1 ± 0.7	70.6 ± 1.0	80.0 ± 0.8	68.2 ± 1.1	86.5 ± 0.4	67.8 ± 1.2	96.4 ± 0.4	65.5 ± 0.8
	f_7	**82.6 ± 0.5**	79.2 ± 0.8	95.9 ± 0.4	79.8 ± 0.5	66.0 ± 1.3	73.7 ± 0.9	80.8 ± 0.7	72.0 ± 1.1	87.4 ± 0.3	69.9 ± 1.1	95.1 ± 0.3	69.4 ± 0.6
ELM	f_6	80.6 ± 0.6	79.5 ± 1.2	96.7 ± 0.2	80.4 ± 0.3	67.2 ± 0.5	75.6 ± 0.5	83.7 ± 0.4	72.2 ± 0.9	87.3 ± 0.4	70.1 ± 0.9	**97.2 ± 0.3**	71.1 ± 0.6
	f_7	82.3 ± 0.5	**81.2 ± 0.7**	**97.0 ± 0.4**	81.8 ± 0.3	**74.0 ± 0.3**	**79.5 ± 0.2**	**85.8 ± 0.3**	**76.7 ± 0.9**	**88.3 ± 0.2**	**72.3 ± 0.9**	96.8 ± 0.3	72.4 ± 0.8
KELM	f_6	82.3 ± 0.5	80.7 ± 0.9	96.5 ± 0.3	**82.6 ± 0.3**	69.5 ± 0.4	77.8 ± 0.4	85.3 ± 0.4	73.8 ± 1.1	88.0 ± 0.4	**72.3 ± 1.0**	**97.6 ± 0.2**	**72.9 ± 0.7**
	f_7	**84.0 ± 0.4**	**82.2 ± 0.9**	**97.3 ± 0.3**	**83.4 ± 0.2**	**75.7 ± 0.3**	**81.1 ± 0.2**	**87.1 ± 0.2**	**78.2 ± 0.8**	**89.1 ± 0.3**	**73.3 ± 0.9**	96.9 ± 0.3	**74.7 ± 0.8**

A Amazon, C Caltech 256, W Webcam, D Dslr

Table 4 Recognition accuracy of each method with **Setting 3**, where the training data is from both source and target domains

Method	CNN_layer	A → D	C → D	W → D	A → C	W → C	D → C	D → A	W → A	C → A	C → W	D → W	A → W
NN	f_6	89.4 ± 0.7	90.1 ± 0.8	97.0 ± 0.4	78.1 ± 0.4	69.0 ± 0.9	72.8 ± 0.8	83.8 ± 0.5	83.3 ± 0.7	85.4 ± 0.4	86.9 ± 0.6	97.2 ± 0.4	86.1 ± 0.8
	f_7	93.0 ± 0.5	90.9 ± 0.9	98.6 ± 0.2	78.9 ± 0.4	73.6 ± 0.6	75.6 ± 0.4	86.7 ± 0.5	84.0 ± 0.5	87.9 ± 0.2	87.8 ± 0.9	96.3 ± 0.2	89.1 ± 0.6
SVM	f_6	94.5 ± 0.4	92.9 ± 0.8	99.1 ± 0.2	84.0 ± 0.3	81.7 ± 0.5	83.0 ± 0.3	90.5 ± 0.2	90.1 ± 0.2	90.0 ± 0.2	91.5 ± 0.6	97.9 ± 0.3	90.4 ± 0.8
	f_7	94.0 ± 0.6	92.7 ± 0.8	98.9 ± 0.2	83.4 ± 0.4	81.2 ± 0.4	82.7 ± 0.4	90.9 ± 0.3	90.6 ± 0.2	90.3 ± 0.2	90.6 ± 0.8	98.0 ± 0.2	91.1 ± 0.8
LSSVM	f_6	92.6 ± 0.5	93.1 ± 0.6	98.8 ± 0.2	82.3 ± 0.5	80.7 ± 0.5	82.3 ± 0.4	90.9 ± 0.2	89.7 ± 0.2	90.3 ± 0.1	90.9 ± 0.6	97.8 ± 0.3	87.7 ± 0.8
	f_7	91.9 ± 0.5	92.4 ± 0.8	98.4 ± 0.2	82.9 ± 0.4	81.7 ± 0.3	82.6 ± 0.5	90.9 ± 0.4	90.0 ± 0.2	90.7 ± 0.2	90.4 ± 0.5	97.2 ± 0.3	89.5 ± 0.7
ELM	f_6	94.6 ± 0.5	93.7 ± 0.6	**99.2 ± 0.2**	83.4 ± 0.3	81.2 ± 0.3	83.5 ± 0.3	91.1 ± 0.2	90.3 ± 0.2	90.5 ± 0.1	91.6 ± 0.7	**98.3 ± 0.2**	90.5 ± 0.6
	f_7	94.9 ± 0.4	93.0 ± 0.6	99.0 ± 0.2	84.1 ± 0.2	82.2 ± 0.4	84.1 ± 0.2	91.7 ± 0.2	**90.8 ± 0.2**	90.9 ± 0.1	91.5 ± 0.7	97.9 ± 0.2	**91.7 ± 0.7**
KELM	f_6	**95.7 ± 0.4**	94.1 ± 0.6	**99.2 ± 0.2**	85.0 ± 0.3	83.0 ± 0.3	84.9 ± 0.2	91.9 ± 0.2	**90.8 ± 0.2**	**91.1 ± 0.1**	**92.2 ± 0.7**	**98.6 ± 0.2**	91.3 ± 0.6
	f_7	95.5 ± 0.4	93.9 ± 0.6	99.1 ± 0.1	**85.4 ± 0.3**	**83.4 ± 0.3**	**85.3 ± 0.3**	**92.1 ± 0.2**	**91.5 ± 0.2**	**91.5 ± 0.1**	91.9 ± 0.6	98.2 ± 0.3	**92.2 ± 0.6**

A Amazon, *C* Caltech 256, *W* Webcam, *D* Dslr

few labeled data from target domain are leveraged in model training with domain adaptation, so the recognition accuracies are much higher than that from Table 3. The average differences between ELMs and SVMs are therefore reduced from 4 % in **Setting 2** to 1.5 % in **Setting 3**.

5 Conclusion

In the paper, we present a systematic comparison between SVMs and ELMs for object recognition with multiple domains based on the deep convolutional activation features trained by CNN on a subset of 1000-category images from ImageNet. We aim at exploring the most appropriate classifiers for high-level deep features in classification. In experiments, the deep features of 10-category object images of 4 domains from the 6-th layer and 7-th layer of CNN are used as the inputs of general classifiers including NN, SVM, LSSVM, ELM and KELM, respectively. The recognition accuracies for each method under three different experimental settings are reported. A number of experimental results clearly demonstrate that ELMs outperform SVM based classifiers in different settings. In particular, KELM shows state-of-the-art recognition performance among the presented 5 popular classifiers.

Acknowledgements This work was supported in part by the National Natural Science Foundation of China under Grant 61401048, in part by the Hong Kong Scholar Program under Grant XJ2013044, and in part by the China Post-Doctoral Science Foundation under Grant 2014M550457.

References

1. Lecun, Y., Bottou, L., Bengio, Y., Haffner, P.: Gradient-based learning applied to document recognition. Proc. IEEE **86**(11), 2278–2324 (1998)
2. Lawrence, S., Giles, C.L., Ah Chung, T., Back, A.D.: Face recognition: a convolutional neural-network approach. IEEE Trans. Neural Netw. **8**(1), 98–113 (1997)
3. Hinton, G.E., Salakhutdinov, R.R.: Reducing the dimensionality of data with neural networks. Science **313**(5786), 504–507 (2006)
4. Hinton, G., Osindero, S., The, Y.: A fast learning algorithm for deep belief nets. Neural Comput. **18**(7), 1527–1554 (2006)
5. Salakhutdinov, R., Hinton, G.: Deep Boltzman machines. In: Proceedings International Conference Artificial Intelligence and Statistics, pp. 448–455 (2009)
6. Huang, G.B., Lee, H., Learned-Miller, E.: Learning hierarchical representations for face verification with convolutional deep belief networks. In: Proceedings IEEE International Computer Vision and Pattern Recognition, pp. 2518–2525 (2012)
7. Sun, Y., Wang, X., Tang, X.: Hybrid deep learning for face verification. In: Proceedings IEEE International Conference Computer Vision (2013)

8. Taigman, Y., Yang, M., Ranzato, M.A., Wolf, L.: DeepFace: closing the gap to human-level performance in face verification. In Proceedinga IEEE International Computer Vision and Pattern Recognition (2014)
9. Zhou, E., Cao, Z., Yin, Q.: Naïve-deep face recognition: touching the limit of LFW benchmark or not? arXiv:1501.04690 (2015)
10. Krizhevsky, A., Sutskever, I., Hinton, G.E.: ImageNet classification with deep convolutional neural networks. NIPS (2012)
11. Ciresan, D., Meier, U., Schmidhuber, J.: Multi-column deep neural networks for image classification. In Proceedings IEEE International Conference Computer Vision and Pattern Recognition, pp. 3642–3649 (2012)
12. Karpathy, A., Toderici, G., Shetty, S., Leung, T.: Large-scale video classification with convolutional neural networks. In Proceedings IEEE International Conference Computer Vision and Pattern Recognition, pp. 1725–1732 (2014)
13. Razavian, A.S., Azizpour, H., Sullivan, J., Carlsson, S.: CNN features off-the-shelf: an astounding baseline for recognition. In Proceedings IEEE International Conference Computer Vision and Pattern Recognition, pp. 512–519 (2014)
14. Girshick, R., Donahue, J., Darrell, T., Malik, J.: Accurate object detection and semantic segmentation. In Proceedings IEEE International Conference Computer Vision and Pattern Recognition, pp. 580–587 (2014)
15. He, K., Zhang, X., Ren, S., Sun, J.: Spatial pyramid pooling in deep convolutional networks for visual recognition. arXiv:1406.4729
16. Jarrett, K., Kavukcuoglu, K., Ranzato, M., LeCun, Y.: What is the best multi-stage architecture for object recognition? ICCV, pp. 2146–2153 (2009)
17. Donahue, J., Jia, Y., Vinyals, O., Hoffman, J., Zhang, N., Tzeng, E., Darrell, T.: DeCAF: a deep convolutional activation feature for generic visual recognition. arXiv:1310.1531 (2013)
18. Cover, T., Hart, P.: Nearest neighbor pattern classification. IEEE Trans. Inf. Theory 13(1), 21–27 (1967)
19. Vapnik, V.: Statistical Learning Theory. Wiley, New York (1998)
20. Suykens, J.A.K., Vandewalle, J.: Least squares support vector machine classifiers. Neural Process. Lett. 9(3), 293–300 (1999)
21. Huang, G.B., Zhu, Q.Y., Siew, C.K.: Extreme learning machine: theory and applications. Neurocomputing 70, 489–501 (2006)
22. Huang, G.B., Zhou, H., Ding, X., Zhang, R.: Extreme learning machine for regression and multiclass classification. IEEE Trans. Syst. Man Cybern. Part B 42(2), 513–529 (2012)
23. Huang, G.B., Ding, X.J., Zhou, H.M.: Optimization method based on extreme learning machine for classification. Neurocomputing 74(1–3), 155–163 (2010)
24. Zhang, L., Zhang, D.: Evolutionary cost-sensitive extreme learning machine and subspace extension. arXiv:1505.04373 (2015)
25. Huang, G.B.: What are extreme learning machines? Filling the gap between Frank Rosenblatt's dream and John von Neumann's puzzle. Cogn. Comput. 7, 263–278 (2015)
26. Kasun, L.L.C., Zhou, H., Huang, G.B., Vong, C.M.: Representational learning with extreme learning machine for big data. IEEE Intell. Syst. 28(6), 31–34 (2013)
27. Huang, G.-B., Bai, Z., Kasun, L.L.C., Vong, C.M.: Local receptive fields based extreme learning machine. IEEE Comput. Intell. Mag. 10(2), 18–29 (2015)
28. Zhang, L., Zhang, D.: Domain adaptation transfer extreme learning machines. Proceedings in Adaptation, Learning and Optimization 3, 103–119 (2015)
29. Zhang, L., Zhang, D.: Domain adaptation extreme learning machines for drift compensation in E-nose systems. IEEE Trans. Instrum. Meas. 64(7), 1790–1801 (2015)
30. Huang, G., Song, S., Gupta, J.N.D., Wu, C.: Semi-supervised and unsupervised extreme learning machines. IEEE Trans. Cybern. 44(12), 2405–2417 (2014)
31. Saenko, K., Kulis, B., Fritz, M., Darrell, T.: Adapting visual category models to new domains. ECCV (2010)
32. Gong, B., Shi, Y., Sha, F., Grauman, K.: Geodesic flow kernel for unsupervised domain adaptation. CVPR, pp. 2066–2073 (2012)

Learning with Similarity Functions: A Novel Design for the Extreme Learning Machine

Federica Bisio, Paolo Gastaldo, Rodolfo Zunino, Christian Gianoglio and Edoardo Ragusa

Abstract This research analyzes the affinities between two well-known learning schemes that apply randomization in the training process, namely, Extreme Learning Machines (ELMs) and the learning framework using similarity functions. These paradigms share a common approach to inductive learning, which combines an explicit data remapping with a linear separator; however, they seem to exploit different strategies in the design of the mapping layer. This paper shows that the theory of learning with similarity functions can stimulate a novel reinterpretation of ELM, thus leading to a common framework. This in turn allows one to improve the strategy applied by ELM for the setup of the neurons' parameters. Experimental results confirm that the new approach may improve over the standard strategy in terms of the trade-off between classification accuracy and dimensionality of the remapped space.

Keywords Feed-forward neural networks · Extreme learning machine · Similarity functions

F. Bisio (✉) · P. Gastaldo · R. Zunino · C. Gianoglio · E. Ragusa
Department of Electrical, Electronic, Telecommunications Engineering
and Naval Architecture, DITEN, University of Genoa, Genoa, Italy
e-mail: federica.bisio@edu.unige.it

P. Gastaldo
e-mail: paolo.gastaldo@unige.it

R. Zunino
e-mail: rodolfo.zunino@unige.it

C. Gianoglio
e-mail: s3502644@unige.it

E. Ragusa
e-mail: s3523687@studenti.unige.it

© Springer International Publishing Switzerland 2016
J. Cao et al. (eds.), *Proceedings of ELM-2015 Volume 1*,
Proceedings in Adaptation, Learning and Optimization 6,
DOI 10.1007/978-3-319-28397-5_21

1 Introduction

This paper wants to provide new insights on ELM, which has recently emerged as a powerful and flexible paradigm in this context. In practice, the ELM framework implements a Single-hidden-Layer Feedforward Network (SLFN), in which all hidden-nodes parameters are set randomly. This configuration simplifies the learning stage with respect to conventional feedforward NNs, since ELM training just requires to solve a linear system. Nonetheless, the literature shows that such framework can obtain effective results on a wide range of applicative domains [1–6].

The present research will analyze *pro et contra* of the ELM model by exploiting the relationship between such model and the framework discussed in [7], where a learning theory based on similarity functions is formalized. In this regard, two are the interesting aspects that emerge from [7]. First, one can tackle inductive learning with similarity functions instead of kernel functions; as a major consequence, one is no longer tied to functions that (1) span high-dimensional spaces implicitly, and (2) rely on positive semi-definite matrixes. Secondly, similarity functions support a two-stage learning algorithm that shares a common formalism with the ELM model. The first stage involves an explicit mapping of data in a new space whose dimensionality corresponds to the number of units in the hidden layer. In the second stage, a conventional learning algorithm sets a linear classifier in the remapped space.

The reinterpretation of ELM can stimulate novel investigations on the learning model itself. This research focuses in particular on the strategy applied to set the free parameter of the activation/similarity function (e.g., the bias in the sigmoid function). In the standard ELM model, the value of such parameter is different for each neuron, and is assigned on a random basis. However this paper shows that—in the view of the parallel between activation and similarity function—such approach may lead to ineffective mapping layers. Accordingly, a novel strategy for the setup of the free parameters in the mapping layer is presented and empirically validated.

The experimental verification involved three real-world benchmarks: Ionosphere [8], Glass Identification [9], and Statlog Landsat Satellite [9]. Experimental results show the effectiveness of the proposed strategy, which allows one to achieve a better performance in terms of the trade-off between size of the mapping layer (i.e., number of neurons) and generalization error scored by the ELM-based predictor.

2 Connecting ELM with Similarity-Function Learning

2.1 Background: Learning with Similarity Functions

Similarity-based classifiers predict the class of an input sample by exploiting the notion of similarity between two generic patterns $\mathbf{x}, \mathbf{x}' \in x$, were x is the sample domain (in general, $x \subset \mathbb{R}^D$, with $D \in N^+$). Let $T = \{(\mathbf{x}, y)_i; \mathbf{x} \in x; y \in \{-1, 1\};$ $i = 1, ..., Z\}$ be a labeled training set and let $K: x \times x \to \mathbb{R}$ be the function that expresses the notion of similarity $K(\mathbf{x}_m, \mathbf{x}_n)$; then, the classifier will assess the label of a new input sample $\mathbf{x}'$ by utilizing (1) the pairwise similarities between the training patterns and (2) the similarities between $\mathbf{x}'$ and the training patterns. Kernel-based learning machines [10] are a popular and powerful family of such classifiers.

In practice, if K satisfies the notion of "good similarity function" formalized in [7] for the learning problem at-hand, then one can exploit the notion similarity to perform an explicit remapping of the original data in a new space in which exists a low-error large-margin separator. As a major result, it is possible to use standard predictors to find the linear separator in the new space.

For the present research, the crucial elements of the theory of learning with similarity functions proposed in [7] can be summarized by the following definition.

Definition [7] A similarity function K is an (ε, γ)—good similarity function for a learning problem P if there exists a bounded weighting function ω over x ($\omega(\mathbf{x}') \in [0, 1]$ for all $\mathbf{x}' \in$ x) such that at least a $(1 - \varepsilon)$ probability mass of example $\mathbf{x}$ satisfy:

$$E_{\mathbf{x}' \sim P}[\omega(\mathbf{x}')K(\mathbf{x}, \mathbf{x}')|y(\mathbf{x}) = y(\mathbf{x}')] \geq E_{\mathbf{x}' \sim P}[\omega(\mathbf{x}')K(\mathbf{x}, \mathbf{x}')|y(\mathbf{x}) \neq y(\mathbf{x}')] + \gamma \quad (1)$$

When considering the eventual learning algorithm, the crucial aspect is that the definition requires the bounded weighting function ω to exist, but it is not required that such function is known a-priori.

As a major result, a similarity function K that is (ε, γ)—good can support the learning algorithm outlined in Fig. 1, which mainly includes two steps [7]. The first

---

Algorithm 1

Inputs:
    a labelled training set $\mathrm{T} = \{(\mathbf{x}, y)_i; i = 1, .., Z\}$, $\mathbf{x} \in$ x, $y \in \{-1, 1\}$, $Z > 0$
    an unlabelled data set $\mathrm{U} = \{\mathbf{u}_j; j = 1, .., Q\}$, $\mathbf{u} \in$ x, $Q \geq 0$
    a similarity function K
    number of landmarks L

0. *Initialize* - extract L random samples $\mathrm{L} = \{\mathbf{l}_d; d = 1, .., L\}$ from the set

    $\{\mathbf{x}_i, \mathbf{u}_j; i = 1, .., Z; j = 1, .., Q \}$

1. *Mapping* - remap all the patterns $\mathbf{x} \in \mathcal{T}$ by using the following mapping function

    $$\phi(\mathbf{x}) = \left\{ \frac{1}{\sqrt{L}}K(\mathbf{x}, \mathbf{l}_1), .., \frac{1}{\sqrt{L}}K(\mathbf{x}, \mathbf{l}_L) \right\}$$

2. *Learning* - train a linear predictor in the transformed space $\phi : \mathrm{x} \to \mathbb{R}^L$

Fig. 1 The learning scheme that exploits the theory of learning with (ε, γ)—good similarity functions [7]

step implements an explicit remapping of the space x into a new space $\mathbb{R}^L$; here L is the number of *landmarks*, i.e., a subset of samples randomly extracted from the domain distribution $p(x)$ that characterizes P. Thus, both labeled and unlabeled patterns provide an admissible resource of landmarks. For a given pattern, $\mathbf{x}$, the remapping requires to compute the similarity K between the pattern itself and each landmark. In the second step, a linear predictor is trained in the new feature space $\mathbb{R}^L$.

Algorithm 1 exploits the notion of (ε, γ)—good similarity function to remap the original space into a new space where data are separated by a (possibly large) margin with error ε. Then, the task of assessing the weighting function ω is assigned to the linear predictor. The learning abilities of this procedure have been formally analyzed in [7]: if one set $L = 16 \cdot \ln(4/\varepsilon_*)/\gamma^2$, then with probability at least $1 - \varepsilon_*/2$ there exists a low-error ($\leq \varepsilon + \varepsilon_*$) large-margin ($\geq \gamma/2$) separator in the new feature space.

2.2 ELM and Similarity-Function Learning: Common Elements

The ELM model and the learning machine that exploits the theory of learning with (ε, γ)—good similarity functions (Fig. 1) feature two distinct affinities:

1. Hypothesis space: both the frameworks realize a single-hidden layer feedforward neural network. In the case of Algorithm 1, one has

$$f_{SIM}(\mathbf{x}) = \sum_{j=1}^{L} w_j \phi_j(\mathbf{x}, \mathbf{l}_j) \qquad (2)$$

Thus, the hidden layer involves L mapping neurons, one for each landmark.
2. Learning model: both the frameworks apply in the hidden layer an explicit remapping of the original input space x. Thus, the training procedure should only address the setting of the weights w_j that connect the hidden layer to the output.

The final aspect that the two learning models share in common is the use of randomization in the implementation of the explicit mapping stage. The two frameworks, though, apparently utilize different strategies. In Algorithm 1, the design of the mapping stage stems from the theoretical background provided in [7]. Hence, by sampling at random the available dataset one tries to take advantage, with a given probability, of the fruitful properties of the underlying (ε, γ)—good similarity function. In the ELM model, on the other hand, the goal is to set the parameterization of the hidden layer (i.e., of the activation functions) independently of the training data, thus simplifying the learning procedure. In this regard, randomness provides a suitable criterion to achieve the goal while preserving effective generalization performance [11, 12].

3 ELM Model and Similarity-Function Learning: A Common Framework

3.1 Activation Functions as Similarity Functions

In ELM, a neuron utilizes an activation function with preset parameters to remap a pattern $\mathbf{x} \in \mathrm{x}$ into $\mathbb{R}$; hence, $a(\mathbf{x}, \mathcal{R})$: $\mathrm{x} \to \mathbb{R}$ defines the activation function for each neuron, where $\mathcal{R}$ is the set of parameters. Let $a(\mathbf{x}, \mathcal{R})$ be, for example, the RBF function. The parameters to be (randomly) set are the centroid $\mathbf{c}$ and the spread factor ζ. From the point of view of the theory of learning with similarity functions, one may indeed affirm that the neuron uses a similarity function K (i.e., the RBF) to remap the original input $\mathbf{x}$. In this case, the "landmark" is the centroid $\mathbf{c}$, which has been randomly selected between all the admissible points that lay on x. In addition, the parameter ζ simply sets the specific shape of the similarity function to be adopted, as K basically involves a family of functions.

A similar reasoning can be extended to all the activation functions commonly used in ELM. In practice, an activation function defines a notion of similarity between the input $\mathbf{x} \in \mathrm{x}$ and a random vector $\mathbf{x}^*$ that lies on the same domain. Indeed, the activation/similarity function may actually involve a family of functions; accordingly: $a(\mathbf{x}, \mathcal{R}) = K(\mathbf{x}, \mathbf{x}^*, \kappa)$, where κ is the free parameter that characterizes the family of functions at hand.

In the ELM framework, κ is set randomly; nonetheless, in general, each neuron defines its κ and its $\mathbf{x}^*$. Hence, one may re-formalize ELM decision function as follows:

$$f_{ELM}(\mathbf{x}) = \sum_{j=1}^{N} w_j K(\mathbf{x}, \mathbf{x}_j^*, \kappa_j) \tag{3}$$

3.2 Landmarks Placement

The strategy applied by ELM for the placement of the landmarks is apparently in contrast with that suggested in [7]. However, a closer look to the definition of (ε, γ) —good similarity function (see Sect. 2) reveals that landmarks may also be obtained from a random sampling of the overall input space, x.

First, to avoid mixing two issues (landmark placement and parametrization of the similarity function), one may temporary work under the hypothesis that ELM applies the following decision function

$$f_{\overline{ELM}}(\mathbf{x}) = \sum_{j=1}^{N} w_j K(\mathbf{x}, \mathbf{x}_j^*, \kappa^*) \tag{4}$$

Second, it is useful to provide a novel definition of (ε, γ)—good similarity function.

Definition I A similarity function K is an (ε, γ)—good similarity function for a learning problem P and a set of landmarks L if there exists a bounded weighting function ω over x $(\omega(\mathbf{x}') \in [0, 1]$ for all $\mathbf{x}' \in$ x) such that at least a $(1 - \varepsilon)$ probability mass of example $\mathbf{x}$ satisfy:

$$E_{\mathbf{x}' \sim L}[\omega(\mathbf{x}')K(\mathbf{x}, \mathbf{x}')|y(\mathbf{x}) = y(\mathbf{x}')] \geq E_{\mathbf{x}' \sim L}[\omega(\mathbf{x}')K(\mathbf{x}, \mathbf{x}')|y(\mathbf{x}) \neq y(\mathbf{x}')] + \gamma$$

Such definition stresses the fact that one, in principle, is interested in the combination (K, L) most suitable for the learning problem P, which in fact is represented by a training set T. The latter aspect is a major concern when considering, for example, that the size of the training set may be modest, or that the training set may be corrupted by some noise. Therefore, T might not be a consistent source of landmarks.

In this regard, the ELM model suggests a possible strategy to deal with the intricacies of Definition I. Given a training set T and a similarity function K, L may be generated by randomly sampling the input space x. Such approach might not fully inherit the theoretical credentials that characterize the original definition of (ε, γ)—good similarity function. However, in practice, one expects this approach to be quite reliable in a probabilistic sense. Hence, one may conclude that, at this stage, the mapping strategy applied in the ELM framework is fully congruent with that suggested in [7].

3.3 Similarity Function

The mapping function $\phi(\mathbf{x})$ (as per Algorithm 1) does not explicitly involve similarity functions with free parameters. In practice, $\phi(\mathbf{x})$ can be also formulated as $\phi(\mathbf{x}) = \{\eta \cdot K(\mathbf{x}, \mathbf{l}_1, \kappa), \ldots, \eta \cdot K(\mathbf{x}, \mathbf{l}_L, \kappa)\}$. This expression points up that in $\phi(\mathbf{x})$ all the mapping units share a common setting for the free parameter κ.

Indeed, to fit the ELM model, one should actually redefine $\phi(\mathbf{x})$ as follows

$$\phi'(\mathbf{x}) = \{\eta \cdot K(\mathbf{x}, \mathbf{l}_1, \kappa_1), \ldots, \eta \cdot K(\mathbf{x}, \mathbf{l}_L, \kappa_L)\}$$

In general, though, given a similarity function K and a set L of landmarks, one has that $\phi_d(\mathbf{x}) = \alpha_d \cdot \phi'_d(\mathbf{x})$, where α_d is a scalar. That is, the output of the dth unit of $\phi(\mathbf{x})$ differs from the output of the dth unit of $\phi'(\mathbf{x})$ only for a scale factor. Thus, in practice, the ELM model, by setting κ for each mapping unit (neuron), actually defines a specific scale factor for each unit. In terms of the theory of learning with similarity functions, this ultimately means that the mapping function $\phi'(\mathbf{x})$ may convert an (ε, γ)—good similarity function into an (ε, γ')—good similarity

function. Such transformation is a direct consequence of the role played by the scale
factors, which set the span covered by each single mapping unit.

Indeed, this analysis confirms that the learning model implemented by ELM is
fully congruent with that outlined in Fig. 1, which implicitly utilizes a single κ for
each mapping unit.

4 An Improved Mapping Strategy for the ELM Model

The reinterpretation of the ELM model as a learning machine that fits within the
framework discussed in [7] allows one to provide novel insights on the model itself.
In particular, it is interesting to analyse the peculiar opportunity offered by ELM of
setting the free parameter κ of the similarity function K for each mapping unit.

4.1 ELM: Activation Function and Free Parameters

According to the conventional learning model that characterizes ELM, each neuron
actually implements an activation function, which may involve a free parameter. In
fact, if the activation function is interpreted as a similarity function, the role played
by κ changes considerably. To fully understand this issue, one can consider the case
in which the RBF is used as activation/similarity function.

Let $T = \{(\mathbf{x}, y)_i;\ i = 1, \ldots, 18;\ \mathbf{x} \in \mathbb{R}^6\}$ be a balanced training set (i.e., patterns are
evenly distributed among class '+1' and class '−1'). Accordingly, Fig. 2 refers to
the mapping $\mathbb{R}^6 \rightarrow \mathbb{R}$ applied by the dth mapping unit of ELM on the patterns in
T when using (a) the RBF as notion of similarity and (b) a landmark $\mathbf{l}'$ randomly
selected on $\mathbb{R}^6$. Thus, in Fig. 2 the 18 patterns are plotted according to their
similarity to $\mathbf{l}'$. Asterisk markers refer to patterns of class '+1', while circle markers
refer to patterns of class '−1. Figure 2a refers to a case in which all the patterns

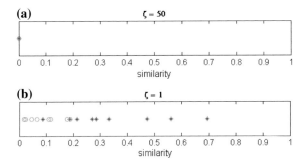

Fig. 2 The mapping applied by the dth unit of ELM on the patterns in T when using the RBF as
notion of similarity: **a** $\zeta = 50$; **b** $\zeta = 1$

collapse in $K(\mathbf{x}, \mathbf{l}') = 0$; that is, all the patterns are very far from the landmark $\mathbf{l}'$ with respect to the standard deviation assigned to the RBF. Figure 2b refers to a set up in which the patterns eventually spread along $\mathbb{R}$; in this case, one is able to distinguish all the patterns.

Overall, this example proves that—given T, K and a landmark—the value assigned to the free parameter κ can heavily influence the behavior of the mapping unit. A mapping unit that collapses all the patterns $\mathbf{x} \in \mathrm{x}$ in one single point $\mathbf{p} \in \mathbb{R}$ is not useful in terms of learning. Actually, the eventual goal of the mapping layer is to project the input samples in a new space in which positive and negative patterns are separable. In practical terms, if μ_+ denotes the barycenter of the positive patterns on $\mathbb{R}$ after the remapping provided by the dth unit, and μ_- denotes the corresponding barycenter of the negative patterns, one expects:

$$\mu_+^{(d)} \neq \mu_-^{(d)} \quad d = 1, \ldots, L \tag{5}$$

where L is the number of mapping units (i.e., the number of landmarks in the mapping layer). Nonetheless, it is clear that—in principle—the best mapping unit is the one that guarantees the largest margin between μ_+ and μ_-.

4.2 Design of a Novel Algorithm for the Setup of Free Parameters

The present research aims at assessing experimentally the consequences of a 'blind' assignment of the free parameters κ_d in ELM. Therefore, it is convenient to compare the conventional strategy applied by ELM for the setup of κ_d with a novel strategy that takes into account the goal of being compliant with (5).

Basically, the objective of the novel strategy is to check the configuration assigned to a mapping unit d; such configuration is considered flawed when

$$\left| \mu_+^{(d)} - \mu_-^{(d)} \right| < \tau \tag{6}$$

where τ is a threshold to be set empirically. In practice, the algorithm should validate the value assigned to κ_d, as both the similarity function K implemented by the unit and the landmark assigned to the unit are considered chosen. Hence, for each mapping unit d, the goal is to find a compliant value κ_d.

Figure 3 outlines the algorithm that implements such strategy. The algorithm receives as input: a labeled training set T = $\{(\mathbf{x}, y)_i;\ i = 1, \ldots, Z\}$, an activation/similarity function K, an admissible range of values for $\kappa = [\kappa_{inf}, \kappa_{sup}]$, the threshold τ, and a target value L for the dimensionality of the mapping layer; the latter parameter corresponds to the number of neurons according to the conventional ELM notation, i.e., the number of landmarks in the theory of learning with similarity functions. For each mapping unit, the algorithm first generates the

Algorithm 2

1. *Landmarks* - generate the set of landmarks $L = \{l_d; d = 1,..,L\}$

   $l_{d,n} = \text{rand}(-1,1)$

2. *Free parameters* - for each mapping unit, set κ_d by applying the following routine

   $X_+ = \{\}; X_- = \{\}$
   $ok = 0$
   while $ok == 0$
   $\quad \kappa_d = \text{rand}(\kappa_{inf}, \kappa_{sup})$
   $\quad$ **for** $i = 1$ **to** Z
   $\quad\quad k_i = K(x_i, l_d, \kappa_d)$
   $\quad\quad$ **if** $y_i == 1$ **then** $k_i \rightarrow X_+$ **else** $k_i \rightarrow X_-$
   $\quad$ **end**
   $\quad \mu_+ = \text{mean}(X_+); \mu_- = \text{mean}(X_-)$
   $\quad$ **if** $|\mu_+ - \mu_-| > \tau$ **then** $ok = 1$
   end

Fig. 3 The proposed algorithm for the setup of free parameters in ELM

corresponding landmark by using the standard ELM strategy; thus, a random number generator is exploited. Without loss of generality, it is assumed that $x = [-1, 1]^T$. Then, the routine for the selection of κ_d starts: (1) choose a random value in the suggested range, and (2) validate such value. The routine stops when the second step completes with success.

The proposed algorithm clearly is less efficient of the original procedure applied by ELM in terms of computational complexity, as the number of attempts required to find a suitable κ for each mapping unit is not predictable. On the other hand, it seems quite interesting to evaluate also the advantages that the strategy applied in Algorithm 2 can offer in terms of the trade-off between generalization performance and size of the mapping layer.

5 Experimental Results

The experimental section aims at evaluating the ability of Algorithm 2 (as per Fig. 3) to improve the overall performance of the ELM model in terms of classification accuracy. To the purpose of robustly assessing such aspect, three different benchmarks have been involved in the experimental evaluation: Ionosphere [8], Glass Identification [9], and Statlog Landsat Satellite [9].

Each experimental session has been designed to provide a fair comparison between the generalization performances of two ELM models: the one that applies the conventional strategy in the setup of free parameters and the one that exploits Algorithm 2. Therefore, in each experiment, the two implementations of ELM have been compared by defining a common configuration for both the range of admissible λs (i.e., the regularization parameter), and the dimensionality, L, of the remapped space (i.e., the number of landmarks/neurons):

- $\Lambda \in \{1 \times 10^{-6}, 1 \times 10^{-5}, 1 \times 10^{-4}, 1 \times 10^{-3}, 1 \times 10^{-2}, 1 \times 10^{-1}, 1, 1 \times 10^{1},$ $1 \times 10^{2}, 1 \times 10^{3}, 1 \times 10^{4}, 1 \times 10^{5}\}$
- $L \in \{50, 100, 200, 500, 1000\}$

5.1 Ionosphere Dataset

The Ionosphere dataset includes a total of 351 patterns, which lie in a 34-dimensional space. In the proposed experimental design, both the training set and the test set include 50 patterns per class; all the 34 features are renormalized in the interval $[-1, 1]$.

Two different activation/similarity functions have been involved in the session: sigmoid function and RBF. Thus, the free parameters are the bias b and the spread factor ζ, respectively. In both cases the enhanced ELM implementation exploits Algorithm 2 with threshold $\tau = 0.3$.

Figure 4 provides the outcomes of the experiments. The graph compares, for each admissible value of L, the performance of the standard ELM (dark markers) with the performance of the enhanced ELM (light markers). In particular, given a value of L (x axis), four predictors are plotted: standard ELM with sigmoid activation/similarity function (marker: asterisk), enhanced ELM with sigmoid activation/similarity function (marker: circle), standard ELM with RBF activation/similarity function (marker: square), enhanced ELM with RBF activation/similarity function (marker: triangle). The y axis gives the classification error: the performance of a predictor is assessed by the average classification error on the test set over 50 different rounds; that is, 50 different randomizations of the mapping layer in the ELM model. Accordingly, the graph also provides the confidence interval $\pm\sigma$.

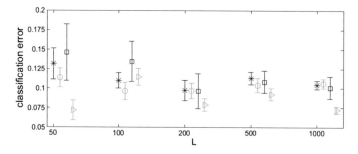

Fig. 4 Results of the experiments involving the Ionosphere dataset

Overall, the graphs clearly show that the enhanced ELM can improve over standard ELM in terms of classification performance. Indeed, two are the interesting outcomes of the experimental session. The first conclusion is that Algorithm 2 offers more improvement when using RBF as activation/similarity function than with the sigmoid as activation function. The second conclusion is that the enhanced ELM based on RBF is actually the best predictor, with classification error of 7.2 % ($L = 50$).

5.2 Glass Dataset

The Glass Identification dataset includes 214 samples that lie in a 9-dimensional space. The benchmark involves a multi-class problem, as six different classes are represented in the dataset; the experiments presented here, though, only addresses a binary classification problem, namely, class 1 versus class 2. In the proposed experimental design, both the training set and the test set include 30 patterns per class randomly extracted from the original dataset. All the 9 features were renormalized in the interval $[-1, 1]$.

The experimental session has been organized by following the same design of the one involving the Ionosphere dataset. As above, both the sigmoid function and the RBF are involved in the experiments; the enhanced ELM exploits Algorithm 2 with threshold $\tau = 0.3$.

Figure 5 provides the outcomes of the experiments. The format adopted for the graphs replicates the one used for the previous experimental session. In general, these results confirm the tendency identified with the Ionosphere dataset. However, this run seems to show that the improvement provided by the enhanced ELM is less sharp in this case. In addition, with this dataset the RBF does not seem able to outperform the sigmoid function, which supported a predictor that scored a classification error of 25.0 % as best result ($L = 200$).

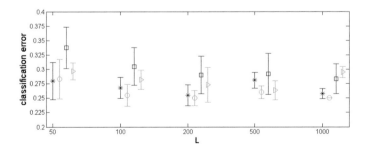

Fig. 5 Results of the experiments involving the Glass dataset

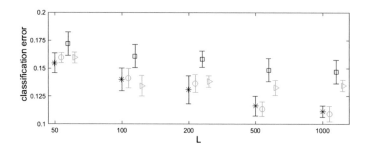

Fig. 6 Results of the experiments involving the Landsat dataset

5.3 *Landsat Dataset*

The Landsat satellite dataset provides a training set including 4435 samples and a test set including 2000 samples; data are drawn from a 36-dimensional space. The original benchmark involves a multi-class problem, but the present experiments only address a binary classification problem: class 4 versus class 7. In the proposed experimental design, the training set includes 300 patterns per class randomly extracted from the original training database; the test set includes 150 patterns per class randomly extracted from the original test database. All the 36 features have been renormalized in the interval $[-1, 1]$.

As above, both the sigmoid function and the RBF are involved in the experiments; the enhanced ELM exploits Algorithm 2 with threshold $\tau = 0.3$. Figure 6 gives the outcomes of the experiments. The format adopted for the graphs replicates the one used for the previous experimental sessions. The experiments confirm that the enhanced ELM can always outperform the standard ELM. In the case of RBF as activation/similarity function, the gap is more relevant. One the other hand, the best results in terms of classification error have been obtained with the sigmoid function as activation/similarity function: 10.9 % ($L = 1000$).

6 Conclusions

This research showed that the theory of learning with similarity functions can provide the basis for the development of novel insights on the ELM model. The crucial outcome is that it is possible to reinterpret the ELM mapping layer by introducing the concepts of similarity function and landmark. As a major result, the learning scheme applied by ELM can be described as a viable strategy to search for a consistent (ε, γ)—good similarity function.

Within this context, the paper suggested a possible enhancement to such strategy by focusing on the peculiar role played by the free parameter of the activation/similarity function. The proposed upgrading addressed a validation of the actual effectiveness of a mapping unit. The basic directive is that each mapping unit should avoid the remapped patterns to collapse in a single point, as such configuration would hamper any distinction between samples belonging to different classes. Empirical evidence supported the proposed solution.

References

1. Decherchi, S., Gastaldo, P., Leoncini, A., Zunino, R.: Efficient digital implementation of extreme learning machines for classification. IEEE Trans. Circ. Syst. **II**(50), 496–500 (2012)
2. Gastaldo, P., Pinna, L., Seminara, L., Valle, M., Zunino, R.: A Tensor-based pattern-recognition framework for the interpretation of touch modality in artificial skin sys-tems. IEEE Sens. **14**, 2216–2225 (2014)
3. Poria, S., Cambria, E., Winterstein, G., Huang, G.-B.: Sentic patterns: dependency-based rules for concept-level sentiment analysis. Knowl. Based Syst. **69**, 45–63 (2014)
4. Chen, H., Peng, J., Zhou, Y., Li, L., Pan, Z.: Extreme learning machine for ranking: generalization analysis and applications. Neural Netw. **53**, 119–126 (2014)
5. Grigorievskiy, A., Miche, Y., Ventelä, A.-M., Séverin, E., Lendasse, A.: Long-term time series prediction using OP-ELM. Neural Netw. **51**, 50–56 (2014)
6. Cambria, E., Gastaldo, P., Bisio, F., Zunino, R.: An ELM-based model for affective ana-logical reasoning. Neurocomputing **149A**, 443–455 (2015)
7. Balcan, M.F., Blum, A., Srebro, N.: A theory of learning with similarity functions. Mach. Learn. **72**, 89–112 (2008)
8. http://www.csie.ntu.edu.tw/~cjlin/libsvmtools/datasets/binary.html
9. http://www.csie.ntu.edu.tw/~cjlin/libsvmtools/datasets/multiclass.html
10. Hofmann, T., Schölkopf, B., Smola, A.J.: Kernel methods in machine learning. Ann. Stat. **36**, 1171–1220 (2008)
11. Liu, X., Lin, S., Fang, J., Xu, Z.: Is extreme learning machine feasible? A theoretical assessment (Part I). IEEE Trans. Neural Netw. Learn. Syst. **26**, 7–20 (2015)
12. Liu, X., Lin, S., Fang, J., Xu, Z.: Is extreme learning machine feasible? A theoretical assessment (Part II). IEEE Trans. Neural Netw. Learn. Syst. **26**, 21–34 (2015)

A Semi-supervised Low Rank Kernel Learning Algorithm via Extreme Learning Machine

Bing Liu, Mingming Liu, Chen Zhang and Weidong Wang

Abstract Semi-supervised kernel learning methods have been received much more attention in the past few years. Traditional semi-supervised *Non-Parametric Kernel Learning* (NPKL) methods usually formulate the learning task as a Semi-Definite Programming (SDP) problem, which is very time consuming. Although some fast semi-supervised NPKL methods have been proposed recently, they usually scale very poorly. Furthermore, many semi-supervised NPKL methods are developed based on the *manifold assumption*. But, such an assumption might be invalid when handling some high-dimensional and sparse data, which has severely negative effect on the performance of learning algorithms. In this paper, we propose a more efficient semi-supervised NPKL method, which can effectively learn a low-rank kernel matrix from must-link and cannot-link constraints. Specially, by virtue of the nonlinear spectral embedded technique based on extreme learning machine (ELM), the proposed method has the ability of coping with data points that do not have a clear manifold structure in a low dimensional space. The proposed method is formulated as a trace ratio optimization problem, which is combined with dimensionality reduction in ELM feature space and aims to find optimal low-rank kernel matrices. The proposed optimization problem can be solved much more efficiently than SDP solvers. Extensive experiments have validated the superior performance of the proposed method compared to state-of-the-art semi-supervised kernel learning methods.

B. Liu (✉) · C. Zhang · W. Wang
School of Computer Science and Technology, China University of Mining and Technology,
Xuzhou, Jiangsu, China
e-mail: liubing@cumt.edu.cn

C. Zhang
e-mail: zc@cumt.edu.cn

W. Wang
e-mail: wdwang@cumt.edu.cn

M. Liu
School of Information and Electrical Engineering, China University of Mining and
Technology, Xuzhou, Jiangsu, China
e-mail: liumingming2004@126.com

© Springer International Publishing Switzerland 2016
J. Cao et al. (eds.), *Proceedings of ELM-2015 Volume 1*,
Proceedings in Adaptation, Learning and Optimization 6,
DOI 10.1007/978-3-319-28397-5_22

Keywords Kernel learning · Low-rank kernel · Spectral embedding · Clustering

1 Introduction

Kernel learning is one of the fundamental topics in machine learning and pattern recognition. It is crucial to choose appropriate kernel functions or kernel matrices in many kernel-based machine learning methods. Unfortunately, kernel learning cannot be addressed effectively in unsupervised settings since it is extremely difficult to construct a well-defined optimization problem in the absence of supervisory information. Since supervisory information is not generally available for learning tasks, side information, such as pairwise constraints, is usually substituted for class labels. The goal of semi-supervised kernel learning is to effectively learn appropriate kernel functions or kernel matrices by utilizing some limited supervisory information as well as a large amount of unlabeled data.

In the last few years, there has been a growing research interest in semi-supervised kernel learning. Some kernel-based nonlinear metric learning methods have been introduced to learn a kernel or equivalently a nonlinear transformation based on the global Mahalanobis metric [1]. Few algorithms [2, 3] are proposed to learn parameters of parametric kernels. Some other studies, such as semi-supervised multiple kernel learning (MKL) [4], aims at learning a convex combination of several predefined base kernels or matrices. Since these methods generally assume the target kernel is of some parametric forms with several fixed kernels or graph Laplacian matrices, the choice of the target kernel matrices is limited, which decreases the flexibility of kernel learning algorithms and limits their capacity of fitting diverse patterns.

Recently, Non-Parametric Kernel Learning (NPKL) methods, which aim to learn a Positive Semi-Definite (PSD) kernel matrix directly from data, have been actively explored. NPKL not only avoids the parametric form of the target kernel, but is easy to incorporate prior/side information into its learning model. Thus, the learned kernel can characterize the data similarity very well. By means of the pairwise constraints and unlabeled data, Hoi et al. [5] and Li et al. [6] have developed NPKL methods early. Nevertheless, these methods need to solve a Semi-Definite Programming (SDP) problem because of the PSD constraint. To improve the efficiency and scalability of semi-supervised NPKL, Zhuang et al. [7] have proposed a Simple NPKL method. By virtue of trace ratio maximization and the pairwise constraints, Baghshah et al. [8] have introduced a more efficient semi-supervised low rank NPKL method by constructing more appropriate optimization model. Nonetheless, these efficient methods heavily depend on the manifold assumption [9, 10], namely, that two nearby data points of a low-dimensional manifold have the same class label. Such an assumption may not hold due to the bias caused by the curse of dimensionality. In other words, NPKL methods lack of dimensionality reduction mechanism to exploit the complex topological structure of data without the clear low-dimensional manifold, which would result in the inaccuracies of the target kernels.

In this paper, we present a novel NPKL method to address the manifold assumption invalidation issue, which still maintains the efficiency and scalability of semi-supervised NPKL methods proposed by Baghshah et al. The proposed method is formulated as a trace ratio optimization problem by incorporating the pairwise constraints and the structure of the data. More importantly, we impose a linearity regularization with the nonlinear spectral embedded technique on the objective function. Motivated by the efficiency of extreme learning machine (ELM), we utilize random activation functions of ELM as nonlinear embedding functions. Thus, the proposed method is much more efficient than those methods with standard SDP solvers by trace ratio optimization algorithms. Compared to the nonlinear distance metric semi-supervised learning methods, the proposed low-rank kernel learning method does not need an initial kernel and directly learns a non-parametric low-rank target kernel matrix. In addition, compared to the existing efficient semi-supervised NPKL methods introduced in Refs. [5–7, 11, 12], our method is more robust to some high-dimensional and sparse data without clear low-dimensional manifold structure and also applicable to large scale problems.

The rest of this paper is organized as follows: in Sect. 2, the proposed algorithm is introduced. In this section, we formulate an optimization problem using constraints and the topological structure of the data and solve this problem to find an appropriate low-rank kernel matrix. Moreover, we show the relation between the spectral clustering methods and low-rank kernel learning. Experimental results are presented in Sect. 4. Finally, we give concluding remarks in Sect. 5. In order to avoid confusion, we give a list of the main notations used in this paper in Table 1.

Table 1 Notations

Notations	Descriptions
$\mathbb{R}^d$	The input d-dimensional Euclidean space
n	The number of total training data points
c	The number of classes that the samples belong to
X	$X = [x_1, \ldots, x_n] \in \mathbb{R}^{d \times n}$ is the training data matrix
Y	$Y = [y_1, \ldots, y_n]^T \in \mathbb{B}^{n \times c}$ is the 0–1 class assignment matrix
ϕ	$\phi(x) = (\psi_1(x_1), \ldots, \psi_n(x_n))$ is the transformed data to the kernel space
$k(x, y)$	Kernel function of variables x and y
K	Kernel matrix $K = [k(x_i, x_j)]_{n \times n} = \phi^T \phi$
e_i	The ith column of the $n \times n$ identity matrix
$\text{tr}(A)$	The trace of the matrix A, that is, the sum of the diagonal elements of the matrix A

2   Related Works

2.1   Extreme Learning Machine

The output function of ELM for generalized SLFNs in the case of one output node is

$$f_L(\mathbf{x}) = \sum_{i=1}^{L} \beta_i h_i(\mathbf{x}) = \mathbf{h}(\mathbf{x})\boldsymbol{\beta}, \tag{1}$$

where $\boldsymbol{\beta} = [\beta_1, \ldots, \beta_L]^T$ is the vector of the output weights between the hidden layer of L nodes and the output node, and $\mathbf{h}(\mathbf{x}) = [h_1(\mathbf{x}), \ldots, h_L(\mathbf{x})]$ is the output (row) vector of the hidden layer with respect to the input x. In fact, $\mathbf{h}(\mathbf{x})$ maps the data from the d-dimensional input space to the L-dimensional hidden-layer feature space (ELM feature space) $\mathbf{H}$. ELM is to minimize the training error as well as the norm of the output weights [13]

$$\min_{\beta} \frac{C}{2} \|\mathbf{H}\boldsymbol{\beta} - \mathbf{T}\|^2 + \frac{1}{2} \|\boldsymbol{\beta}\|^2, \tag{2}$$

where C is a tradeoff parameter between the complexity and fitness of the decision function and $\mathbf{H}$ is the hidden-layer output matrix denoted by

$$\mathbf{H} = \begin{bmatrix} \mathbf{h}(\mathbf{x}_1) \\ \mathbf{h}(\mathbf{x}_2) \\ \vdots \\ \mathbf{h}(\mathbf{x}_n) \end{bmatrix} = \begin{bmatrix} h_1(\mathbf{x}_1) & \ldots & h_L(\mathbf{x}_1) \\ h_1(\mathbf{x}_2) & \ldots & h_L(\mathbf{x}_2) \\ \vdots & \vdots & \vdots \\ h_1(\mathbf{x}_n) & \ldots & h_L(\mathbf{x}_n) \end{bmatrix}. \tag{3}$$

Similar to support vector machine (SVM), to minimize the norm of the output weights $\|\boldsymbol{\beta}\|$ is actually to maximize the distance of the separating margins of the two different classes in the ELM feature space: $2/\|\boldsymbol{\beta}\|$, which actually controls the complexity of the function in the ELM feature space.

2.2   Locality Preserving Regularization

In this section, we use graph Laplacian to represent the geometrical structure of data in input space. To preserve the intrinsic geometric structure of the data in transformed feature space, we define $Q(\boldsymbol{\phi}, \mathbf{X})$ to measure how closely the mapping $\boldsymbol{\phi}$ can preserve the topological structure of the data points in $\mathbf{X}$.

Specifically, denote an undirected weighted graph by $\mathbf{G} = \{\mathbf{X}, \mathbf{W}\}$, where $\mathbf{X}$ is a vertex set and $\mathbf{W} \in \mathbb{R}^{n \times n}$ represents an affinity matrix. Each entry W_{ij} of the symmetric matrix $\mathbf{W}$ is used to record the edge weights that characterize the similarity relationship between a pair of vertices of G. In this paper, we use the k-nearest neighbor method to define W [14]:

$$W_{ij} = \begin{cases} 1 & \text{if} (x_i \in N_k(x_j) \bigwedge x_j \in N_k(x_i)) \\ 0 & \text{otherwise} \end{cases} \tag{4}$$

where $N_k(x_i)$ denotes the k-nearest neighbors set of x_i.

The smoothness of the mapping $\phi(x)$ on the data points, which captures the local dependency between ψ_i and ψ_j, can be defined as

$$Q(\phi, X) = \frac{1}{2} \sum_{i,j=1}^{n} w_{ij} \left\| \frac{\psi_i}{\sqrt{D_i}} - \frac{\psi_j}{\sqrt{D_j}} \right\|_2^2 = tr(\phi L \phi^T) = tr(KL), \tag{5}$$

where L is the normalized graph Laplacian matrix defined as:

$$L = I - D^{-1/2} W D^{-1/2}, \tag{6}$$

where $D = diag(D_1, \ldots, D_n)$ is a diagonal matrix with the diagonal elements defined as $D_i = \sum_{j=1}^{n} w_{ij} \cdot Q(\phi, X)$ essentially measures the weighted sum of the squared distances between the neighboring data points in the kernel space.

3 Semi-supervised Nonlinear Embedded Low Rank Kernel Learning

3.1 Formulation

We first denote ML and CL as the must-link and cannot-link pairwise constraints set respectively. To effectively utilize pairwise constraints information, we define $S_{ML}(K)$ as the sum of the squared distances between pairs of similar data in the kernel space as follows.

$$\begin{aligned} S_{ML}(K) &= \sum_{(x_i, x_j) \in ML} \left\| \psi(x_i) - \psi(x_j) \right\|_2^2 \\ &= \sum_{(i,j) \in ML} K_{ii} + K_{jj} - 2K_{ij} \\ &= \sum_{(e_i, e_j) \in ML} (e_i - e_j)^T K (e_i - e_j) \\ &= \sum_{(e_i, e_j) \in ML} tr\left((e_i - e_j)(e_i - e_j)^T K \right) = tr(K E_{ML}) \end{aligned} \tag{7}$$

where $E_{ML} = (e_i - e_j)(e_i - e_j)^{\mathrm{T}}$, and e_i denotes a column vector, in which the ith element is one when the ith sample point has been chosen into set ML (must-link), while others are zero. Similarly, $S_{CL}(K)$, which denotes the sum of the squared distances between pairs of dissimilar data in the kernel space, is defined as:

$$S_{CL}(K) = \sum_{(x_i, x_j) \in CL} \left\| \psi(x_i) - \psi(x_j) \right\|_2^2 = \mathrm{tr}(KE_{CL}) \tag{8}$$

where $E_{CL} = (e_i - e_j)(e_i - e_j)^{\mathrm{T}}$.

It is desirable that the target kernel matrix should be consistent with small distances between similar pairs and large distances between dissimilar pairs in the kernel space. Furthermore, it should preserve the intrinsic geometric structure of the data in the kernel space. In addition, we make good use of the nonlinear spectral embedded method to improve the performance of semi-supervised NPKL methods and enhance the scalability of them through optimizing a low rank kernel matrix. We can present the kernel matrix in terms of a decomposition $K = FF^{\mathrm{T}}$, where F is a $n \times r$ matrix and $r < n$. It should be noted that F can be considered as a class assignment matrix if we set r as the number of class. The semi-supervised low rank kernel learning optimization problem is defined as follows:

$$\max_{K \geqslant 0} \frac{S_{CL}(K)}{S_{ML}(K) + \alpha Q(\phi, X) + \mu \left(\|H\beta - F\|^2 + \gamma_g \mathrm{tr}(\beta^T \beta) \right)} \tag{9}$$

where α and μ are adjusting parameters. The term $Q(\phi, X)$ characterizes the topological structure of the data in the kernel space. The term $\|H\beta - F\|^2 + \gamma_g \mathrm{tr}(\beta^T \beta)$ uses the nonlinear transformation of ELM to control the error between the cluster assignment matrix and the low-dimensional embedding of the data. The low rank kernel learning method proposed in Ref. [15] has added the constraint item rank $(K) \leq r$. By substituting (5), (7) and (8) into (9), the problem (9) can be rewritten as:

$$\max_{F \in \mathbb{R}^{n \times r}} \frac{\mathrm{tr}(F^{\mathrm{T}} E_{CL} F)}{\mathrm{tr}(F^{\mathrm{T}}(E_{ML} + \alpha L)F) + \mu(\|H\beta - F\|^2 + \gamma_g \mathrm{tr}(\beta^T \beta))} \tag{10}$$

Similar to the method in Ref. [11], we add the orthogonal constraint $F^{\mathrm{T}}F = I_r$ to avoid the trivial solution. The problem (10) can be reformulated as

$$\max_{F^{\mathrm{T}}F = I_r} \frac{\mathrm{tr}(F^{\mathrm{T}} E_{CL} F)}{\mathrm{tr}(F^{\mathrm{T}}(E_{ML} + \alpha L)F) + \mu \left(\|H\beta - F\|^2 + \gamma_g \mathrm{tr}(\beta^T \beta) \right)} \tag{11}$$

3.2 Algorithm

To solve the optimization (11), we need to transform it into another simple form. We have the following theorem.

Theorem 1 *The optimization problem (11) can be transformed into the following minimization problem:*

$$\max_{F^{\mathrm{T}}F=I_r} \frac{\mathrm{tr}(F^{\mathrm{T}}E_{CL}F)}{\mathrm{tr}(F^{\mathrm{T}}(E_{ML}+\alpha L+\mu L_H)F)} \tag{12}$$

where $L_H = I_n - HH^T(\gamma_g I_n + HH^T)^{-1}$ or $L_H = I_n - H(\gamma_g I_L + H^T H)^{-1} H^T \cdot I_L$ represents the identity matrix of size c by c and L is the number of hidden layer nodes in ELM.

Proof By setting the derivatives of the objective function (11) with respect to β to zero, we have

$$\beta = H^T(\gamma_g I_n + HH^T)^{-1}F \tag{13}$$

By substituting β in (11) by (13), the optimization problem (11) becomes

$$\max_{F^{\mathrm{T}}F=I_r} \frac{\mathrm{tr}(F^{\mathrm{T}}E_{CL}F)}{\mathrm{tr}(F^{\mathrm{T}}(E_{ML}+\alpha L)F) + \mu tr\left(F^T\left(I_n - HH^T(\gamma_g I_n + HH^T)^{-1}\right)F\right)} \tag{14}$$

which can be denoted as follows:

$$\max_{F^{\mathrm{T}}F=I_r} \frac{\mathrm{tr}\left(F^{\mathrm{T}}E_{CL}F\right)}{\mathrm{tr}\left(F^{\mathrm{T}}(E_{ML}+\alpha L+\mu L_H)F\right)} \tag{15}$$

where $L_H = I_n - HH^T(\gamma_g I_n + HH^T)^{-1}$ and L_H can be transformed into another form as follows:

$$L_H = I_n - H(\gamma_g I_L + H^T H)^{-1} \cdot (\gamma_g I_L + H^T H) \cdot H^T(\gamma_g I_n + HH^T)^{-1}$$
$$= I_n - H(\gamma_g I_L + H^T H)^{-1}H^T$$

This completes the proof of Theorem 1.   □

Obviously, problem (12) is the orthogonally constrained trace ratio optimization problem. Although it can be solved by using generalized eigenvalue decomposition. But, prior woks in Ref. [16] suggest that it is more reasonable to solve it directly. Since E_{CL} and $E_{ML} + \alpha L + \mu L_H$ are PSD matrices, the algorithm proposed in Ref. [17] can be used to solve problem (12). Table 2 summarizes the proposed

Table 2 LRKL-ESE
algorithm

Input: A dataset $\mathcal{X} = \{x_i\}_{i=1}^n \in \mathbb{R}^{n \times d}$, must-link set ML and cannot-link set, the number of class c, the number of the nearest k, parameters α, μ and γ_g, an error constant ε.
Output: $K^* = F^* F^{*T} \in \mathbb{R}^{n \times n}$, where $F^* = \text{argmax}_{F^T F = I_c} \text{tr}(F^T E_{CL} F) / \text{tr}(F^T (E_{ML} + \alpha L + \mu L_H) F)$.
Step 1: Construct the weighted matrix W using the k-nearest neighbors method.
Step 2: Calculate Laplacian matrix $L = I - D^{-1/2} W D^{-1/2}$.
Step 3: Randomly generate input weights $\{(a_i, b_i)\}_{i=1}^L$ and initiate an ELM network of L hidden neurons, calculate the output matrix of the hidden layer.
Step 4: if $L \le n$ let $L_H = I_n - H(\gamma_g I_L + H^T H)^{-1} H^T$ else $L_H = I_n - H H^T (\gamma_g I_n + H H^T)^{-1}$.
Step 5: Perform the eigen-decomposition on E_{CL} and $E_{ML} + \alpha L + \mu L_H$. $k = [k_1, k_2, \ldots, k_c]$ is a matrix composed of the first c largest eigenvalues of E_{CL}, while $t = [t_1, t_2, \ldots, t_c]$ composed of the first c smallest eigenvalues of $E_{ML} + \alpha L + \mu L_H$.
Step 6: Let $\lambda_1 = \text{tr}(T_1) / \text{tr}(T_2)$ and $\lambda_2 = \sum_{i=1}^c k_i / \sum_{i=1}^c t_i, \lambda = (\lambda_1 + \lambda_2)/2$.
Step 7: repeat Compute r(λ) as the sum of the first c largest eigenvalues of $E_{CL} - \lambda(E_{ML} + \alpha L + \mu L_H)$ if r(λ) > 0 then $\lambda = \lambda_1$, otherwise $\lambda = \lambda_2$ $\lambda = (\lambda_1 + \lambda_2)/2$ until $\lambda_1 - \lambda_2 > \varepsilon$.
Step 8: $F^* = [z_1, z_2, \ldots, z_c]$ constructed by c eigenvectors corresponding the first c largest eigenvalues of $E_{CL} - \lambda(E_{ML} + \alpha L + \mu L_H)$. **return** the kernel matrix $K^* = F^* F^{*T}$.

algorithm. Since the proposed semi-supervised nonlinear spectral embedded low rank kernel learning method is based on ELM, we call it as LRKL-ESE in the following sections.

3.3 Computational Complexity

From the LRKL-ESE algorithm, we can see that the most costly computation is carrying out the eigen-decomposition of E_{CL} and $E_{ML} + \alpha L + \mu L_H$. Since the number of constraints is usually very low, the matrix E_{CL} is sparse and its eigen-decomposition can be performed more quickly than the matrix $E_{ML} + \alpha L + \mu L_H$ [7]. Consequently, the computational complexity of the proposed algorithm mainly depends on the time cost of obtaining the c eigenvectors corresponding to c largest eigenvalues of $E_{ML} + \alpha L + \mu L_H$, which is $O(n^3)$. In addition, the computational complexity of computing L_H is $O(L^3)$, where $L \ll n$. Therefore, the total computational complexity of the proposed algorithm is $O(n^3)$, which is much lower than that of the SDP problem solvers that can be as high as $O(n^{6.5})$.

4 Experiments

In this section, we test the proposed algorithm on some benchmark datasets to evaluate the performance of our method. To compare the performance of different kernel learning methods, we apply the kernel k-means algorithm on the obtained kernels. Thus, the performance of kernel learning methods can be evaluated by the clustering results. Specifically, we first perform the kernel k-means clustering algorithm on the kernels learned by our method, termed as (LRKL-ESEC). Then, LRKL-ESEC is compared with the traditional K-means without metric learning algorithm (Euclidean), kernel k-means with the low-rank kernel learned by Baghshah et al.'s method (LRKL) [11], kernel k-means with the kernel learned by Hoi et al.'s method (NPK) [5] and spectral clustering based on the LRKL algorithm (LRKL + SC) [18]. All the experiments have been performed in MATLAB R2013a running in a 3.10 GHZ Intel CoreTMi5-2400 with 4-GB RAM. In the experiments, we set the number of clusters as the number of classes c in each dataset. Normalized mutual information (NMI) and the clustering accuracy (ACC) [19] are used to evaluate the clustering performance.

4.1 Experimental Datasets and Parameter Settings

The basic information of the datasets is listed in Table 3. All of the data are normalized before conducting experiments. For fair comparison, we set the number of nearest neighbors to $k = 5$ and the regularization parameter to $\alpha = 0.2$. For LRKL-ESEC, the parameters γ_g varies from 0.01 to 0.9 and the parameter μ ranges from 0 to 1. They are specified using 5-fold cross validation. The error constant ε is

Table 3 Properties of datasets

Dataset	Samples	Dimensions	Clusters
Iris	150	4	3
Wine	178	13	3
USPS	900	256	3
Balance	625	4	3
Glass	214	10	6
Yale	165	1024	15
ORL	100	4096	10
Isolet	390	1234	13
COIL20	1440	1024	20

set as 0.01. The RBF kernel is selected as the hidden node function and a grid search of the number of hidden nodes L on $\{100, 150, 200, \ldots, 1000\}$ is conducted in seek of the optimal result by using 5-fold cross validation. The number of must-link pairwise constraints is equal to that of cannot-link. All clustering algorithms are independently repeated 50 times with different random initializations for each set of constraints.

4.2 Experiment Results

We first compare LRKL and LRKL-ESEC which use the same trace ratio maximization algorithm to obtain the target kernel. Tables 4 and 5 report the mean ACC results on Iris and the NMI results on Yale, respectively. As can be seen from Tables 4 and 5, LRKL-ESEC significantly outperforms LRKL with the increasing of the number of pairwise constraints. For the high-dimensional Yale dataset, LRKL-ESEC performs better than LKRL by using the nonlinear spectral embedded technique based on ELM. For the low dimensional Iris dataset, LRKL-ESEC still achieves superior performance compared to LRKL, which is due to the fact that LRKL-ESEC introduces the nonlinear spectral embedded method into its model, which is essentially a kind of regularization methods and contributes to the performance improvement of kernel learning algorithms.

The average NMI of each method versus the number of constraints on UCI datasets is displayed in Fig. 1. From Fig. 1, we can see that the proposed LRKL-ESEC method achieves much better clustering results than the other methods. Only LRKL is comparable to our method on the Glass dataset. The experimental results on UCI datasets demonstrate that LRKL-ESEC is also applicable to handle low-dimensional datasets. The NMI performance curves for high-dimensional data sets have also been shown in Fig. 2. As can be seen from

Table 4 The ACC results of LRKL and LRKL-ESEC on the Iris dataset

Algorithm	The number of pairwise constraints							
	30	50	70	90	110	130	150	170
LRKL	35.7333	35.7466	45.4666	51.7066	56.8400	64.1333	66.6000	70.0266
LRKL-ESEC	88.4222	88.8667	93.1111	87.4889	90.3111	90.5778	89.7111	89.5778

Table 5 The NMI results of LRKL and LRKL-ESEC on the Yale dataset

Algorithm	The number of pairwise constraints							
	30	50	70	90	110	130	150	170
LRKL	0.1868	0.2736	0.3791	0.6005	0.6841	0.7534	0.8309	0.8429
LRKL-ESEC	0.3312	0.3373	0.3829	0.6500	0.7151	0.7892	0.8297	0.8526

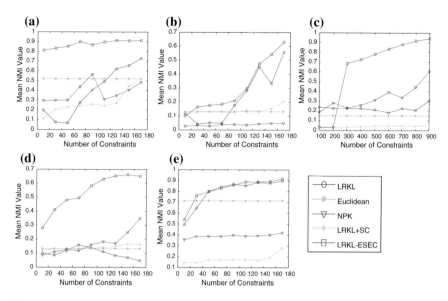

Fig. 1 Clustering results of different algorithms on UCI datasets (the mean NMI values versus the number constraints) **a** iris, **b** wine, **c** USPS, **d** balance, **e** glass

Fig. 2, LRKL-ESEC significantly performs better than the other algorithm since it can cope with the high-dimensional data which do not exhibit a clear low-dimensional manifold structure. Thus, these clustering results demonstrate that our method has better kernel learning performance than the other methods and validate the effectiveness of the proposed algorithm for both low-dimensional and high-dimensional datasets.

To compare the learned kernel matrices, we take the Iris and USPS dataset for example and transform the learned kernel matrices into gray images, which are shown in Fig. 3. Obviously, the block structure of the kernel matrix learned by our method is much clearer than that obtained by LRKL, which shows that data points in the same cluster are compact and those in different clusters are separated very well. Consequently, the better clustering results can be achieved based on the proper kernel matrices learned by our method. In UCI datasets demonstrate that LRKL-ESEC is also applicable to handle low-dimensional datasets. The NMI performance curves for high-dimensional datasets have also been shown in Fig. 2. As can be seen from Fig. 2, LRKL-ESEC significantly performs better than the other algorithm since it can cope with the high-dimensional data which do not exhibit a clear low-dimensional manifold structure. Thus, these clustering results demonstrate that our method has better kernel learning performance than the other methods and validate the effectiveness of the proposed algorithm for both low-dimensional and high-dimensional datasets.

To compare the learned kernel matrices, we take the Iris and USPS dataset for example and transform the learned kernel matrices into gray images, which are

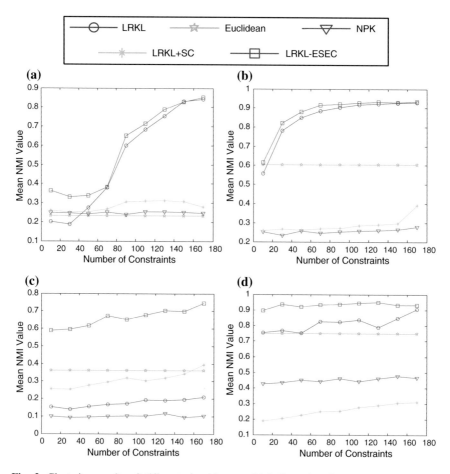

Fig. 2 Clustering results of different algorithms on high-dimension datasets (the mean NMI values versus the number constraints) **a** Yale, **b** ORL, **c** Isolet, **d** COIL20

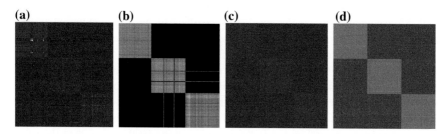

Fig. 3 Kernel matrices learned by LRKL and LRKL-ESEC on Iris and USPS: **a** the kernel matrix learned by LRKL on Iris, **b** the kernel matrix learned by LRKL-ESEC on iris, **c** the kernel matrix learned by LRKL on USPS and **d** the kernel matrix learned by LRKL-ESEC on USPS

shown in Fig. 3. Obviously, the block structure of the kernel matrix learned by our method is much clearer than that obtained by LRKL, which shows that data points in the same cluster are compact and those in different clusters are separated very well. Consequently, the better clustering results can be achieved based on the proper kernel matrices learned by our method.

5 Conclusions

In this paper, we have present a semi-supervised low rank kernel learning method combined with nonlinear spectral embedded regularization method and ELM. The proposed method is formulated as an orthogonally constrained trace ratio maximization problem by virtue of pairwise constraints and the structure of the data. The proposed model can be can be solved efficiently by using the popular trace ratio optimization algorithm. Experimental results show that the performance of our method is much better than that of some existing NPKL methods for both low-dimensional and high-dimensional datasets. In the future, we will find a way to automatically specify the optimal rank of the target kernel matrix.

Acknowledgments This work was supported by the National Natural Science Foundation of China (NO. 61403394) and the Fundamental Research Funds for the Central Universities (NO. 2014QNA46).

References

1. Li, F., Yang, J., Wang, J.: A transductive framework of distance metric learning by spectral dimensionality reduction. In: Proceedings of the 24th International Conference on Machine Learning (ICML), pp. 513–520, Corvallis, OR, USA (2007)
2. Zhong, S., Chen, D., Xu, Q., et al.: Optimizing the Gaussian kernel function with the formulated kernel target alignment criterion for two-class pattern classification. Pattern Recogn. **46**(7), 2045–2054 (2013)
3. Yin, X., Chen, S., Hu, E., Zhang, D.: Semi-supervised clustering with metric learning: an adaptive kernel method. Pattern Recogn. **43**, 1320–1333 (2010)
4. Wang, S., Jiang, S., Huang, Q., et al.: S3MKL: scalable semi-supervised multiple kernel learning for image data mining. In: Proceedings of the International Conference on Multimedia. ACM, pp. 1259–1274 (2010)
5. Hoi, S.C.H., Jin, R., Lyu, M.R.: Learning nonparametric kernel matrices from pairwise constraints. In: Proceedings of the 24th International Conference on Machine Learning (ICML), pp. 361–368, New York, USA (2007)
6. Li, Z., Liu, J., Tang, X.: Pairwise constraint propagation by semidefinite programming for semi-supervised classification. In: Proceedings of the 25th International Conference on Machine Learning (ICML), pp. 576–583 (2008)
7. Zhuang, J., Tsang, I.W., Hoi, S.C.H.: A family of simple non-parametric kernel learning algorithms. J. Mach. Learn. Res. **12**, 1313–1347 (2011)
8. Baghshah, M.S., Shouraki, S.B.: Learning low-rank kernel matrices for constrained clustering. Neurocomputing **74**(12), 2201–2211 (2011)

9. Belkin, M., Niyogi, P.: Laplacian eigenmaps for dimensionality reduction and data representation. Neural Comput. **15**(6), 1373–1396 (2003)
10. Belkin, M., Niyogi, P., Sindhwani, V.: Manifold regularization: a geometric framework for learning from labeled and unlabeled examples. J. Mach. Learn. Res. **7**, 2399–2434 (2006)
11. Soleymani Baghshah, M., Bagheri Shouraki, S.: Kernel-based metric learning for semi-supervised clustering. Neurocomputing **73**, 1352–1361 (2010)
12. Yeung, D.Y., Chang, H.: A kernel approach for semi-supervised metric learning. IEEE Trans. Neural Netw. **18**(1), 141–149 (2007)
13. Huang, G.B., Zhou, H., Ding, X., Zhang, R.: Extreme learning machine for regression and multi-class classification. IEEE Trans. Syst. Man Cybern. **42**(2), 513–529 (2012)
14. Kulis, B., Basu, S., Dhillon, I.: Semi-supervised graph clustering: a kernel approach. Mach. Learn. **74**(1), 1–22 (2009)
15. Kulis, B., Sustik, M., Dhillon, I.: Learning low-rank kernel matrices. In: Proceedings of the 23th International Conference on Machine Learning (ICML), pp. 505–512, Pittsburg, PA (2006)
16. Jia, Y., Nie, F., Zhang, C.: Trace ratio problem revisited. IEEE Trans. Neural Netw. **20**(4), 729–735 (2009)
17. Xiang, S., Nie, F., Zhang, C.: Learning a Mahalanobis distance metric for data clustering and classification. Pattern Recogn. **41**(12), 3600–3612 (2008)
18. Shi, J., Malik, J.: Normalized cuts and image segmentation. IEEE Trans. Pattern Anal. Mach. Intell. **22**(8), 888–905 (2000)
19. Chen, W., Feng, G.: Spectral clustering: a semi-supervised approach. Neurocomputing **77**(1), 229–242 (2012)

Application of Extreme Learning Machine on Large Scale Traffic Congestion Prediction

Xiaojuan Ban, Chong Guo and Guohui Li

Abstract Short-Term prediction aimed at the urban traffic congestion is an important goal for Intelligent Transport Systems (ITS). Short-term traffic prediction module tries to predict Traffic Congestion Index accurately and in approximately real-time. Up to date, there have been three basic methods in short-Term traffic prediction research namely the Kalman Filtering (KF) (Okutani and Stephanedes in Transport Res Part B 18B:1–11, 1984, [1]) method, Time Series models (Williams and Hoel in ASCE J Transport Eng 129:664–672, 2003, [2]) and Neural Network (NN) models (Smith and Demetsky in Transport Res Rec J Transport Res Board 1453:98–104, 1994, [3]). The Neural Networks based methods have proven to give good accuracy rate but they are time consuming in training. In our paper, we have implemented the new Neural Networks based algorithm called Extreme Learning Machine (ELM) (Huang and Siew in ICARCV, pp. 1029–1036, 2004, [4]) to design a real-time traffic index in the data of a real world city of Nanning in South China. Our experiment results show that ELM algorithm provides good generalization performance at extremely fast learning speed compared with other state-of-art algorithms. The algorithm obtains high accuracy in practical prediction application. In addition, quick training and good fitting results on our own large scale traffic data set proves ELM algorithm works well on large data sets.

Keywords Extreme learning machine · ELM · Traffic congestion prediction · Intelligent transport system · Large-scale computing · Real-time predicting · High-speed

X. Ban (✉) · C. Guo
School of Computer and Communication Engineering, University of Science and Technology Beijing, Beijing 100083, China
e-mail: banxj@ustb.edu.cn

C. Guo
e-mail: kotrue2015@gmail.com

G. Li
Ao Jin Tech Co., Ltd, Tianjin 300072, China
e-mail: 951859818@qq.com

© Springer International Publishing Switzerland 2016
J. Cao et al. (eds.), *Proceedings of ELM-2015 Volume 1*,
Proceedings in Adaptation, Learning and Optimization 6,
DOI 10.1007/978-3-319-28397-5_23

293

1 Introduction

Nanning City in China has successfully constructed a complete Taxi Read-time Monitoring System (TRMS). Taxis can continuously transmit real-time information including GPS, driving speed and direction to Information Processing Center via TRMS. We are responsible for building a real-time evaluation and short-term prediction system of the urban traffic congestion based on TRMS. The system describes the urban traffic condition by Traffic Congestion Index (continuous integer range from 0 to 100). High dimensions of features and large scale of training data drive us to look for a faster neural network algorithm. Extreme learning machine (ELM) is an emerging learning algorithm for the generalized single hidden layer feed-forward neural networks (SLFNs), of which the hidden node parameters are randomly generated and the output weights are analytically computed. ELM has been proven to provide good generalization performance at extremely fast learning speed [4].

The traffic conditions evaluation and prediction are vital components of Intelligent Transport System (ITS) which aim to impact travel routes selection and reduce traffic congestion. Real-time traffic congestion evaluation and accurate predictions of future traffic condition can offer to citizens advices to determine appropriate travel routes and enhance performance of urban road network. Nanning city in China has established a relatively complete Real-time Taxi Monitoring System. About 8000 taxis generate voluminous real-time traffic information everyday and more than 20,000 messages are transmitted to data processing center per minute. The information contains speed, taxi location GPS, taxi direction and more. Actually, the voluminous data covers nearly 75 % roads of Nanning city every five minutes. These huge amount of data, and the traffic condition information they represent are essential to assess the urban traffic condition. Nanning has constructed Traffic Congestion Index Evaluation and Prediction System to offer to its citizens traffic congestion index which range from 0 to 100 and indicate the traffic congestion condition. The higher congestion index is, the worse traffic condition is. Traffic congestion index in the system are divided into three categories: historic index, current index and predictive index. Historic index describes the state of traffic condition in the previous time periods. Current index is the most up-to-date status about traffic. Considering taxi density, current index is produced every 5 min by using aggregate data collected in past 5 min, and it turns to be historic with time going on. Predictive index, to be exact the short-term predictive index, is the predicting traffic status for next 30 min. Like current index, the predictive index is forecasted every 5 min. This paper mainly focuses on the prediction implementation.

Over the last four decades, many researchers developed or proposed a considerable number of algorithms of short-term traffic prediction and numerous publications can be found in the open literature. A lot of tools and algorithms have been applied, most of which focused on developing new or improving existing prediction models. For instance, the Kalman Filtering (KF) method was first used for traffic volume prediction by Okutani and Stephanedes [1]. Recently, in a study by Yang et al. [5], a Recursive Least Square (RLS) approach was proposed for short-term traffic speed prediction by means of KF to adapt to changing patterns quickly, based on the maximum likelihood method and Bayesian rule.

Another group of short-term traffic prediction models are based on Time Series models, which predict different types of traffic parameters. For instance, some studies developed a time series model to predict future traffic volumes, such as Williams and Hoel [2], while others commonly used time series for short-term traffic speed prediction, such as Farokhi et al. [6].

Apart from Time Series models, Neural Network (NN) models are another class of models successfully used for short term traffic prediction. Smith and Demetsky [3] introduced the back-propagation neural network model for traffic volume prediction. Park et al. [7] conducted a study to involve prediction short-term freeway traffic volumes with a Radial Basis Function (RBF) neural network. Yin et al. [8] realized a fuzzy-neural model (FNM) to predict the traffic flows in an urban street network. The model contained two modules: a gate network (GN) and an expert network (EN). Ishak [9] conducted a study to optimize short-term traffic prediction performance via utilizing multiple topologies of dynamic neural networks under numerous parameters and traffic-condition settings. With the purpose of improving the performance of the NN models, many hybrid models were proposed. For example, Abdulhai et al. [10] developed a system on the basis of Time Delay Neural Network (TDNN) model synthesized using Genetic Algorithm (GA) for short-term traffic prediction. Alecsandru and Ishak [11] put forward a hybrid model-based and memory-based methodology to advance predictions under both recurrent and non-recurrent conditions. The model-based approach depended on a combination of static and dynamic neural network architectures to attain optimal prediction performance under various input and traffic condition settings. Vlahogianni et al. [12] incorporated GA to optimize the learning rule along with the network structure. The GA approach consisted of three steps: selection, crossover, and mutation, founded on the principles of genetics. Many other NN models on short-term traffic prediction were found in the literature (see for instance Jiang and Adeli [14]).

The literature review shows that the Time Series and NN models were the most commonly used models for short-term traffic prediction. Although NN models have good fitting results, large scale training data make it's training time consuming. In this paper, we used a new NN networks called ELM, which can produce good generalization at extremely fast speed, to predict traffic status.

The rest of this paper is organized as follows. Section 2 introduces the fundamental theory and implementation of ELM. The output of traffic evaluation is the source of traffic congestion prediction training data. Therefore it's necessary to make a brief introduction on traffic evaluation and make it easy to understand the following. Section 3 will give a very brief description on traffic evaluation strategies. Section 4 illustrates how to pre-process input feature by clustering feature to reduce dimensions. Performance evaluation is presented in Sect. 5. Discussions and conclusions are given in Sect. 6.

2 Brief Review of ELM Algorithm

Learning speed of feedforward neural networks is in general far slower than required. Extreme learning machine (ELM) for single-hidden layer feedforward neural networks (SLFNs) is proposed which randomly chooses hidden nodes and analytically determines the output weights of SLFNs. For N arbitrary distinct samples (x_i, t_i), where $x_i = [x_{i1}, x_{i2}, \ldots, x_{in}]^T \in R^n$ and $t_i = [t_{i1}, t_{i2}, \ldots, t_{im}]^T \in R^m$, standard SLFNs with Ñ hidden nodes and activation function $g(x)$ are mathematically modeled as

$$\sum_{i=1}^{\tilde{N}} \beta_i g_i(x_j) \sum_{i=1}^{\tilde{N}} \beta_i g_i(w_i * x_j + b_i) = O_j, \quad j = 1, \ldots, N \qquad (1)$$

where $w_i = [w_{i1}, w_{i2}, \ldots, w_{in}]^T$ is the weight vector connecting the ith hidden node and the input nodes, $\beta_i = [\beta_{i1}, \beta_{i2}, \ldots, \beta_{im}]^T$ is the weight vector connecting the ith hidden node and the output nodes, and b_i is the threshold of the ith hidden node. $w_i * x_j$ denotes the inner product of w_i and x_j. The output nodes are chosen linear.

That standard SLFNs with Ñ hidden nodes with activation function $g(x)$ can approximate these N samples with zero error means that $\sum_{i=1}^{\tilde{N}} \|O_j - t_j\| = 0$, i.e., there exist β_i, w_i and b_i such that

$$\sum_{i=1}^{\tilde{N}} \beta_i g_i(w_i * x_j + b_i) = t_j, \quad j = 1, \ldots, N. \qquad (2)$$

The above N equations can be written compactly as

$$H\beta = T \qquad (3)$$

According to ELM theory [4], input weights and hidden layer biases can be randomly assigned if only the activation function is infinitely differentiable. For fixed input weights w_i and the hidden layer biases b_i, to train an SLFN is simply equivalent to finding a least-squares solution $\widehat{\beta}$ of the linear system $H\beta = T$:

$$\left\| H(w_1, \ldots, w_{\tilde{N}}, b_1, \ldots, b_{\tilde{N}}) \check{\beta} - T \right\| = \min_{\beta} \left\| H(w_1, \ldots, w_{\tilde{N}}, b_1, \ldots, b_{\tilde{N}}) \check{\beta} - T \right\| \qquad (4)$$

If the number Ñ of hidden nodes is equal to the number N of distinct training samples, Ñ = N, matrix H is square and invertible when the input weight vectors w_i

and the hidden biases b_i are randomly chosen, and SLFNs can approximate these training samples with zero error.

However, in most cases the number of hidden nodes is much less than the number of distinct training samples, $\tilde{N} \ll N$, H is a nonsquare matrix and there may not exist w_i, b_i, $\beta_i (i = 1, \ldots, \tilde{N})$ such that $H\beta = T$. The smallest norm least-squares solution of the above linear system is

$$\widehat{\beta} = H^+ T \tag{5}$$

where H^+ is the Moore–Penrose generalized inverse of matrix H.

ELM algorithm have the following important properties: minimum training error; smallest norm of weights [4].

3 Traffic Evaluation

3.1 Average Speed

In practice, the roads are split into small road pieces as basic units of evaluation and prediction. There are 18,041 pieces in Nanning City. Figure 1a show the road pieces in Nanning City and Fig. 1b is the enlarged view of road pieces. Complex algorithms are used to bind taxi with road by using taxi GPS and driving direction information, thus it's easy to obtain the current running speed of the car on this road piece. Taxis cover at least 75 % road pieces in previous 5-min and transmit

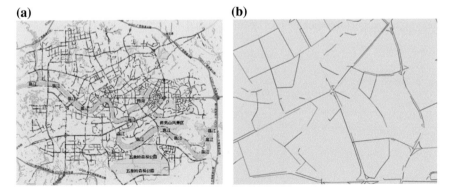

(a)   **(b)**

Fig. 1 Roads are split into many small pieces. Road pieces are the basic unit to evaluate or predict. **a** Overall view of road pieces in Nanning city. **b** Enlarged view of part of road pieces in Nanning City

100,000 messages. It's easy to get average driving speed by aggregate 5-min speed.

3.2 Speed-TCI Model

Driving speed is an important indicator to judge road condition. It is easy to get average driving speed by collecting 5-min speed. We mainly use speed to determine the Traffic Congestion Index (TCI). Equation 6 is the model used to evaluate transport condition in the system and its graph is shown in Fig. 2.

$$f(x) = 100 - 200 \cdot \left(\frac{1}{1 + e^{-0.46x}} - \frac{1}{2} \right) \qquad (6)$$

where x is the speed of road piece (unit: km/h) while f(x) represents TCI. As described in the graph, road condition reaches worst when the running speed is zero, thus TCI equals 100. Conversely, traffic condition turns to be better when speed grows. Therefore when speed tends to be infinite, TCI tends to be 0. $y = 0$ is the horizontal asymptote of $f(x)$.

The output of evaluation is a continuous value which varies between 0 and 100. The practical result shows that the speed-TCI model is reasonable and can describe traffic status properly. The evaluation results meet the real situation as shown on Fig. 3. The system generates evaluation output every 5 min to describe the current urban traffic status of each road piece. These data becomes historical with time going on. The historical evaluation data can be used for training predicting model and obviously there are a huge amount of training samples every day. Actually almost 1,000,000 useful samples are produced every day. Big data calls for a fast model. In this paper we have chosen ELM algorithm which tends to provide good generalization performance at fast learning speed.

Fig. 2 Graph of Eq. 6

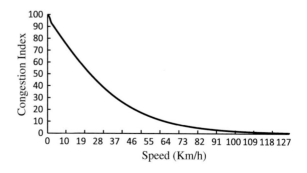

Fig. 3 The comparison of camera results and evaluation results shows that the evaluation model is reasonable

4   Feature Extraction

For a machine learning application, feature extraction is a very important part and will directly affect the predicting results. Apparently, factors such as if there are traffic lights on the road, which region the road is in, whether there is a business district or a school nearby and more strongly influence the traffic condition of roads. But in our system, these features are difficult to get automatically. Therefore these features are not considered. After a lot of trials with feature select tools, we selected the following five features.

- Current time: categorical, 06:05, 08:05, ..., 21:55. 191 in total.
- Road logical region: categorical feature, 1, 2, 3, ..., 50, 50 in total.
- Traffic congestion index of last time: continuous, vary from 0 to 100.
- Road type: categorical, express way, backbone road, side road.
- Number of adjacent roads: continuous, positive integer.

Physical region, which to some extent describes the environment of roads, can play a significant role in road congestion. For instance, roads near the school may always be in good condition except when children go to school in the morning, and when they go home in the afternoon. Meanwhile, traffic condition has a strong spatiotemporal periodicity and region feature should be taken into account. Unfortunately, physical region information are not automatically provided by the TRMS and are very sensitive to change, especially under the circumstance that there is a new road constructed when system runs. Besides, the region labeled by people may not be accurate sometimes. This paper tries to use a logical region instead of physical region to describe roads' environment information.

We define $x_{i,j}(0 \leq x_{i,j} < 100)$ as the traffic congestion index of road i at time j (is evaluated every 5-min and 191 indexes a day in total, from 6:00 to 22:00). For a specific road i, $[x_{i,0}, x_{i,1}, x_{i,2}, \ldots, x_{i,190}]^T$ is the vector of congestion index of a day. Thus vectors of all roads can be written as following:

$$x_0 = [x_{0,0}, x_{0,1}, x_{0,2}, \ldots, x_{0,190}]^T$$
$$x_1 = [x_{1,0}, x_{1,1}, x_{1,2}, \ldots, x_{1,190}]^T$$
$$\cdots$$
$$x_n = [x_{n,0}, x_{n,1}, x_{n,2}, \ldots, x_{n,190}]^T$$

K-means algorithms can be used to cluster vectors above, thus roads are clustered into k clusters, which is called logical region in this paper. Roads in same cluster have the similar traffic condition in a day, and roads from different clusters differ a lot. Roads cluster or so-called road logical region can distinguish roads logically. The value of k is tuned and set k = 50 in this paper.

In general, variables either indicate measurements on some continuous scale like traffic congestion index of last time in this case, or represent information about some categorical or discrete characteristics like current time in the above features. The features used in this paper mix categorical features with real-valued features, and should be transformed into all categorical or real-valued features. In our cases, all categorical features are subdivided into several single real-valued features, whose value is set to be either 1 or 0. For instance, feature current time will be transformed into 192 new features. Each of them is 0 or 1 to indicate if it appears or not. Thus all the features become real-valued and can be equally evaluated. According to the experiments in practice, such transformation increase the accuracy rate by 2.2 % compared with using original feature directly (limit to the words this paper won't give the details).

5 Experimental Result

In this section, the performance of ELM on real world traffic data is evaluated. Comparisons are made with other popular regression algorithms used in practice, e.g., Ridge Regression, Support Vector Machine for Regression, Lasso Regression and Gradient Boosting Decision Tree Regression. This section mainly contains 3 parts. Part 1 shows the performance of ELM when hidden nodes change. The performance of ELM is compared with the popular regression algorithms in part 2. Part 3 gives the performance of our prediction system based on ELM. All the simulations are carried out in Ubuntu 12.04 environment running in a Core Duo i7 2.50 GHZ CPU.

Although 2,400,000 training data are produced a day, there are only 1,515,446 distinct training samples, the redundancy of the training data has been removed. All the simulations run 4-fold cross validation except for special explanation. The training samples is randomly partitioned into 4 equal sized subsamples. Of the 4 subsamples, a single subsample is remained as the validation data for testing the

Table 1 Information of the simulation training and testing data

Training samples	Testing samples	Attributes
1,136,585	378,861	243

model, and the remaining 3 subsamples are used as training data each time. The details of data set are shown in Table 1.

The regression results are continuous integer varying between 0 and 100. In this paper, prediction is correct when $|y_{truth} - y_{estimate}| \leq 25, y_{truth}(0 \leq y_{truth} < 100)$, is the actual congestion index in testing data while $y_{estimate}(0 \leq y_{estimate} < 100)$ is the estimate result predicted by models.

5.1 ELM Performance on Traffic Data Set

The ELM is implemented using Python 2.7. In this experiment, activation function g is tuned and set as defined in Eq. 7, the other parameters are given in default. The number of hidden nodes are gradually increased from 2 to 800. Figure 4 shows the performance of ELM with different number of neurons.

$$g(x) = \frac{1}{\left(1 + x^2\right)^2} \tag{7}$$

As observed from Fig. 4, overall speaking, the more hidden nodes ELM has, the better fitting result it produces, but more time it cost. Figure 4a is the accuracy performance with growth of nodes. At the beginning, with the growth of hidden nodes, prediction accuracy increases dramatically. Prediction accuracy can increase from 86.86 to 91.18 % while nodes grow from 2 to 200. Finally, the accuracy increases slightly even stabilizes when hidden layer nodes reach to 200. Although ELM runs extremely fast compared with other SLFNs algorithms, number of hidden nodes has a great influence on training time. Figure 4b shows the training time cost when hidden nodes grow. When number of nodes grow to 400, training time becomes obvious and tend to be increase dramatically. Taking into account cost and benefit, we set L = 200 when ELM compared with other algorithms.

The part of prediction result (L is set to 400) is shown Fig. 5. For a good visualization, we just randomly select 50 test samples and draw true values and ELM predictions respectively. Figure 5 shows the ELM algorithms has great fitting results on traffic training data. The difference between the true one and the estimated values are so small for most the cases, some even completely same!

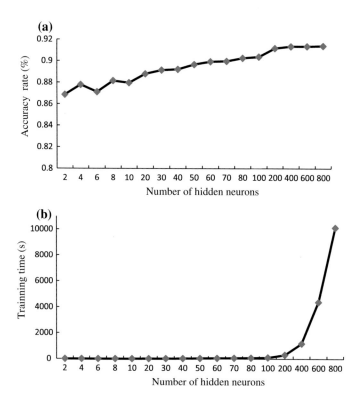

Fig. 4 Performance of ELM when hidden neurons increase

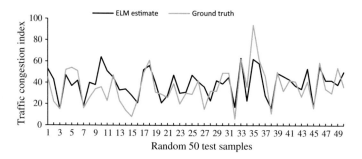

Fig. 5 Part of regression prediction shows ELM (L = 400) has a great fitting result

5.2   *Comparison with Other Algorithms*

ELM is compared with some of other the-state-of-art algorithms in this section. The algorithms to compare include Ridge Regression, Support Vector Machine for Regression, Lasso Regression and Gradient Boosting Decision Tree Regression. All

Table 2 Performance comparison in traffic congestion index predict application

Algorithms	Train time (s)	Best success rate (%)
ELM	309.39	91.08
GBDT	4204.86	92.26
Ridge	1006.18	90.68
Lasso	9.80	88.77
LR	6.78	75.64

the algorithms in experiments except for ELM are implemented by Sklearn 1.4 machine learning package. There are a lot of parameters in these algorithms, they are set to be default exception for specific explanation. For ELM algorithm, the number of hidden nodes L is 200 and activation function is Eq. 7. Depth in Gradient Boosting Decision Tree is set to 9. The results of comparison is shown in Table 2.

As observed in Table 2, the ELM and GBDT have almost the same and best success rate but the ELM training time is 10 times better than the GBDT. Though Lasso regression has considerably a short training time comparing to ELM but it is accuracy is less reliable than the ELM accuracy in the same accuracy condition the ELM training time is 6.48 s (where L = 8 and accuracy is 88.15).

It's worth to mention that we actually did the Linear SVM simulations. Unfortunately, SVM seems not able to give the output result in a short time on this full dataset. The simulation has run 2080 min (about 35 h) before we kill it on this full data set.

5.3 Performance of Prediction System Based on ELM

The model used to predict traffic condition is trained everyday with the data of previous week. The next 30 min traffic status will be updated every 5 min. For instance, the predicting results of period from 8:00 to 8:35 are produced at 7:55 while period from 8:05 to 8:40 are updated at 8:00. Figure 6 are part of predicting results in the morning, in the middle of the day and in the evening respectively.

Figure 6a shows the accuracy of prediction produced at 7:55 am. The accuracy of 8:00 am is 92.12 % while 8:05 is 83.91 %. We can see that forecast accuracy decreases with the increasing of time. Actually, this change is mainly caused by feature *traffic congestion index of last time*. The 8:00 prediction is based on the actual evaluation congestion index value at 7:55, while 8:05 prediction is based on the previous predicted value at 8:00, Therefore, prediction accuracy is gradually decreasing over time.

Figure 6b is the performance result of prediction at 10:55 and Fig. 6c is the result at 16:55. The former represent the performance in normal period while the later represents the result in evening run hours. Both of them have the similar performance compared with Fig. 6a. The prediction accuracy in all day ranges from 76 to 92 % and our system have a good performance at short-term traffic status prediction.

Fig. 6 Performance of
prediction system based on
ELM, which predicts the
traffic status for next 30 min.
a shows the prediction
accuracy in the morning rush
hours (predicted at 7:55 am
on March 15th, 2015). b and
c are the predicting results
estimated at 10:55 am and
16:55 pm respectively

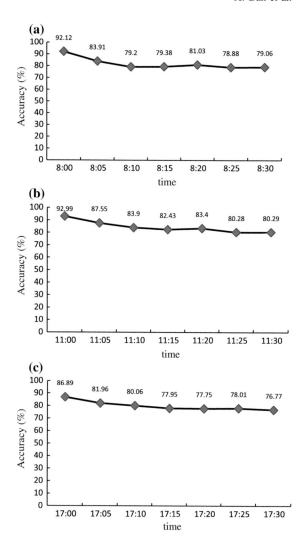

6   Discussion and Conclusion

The experimental results in 5.3 shows the more hidden nodes ELM has, the better
fitting result it produces, but more time it cost. The number of hidden nodes is
adjusted to 200 in our prediction system for a good performance both in training
and predicting. Our simulations also show that ELM algorithm provides good
generalization performance at extremely fast learning speed compared with other
state-of-art algorithms. In addition, successful application in real world traffic
congestion prediction proves ELM algorithm works well on large data sets.

Acknowledgement This work was supported by National Nature Science Foundation of P. R. China (No. 61272357, 61300074, 61572075).

References

1. Okutani, I., Stephanedes, Y.J.: Dynamic prediction of traffic volume through Kalman filtering theory. Transport. Res. Part B **18B**, 1–11 (1984)
2. Williams, B.M., Hoel, L.A.: Modeling and forecasting vehicular traffic flow as a seasonal ARIMA process: theoretical basis and empirical results. ASCE J. Transport. Eng. **129**(6), 664–672 (2003)
3. Smith, B.L., Demetsky, M.J.: Short-term traffic flow prediction: neural network approach. Transport. Res. Rec. J. Transport. Res. Board **1453**, 98–104 (1994)
4. Huang, G.B., Siew, C.K.: Extreme learning machine: RBF network case. In: ICARCV, pp. 1029–1036 (2004)
5. Yang, F., Yin, Z., Liu, H.X., Ran, B.: Online recursive algorithm for short-term traffic prediction. Transport. Res. Rec. J. Transport. Res. Board **1879**, 1–8 (2004)
6. Farokhi, K.S., Masoud, H., Haghani, A.: Evaluating moving average techniques in short-term travel time prediction using an AVI data set. Paper presented at the transportation research board 89th annual meeting, Washington, DC (2010)
7. Park, B., Messer, C.J., Urbanik II, T.: Short-term freeway traffic volume forecasting using radial basis function neural network. Transport. Res. Rec. J. Transport. Res. Board **1651**, 39–47 (1998)
8. Yin, H., Wong, S.C., Xu, J., Wong, C.K.: Urban traffic flow prediction using a fuzzy-neural approach. Transport. Res. Part C Emerg. Technol. **10**(2), 85–98 (2002)
9. Ishak, S.: Deriving traffic-performance measures and levels of service from second-order statistical features of spatiotemporal traffic contour maps. Transport. Res. Rec. J. Transport. Res. Board **1858**, 148–157 (2003)
10. Abdulhai, B., Porwal, H., Recker, W.W.: Short-term freeway traffic flow prediction using genetically-optimized time-delay-based neural networks. Paper presented at transportation research board 78th annual meeting, Washington, DC (1999)
11. Alecsandru, C., Ishak, S.: Hybrid model-based and memory-based traffic prediction system. Transport. Res. Rec. J. Transport. Res. Board **1879**, 59–70 (2004)
12. Vlahogianni, E.I., Karlaftis, M.G., Golias, J.C.: Optimized and meta-optimized neural networks for short-term traffic flow prediction: a genetic approach. Transport. Res. Part C Emerg. Technol. **13**(3), 211–234 (2005)
13. Jiang, X., Adeli, H.: Dynamic wavelet neural network model for traffic flow forecasting. ASCE J. Transport. Eng. **131**(10), 771–779 (2005)
14. Ishak, S., Alecsandru, C.: Optimizing traffic prediction performance of neural networks under various topological, input, and traffic condition setting. ASCE J. Transport. Eng. **130**(7), 452–465 (2004)
15. Shen, L., Hadi, M.: Freeway travel time prediction with dynamic neural networks. Paper presented at transportation research board 89th annual meeting, Washington, DC (2010)

Extreme Learning Machine-Guided Collaborative Coding for Remote Sensing Image Classification

Chunwei Yang, Huaping Liu, Shouyi Liao and Shicheng Wang

Abstract Remote sensing image classification is a very challenging problem and covariance descriptor can be introduced in the feature extraction process for remote sensing image. However, covariance descriptor lies in non-Euclidean manifold, and conventional extreme learning machine (ELM) cannot effectively deal with this problem. In this paper, we propose an improved ELM framework incorporating the collaborative coding to tackle the covariance descriptor classification problem. First, a new ELM-guided dictionary learning and coding model is proposed to represent the covariance descriptor. Then the iterative optimization algorithm is developed to solve the model. By evaluating the proposed approach on the public dataset, we show the effectiveness of the proposed strategy.

Keywords Extreme learning machine · Collaborative coding · Covariance descriptor

1 Introduction

Extreme Learning Machine (ELM), which was firstly proposed by Huang [1], has become an effective learning algorithm for various classification tasks. It works on a simple structure named Single-hidden Layer Feed-forward Neural networks (SLFNs) and randomly applies computational hidden nodes. This mechanism is different from the conventional gradient descent learning of SLFNs. It provides a good generalization and a highly accurate learning solution for both classification and regression problems [2, 3]. ELM yields better performance than other conventional learning algorithms in application with higher noise. It also has an extremely fast

C. Yang · S. Liao · S. Wang
High-Tech Institute of Xi'an, Xi'an 710025, Shaanxi, China

C. Yang · H. Liu (✉)
Department of Computer Science and Technology, Tsinghua University,
Beijing 100084, China
e-mail: hpliu@tsinghua.edu.cn

© Springer International Publishing Switzerland 2016
J. Cao et al. (eds.), *Proceedings of ELM-2015 Volume 1*,
Proceedings in Adaptation, Learning and Optimization 6,
DOI 10.1007/978-3-319-28397-5_24

307

learning speed compared with traditional gradient-based algorithms. Furthermore, ELM technique successfully overcomes the difficulty of the curse of dimensionality. Currently, the application scope of ELM covers classification [4, 5], detection [6, 7], recognition [8, 9], and so on. Among these applications, remote sensing image classification using ELM has been attracting more and more attentions in recent years [5, 10]. For example, Ref. [5] presented a remote sensing image classification method based on NMF and ELM ensemble (NMF-ELM). Reference [10] fused ELM and graph-based optimization methods and proposed a multiclass active learning method for remote sensing image classification.

On the other hand, Covariance Descriptor (CovD) can be introduced in the feature extraction and representation processes for complicated image, and has been widely adopted in many computer vision applications [11]. One key problem of CovD method is the model and computation of CovD. It is well known that the space of CovD is not a linear space, but forms a Lie group that is a Riemannian manifold [11]. Hence, the mathematical modeling in this space is different from what is commonly done in the Euclidean space. This results in great challenges for application of ELM. To tackle this problem, we propose a collaborative coding approach incorporating ELM supervised term which can jointly optimize the reconstruction error and the ELM classifier. The main contributions are listed as follows:

1. A new ELM-guided dictionary learning and collaborative coding method is proposed which can learn a dictionary, represent the CovD as a more discriminative feature vector and derive the ELM classifier simultaneously.
2. An iterative optimization algorithm is developed to solve the ELM-guided collaborative coding model.
3. Extensive experimental results on UCMERCED dataset show the proposed ELM-guided collaborative coding strategy performs well on the representation task for CovD.

The rest of this paper is organized as follows: Sects. 2 and 3 give brief introductions of ELM and CovD, respectively. Sections 4 and 5 show the proposed model and corresponding optimization algorithm. Experiments and conclusions are given in Sects. 6 and 7.

2   Brief Introduction of ELM

ELM was initially proposed for SLFNs and then extended to the "generalized" SLFNs with wide types of hidden neurons [3, 12, 13]. ELM stands out from other learning methods with the following characteristics: extremely fast training, good generalization, and universal approximation capability and has been demonstrated to have excellent learning accuracy and speed in various applications [14, 15].

Suppose SLFNs with L hidden nodes can be represented by the following equation

$$f(\mathbf{x}) = \mathbf{h}(\mathbf{x})\boldsymbol{\beta} = \sum_{i=1}^{L} h_i(\mathbf{x}) * \boldsymbol{\beta}_i = \sum_{i=1}^{L} G_i(\mathbf{x}, \mathbf{c}_i, b_i) * \boldsymbol{\beta}_i. \tag{1}$$

where $\boldsymbol{\beta}_i = [\beta_1, \dots, \beta_L]^T$ is the vector of the output weights between the hidden layer with L nodes to the output layer with m nodes, and $\mathbf{h}(\mathbf{x}) = [h_1(\mathbf{x}), \dots, h_L(\mathbf{x})]$ is the output vector of the hidden layer. $G_i(\cdot)$ denotes the ith hidden node activation function which is a nonlinear piecewise continuous function, and $\mathbf{c}_i, b_i$ are the input weight vector connecting the input layer to the ith hidden layer and the bias weight of the ith hidden layer, respectively.

For additive nodes with activation function g, G_i is defined as follows

$$G(\mathbf{x}, \mathbf{c}_i, b_i) = g(\mathbf{c}_i \cdot \mathbf{x} + b_i). \tag{2}$$

From the learning point of view, unlike traditional learning algorithms, ELM theories emphasize that the hidden neurons need not be adjusted, and ELM solutions aim to simultaneously reach the smallest training error and the smallest norm of output weights [1]:

$$Minimize : ||\boldsymbol{\beta}||_p^{\sigma_1} + C||\mathbf{H}\boldsymbol{\beta} - \mathbf{T}||_q^{\sigma_2}. \tag{3}$$

where $\sigma_1 > 0$, $\sigma_2 > 0$, $p, q = 0, 1/2, 1, 2, \dots, +\infty$. $\mathbf{H}$ is the hidden layer output matrix, and $\mathbf{T}$ is the training data target matrix:

The output weight $\boldsymbol{\beta}$ is calculated by

$$\hat{\boldsymbol{\beta}} = \mathbf{H}^\dagger \mathbf{T} \tag{4}$$

where $\mathbf{H}^\dagger$ is the Moore-Penrose generalized inverse of matrix $\mathbf{H}$.

3 Covariance Descriptor

CovD was first proposed as a compact region descriptor by Tuzel et al. [11]. Formally, let $\{\mathbf{f}_i\}_{i=1,\dots,d}$ be a feature vector denoting the p-dimensional feature points such as intensity, color orientation, spatial attributes, etc. Then, a $p \times p$ CovD $\mathbf{R}$ of an image can be represented as:

$$\mathbf{R} = \frac{1}{d-1} \sum_{i=1}^{d} (\mathbf{f}_i - \boldsymbol{\mu})(\mathbf{f}_i - \boldsymbol{\mu})^T. \tag{5}$$

where d is the number of pixels of the image and $\boldsymbol{\mu}$ is the mean feature vector.

As is indicated in [11], CovD has several advantages. First, a CovD extracted from an image is usually enough to match the image in different views and poses. Second, it proposes a natural way of fusing multiple features which might be correlated. The diagonal entries of the CovD represent the variance of each feature and the nondiagonal entries represent the correlations. Third, it is low-dimensional compared to other descriptors and due to symmetry $\mathbf{R}$ has only $(p^2 + p)/2$ different values.

However, the CovD is a symmetric positive definite (SPD) matrix, and one key problem of SPD matrix-based learning methods is the model and computation of SPD matrices. We know that the space of $p \times p$ SPD matrices $\mathbf{R}$ is not a linear space but a Lie group which is a Riemannian manifold [16]. Hence, the mathematical modeling in this space is different from what is commonly done in the Euclidean space. Here, we approximate their distances using the log-Euclidean metric [17],

$$d(\mathbf{R}_1, \mathbf{R}_2) = ||logm(\mathbf{R}_1) - logm(\mathbf{R}_2)||_F. \tag{6}$$

where $\mathbf{R}_1$ and $\mathbf{R}_2$ are two SPD matrices, $logm$ is the matrix logarithm and $|| \cdot ||_F$ is the Frobenius norm.

4   ELM-guided Collaborative Coding Model

In this section, we propose an ELM-guided collaborative coding model, which consists of two components working jointly as dictionary learning and ELM classification which is shown in Fig. 1. Firstly, in the dictionary learning phase, the ELM training is incorporated into the process, making the resulting coding vector $\mathbf{X}$ more discriminative. Then, in the classification phase, the ELM classifier β is obtained in terms of coding vector $\mathbf{X}$. The objective functions in each phase are combined in one unified optimization problem so that a collaborative coding strategy and the matching ELM classifier can be jointly found. Finally, based on the obtained dictionary coefficients $\mathbf{A}$, the testing signal $\mathbf{s}$ is converted to a linear representation $\mathbf{z}$, which is used for ELM classification.

Given the N training samples $\{\mathbf{y}_i\}_{i=1}^{N} \subset \mathcal{M}$, where $\mathcal{M}$ is a Riemannian manifold. We map the training samples into a higher dimensional space by a proper mapping function. In other words, we denote $\Phi(\cdot) : \mathcal{M} \to \mathcal{H}$ to be the implicit nonlinear mapping from $\mathcal{M}$ into a high-dimensional (maybe infinite dimensional) dot product space $\mathcal{H}$. For convenience, we denote the dimension of $\mathcal{H}$ as $\tilde{n}$. This mapping function is associated with some kernel $\kappa(\mathbf{y}_i, \mathbf{y}_j) = \Phi^T(\mathbf{y}_i)\Phi(\mathbf{y}_j)$, where $\mathbf{y}_i, \mathbf{y}_j \in \mathcal{M}$. Here, for CovD, we choose Gaussian kernel due to its excellent performance in many works [18]

$$\kappa(\mathbf{y}_i, \mathbf{y}_j) = exp(-\beta||logm(\mathbf{y}_i) - logm(\mathbf{y}_j)||) \tag{7}$$

where β is a decay parameter which is empirically set as 0.2 here.

The aim of dictionary learning is to empirically learn a dictionary adapted to the training sample set $\{\mathbf{y}_i\}_{i=1}^{N}$. Therefore we need to determine some atoms $\mathbf{d}_1, \mathbf{d}_2, \dots, \mathbf{d}_K \in \mathcal{H}$, where $K < N$ is the size of the dictionary, to represent each training sample in the feature space. By denoting $\Phi(\mathbf{Y}) = [\Phi(\mathbf{y}_1), \ldots, \Phi(\mathbf{y}_N)] \in R^{\tilde{n} \times N}$ and $\mathbf{D} = [\mathbf{d}_1, \ldots, \mathbf{d}_K] \in R^{\tilde{n} \times K}$, we can formulate the kernel dictionary learning problem as

$$\min_{\mathbf{X}, \mathbf{D}} ||\Phi(\mathbf{Y}) - \mathbf{DX}||_2^2 + \alpha ||\mathbf{X}||_2^2, \qquad (8)$$

where $\mathbf{X} \in R^{K \times N}$ is the coding matrix and α is the penalty parameter.

By using the mapping function $\Phi(\cdot)$, we can transform the problem on Riemannian manifold to the collaborative coding problem in feature space. This is the great advantage of the kernel trick [19]. However, such a formulation admits challenge to the dictionary learning since the dictionary atoms $\mathbf{d}_j$ may be in infinite dimensional space. Fortunately, some recent literatures pointed that the dictionary can be represented by $\mathbf{D} = \Phi(\mathbf{Y})\mathbf{A}$, where $\mathbf{A} \in R^{N \times K}$ is a coefficient matrix. This means that the dictionary atoms can be linearly reconstructed by the training samples in the feature space. This conclusion was proved in [20] and [21]. Based on this formulation, the dictionary learning problem becomes

$$\min_{\mathbf{A}, \mathbf{X}} ||\Phi(\mathbf{Y}) - \Phi(\mathbf{Y})\mathbf{A}\mathbf{X}||_2^2 + \alpha ||\mathbf{X}||_2^2. \qquad (9)$$

Such a formation provides significant convenience since the learning of dictionary becomes the search of the matrix $\mathbf{A}$ and provides a principled derivation for nonlinear dictionary learning and coding that essentially reduces to linear problems for any type of kernel function.

Based on the above, we propose to combine the dictionary learning phase and classification phase into a single optimization problem as

$$\min_{\mathbf{A}, \mathbf{X}, \beta} ||\Phi(\mathbf{Y}) - \Phi(\mathbf{Y})\mathbf{A}\mathbf{X}||_2^2 + \lambda ||\mathbf{X}||_2^2 + \alpha ||\mathbf{T} - (\mathbf{X}^{\mathbf{T}}\mathbf{C}^{\mathbf{T}} + \mathbf{B})\beta||_2^2 + \gamma ||\beta||_2^2. \qquad (10)$$

where the first and second term denote dictionary learning term, and the remaining stands for classification term. The parameter α controls the trade-off between the rep-

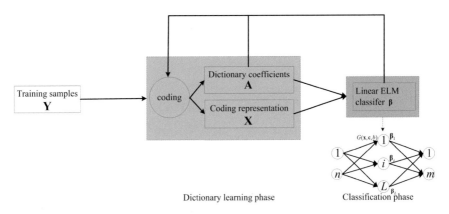

Fig. 1 The illustration of the proposed model

resentation of the signal and classification, and λ and γ are penalty parameters. Note that for the ELM classification stage, we use the additive nodes, and the activation function g is set to 1, which means

$$G(\mathbf{x}, \mathbf{c}_i, b_i) = g(\mathbf{c}_i \cdot \mathbf{x} + b_i) = \mathbf{c}_i \cdot \mathbf{x} + b_i. \tag{11}$$

After the optimal solution of variables $\mathbf{A}$ and $\boldsymbol{\beta}$ are found, as to a testing sample $\mathbf{s}$, the linear representation $\mathbf{z}$ is obtained through

$$\min_{\mathbf{z}} ||\varPhi(\mathbf{s}) - \varPhi(\mathbf{Y})\mathbf{A}\mathbf{z}||_2^2 + \alpha||\mathbf{z}||_2^2. \tag{12}$$

Then, $\mathbf{z}$ is used for ELM classification.

5   Optimization Algorithm

The optimization problem (10) includes three variables: $\mathbf{A}, \mathbf{X}$ and $\boldsymbol{\beta}$. Here, we present an iterative solution for one of the three variables at a time by fixing the others and repeating for a certain number of iterations.

First, fixing $\mathbf{X}$ and $\boldsymbol{\beta}$, taking a derivative with respect to $\mathbf{A}$ (here, (10) is denoted as $\mathbf{F}(\mathbf{A}, \mathbf{X}, \boldsymbol{\beta})$, and the subscripts (k) and $(k+1)$ mean that the variables obtained from the kth and $(k+1)$th iteration, respectively)

$$\frac{\partial \mathbf{F}(\mathbf{A}, \mathbf{X}_{(k)}, \boldsymbol{\beta}_{(k)})}{\partial \mathbf{A}} = 0. \tag{13}$$

And the optimal solution of $\mathbf{A}$ is

$$\mathbf{A}_{(k+1)} = \mathbf{X}_{(k)}^T (\mathbf{X}_{(k)} \mathbf{X}_{(k)}^T)^{-1}. \tag{14}$$

Second, fixing $\mathbf{A}$ and $\boldsymbol{\beta}$, taking a derivative with respect to $\mathbf{X}$

$$\frac{\partial \mathbf{F}(\mathbf{A}_{(k+1)}, \mathbf{X}, \boldsymbol{\beta}_{(k)})}{\partial \mathbf{X}} = 0. \tag{15}$$

Then, the optimal solution of $\mathbf{X}$ is

$$\begin{aligned} \mathbf{X}_{(k+1)} = &(\alpha \mathbf{C}^T \boldsymbol{\beta}_{(k)} \boldsymbol{\beta}_{(k)}^T \mathbf{C} + \mathbf{A}_{(k+1)}^T \mathbf{K}(\mathbf{Y}, \mathbf{Y}) \mathbf{A}_{(k+1)} + \lambda \mathbf{I})^{-1} \\ &(\alpha \mathbf{C}^T \boldsymbol{\beta}_{(k)} \mathbf{T} + \mathbf{A}_{(k+1)}^T \mathbf{K}(\mathbf{Y}, \mathbf{Y}) - \alpha \mathbf{C}^T \boldsymbol{\beta}_{(k)} \boldsymbol{\beta}_{(k)}^T \mathbf{B}^T). \end{aligned} \tag{16}$$

where $\mathbf{K}(\mathbf{Y}, \mathbf{Y})$ is a $N \times N$ square matrix of which (i, j)th element is $\kappa(\mathbf{y}_i, \mathbf{y}_j)$.

Finally, fixing $\mathbf{A}$ and $\mathbf{X}$, taking a derivative with respect to $\boldsymbol{\beta}$ and the optimal solution of $\boldsymbol{\beta}$ is obtained.

$$\frac{\partial F(\mathbf{A}_{(k+1)}, \mathbf{X}_{(k+1)}, \boldsymbol{\beta})}{\partial \boldsymbol{\beta}} = 0. \tag{17}$$

$$\boldsymbol{\beta}_{(k+1)} = (\alpha(\mathbf{C}\mathbf{X}_{(k+1)} + \mathbf{B}^T)(\mathbf{X}_{(k+1)}^T \mathbf{C}^T + \mathbf{B}) + \gamma \mathbf{I})^{-1} \alpha(\mathbf{C}\mathbf{X}_{(k+1)} + \mathbf{B}^T)\mathbf{T}. \tag{18}$$

Once the optimal $\mathbf{A}$ is found, as to (12), the optimal solution $\mathbf{z}$ can be obtained as

$$\mathbf{z}_i = (\mathbf{A}^T \mathbf{K}(\mathbf{Y}, \mathbf{Y})\mathbf{A} + \alpha \mathbf{I})^{-1} \mathbf{A}^T \mathbf{K}^T(\mathbf{s}_i, \mathbf{Y})) \tag{19}$$

where $\mathbf{K}(\mathbf{s}_i, \mathbf{Y}) = [\kappa(\mathbf{s}_i, \mathbf{y}_1), \ldots, \kappa(\mathbf{s}_i, \mathbf{y}_N)]$.
Then, $\mathbf{z}_i$ is used for classification in terms of the optimal $\boldsymbol{\beta}$.

6 Experimental Results

6.1 Dataset and Baseline Methods

In this section, we demonstrate the application of our framework in the classification experiments using the UCMERCED high-resolution aerial image dataset [22]. This dataset includes 21 challenging scene categories with 100 samples per class. Three samples of each category are shown in Fig. 2.

For each class, we randomly partition into five subsets, each of which contains 20 images. Four subsets are used as training and the rest one subset is used as testing. The experiments are repeated five times by selecting one of the five subsets as testing, and the average classification is reported in this paper. At each pixel (u, v) of an image, we compute the 15-dimensional feature vector $\mathbf{f}_{u,v} = [\mathbf{c}_{R,u,v}^T, \mathbf{c}_{G,u,v}^T, \mathbf{c}_{B,u,v}^T]^T$, where $\mathbf{c}_{C,u,v} = [I_{C,u,v}, |\partial I_C/\partial u|, |\partial I_C/\partial v|, |\partial^2 I_C/\partial u^2|, |\partial^2 I_C/\partial v^2|]^T$, where I_C is the intensity image for the C channel and $C \in \{R, G, B\}$ represents one of the color channel.

For comparison, we designed the following dictionary learning methods:

1. RandDict: This method just randomly selects K atoms subset $\{\mathbf{y}_1, \ldots, \mathbf{y}_K\}$ from the training sample set $\{\mathbf{y}_i\}_{i=1}^N$ to construct the dictionary.
2. K-Medoids method: For RCovD, we just use the conventional K-Medoids clustering method to get the K dictionary atoms. As a result, we also get the dictionary as $\{\mathbf{y}_1, \ldots, \mathbf{y}_K\}$.
3. Conventional ELM: This method separately train the dictionary and the ELM classifier.
4. Proposed method: This method jointly optimizes the reconstruction error in the kernel space and the classification performance using ELM.

Fig. 2 Samples from the UCMERCED. Example images associated with 21 land-use categories are shown here

6.2 Parameter Analysis

There are five parameters needing to be tuned: λ, α, γ, number of hidden nodes L, and dictionary size K. Fixing the other parameters, we investigate the influence the each parameter. Note that we initialize the above five parameters as 0.1, 0.1, 0.1, 600 and 210.

Firstly, we range parameter λ, α, and γ from 0.1 to 1, and the classification accuracy versus each parameter is shown in Fig. 3a–c. We can see that when $\lambda = 0.2$, $\alpha = 0.2$, $\gamma = 0.2$ (or 0.4), the method obtains the best performance (84.76 %). Mean-

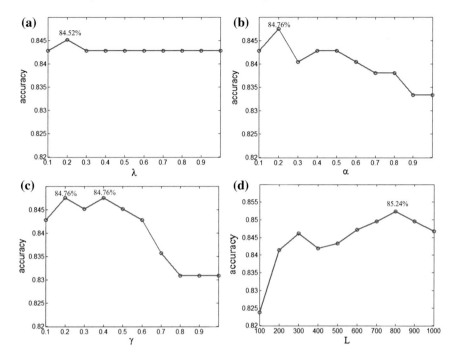

Fig. 3 Evaluation of the effect on the classification accuracy for parameters **a** λ, **b** α, **c** γ and **d** number of hidden nodes L

while, our method is not sensitive to these three parameters, which indicates the robustness of our method.

Then, fixing $\lambda = 0.2$, $\alpha = 0.2$, $\gamma = 0.2$, we investigate the effect of the number of hidden nodes of ELM shown in Fig. 3d, which indicates that the best classification accuracy (85.24 %) occurs when $L = 800$.

6.3 Experiment Results and Comparison

To validate the effectiveness of our method, we show the classification accuracy compared with other methods in Fig. 4. From this figure we see that the proposed method always performs better than other methods, and the proposed method obtains best performance (88.10 %) when $K = 336$. From this figure we also find that the conventional ELM, which neglects the intrinsic structure of the CovD and the ELM classifier, cannot obtain the satisfactory performance.

Figures 5 and 6 show the confusion matrices of the conventional ELM and our proposed ELM-guided method, respectively. Compared with conventional ELM, the classification accuracy of 20 out of 21 categories of our method is higher. From Fig. 6

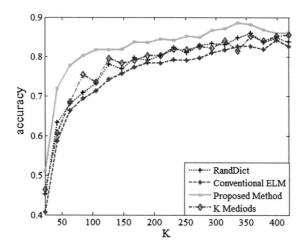

Fig. 4 Comparison of different methods

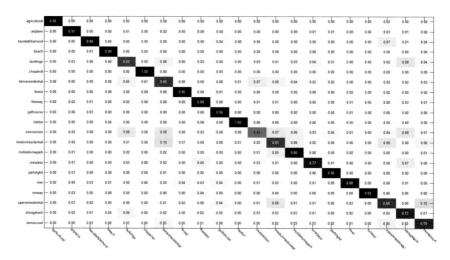

Fig. 5 The average confusion matrix of conventional ELM

we can find that the classification accuracy of 15 out of 21 categories is more than 80 %, among which 11 categories are more than 90 %. However, the classification accuracy of 4 categories is less than 70 %, which are buildings, dense residential, intersection and medium density residential.

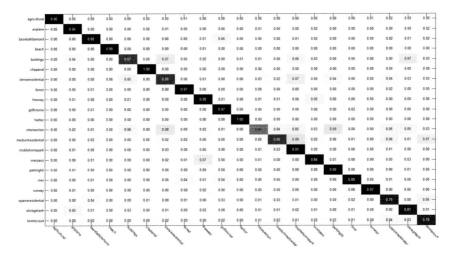

Fig. 6 The average confusion matrix of proposed method

7 Conclusions

In this paper, the ELM classifier is developed to tackle the covariance descriptor classification problem. Since covariance descriptor lies in non-Euclidean space, we propose a new ELM-guided coding strategy incorporating dictionary learning and ELM classifier design. Such a coding strategy can jointly obtain the dictionary and ELM classifier. Experiments on the public dataset show the effectiveness of our method.

Acknowledgments This work was supported in part by the National Key Project for Basic Research of China under Grant 2013CB329403; in part by the National Natural Science Foundation of China under Grant 61210013 and in part by the Tsinghua University initiative Scientific Research Program under Grant 20131089295.

References

1. Huang, G., Zhu, Q., Siew, C.: Extreme learning machine: theory and applications. Neurocomputing **70**, 489–501 (2006)
2. Huang, G., Chen, L., Siew, C.: Universal approximation using incremental constructive feedforward networks with random hidden nodes. IEEE Trans. Neural Netw. **17**(4), 879–892 (2006)
3. Huang, G., Chen, L.: Enhanced random search based incremental extreme learning machine. Neurocomputing **71**, 3460–3468 (2008)
4. Alexandre, E., Cuadra, L., Salcedo-Sanz, S., Pastor-Sanchez, A., Casanova-Matero, C.: Hybridizing extreme learning machines and genetic algorithms to select acoustic features in vehicle classification applications. Neurocomputing **152**, 58–68 (2015)
5. Han, M., Liu, B.: Ensemble of extreme learning machine for remote sensing image classification. Neurocomputing **149**, 65–70 (2015)

6. Zhu, W., Miao, J., Hu, J., Qing, L.: Vehicle detection in driving simulation using extreme learning machine. Neurocomputing **128**, 160–165 (2014)

7. An, L., Yang, S., Bhanu, B.: Efficient smile detection by extreme learning machine. Neurocomputing **149**, 354–363 (2015)

8. Rong, H., Jia, Y., Zhao, G.: Aircraft recognition using modular extreme learning machine. Neurocomputing **128**, 166–174 (2014)

9. Chen, X., Koskela, M.: Skeleton-based action recognition with extreme learning machine. Neurocomputing **149**, 387–396 (2015)

10. Bencherif, M.A., Bazi, Y., Guessoum, A., Alajlan, N., Melgani, F., Alhichri, H.: Fusion of extreme learning machine and graph-based optimization methods for active classification of remote sensing images. IEEE Geosci. Remote Sens. Lett. **12**(3), 527–531 (2015)

11. Tuzel, O., Porikli, F., Meer, P.: Region covariance: a fast descriptor for detection and classification. In: European Conference on Computer Visoin, pp. 589–600 (2006)

12. Huang, G., Chen, L.: Convex incremental extreme learning machine. Neurocomputing **70**, 3056–3062 (2007)

13. Huang, G., Zhou, H., Ding, X., Zhang, R.: Extreme learning machine for regression and multiclass classification. IEEE Trans. Syst. Man Cybern. **42**(2), 513–529 (2012)

14. Tang, J., Deng, C., Huang, G., Zhao, B.: Compressed-domain ship detection on spaceborne optical image using deep neural network and extreme learning machine. IEEE Trans. Geosci. Remote Sens. **53**(3), 1174–1185 (2015)

15. Decherchi, S., Gastaldo, P., Cambria, E., Redi, J.: Circular-ELM for the reduced-reference assessment of perceived image quality. Neurocomputing **102**, 78–89 (2013)

16. Arsigny, V., Fillard, P., Pennec, X., Ayache, N.: Geometric means in a novel vector space structure on symmetric positive-definite matrices. SIAM J. Matrix Anal. Appl. **29**(1), 328–347 (2006)

17. Li, P., Wang, Q., Zuo, W., Zhang, L.: Log-Euclidean kernels for sparse representation and dictionary learning. In: IEEE International Conference on Computer Vision, pp. 1601–1608 (2013)

18. Bo, L., Sminchisescu, C.: Efficient match kernels between sets of features for visual recognition. In: Advances in Neural Information Processing Systems, pp. 135–143 (2009)

19. Gao, S., Tsang, I.W., Chia, L.: Sparse representation with kernels. IEEE Trans. Image Process. **22**(2), 423–434 (2013)

20. Nguyen, H., Patel, V., Nasrabadi, N., Chellappa, R.: Design of non-linear kernel dictionaries for object recognition. IEEE Trans. Image Process. **22**(12), 5123–5135 (2013)

21. Kim, M.: Efficient kernel sparse coding via first-order smooth optimization. IEEE Trans. Neural Netw. Learn. Syst. **25**(8), 1447–1459 (2014)

22. Yang, Y., Shawn, N.: Bag-of-visual-words and spatial extensions for land-use classification. In: Proceedings of the 18th SIGSPATIAL International Conference on Advances in Geographic Information Systems, pp. 270–279 (2010)

Distributed Weighted Extreme Learning Machine for Big Imbalanced Data Learning

Zhiqiong Wang, Junchang Xin, Shuo Tian and Ge Yu

Abstract Extreme Learning Machine (ELM) and its variants have been widely used in many big data learning applications where raw data with imbalanced class distribution can be easily found. Although there have been several works solving the machine learning and robust regression problems using MapReduce framework, they need multi-iterative computations. Therefore, in this paper, we propose a novel Distributed Weighted Extreme Learning Machine based on MapReduce framework, named DWELM, which can learn the big imbalanced training data efficiently. Firstly, after indepth analyzing the properties of centralized Weighted ELM (WELM), it can be found out that the matrix multiplication operators in WELM are decomposable. Next, a DWELM based on MapReduce framework can be developed, which can first calculate the matrix multiplications effectively using two MapReduce Jobs in parallel, and then calculate the corresponding output weight vector with centralized computing. Finally, we conduct extensive experiments on synthetic data to verify the effectiveness and efficiency of our proposed DWELM in learning big imbalanced training data with various experimental settings.

Keywords Weighted ELM · Big imbalanced data · MapReduce framework · In-mapper combining

Z. Wang (✉) · S. Tian
Sino-Dutch Biomedical & Information Engineering School, Northeastern
University, Shenyang, China
e-mail: wangzq@bmie.neu.edu.cn

S. Tian
e-mail: xinjunchang@ise.neu.edu.cn

J. Xin · G. Yu
College of Information Science & Engineering, Northeastern University,
Shenyang, China
e-mail: dyhswdza@sina.com

G. Yu
e-mail: yuge@ise.neu.edu.cn

© Springer International Publishing Switzerland 2016
J. Cao et al. (eds.), *Proceedings of ELM-2015 Volume 1*,
Proceedings in Adaptation, Learning and Optimization 6,
DOI 10.1007/978-3-319-28397-5_25

1 Introduction

With the proliferation of mobile devices, artificial intelligence, web analytics, social media, internet of things, location based services and other types of emerging technologies, the amount of data, and the rate at which it's being accumulated, is rising exponentially. For examples, Facebook users share 2.5 billion unique pieces of content, hit the "like" button 2.7 billion times and upload 300 million photos a day. Thus, the era of big data has arrived [1, 2].

Extreme Learning Machine (ELM) [3–8] has recently attracted increasing attention from more and more researchers due to the characteristics of excellent generalization performance, rapid training speed and little human intervene [9]. ELM and its variants have been extensively used in many fields, such as text classification, image recognition, handwritten character recognition, mobile object management and bioinformatics [10–21].

Recently, as important variants of ELM, some Distributed ELM (DELM) [22–25] have been proposed to resolve the problem of big data learning, and a centralized Weighted ELM (WELM) [26] has been proposed to deal with data with imbalanced class distribution. However, neither DELM nor WELM can cope with big imbalanced training data efficiently since they only consider one aspect of big imbalanced data, though raw data with imbalanced class distribution can be found in many big data learning applications [26]. Therefore, in this paper, a Distributed Weighted Extreme Learning Machine (DWELM) which combines the advantages of both DELM and WELM based on distributed MapReduce framework [27–29] is proposed, to improve the scalability of centralized WELM and make it learn the big imbalanced data efficiently. The contributions of this paper are as follows.

- We prove theoretically that the matrix multiplication operators in centralized WELM are decomposable.
- A novel Distributed Weighted Extreme Learning Machine based on MapReduce framework (DWELM) is proposed to learn big imbalanced data efficiently.
- Last but not least, our extensive experimental studies using synthetic data show that our proposed DWELM can learn big imbalanced data efficiently, which can fulfill the requirements of many real-world big data applications.

The rest of the paper is organized as follows. Section 2 briefly reviews the background for our work. The theoretical foundation and the computational details of the proposed DWELM are introduced in Sect. 3. The experimental results to show the effectiveness of the proposed approaches are reported in Sect. 4. Finally, Sect. 5 concludes this paper.

2 Background

2.1 Weighted Extreme Learning Machine

ELM [3, 4] has been originally developed for single hidden-layer feedforward neural networks (SLFNs) and then extended to the "generalized" SLFNs where the hidden layer need not be neuron alike [5, 6]. ELM first randomly assigns the input weights and the hidden layer biases, and then analytically determines the output weights of SLFNs. It can achieve better generalization performance than other conventional learning algorithms at an extremely fast learning speed. Besides, ELM is less sensitive to user-specified parameters and can be deployed faster and more conveniently [7, 8]. Recently, a centralized Weighted ELM (WELM) [26] has been proposed to deal with data with imbalanced class distribution.

For N arbitrary distinct samples $(\mathbf{x}_j, \mathbf{t}_j)$, where $\mathbf{x}_j = [x_{j1}, x_{j2}, \dots, x_{jn}]^T \in \mathbb{R}^n$ and $\mathbf{t}_j = [t_{j1}, t_{j2}, \dots, t_{jm}]^T \in \mathbb{R}^m$, standard SLFNs with L hidden nodes and activation function $g(x)$ are mathematically modeled as

$$\sum_{i=1}^{L} \beta_i g_i(\mathbf{x}_j) = \sum_{i=1}^{L} \beta_i g(\mathbf{w}_i \cdot \mathbf{x}_j + b_i) = \mathbf{o}_j \qquad (j = 1, 2, \dots, N) \qquad (1)$$

where $\mathbf{w}_i = [w_{i1}, w_{i2}, \dots, w_{in}]^T$ is the weight vector connecting the ith hidden node and the input nodes, $\beta_i = [\beta_{i1}, \beta_{i2}, \dots, \beta_{im}]^T$ is the weight vector connecting the ith hidden node and the output nodes, b_i is the threshold of the ith hidden node, and $\mathbf{o}_j = [o_{j1}, o_{j2}, \dots, o_{jm}]^T$ is the jth output of the SLFNs [3].

The standard SLFNs with L hidden nodes and activation function $g(x)$ can approximate these N samples with zero error. It means $\sum_{j=1}^{L} ||\mathbf{o}_j - \mathbf{t}_j|| = 0$ and there exist β_i, $\mathbf{w}_i$ and b_i such that

$$\sum_{i=1}^{L} \beta_i g(\mathbf{w}_i \cdot \mathbf{x}_j + b_i) = \mathbf{t}_j \qquad (j = 1, 2, \dots, N) \qquad (2)$$

The equation above can be expressed compactly as follows:

$$\mathbf{H}\beta = \mathbf{T} \qquad (3)$$

where $\mathbf{H}$ is called the hidden layer output matrix of the neural network and the ith column of $\mathbf{H}$ is the ith hidden node output with respect to inputs $\mathbf{x}_i, \mathbf{x}_2, \dots, \mathbf{x}_N$.

To maximize the marginal distance and to minimize the weighted cumulative error with respect to each sample, we have an optimization problem mathematically written as

$$Minimize : \frac{1}{2} \|\beta\|^2 + CW \frac{1}{2} \sum_{i=1}^{N} \|\xi_i\|^2$$
$$Subject\ to\ :\ \mathbf{h}(\mathbf{x}_i)\beta = \mathbf{t}_i^T - \xi_i^T \tag{4}$$

where C is the regularization parameter to represent the trade-off between the minimization of weighted cumulative error and the maximization of the marginal distance. ξ_i, the training error of sample $\mathbf{x}_i$, is caused by the difference of the desired output $\mathbf{t}_i$ and the actual output $\mathbf{h}(\mathbf{x}_i)\beta$. $\mathbf{W}$ is a $N \times N$ diagonal matrix associated with every training sample $\mathbf{x}_i$, and

$$W_{ii} = 1/\#(\mathbf{t}_i) \tag{5}$$

or

$$W_{ii} = \begin{cases} 0.618/\#(\mathbf{t}_i) & \text{if } \#(\mathbf{t}_i) > AVG \\ 1/\#(\mathbf{t}_i) & \text{if } \#(\mathbf{t}_i) \leq AVG \end{cases} \tag{6}$$

where $\#(\mathbf{t}_i)$ is the number of samples belonging to class $\mathbf{t}_i$, and AVG is the average number of samples per class.

According to Karush-Kuhn-Tucker (KKT) theorem [30], we have the following solutions for Weighted ELM (WELM):

$$\beta = \left(\frac{\mathbf{I}}{\lambda} + \mathbf{H}^T \mathbf{W} \mathbf{H}\right)^{-1} \mathbf{H}^T \mathbf{W} \mathbf{T} \tag{7}$$

when N is large or

$$\beta = \mathbf{H}^T \left(\frac{\mathbf{I}}{\lambda} + \mathbf{W} \mathbf{H} \mathbf{H}^T\right)^{-1} \mathbf{W} \mathbf{T} \tag{8}$$

when N is small.

2.2 MapReduce Framework

MapReduce is a simple and flexible parallel programming model initially proposed by Google for large scale data processing in a distributed computing environment [27–29], with one of its open source implementations Hadoop.[1] The typical procedure of a MR job is as follows: First, the input to a MR job starts as the dataset stored on the underlying distributed file system (e.g. GFS [31] and HDFS [32]), which is split into a number of files across machines. Next, the MR job is partitioned into many independent map tasks. Each map task processes a logical split of the input dataset. The map task reads the data and applies the user-defined map function on each record, and then buffers the resulting intermediate output. This intermediate data is sorted and partitioned for reduce phase, and written to the local disk of the machine executing the corresponding map task. After that, the intermediate data

---

[1] http://hadoop.apache.org/.

files from the already completed map tasks are fetched by the corresponding reduce task following the "pull" model (Similarly, the MR job is also partitioned into many independent reduce tasks). The intermediate data files from all the map tasks are sorted accordingly. Then, the sorted intermediate data is passed to the reduce task. The reduce task applies the user-defined reduce function to the intermediate data and generates the final output data. Finally, the output data from the reduce task is generally written back to the corresponding distributed file system.

3 Distributed Weighted Extreme Learning Machine

3.1 Preliminaries

In big imbalanced data learning applications, the number of training records is much larger than the dimensionality of the feature space, that is to say, $N \gg L$. According to $N \gg L$, the size of $\mathbf{H}^T \mathbf{W} \mathbf{H}$ is much smaller than that of $\mathbf{W} \mathbf{H} \mathbf{H}^T$. Therefore, it is a better choice of using Eq. (7) to calculate the output weight vector β in WELM. Similar with ELM* [23], we analyze the properties of centralized WELM, and find the part that can be processed in parallel, and then transplant it into MapReduce framework. In this way, we can make WELM extend to the scale of big imbalance data efficiently. Let $\mathbf{U} = \mathbf{H}^T \mathbf{W} \mathbf{H}$, $\mathbf{V} = \mathbf{H}^T \mathbf{W} \mathbf{T}$, and we can get,

$$\beta = \left(\frac{\mathbf{I}}{\lambda} + \mathbf{U} \right)^{-1} \mathbf{V} \tag{9}$$

According to the matrix multiplication operator, we have

$$\mathbf{U} = \mathbf{H}^T \mathbf{W} \mathbf{H} = \sum_{k=1}^{N} h(\mathbf{x}_k)^T W_{kk} h(\mathbf{x}_k) \tag{10}$$

Then, we can further get,

$$u_{ij} = \sum_{k=1}^{N} W_{kk} \times g(\mathbf{w}_i \cdot \mathbf{x}_k + b_i) \times g(\mathbf{w}_j \cdot \mathbf{x}_k + b_j) \tag{11}$$

Similarly, according to the matrix multiplication operator, we also have

$$\mathbf{V} = \mathbf{H}^T \mathbf{W} \mathbf{T} = \sum_{i=1}^{N} h(\mathbf{x}_k)^T W_{kk} \mathbf{t}_k \tag{12}$$

Then, we can further get,

$$v_{ij} = \sum_{k=1}^{N} W_{kk} \times g(\mathbf{w}_i \cdot \mathbf{x}_k + b_i) \times t_{kj} \qquad (13)$$

According to Eq. (11), we know that the item u_{ij} in matrix $\mathbf{U}$ can be expressed by the summation of $W_{kk} \times g(\mathbf{w}_i \cdot \mathbf{x}_k + b_i) \times g(\mathbf{w}_j \cdot \mathbf{x}_k + b_j)$. Here, W_{kk} is the weight of training sample $(\mathbf{x}_k, \mathbf{t}_k)$, and $h_{ki} = g(\mathbf{w}_i \cdot \mathbf{x}_k + b_i)$ and $h_{kj} = g(\mathbf{w}_j \cdot \mathbf{x}_k + b_j)$ are the ith and jth elements in the kth row $h(\mathbf{x}_k)$ of the hidden layer output matrix $\mathbf{H}$, respectively. Similarly, according to Eq. (13), we know that item v_{ij} in matrix $\mathbf{V}$ can be expressed by the summation of $W_{kk} \times g(\mathbf{w}_i \cdot \mathbf{x}_k + b_i) \times t_{kj}$. Here, W_{kk} is the weight of training sample $(\mathbf{x}_k, \mathbf{t}_k)$, $h_{ki} = g(\mathbf{w}_i \cdot \mathbf{x}_k + b_i)$ is the ith element in the kth row $h(\mathbf{x}_k)$ of the hidden layer output matrix $\mathbf{H}$, and t_{kj} is the jth element in the kth row $\mathbf{t}_k$ of matrix $\mathbf{T}$ which related to $(\mathbf{x}_k, \mathbf{t}_k)$.

The variables involved in equations of matrices $\mathbf{U}$ and $\mathbf{V}$ include: W_{kk}, h_{ki}, h_{kj} and t_{kj}. According to Eqs. (5) and (6), to calculate the corresponding weight W_{kk} related to training sample $(\mathbf{x}_k, \mathbf{t}_k)$, we must first get the number #$(\mathbf{t}_k)$ of training samples which belongs to the same class as $\mathbf{t}_k$. The numbers of training samples in all classes can be easily calculated in one MR job. At the same time, the remaining three variables h_{ki}, h_{kj} and t_{kj} only have relationship with training sample $(\mathbf{x}_k, \mathbf{t}_k)$ itself, and have nothing to do with the other training samples, so the calculation of matrices $\mathbf{U}$ and $\mathbf{V}$ can be done in another MR Job.

To sum up, the calculation process of matrices $\mathbf{U}$ and $\mathbf{V}$ is decomposable, therefore, similar to ELM* [23], we can realize the parallel computation of matrices $\mathbf{U}$ and $\mathbf{V}$ by using MapReduce framework, to break through the limitation of single machine, so as to improve the efficiency of which WELM learns big imbalanced training data.

3.2 DWELM

The process of DWELM is shown in Algorithm 1. Firstly, we randomly generate L pairs of hidden node parameters $(\mathbf{w}_i, b_i)$ (Lines 1–2). And then, using a MR Job to count the number of training samples contained in each class (Line 3). Next, using another MR Job to calculate matrices $\mathbf{U}$ and $\mathbf{V}$ according to the input parameters and randomly generate parameters (Line 4). Finally, we solve output weight vector β according to the Eq. 7 (Line 5).

Algorithm 1 DWELM

---

for $i = 1$ to L **do**
    Randomly generate hidden node parameters $(\mathbf{w}_i, b_i)$
Calculate all #$(\mathbf{t}_k)$ using Algorithm 2
Calculate $\mathbf{U} = \mathbf{H}^T \mathbf{WH}$, $\mathbf{V} = \mathbf{H}^T \mathbf{WT}$ using Algorithm 3
Calculate the output weight vector $\beta = (\mathbf{I}/\lambda + \mathbf{U})^{-1} \mathbf{V}$

Here are the specific processes of two MR Jobs involved in DWELM:

The process of the 1st MR Job is shown in Algorithm 2. The algorithm includes two classes, Class Mapper (Lines 1–10) and Class Reducer (Lines 11–16). Class Mapper contains three methods, Initialize (Lines 2–3), Map (Lines 4–7) and Close (Line 8–10), while Class Reducer only contains one method, Reduce (Lines 12–16). In the Initialize method of Mapper, we initialize one array, c, which is used to store the intermediate summation of training samples contained in each class (Line 3). In the Map method of Mapper, firstly, we analyze the training sample s, and resolve the class which sample s belongs to (Lines 5–6). Then, adjust the corresponding value in the array c (Line 7). In the Close method of Mapper, the intermediate summations stored in c are emitted by the mapper (Lines 9–10). In the Reduce method of Reducer, firstly, we initialize a temporary variable *sum* (Line 13). And then, we combine the intermediate summations of different mappers which have the same Key, and furthermore, get the final summation of the corresponding element of the Key (Lines 14–15). Finally, we store the results into the distributed file system (Line 16).

Algorithm 2 The 1st MR Job of DWELM

---

class MAPPER
   method INITIALIZE()
     c = new ASSOCIATIVEARRAY
   method MAP(sid *id*, sample *s*)
     $\mathbf{t}$ =ParseT(s)
     num =Class($\mathbf{t}$)
     $c[num] = c[num] + 1$
   method CLOSE()
     for $i = 1$ to c.Length() **do**
       context.write(cid i, count $c[i]$)
class REDUCER
   method REDUCE(cid *id*, counts $[c_1, c_2, \dots]$)
     sum = 0
     for all count $c \in [c_1, c_2, \dots]$ **do**
       sum = *sum* + c
     context.write(cid *id*, count *sum*)

The process of the 2nd MR Job is shown in Algorithm 3. The algorithm includes two classes, Class Mapper (Lines 1–21) and Class Reducer (Lines 22–27). Class Mapper contains three methods, Initialize (Lines 2–4), Map (Lines 5–15) and Close (Line 16–21), while Class Reducer only contains one method, Reduce (Lines 23–27). In the Initialize method of Mapper, we initialize two arrays, u and v, which are used to store the intermediate summations of the elements in matrices $\mathbf{U}$ and $\mathbf{V}$ respectively. In the Map method of Mapper, firstly, we initialize a local variable h (Line 6). Then, we resolve the input training sample s, dividing s into training feature $\mathbf{x}$ and its corresponding training result $\mathbf{t}$ (Line 7). Again, according to training result $\mathbf{t}$ and the result of Algorithm 2, we get the corresponding weight w of s (Line8). And then calculate the corresponding hidden layer output vector $h(\mathbf{x})$ (Lines 9–10). Finally, separately calculate local summations of the elements in matrices $\mathbf{U}$ and $\mathbf{V}$, and save the result to local variables u and v (Lines 11–15). In the Close method of Mapper, the intermediate summations stored in u and v are emitted by the mapper (Lines 17–21). In the Reduce method of Reducer, firstly, we initialize a temporary variable uv (Line 24). And then, we combine the intermediate summations which have the same Key, and furthermore, get the final summation of the corresponding element of the Key (Lines 25–26). Finally, we store the results into the distributed file system (Line 27).

Algorithm 3 The 2nd MR Job of DWELM

---

class MAPPER
    method INITIALIZE()
        u = new ASSOCIATIVEARRAY
        v = new ASSOCIATIVEARRAY
    method MAP(sid *id*, sample *s*)
        h = new ASSOCIATIVEARRAY
        $(\mathbf{x}, \mathbf{t})$ =ParseAll(s)
        w =Weight(*Counts*[Class($\mathbf{t}$)])
        for $i = 1$ to L **do**
          $h[i] = g(\mathbf{w}_i \cdot \mathbf{x} + b_i)$
        for $i = 1$ to L **do**
          for $j = 1$ to L **do**
            $u[i,j] = u[i,j] + w \times h[i] \times h[j]$
          for $j = 1$ to m **do**
            $v[i,j] = v[i,j] + w \times h[i] \times \mathbf{t}[j]$
    method CLOSE()
        for $i = 1$ to L **do**
          for $j = 1$ to L **do**
            context.write(triple ($'U'$, i, j), sum $u[i,j]$)
          for $j = 1$ to m **do**
            context.write(triple ($'V'$, i, j), sum $v[i,j]$)
class REDUCER
    method REDUCE(triple p, sum $[s_1, s_2, \dots]$)
        $uv = 0$
        for all sum $s \in [s_1, s_2, \dots]$ **do**
          $uv = uv + s$
        context.write(triple p, sum uv)

---

4 Performance Evaluation

4.1 Experimental Platform

All the experiments are running on a cluster with 9 computers which are connected in a high speed Gigabit network. Each computer has an Intel Quad Core 2.66 GHZ CPU, 4 GB memory and CentOS Linux 5.6. One computer is set as the Master node

Table 1 Experimental parameters

Parameter	Range and default
Dimensionality (D)	10, 20, 30, 40, **50**
Number of hidden nodes (N_h)	100, 150, **200**, 250, 300
Number of records (N_r)	3M(1.4G), 4M(1.86G), **5M(2.3G)**, 6M(2.8G), 7M(3.27G)
Number of classes (N_c)	5, 10, **15**, 20, 25
Imbalance ratio (R)	0.3, 0.4, **0.5**, 0.6, 0.7
Number of nodes (N_n)	1, 2, 3, 4, 5, 6, 7, **8**

and the others are set as the Slave nodes. We use Hadoop version 0.20.2 and configure it to run up to 4 map tasks or 4 reduce tasks concurrently per node. Therefore, at any point in time, at most 32 map tasks or 32 reduce tasks can run concurrently in our cluster.

Because DWELM is MapReduce-based implementation of centralized WELM, and it does not change any formula in WELM, so it does not have any effect on the classification accuracy rate. In addition, the other learning algorithms of MapReduce solutions such as SVM needs many iterations to obtain the final results. Our DWELM only use two MapReduce job to gain the results. So, the performance of two MapReduce jobs is obviously optimal to several MapReduce computations. Even though we compare the SVM and DWELM, the results of our DWELM are better than SVM. Therefore, we only evaluate the training time of DWELM in the experiments. Table 1 summarizes the parameters used in our experimental evaluation, along with their ranges and default values shown in bold. In each experiment, we vary a single parameter, while setting the remainders to their default values. The imbalance ratio which quantitatively measure the imbalance degree of a dataset is defined as $Min(\#(\mathbf{t}_i))/Max(\#(\mathbf{t}_i))$ [26].

4.2 Experimental Results

Firstly, we investigate the influence of the training data dimensionality. As shown in Fig. 1, with the increase of training data dimensionality, the training time of DWELM increase slightly. Increase of training data dimensionality leads to the running time for calculating the corresponding row h_k of hidden layer output matrix $\mathbf{H}$ in Mapper slightly increases, then leads to the training time of DWELM slightly increases.

Secondly, we investigate the influence of the number of hidden nodes. As shown in Fig. 2, with the increase of the number of hidden nodes, the training time of DWELM increases. Increasing of the number of hidden nodes leads to an increase of the dimensionality of hidden layer output matrix $\mathbf{H}$, and indirectly leads to the increase of the dimensionality of the intermediate matrices $\mathbf{U}$ and $\mathbf{V}$. This not only

Fig. 1 The influence of D

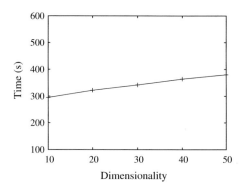

Fig. 2 The influence of N_h

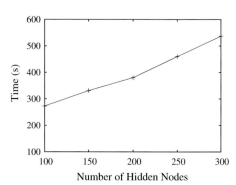

makes the computation time of the local accumulated sum of **U** and **V** increase, but also makes the transmission time of intermediate results in MR Job increase. Therefore, the training time of DWELM increases with the number of hidden nodes.

Again, we investigate the influence of the number of training records. As shown in Fig. 3, with the increase of the number of records, the training time of DWELM increases obviously. Increasing of the number of records means that the number that MR Job needs to deal with increases, leading to the amount of Mapper and Reducer

Fig. 3 The influence of N_r

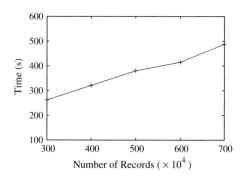

which need to be launched increase. On the other hand, it increases the number of corresponding local accumulated sum of **U** and **V** which need to be transmitted, leading to transmission time of intermediate results increases. Therefore, the training time of DWELM increases with the increasing of the number of training records.

Then, we investigate the influence of the number of classes. As shown in Fig. 4, along with the increase of the number of classes, the training time of DWELM is basically stable. The number of classes increases, which only increases the number of statistical values in the 1st MR Job and the number of input values in the 2nd MR Job of DWELM, which has limited impact on the overall training time, so the training time is relatively stable.

Next, we investigate the influence of imbalance ratio. As shown in Fig. 5, with the increase of imbalance ratio, the training time of DWELM is basically stable. Increasing of imbalance ratio did not produce any substantial effects on the calculation process of MR Job, so the training time is relatively stable.

Finally, we discuss the influence of the number of working slave nodes in the Cluster. As shown in Fig. 6, with the number of slave nodes increasing, the training time of DWELM decreased significantly. Increasing of number of slave nodes implies that increasing of the amount of Mapper/Reducers that be launched at the same time, it also means that the work can be completed in unit time increasing.

Fig. 4 The influence of N_c

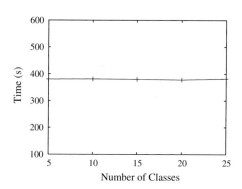

Fig. 5 The influence of R

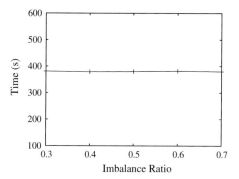

Fig. 6 The influence of N_n

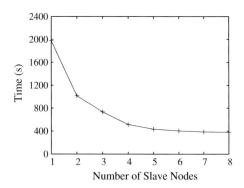

Therefore, in the premise of constant total workload, the training time of DWELM decreases.

In summary, no matter how the experimental parameters change, DWELM can always deal with large-scale data (millions of data) effectively and rapidly (several minutes). At the same time, DWELM has better scalability, through the expansion of the hardware platform, they can easily handle billions and even hundreds of billion of the big imbalanced training data, thereby improve the processing efficiency of big data learning applications significantly.

5 Conclusions

Neither WELM nor DELM can cope with big imbalanced training data efficiently since they only consider either "big" or "imbalanced" aspect of big imbalanced training data. In this paper, we combine the advantages of WELM and DELM, and propose a Distributed Weighted Extreme Learning Machine based on MapReduce framework (DWELM). Specifically, through analyzing the characters of centralized WELM, we found that the matrix multiplication operators (i.e. $\mathbf{H}^T\mathbf{W}\mathbf{H}$ and $\mathbf{H}^T\mathbf{W}\mathbf{T}$) in WELM are decomposable. Then, we transform the corresponding matrix multiplication operators into summation forms, which suit MapReduce framework well, and propose a DWELM which calculates the matrix multiplications using two MapReduce Jobs. Finally, in the Cluster environment, we use synthetic data to do a detailed validation of the performance of DWELM with various experimental settings. The experimental results show that DWELM can learn big imbalanced training data efficiently.

Acknowledgments This research was partially supported by the National Natural Science Foundation of China under Grant Nos. 61402089 and 61472069; the 863 Program under Grant No. 2012AA011004, and the Fundamental Research Funds for the Central Universities under Grant Nos. N141904001 and N130404014.

References

1. Chen, M., Mao, S., Liu, Y.: Big data: a survey. Mob. Netw. Appl. **19**(2), 171–209 (2014)
2. Chen, J., Chen, Y., Du, X., Li, C., Lu, J., Zhao, S., Zhou, X.: Big data challenge: a data management perspective. Front. Comput. Sci. **7**(2), 157–164 (2013)
3. Huang, G.-B., Zhu, Q.-Y., Siew, C.-K.: Extreme learning machine: theory and applications. Neurocomputing **70**(1–3), 489–501 (2006)
4. Huang, G.-B., Chen, L., Siew, C.-K.: Universal approximation using incremental constructive feedforward networks with random hidden nodes. IEEE Trans. Neural Netw. **17**(4), 879–892 (2006)
5. Huang, G.-B., Chen, L.: Convex incremental extreme learning machine. Neurocomputing **70**(16–18), 3056–3062 (2007)
6. Huang, G.-B., Chen, L.: Enhanced random search based incremental extreme learning machine. Neurocomputing **71**(16–18), 3460–3468 (2008)
7. Huang, G.-B., Ding, X., Zhou, H.: Optimization method based extreme learning machine for classification. Neurocomputing **74**(1–3), 155–163 (2010)
8. Huang, G.-B., Zhou, H., Ding, X., Zhang, R.: Extreme learning machine for regression and multiclass classification. IEEE Trans. Syst. Man Cybern. Part B: Cybern. **42**(2), 513–529 (2012)
9. Huang, G.-B., Wang, D.H., Lan, Y.: Extreme learning machines: a survey. Int. J. Mach. Learn. Cybern. **2**(2), 107–122 (2011)
10. Zhang, R., Huang, G.-B., Sundararajan, N., Saratchandran, P.: Multi-category classification using an extreme learning machine for microarray gene expression cancer diagnosis. IEEE/ACM Trans. Comput. Biol. Bioinf. **4**(3), 485–495 (2007)
11. Zhu, Q.-Y., Qin, A.K., Suganthan, P.N., Huang, G.-B.: Evolutionary extreme learning machine. Pattern Recogn. **38**(10), 1759–1763 (2005)
12. Wang, G., Zhao, Y., Wang, D.: A protein secondary structure prediction framework based on the extreme learning machine. Neurocomputing **72**(1–3), 262–268 (2008)
13. Zhao, X., Wang, G., Bi, X., Gong, P., Zhao, Y.: XML document classification based on ELM. Neurocomputing **74**(16), 2444–2451 (2011)
14. Sun, Y., Yuan, Y., Wang, G.: An OS-ELM based distributed ensemble classification framework in P2P networks. Neurocomputing **74**(16), 2438–2443 (2011)
15. Wang, B., Wang, G., Li, J., Wang, B.: Update strategy based on region classification using elm for mobile object index. Soft Comput. **16**(9), 1607–1615 (2012)
16. Huang, G.-B., Liang, N.-Y., Rong, H.-J., Saratchandran, P., Sundararajan, N.: On-line sequential extreme learning machine. In: Proceedings of CI, pp. 232–237 (2005)
17. Liang, N.-Y., Huang, G.-B., Saratchandran, P., Sundararajan, N.: A fast and accurate on-line sequential learning algorithm for feedforward networks. IEEE Trans. Neural Netw. **17**(6), 1411–1423 (2006)
18. Rong, H.-J., Huang, G.-B., Sundararajan, N., Saratchandran, P.: On-line sequential fuzzy extreme learning machine for function approximation and classification problems. IEEE Trans. Syst. Man Cybern.: Part B **39**(4), 1067–1072 (2009)
19. Wang, X., Shao, Q., Miao, Q., Zhai, J.: Architecture selection for networks trained with extreme learning machine using localized generalization error model. Neurocomputing **102**(1), 3–9 (2013)
20. Zhai, J., Xu, H., Wang, X.: Dynamic ensemble extreme learning machine based on sample entropy. Soft Comput. **16**(9), 1493–1502 (2012)
21. Wang, Z., Yu, G., Kang, Y., Zhao, Y., Qu, Q.: Breast tumor detection in digital mammography based on extreme learning machine. Neurocomputing **128**(3), 175–184 (2014)
22. He, Q., Shang, T., Zhuang, F., Shi, Z.: Parallel extreme learning machine for regression based on mapreduce. Neurocomputing **102**(2), 52–58 (2013)
23. Xin, J., Wang, Z., Chen, C., Ding, L., Wang, G., Zhao, Y.: ELM*: Distributed extreme learning machine with mapreduce. World Wide Web **17**(5), 1189–1204 (2014)

24. Bi, X., Zhao, X., Wang, G., Zhang, P., Wang, C.: Distributed extreme learning machine with kernels based on mapreduce. Neurocomputing **149**(1), 456–463 (2015)
25. Xin, J., Wang, Z., Qu, L., Wang, G.: Elastic extreme learning machine for big data classification. Neurocomputing **149**(1), 464–471 (2015)
26. Zong, W., Huang, G.-B., Chen, Y.: Weighted extreme learning machine for imbalance learning. Neurocomputing **101**(3), 229–242 (2013)
27. Dean, J., Ghemawat, S.: MapReduce: simplified data processing on large clusters. In: Proceedings of OSDI, pp. 137–150 (2004)
28. Dean, J., Ghemawat, S.: MapReduce: simplified data processing on large clusters. Commun. ACM **51**(1), 107–113 (2008)
29. Dean, J., Ghemawat, S.: MapReduce: a flexible data processing tool. Commun. ACM **53**(1), 72–77 (2010)
30. Fletcher, R.: Practical Methods of Optimization, Volume 2: Constrained Optimization. Wiley, Hoboken (1981)
31. Ghemawat, S., Gobioff, H., Leung, S.-T.: The google file system. In: Proceedings of SOSP, pp. 29–43 (2003)
32. Shvachko, K., Kuang, H., Radia, S., Chansler, R.: The Hadoop distributed file system. In: Proceedings of MSST, pp. 1–10 (2010)

NMR Image Segmentation Based on Unsupervised Extreme Learning Machine

Junchang Xin, Zhongyang Wang, Shuo Tian and Zhiqiong Wang

Abstract NMR image is often used in medical diagnosis. And image segmentation is one of the most important steps in the NMR image analysis, which is valuable for the computer-aided detection (CADe) and computer-aided diagnosis (CADx). As traditional image segmentation methods based on supervised learning required a lot of manual intervention. Thus, segmentation methods based on unsupervised learning have been received much concern, and unsupervised extreme learning machine (US-ELM)'s performance is particularly outstanding among the unsupervised learning methods. Therefore, in this paper, we proposed a NMR image segmentation method based on US-ELM, named NS-UE. Firstly, a NMR image feature model is established for the input NMR image; Secondly, the clustering based on US-ELM is proposed to separate the various regions of NMR image. Finally, a large number of experimental evaluation results demonstrated the effectiveness and efficiency of the proposed algorithms for the NMR image segmentation.

Keywords NMR image · Segmentation · US-ELM · NS-UE

J. Xin (✉)
College of Information Science & Engineering, Northeastern University,
Shenyang, China
e-mail: xinjunchang@ise.neu.edu.cn

Z. Wang · S. Tian · Z. Wang
Sino-Dutch Biomedical & Information Engineering School,
Northeastern University, Shenyang, China
e-mail: zanezhongyang@163.com

S. Tian
e-mail: dyhswdza@sina.com

Z. Wang
e-mail: wangzq@bmie.neu.edu.cn

© Springer International Publishing Switzerland 2016
J. Cao et al. (eds.), *Proceedings of ELM-2015 Volume 1*,
Proceedings in Adaptation, Learning and Optimization 6,
DOI 10.1007/978-3-319-28397-5_26

1   Introduction

Nuclear magnetic resonance (NMR) is one of the most frequently used methods in medical diagnosis. As NMR images contain the information of a large number of lesions used in clinical treatment. The purpose of segmentation of NMR image is to extract more diagnostic auxiliary meaningful information effectively, which refers to that the NMR image is divided into a number of different areas, or divide the region of interest (ROI), which is valuable for computer-aided detection (CADe) and computer-aided diagnosis (CADx) [1, 2]. As NMR image segmentation is one of important components of the NMR image processing system [3], has become a hot issue in image processing technology, which attracted the attention of many researchers [4–6].

Current clustering methods for image segmentation based on machine learning approaches have gained great success. Existing image segmentation methods are divided into two major categories, which are respectively based on supervised segmentation and unsupervised segmentation. Supervised learning methods have been widely used in image segmentation. Erbas et al. [7] proposed that Bayesian classifier and neural network classifier, and get extensive promotion and application in medical image segmentation research. Jyoti et al. [8] proposed a method that both the concept of clustering and thresholding technique with edge based segmentation methods like sobel, prewitt edge detectors is applied. Further the segmented result is passed through a gaussian filter to obtain a smoothed image. Pham et al. [9] proposed for an image segmentation that can be used in fruit defect detection. The main shortcomings of the method based on the supervision are the need of large amount of prior information, the image segmentation process is limited by manual intervention.

Since the unsupervised segmentation method does not need manual intervention, it attracts a lot of researchers' attention in recent years. Ahmadvand et al. [10] uses a proper combination of clustering methods and MRF and proposes a preprocessing step for MRF method for decreasing the computational burden of MRF for segmentation. Hocking et al. [11] presents a novel unsupervised learning approach to automatically segment and label images in astronomical surveys. Maji et al. [12] introduced an unsupervised feature selection method, based on maximum relevance-maximum significance criterion, to select relevant and significant textural features for segmentation problem, along with a comparison with related approaches, is demonstrated on a set of synthetic and real brain MR images using standard validity indices. Halim et al. [13] applied an unsupervised moving k-means clustering algorithm on the various colour components of RGB and HSI colour models for segmentation of blast cells from the red blood cells and background regions in leukemia image, and proved its effectiveness and efficiency.

No prior information of the image in the process of segmentation results that large amount of computation is produced in the image segmentation process, led to the clustering is difficult to achieve satisfactory accuracy. Moreover, extreme Learning Machine (ELM) requires fewer optimization constraints and results in simpler implementation, faster learning, and better generalization performance. Unsupervised extreme learning machine (US-ELM)'s performance is proved particularly outstanding [15]. Therefore, in this paper, a segmentation method based US-ELM is proposed (NS-UE). Firstly, feature model is established for the input NMR image; Secondly, using unsupervised extreme learning machine for data processing, data in the embedded space of US-ELM are clustered by K-means. With the clustered date, we can get the segmentation images. Finally, experimental evaluation proved the effectiveness and efficiency of the algorithm. The contributions of this paper can be summarized as follows.

- A NMR image pixels feature model based on the N neighborhood pixels is proposed, which is prepared for data clustering.
- Based the feature model we established, the segmentation method based on US-ELM is proposed, named NS-UE, for unsupervised clustering segmentation of NMR image.
- Last but not least, the experimental evaluation shows that our proposed approach can separate the various regions of NMR image effectively.

The remainder of the paper is organized as follows. Section 2 introduces unsupervised extreme learning machine. The details description of the proposed NMR image segmentation based on US-ELM, named NS-UE are introduced in Sect. 3. The experimental evaluation results show the effectiveness and efficiency of the proposed approach are reported in Sect. 4. Finally, we conclude this paper in Sect. 5.

2   Unsupervised Extreme Learning Machine

ELM [17, 18] as been originally developed for single hidden-layer feedforward neural networks (SLFNs) and then extended to the "generalized" SLFNs where the hidden layer need not be neuron alike [19, 20]. ELM first randomly assigns the input weights and the hidden layer biases, and then analytically determines the output weights of SLFNs. It can achieve better generalization performance than other conventional learning algorithms at an extremely fast learning speed. Besides, ELM is less sensitive to user-specified parameters and can be deployed faster and more conveniently [21, 22].

Unsupervised Extreme Learning Machine is proposed by Huang et al. [15]. A number of hidden neurons which map the data from the input space into a n_h-dimensional feature space (n_h is the number of hidden neurons) are randomly generated. Denote $h(x_i) \in \mathbb{1 \times n_h}$ to the output vector of the hidden layer with respect to x_i, and $\beta \in \mathbb{n_h \times n_0}$ the output weights that connect the hidden layer with the output layer. Then, the outputs of the network are given by

$$f(x_i) = h(x_i)\beta, \ i = 1, \dots, N. \tag{1}$$

ELMs aim to solve the output weights by minimizing the sum of the squared losses of the prediction errors, which leads to the following formulation

$$\min_{\beta \in^{n_h \times n_0}} \frac{1}{2} \|\beta\|^2 + \frac{C}{2} \sum_{i=1}^{l} \|e_i\|^2 \tag{2}$$
$$s.t. \ h(x_i)\beta = y_i^T - e_i^T, \quad i = 1, \dots, N$$

where the first term in the objective function is a regularization term which controls the complexity of the model, $e_i \in^{n_0}$ is the error vector with respect to the ith training pattern, and C is a penalty coefficient on the training errors. By substituting the constraints into the objective function, we obtain the following equivalent unconstrained optimization problem:

$$\min_{\beta \in^{n_h \times n_0}} \nabla L_{ELM} = \frac{1}{2} \|\beta\|^2 + \frac{C}{2} \|Y - \mathbf{H}\beta\|^2 \tag{3}$$

If $\mathbf{H}$ has more rows than columns and is of full column rank, which is usually the case where the number of training patterns are more than the number of the hidden neurons, the above equation is over determined, the following closed form solution for:

$$\beta^* = \left(\mathbf{H}^T \mathbf{H} + \frac{I_{n_h}}{C} \right)^{-1} \mathbf{H}^T Y \tag{4}$$

If the number of training patterns are less than the number of hidden neurons, then $\mathbf{H}$ will have more columns than rows, which often leads to an under-determined least squares problem.

$$\beta^* = \mathbf{H}^T \left(\mathbf{H}^T \mathbf{H} + \frac{I_{n_h}}{C} \right)^{-1} Y \tag{5}$$

The US-ELM algorithm [15] for unsupervised learning is introduced. In an unsupervised setting, the entire training data $X \in^{n \times n_i}$ are unlabeled. The formulation of US-ELM is reduced to

$$\min_{\beta \in^{n_h \times n_0}} \|\beta\|^2 + \lambda Tr(\beta^T \mathbf{H}^T L \mathbf{H} \beta) \tag{6}$$

Notice that the above formulation always attains its minimum at $\beta = 0$, the formulation of US-ELM is given by

$$\min_{\beta \in^{n_h \times n_0}} \|\beta\|^2 + \lambda Tr(\beta^T \mathbf{H}^T L \mathbf{H} \beta) \tag{7}$$
$$s.t.(\mathbf{H}\beta)^T \mathbf{H}\beta = I_{n_0}$$

An optimal solution to problem (7) is given by choosing β as the matrix whose columns are the eigenvectors (normalized to satisfy the constraint) corresponding to the first n_0 smallest eigenvalues of the generalized eigenvalue problem:

$$I_0 + l\mathbf{H}^T L \mathbf{H} v = \gamma \mathbf{H}^T \mathbf{H} v \tag{8}$$

The problem (7) can be rewrited as

$$\min_{\beta \in {}^{n_h \times n_0}, \beta^T B \beta = I_{n_0}} Tr(\beta^T A \beta) \tag{9}$$

where $A = I_{n_h} + \lambda \mathbf{H}^T L \mathbf{H}$ and $B = \mathbf{H}^T \mathbf{H}$. It is easy to verify that both A and B are Hermitian matrices. Thus, the above trace minimization problem attains its optimum if and only if the column span of β is the minimum span of the eigenspace corresponding to the smallest no eigenvalues of Eq. (8). Therefore, by stacking the normalized eigenvectors of Eq. (8) corresponding to the smallest no generalized eigenvalues, which can obtain an optimal solution to Eq. (7). In the algorithm of Laplacian eigenmaps, the first eigenvector is discarded since it is always a constant vector proportional to 1 (corresponding to the smallest eigenvalue 0), In the US-ELM algorithm, the first eigenvector of Eq. (8) also leads to small variations in embedding and is not useful for data representation.

Let $\gamma_1 \gamma_2 \ldots \gamma_{n_0+1} \gamma_1 \leq \gamma_2 \leq \cdots \leq \gamma_{n_o+1}$ be the $(n_o + 1)$ smallest eigenvalues of (8) and $v_1 v_2 \ldots v_{n_0+1}$ be their corresponding eigenvectors. Then, the solution to the output weights β is given by

$$\beta^* = [\tilde{v}_1, \tilde{v}_2, \ldots, \tilde{v}_{n_0+1}] \tag{10}$$

where $\tilde{v} = v_i / \|\mathbf{H} v_i\|, \quad i = 2, \ldots, n_0 + 1$ are the normalized eigenvectors.

3 NMR Image Segmentation Based on US-ELM

In this section, we proposed the method that NMR image segmentation based on US-ELM. Firstly, feature model are established by image preprocessing; Secondly, using US-ELM for data processing, we can get the segmentation images [15]. Image Feature Model process is introduced in Sect. 3.1, and the main segmentation method based on US-ELM (NS-UE) is introduced in Sect. 3.2.

3.1 NMR Image Feature Model

The data are processed by median filtering [23] to reduce the noise, and to improve the smoothness of the image. The output image I_f of the Median filtering is shown in Eq. (11), where $g(x, y)$ is the output pixel of the median filtering.

$$I_f = \begin{bmatrix} g(x_1, y_1) & g(x_1, y_2) & \cdots & g(x_1, y_n) \\ g(x_2, y_1) & g(x_2, y_2) & \cdots & g(x_2, y_n) \\ \vdots & \vdots & \ddots & \vdots \\ g(x_m, y_1) & g(x_m, y_2) & \cdots & g(x_m, y_n) \end{bmatrix}_{m \times n} \tag{11}$$

Fig. 1 The image feature
model

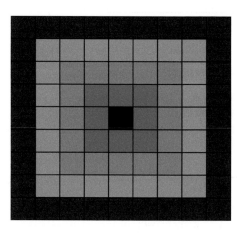

At present, in most image processing methods, extracting image features to achieve
the establishment of image segmentation based on learning method is adopted, however, the feature extraction of image is extracted by artificial methods, which affects
the segmentation results because of the differences and changes of the computation
when the features are extracted artificially. However, different features can cause a
direct impact on the results of segmentation. In order to avoid the occurrence of
this problem, and make full use of the advantage of US-ELM, the method in this
paper establish a feature model directly using the pixels and the information of multiple pixels within its neighborhood as the feature of image. When the Image feature
model is set, pixels are used as a feature directly, it usually selects the neighborhood
of a certain size, and can select all elements with the neighborhood of 3×3 or 5×5
or higher as the corresponding image pixels feature as shown in Fig. 1. The process
of the image feature extraction can be shown in the Eq. (12), F is the feature scale of
the pixel of input image.

$$
F = \begin{bmatrix}
f(x - \frac{n_k-1}{2}, y - \frac{n_k-1}{2}) & \cdots & f(x - \frac{n_k-1}{2}, y) & \cdots & f(x - \frac{n_k-1}{2}, y + \frac{n_k-1}{2}) \\
\vdots & \ddots & \vdots & \ddots & \vdots \\
f(x, y - \frac{n_k-1}{2}) & \cdots & f(x, y) & \cdots & f(x, y + \frac{n_k-1}{2}) \\
\vdots & \ddots & \vdots & \ddots & \vdots \\
f(x + \frac{n_k-1}{2}, y - \frac{n_k-1}{2}) & \cdots & f(x + \frac{n_k-1}{2}, y) & \cdots & f(x + \frac{n_k-1}{2}, y + \frac{n_k-1}{2})
\end{bmatrix}_{n_k \times n_k}
\tag{12}
$$

where $f(x, y)$ is the pixel of the input image, n_k is the size of the neighborhood.

3.2   NMR Image Segmentation Based on US-ELM

In this paper, we presents an image segmentation method based on US-ELM. Firstly
the process of this method is to use the US-ELM to process the image data, to get

---

Algorithm 1 MRI Segmentation based on US-ELM

Input:

    NMR image I;

    Parameters of US-ELM λ;

    Number of hidden neurons n_0;

    Parameters C for clustering;

Output:

    Clustering results: I_1, I_2, I_3, I_4 ;

1   Load input image I: $f(x, y) \in I$;

2   For $x = 1 : m$ do

      for $y = 1 : n$ do

        $F = Denoising(f(x, y))$;

3   For $x = 1 : m$ do

      for $y = 1 : n$ do

        $F = Feature(f(x, y))$;

4    $X \in R^{n \times n_i} = F$;

7    $L = Laplace(X), X \in R^{n \times n_i}$;

6   for $i = 1$ to n_0 do

      Randomly generated hidden layer node parameters $(\mathbf{w_i}, b_i)$;

7   Calculate hidden layer node output matrix H;

8   Calculate $A = I_{n_h} + \lambda H^T L H$;

9   Calculate $B = H^T H$;

10   As $\mathbf{A}v = \gamma \mathbf{B}v$

      Calculate $v_i, i = 2, 3, \dots n_0 + 1$;

11   As $\tilde{v}_i = v_i / \|Hv_i\|, i = 2, 3, \dots n_0 + 1$;

      Calculate $\beta = [\tilde{v}_2, \tilde{v}_3, \dots \tilde{v}_{n_0+1}]$;

12   Calculate the embedded matrix $E = H\beta$:

15   Set the parameters $C = 4$ for clustering;

14   Get clustered results y by $K\text{-}means$;

15   Initializing clustering center C_i;

16   Divided data into clusters;

17   Recalculate the new clustering center C_i';

18   If $C_i' \neq C_{i-1}'$, return to step 16;

19   Get I_1, I_2, I_3, I_4 by y.

---

new embedding space, and then the clustering operation on the data is in the new embedding space, the clustering segmentation of the image is realized by the one-to-one mapping relationship between the new embedding space and the original one, and restore segmented image according to the results of clustering.

Taking a concrete NMR image as an example the process is shown in Algorithm 1, for the input NMR image I_0. Feature model is established for I_0 to get $X \in R^{n \times n_i}$. Firstly, constructing the Laplacian matrix L according to the $X \in R^{n \times n_i}$, then setting the number of conceal layer node of US-ELM, and the hidden node parameters $(\mathbf{w_i}, b_i)$ are generated randomly according to the number of hidden nodes, and the output matrix $H \in R^{n \times n_i}$ of the hidden layer is calculated according to the parameters of the hidden layer nodes in US-ELM. According to the principle of US-ELM, $A = I_{n_h} + \lambda H^T L H$, $B = H^T H$, calculating the generalized eigenvalues and eigenvectors, and the generalized eigenvector $v_2, v_3, \dots v_{n_0+1}$ corresponding to the minimum generalized eigenvalue from 2 to $n_0 + 1$ is found in them, then getting $\beta = [\tilde{v}_2, \tilde{v}_3, \ldots \tilde{v}_{n_0+1}]$ after uniting them, here $\tilde{v}_i = v_i / \|Hv_i\|, i = 2, 3, \ldots n_0 + 1$; and calculating the feature space E according to β, the method of Kmeans based on spatial neighborhood is used to cluster $\mathbf{E} = (e_1, e_2, \ldots e_n)^T$, Here set the number of categories to 4, and the corresponding image elements in the original data are clustered according to the clustering center C_i', get the image after segmentation I_1, I_2, I_3, I_4.

4 Experiments Evaluation

In this section, the experimental setting is introduced in Sect. 4.1, and the experimental results is introduced in Sect. 4.2.

4.1 Experimental Setting

In this paper, US-ELM is applied to the segmentation of MRI according to the way above. The DICOM format images are used for segmentation processing. This experiment are run on the environment of MATLAB2013a, with the Intel Core i3 2.7 GHz CPU and 6GB RAM for simulation and data processing. In the experiment, after discussion, 200 hidden layer nodes are adopted. Excitation function of ELM is "sigmoid", according to the principle of US-ELM above, the value of λ is set to 1 and according to the different requirements to the dimension in US-ELM training, different dimensions of the experiments were carried out. In the experiment, the results of different image segmentation methods are firstly compared in the experiment. And to compare them through a number of evaluation criteria. The including contrast method is:

- K-means: Original K-means clustering approach;
- NS-UE: NMR Image Segmentation based on US-ELM;

Since there is no gold standards for the experimental data and segmentation results so the unsupervised evaluation program proposed by Zhang et al. [24] is adopted in the experiment, the images segmented are evaluated by the regional uniformity and gray uniformity.

4.2 Experimental Results

The effect of the number of hidden nodes on the experimental results is evaluated. With the increase of the hidden layer nodes of data as shown in the Fig. 2a, b, we can see that the uniform does not increase. In contrast, the increase in the number

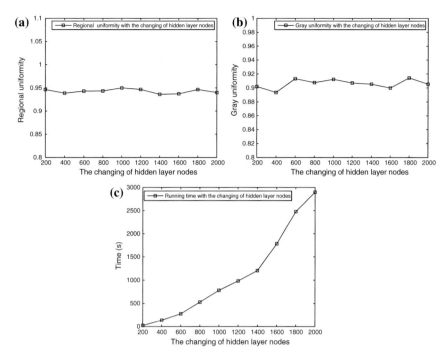

Fig. 2 Changing of the number of hidden layer nodes. **a** Regional uniformity. **b** Gray uniformity. **c** Running time

of nodes in the hidden layer increases the running time, which is shown in Fig. 2c. Therefore, the number of hidden layer nodes in this paper is set to 200.

It can be seen from the four maps, NS-UE can separate different regions of the NMR image, but it still exists phenomenon that the edge is not complete in the process of segmentation (Fig. 3). K-means can not get the segmentation distinguished accurately. According to the evaluation method described in the last section, the segmentation effects of the two methods are compared. In the case of the white matter section, the results are compared with the results of the experiments, the results of the uniformity are compared with the results of the following experiments in the Fig. 4a. By the contrast of the region uniformity, it can be seen that the uniform similarity of the image based on NS-UE is better than others. After comparing the uniformity of the region, we make the evaluation of the effect of the whole gray level, and the evaluation is as follows in the Fig. 4b, c: By the contrast of the gray uniformity in the Fig. 4a, it can be seen that the gray uniformity in the Fig. 4b, the image based on NS-UE is better than K-means. The efficiency of the algorithm is also one of the evaluation indexes of the algorithm, so the running time of the algorithm is compared below: From the Fig. 4c, it can be seen that the operating efficiency of the image based on NS-UE is obviously higher. Take the two evaluation indexes above as an example, in the calculation of US-ELM, when the dimension of the output data

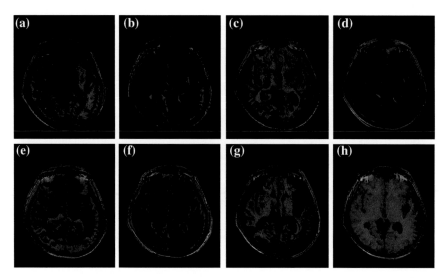

Fig. 3 Segmentation results. **a** Background of K-means. **b** Background of NS-UE. **c** Edge of K-means. **d** Edge of NS-UE. **e** *Gray* matter of K-means. **f** *Gray* matter of NS-UE. **g** *White* matter of K-means. **h** *White* matter of NS-UE

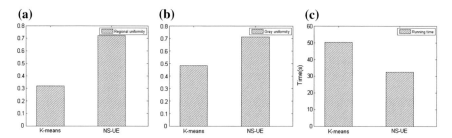

Fig. 4 The compared with the results of *white* matter section. **a** Regional uniformity. **b** Gray uniformity. **c** Running time

changes [15], the segmentation effect can also be affected, the following image is the impact of the change of dimension on the three evaluation indexes above:

With the increase of the dimensions of data as shown in the Fig. 5a, we can see that the uniform does not increase, on the contrary, when the dimension is higher than a certain threshold, the segmentation effect will decline.It can be seen that the segmentation effect is the best when the dimension is about 25. With the increase of the dimensions of data as shown in the Fig. 5b, you can see that gray uniform does not increase, on the contrary, when the dimension is higher than a certain threshold, the segmentation effect will decline.It can be seen that the segmentation effect is the best when the dimension is about 25. From the image shown in the Fig. 5c, because of the increase of dimension, the clustering data increase, so the running time will increase according to the increase of the dimension. According to the image pre-processing

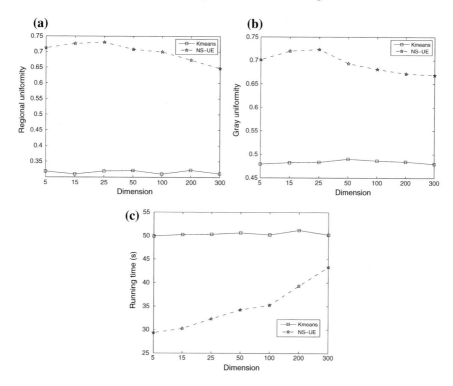

Fig. 5 The contrast of dimension. **a** Regional uniformity. **b** Gray uniformity. **c** Running time

described in last section, this paper established a feature model for adjacent area pixel set as the image feature, the above experimental results select size of neighborhood of 5, in order to verify the effects of the size of selected on the experimental results,and start from 3 respectively, gradually increasing the size of its neighborhood pixels, and analyze its effect on segmentation.

From the Fig. 6a, when the size of neighborhood increases, the regional uniformity increase at first, however when the neighborhood continue increases, the regional uniformity of segmentation results will decrease. Because when the pixels in region increase, the pixels of different classification are introduced as their features, and the effect of the segmentation is disturbed. It can be seen from the in the Fig. 6b that when the size of neighborhood increases, the gray uniformity reduces, which is the same reason as described above. From the Fig. 6c, the running time increases continuously for different neighborhood sizes, because with the increase of the data dimension, it increases the computational complexity of the learning machine, resulting the increase of computation time.

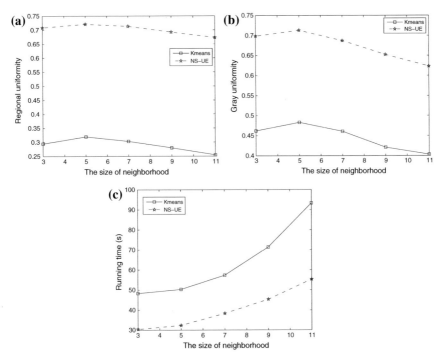

Fig. 6 The contrast of changing of neighborhood. **a** Regional uniformity. **b** Gray uniformity. **c** Running time

5 Conclusions

In this paper, we proposed an MRI segmentation method based on unsupervised extreme learning machine, named NS-UE. As NMR image is an important part in the disease diagnosis, image segmentation is one of the first the most important step in image analysis. And unsupervised learning has been a great success in application. In which US-ELM's performance is particularly outstanding. Firstly, feature model is established for the input NMR image; Secondly, using unsupervised extreme learning machine for data processing. Finally, through a large number of experimental data and calculation proved the effectiveness and efficiency of the algorithms.

Acknowledgments This research was partially supported by the National Natural Science Foundation of China under Grant Nos. 61472069 and 61402089, and the Fundamental Research Funds for the Central Universities under Grant Nos. N130404014.

References

1. Kallergi, M., Clark, R., Clarke, L.: Medical image databases for CAD applications in digital mammography: design issues. Stud. Health Technol. Inf. **43**, 601–605 (1997)
2. Vijaya, G., Suhasini, A.: Synergistic clinical trials with CAD systems for the early detection of lung cancer. Adv. Intell. Syst. Comput. **324**, 561–567 (2015)
3. Pham, D., Xu, C., Prince, J.: Current methods in medical image segmentation. Annu. Rev. Biomed. Eng. **2**(1), 175–272 (2000)
4. Clarke, L., Velthuizen, R., Camacho, M., Heine, J., Vaidyanathan, M.: MRI segmentation: methods and applications. Magn. Reson. Imag. **13**(3), 343–368 (1995)
5. Scott, A., Macapinlac, H., Divgi, C.D., Zhang, J., Kalaigian, H.: Clinical validation of SPECT and CT/MRI image registration in radiolabeled monoclonal antibody studies of colorectal carcinoma. J. Nuclear Med. Official Publication Society of Nuclear Medicine **35**(12), 1976–1984 (1994)
6. Mauri, G.: Real-time US-CT/MRI image fusion for guidance of thermal ablation of liver tumors undetectable with US: results in 295 cases. Cardiovasc. Intervent. Radiol. **38**(11), 143–151 (2014)
7. Erb, R.J.: The backpropagation neural network-a Bayesian classifier. Introduction and applicability to pharmacokinetics. Clin. Pharmacokinet. **29**(2), 69–79 (1995)
8. Jyoti, A., Mohanty, M.N., Kar, S.K.: Optimized clustering method for CT brain image segmentation. In: FICTA, vol. 327, pp. 317–324 (2015)
9. Pham, V.H., Lee, B.R.: An image segmentation approach for fruit defect detection using k-means clustering and graph-based algorithm. Vietnam J. Comput. Sci. **2**(1), 25–33 (2015)
10. Ahmadvand, A., Daliri, M.R.: Improving the runtime of MRF based method for MRI brain segmentation. Appl. Math. Comput. **256**, 808–818 (2015)
11. Hocking, A., James, E.G., Davey, N., Sun, Y.: Teaching a machine to see: unsupervised image segmentation and categorisation using growing neural gas and hierarchical clustering. Instrumentation and Methods for Astrophysics (astro-ph.IM); Cosmology and Nongalactic Astrophysics (astro-ph.CO), vol. 1507 (2015)
12. Maji, P., Roy, S.: Rough-fuzzy clustering and unsupervised feature selection for wavelet based MR image segmentation. PLoS ONE **10**(4), e0123677 (2015)
13. Abd, H.H., Mashor, M.Y., Abdul, A.S., Mustafa, N., Hassan, R.: Colour image segmentation using unsupervised clustering technique for acute leukemia images. In: ICoMEIA 2014, vol. 1660 (2014)
14. Vargas, R.R., Bedregal, B.R., Palmeira, E.S.: A comparison between K-Means, FCM and ckMeans Algorithms. 2011 Workshop-School on Theoretical Computer Science, pp. 32–38 (2011)
15. Huang, G., Song, S., Gupta, J.N.D., Wu, C.: Semi-supervised and unsupervised extreme learning machines. IEEE Trans. Cybern. **44**(12), 2405–2417 (2014)
16. Adhikari, S.K., Sing, J.K., Basu, D.K.: Conditional spatial fuzzy C-means clustering algorithm with application in MRI image segmentation. Find out how to access preview-only content. Inf. Syst. Des. Intell. Appl. **340**, 539–547 (2015)
17. Huang, G.-B., Zhu, Q.-Y., Siew, C.-K.: Extreme learning machine: theory and applications. Neurocomputing **70**(1–3), 489–501 (2006)
18. Huang, G.-B., Chen, L., Siew, C.-K.: Universal approximation using incremental constructive feedforward networks with random hidden nodes. IEEE Trans. Neural Netw. **17**(4), 879–892 (2006)
19. Huang, G.-B., Chen, L.: Convex incremental extreme learning machine. Neurocomputing **70**(16–18), 3056–3062 (2007)
20. Huang, G.-B., Chen, L.: Enhanced random search based incremental extreme learning machine. Neurocomputing **71**(16–18), 3460–3468 (2008)
21. Huang, G.-B., Ding, X., Zhou, H.: Optimization method based extreme learning machine for classification. Neurocomputing **74**(1–3), 155–163 (2010)

22. Huang, G.-B., Zhou, H., Ding, X., Zhang, R.: Extreme learning machine for regression and multiclass classification. IEEE Trans. Syst. Man Cybern. Part B: Cybern. **42**(2), 513–529 (2012)
23. Yong, H.-L., Kassam, S.-A.: Generalized median filtering and related nonlinear filtering techniques. IEEE Trans. Acoust. Speech Signal Process. **33**(3), 673–683 (2012)
24. Zhang, S., Dong, J.W., She, L.H.: The methodology of evaluating segmentation algorithms on medical image. J. Image Graphics **14**(9), 1972–1880 (2009)

Annotating Location Semantic Tags in LBSN Using Extreme Learning Machine

Xiangguo Zhao, Zhen Zhang, Xin Bi, Xin Yu and Jingtao Long

Abstract In recent years, location-based social networks have become very popular. However, it is difficult to extract proper location features from constrained users' check-in activities datasets. In this paper, by capturing the check-in activities of similar users, we propose a new extracting location feature method similar user pattern (SUP) to automatically annotate category tags for all locations lacking semantic tags. Extreme learning machine (ELM) is well-known for having a faster learning speed and good generalization performance, so we apply a binary ELM to train extracting location features for each tag in the tag space in order to support multi-label classification. We also combined with other existing feature extraction methods to train ELM aimed for finding the most effective feature combinations for annotating semantic tags. Finally, to verify the effectiveness and efficiency, we conduct experimental study based on a real dataset collected from Foursquare, which is popular LBSN service. The results show that our proposed method SUP is effective in annotating semantic tag for locations.

Keywords Location-based social networks · Extreme learning machine · Multi-label classification · Semantic annotation of locations

1 Introduction

With the development of GPS technology and the wide use of intelligent terminal, location-based social networks (LBSN) have become very popular. There are lots of LBSN services such as Foursquare, Gowalla, Facebook Places, Whrrl. These services allow users to share their locations and location-related content, such as geo-tagged photos and notes [2]. Based on real traces of users' locations and activities, LBSNs have attracted the attention of many researchers. One of them is

X. Zhao (✉) · Z. Zhang · X. Bi · X. Yu · J. Long
College of Information Science and Engineering, Northeastern University,
Shenyang 110819, Liaoning, China
e-mail: zhaoxiangguo@mail.neu.edu.cn

© Springer International Publishing Switzerland 2016
J. Cao et al. (eds.), *Proceedings of ELM-2015 Volume 1*,
Proceedings in Adaptation, Learning and Optimization 6,
DOI 10.1007/978-3-319-28397-5_27

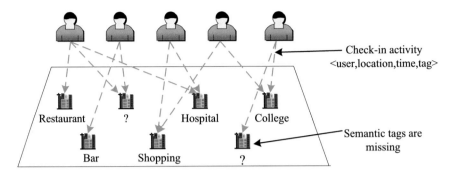

Fig. 1 Users and locations in LBSN

automatically annotating category for all locations lacking any semantic descriptions [17]. Ye et.al proposed many locations have been annotated with useful tags such as *restaurant* or *cinema*. However, based on analysis of data collected from Whrrl and Foursquare, about 30 % of all locations are lacking any semantic tags. So, it is significant to research how to automatically annotate semantic tags for locations. In LBSN, a location also may be associated with multiple tags. Such as, a location associated with a tag *restaurant* may also be tagged with *bar*. Hence, location semantic annotation in LBSN may be addressed as a muti-lable classification problems [3, 18].

Figure 1 shows some users' check-in activity records. Users and locations are connected through a set of check-in activities $C = \{\langle u, l, h, t \rangle | u \in U \wedge l \in L \wedge h \in H \wedge t \in T\}$, in which U denotes the set of users, L denotes the set of locations, H denotes the set of time stamps and T denotes the tag space. Every check-in activity $c \in C$ describes that a user u has checked in a location l at time h in which the semantic tag of the location is t. The locations of majority records have existed semantic tags $t \in T$. Such as *bar, restaurant*. However a few locations (the question marks in Fig. 1) lack any semantic tags.

There are two phases to automatically annotate category tags for all locations lacking semantic tags: (1) *extracting the features*; (2) *the features to train extreme learning machine (ELM)*. For the first phase, we propose a new method similar user pattern (SUP) to extract features of locations and improve the semantic annotation technique in LBSN. For every check-in activity in which the locations lack any semantic tags, we obtain the user information, calculate the top-k similar users of the user and extract those similar users' the location tags of check-in activities within a time period around the the user' check-in time. Finally, we approximately take the statistical result as probability that the location tag of this check-in activity belongs to every category tag in the tag space.

For the second phase, we use the SUP combined with other existing feature extraction methods to train a set of binary ELM classifiers in order to automatically annotate category tags for all locations lacking semantic tags. Location semantic annotation in LBSN has been considered as a muti-lable classification problems.

However, most muti-lable classification learning methods still are based on transformation of the multi-label classification problem to a binary classification problem and then use binary classification learning methods to classify the input data.

Specifically, our research work main contributions can be summarized as:

- we propose a new feature extraction method SUP to annotate semantic tags for the locations which are lacking any semantic tags. It is proved to be a very effective feature extracting method.
- We also combined with other existing feature extraction methods to train ELM aimed for finding the most effective feature combinations for annotating semantic tags of locations.
- By a comprehensive experimental study, using a real dataset collected from Foursquare. We confirm that our proposed method SUP is effective in annotating semantic tag for locations.

The remainder of this paper is organized as follow. In Sect. 2, we review related works. Next, in Sect. 3, according to the information of users' check-in the activities, we summarize existing existing feature extraction methods and introduce SUP in detail. In Sect. 4, we introduce the ELM and further discuss the issue that the extracting the features of locations to train ELM. In Sect. 5, we conduct an empirical study using the collected Foursquare and analyze our results. Finally, in the Sect. 6, we conclude our work.

2   Related Work

In this section, we review a number of existing works in the areas of *extreme learning machine (ELM)*, and *multi-label classification*.

Previous studies on multi-label classification have been conducted in the domains of text classification [14, 16], protein function classification [5], music categorization [13], and semantic scene classification [3]. In [16], BoosTexer has been developed to handle multi-label text categorization. In [14], a mixture model derived by expectation maximization (EM) has been trained to select the most probable set of tags. In [11], a set of binary SVM classifiers have been developed to realize multi-label classification for text classification.

In ELM, hidden node parameters are chosen randomly. ELM generally requires much less training time than the conventional learning machines and tends to reach the smallest training error. ELM has been originally developed based on single-hidden layer feed forward neural networks (SLFNs) [7–10]. It is well-known for having a faster learning speed than other traditional learning methods, and it not only tends to reach the smallest training error but also the smallest norm of weights. Its classification performance is also better than the gradient based learning method. The simulation aiso results proved that ELM can achieve better generation performance than the traditional support vector machine (SVM) [6].

3   Semantic Annotation of Places

In this section we first present a brief introduction about existing feature extraction method, then present our similar user pattern (SUP) method to extract features of locations.

3.1   The Existing Feature Extraction Methods

So far, there have been some the feature extraction methods to annotate category tags for all locations lacking semantic tags. Based on those, we propose a new feature extraction method SUP and combine with other other existing feature extraction methods to train extreme learning machine (ELM). We first briefly introduce the existing feature extraction methods involved in this paper.

Ye et al. proposed to extract explicit pattern (EP) feature f_{ed} from users' check-in records to train SVM classifiers [17]. The EP features contain five aspects: (1) total number of check-ins; (2) total number of unique visitors; (3) maximum number of check-ins by a single visitor; (4) distribution of check-in time in a week; (5) distribution of check-in time in 24-h scale. The features extracted from EP are summarized from all check-ins at a specific location which is different angles with SUP. So, EP will be combined with SUP in Sect. 5 to find the most effective feature combinations for annotating semantic tags for the locations. Ye et al. also proposed implicated relatedness(IR) features which is a major contribution to this paper. The regularity appears in certain users to be used for correlating similar locations. The locations checked in by the same user at around the same time show strong relatedness. To capture the relatedness among locations and extract discriminative features from IR. It builds a network of related places(NRP). The feature extracted from IR are summarized from the check-in activities of users do exhibits a strong regularity. Our SUP is improved based on IR. So, The SUP will be compared with IR in Sect. 5. Ou et al. proposed to extract a demographic feature f_{de}(DFs) that is complementary to EP based on the interest features [15]. So, DFs also will be combined with SUP in Sect. 5 to find the most effective feature combinations for annotating semantic tags for the locations.

3.2   The SUP Method to Extract Features

we propose a new SUP method to extract features of location. As mentioned earlier, the only data resource we have is the users' check-in activities at various locations and time. Fortunately, human behaviors are not completely random. We discover two regulars from Foursquare datasets: (1) *regularity of the certain user*; (2) *regularity of between users*. Regularity of the certain user reflects on the check-in activities of the certain user do exhibits a strong regularity. Regularity of the between users reflects on the some users daily behaviors have similarities and possible go to the locations which have similar semantic tags at the same time or closer time. Moreover. According to it,we formulate the similar users problem.

Definition 1 (*Similar Users*) Given the check-in activities of two users: $C_{11} = \{\langle u_1, l_1, h_1, t_1 \rangle | u_1 \in U \wedge l_1 \in L \wedge h_1 \in H \wedge t_1 \in T\}$ and $C_{21} = \{\langle u_2, l_2, h_2, t_2 \rangle | u_2 \in U \wedge l_2 \in L \wedge h_2 \in H \wedge t_2 \in T\}$. t_1, t_2 are the similar location tags and h1, h2 are closer time, we can call $u_1 \sim u_2$. P is similar rate between the two users. Here we think the same user is the greatest similarity, and the similar rate P is 1.

Before introducing SUP method, we explain how to calculate similar users. First, it assumes that we regroup locations by functions into n general tags. For example, Coffee, Snacks, Cafe, Delis and etc., all belong to the same category '*Restaurant*'. And it assumes that we divide 24 h one day into m time periods (In the Sect. 5, we will find the best time period partition method by the experiment). Then we put check-in activities having location semantic tags of all the users into different groups according to both semantic tags of location and check-in period of time. Such as, we assume that S_a is one of the general tags and k is one of the periods of time. So, all check-in records with S_a location tags and also within the k time period will be classified into one group. Thus, we can put check-in activities having location semantic tags of all the users into $m \times n$ groups.

After the end of the grouping of the check-in activities, we use the *Vector Space Model (VSM)* [4] to calculate the Similar Users. The grouping of each user check-in is considered as a vector. In order to reflect the users repeated access to the same group and the similar behavior of visiting users in the groups. The groups of all users consists of the user-group Matrix, as follows:

$$V_{m\times n} = \begin{bmatrix} V_{1,1} & V_{1,2} & \cdots & V_{1,n-1} & V_{1,n} \\ V_{2,1} & V_{2,2} & \cdots & V_{2,n-1} & V_{2,n} \\ \vdots & \vdots & \ddots & \vdots & \vdots \\ V_{m,1} & V_{m,2} & \cdots & V_{m,n-1} & V_{m,n} \end{bmatrix} \quad (1)$$

where m is the number of visiting users, n is the number of groups. V_{ij} is the number of check-ins by the user i in the group region j.

When obtaining the user-group Matrix, we use the cosine angle [1] between two vectors to measure Similar Users. Suppose that the user a and the user b are represented as vector U_a and U_b in n-dimensional the user-group Matrix, the similarity between user a and the user b is defined as follows:

$$sim(a,b) = \cos(U_a, U_b) = \frac{U_a \bullet U_b}{\|U_a\| \|U_b\|} \quad (2)$$

When understanding the method of similar users calculation, We present SUP method to extract location features. First, we find the check-in activities $C_m = \{c_{m1}, c_{m2}, \ldots, c_{mn}\}$ in which the location is no semantic tagging. Here it assumes

that we will give the check-in activity $c_{m1} = \{\langle u_{m1}, l_{m1}, h_{m1}, ?\rangle | u_{m1} \in U \wedge l_{m1} \in L \wedge$
$h_{m1} \in H \wedge ? \in T\}$ to annotate tags. First we find the user of check-in activity u_{m1} and
calculate the most similar k users $\{u_{m1,1}, u_{m1,2}, \dots, u_{m1,k}\}$ with u_{m1} by the above introduction of similar users calculation method. Then we extract all the check-in activities records of existing location tag from $\{u_{m1,1}, u_{m1,2}, \dots, u_{m1,k}\}$ which is around
h_{m1} within a certain period. For example, $[h_{m1} - i, h_{m1} + i]$ in which i can be valued 0.5, 1, 1.5, 2 h. We define those the check-in activities as $C_{\{[m1,1],[m1,2],\dots,[m1,n]\}}$.
we find all location semantic tagging $\{T_{m1,1}, T_{m1,2}, \dots, T_{m1,n}\}$ from check-in records
$C_{\{[m1,1],[m1,2],\dots,[m1,n]\}}$ .Then we do probability statistics for each tag in order to adapt
to ELM. As mentioned earlier, we still assume that it takes the location of semantic tags into n groups as $\{T_1, T_2, \dots, T_n\}$. Then we calculate the probability of
$\{T_{m1,1}, T_{m1,2}, \dots, T_{m1,n}\}$ belonging each category which approximate as probability
that the location tag T_{m1} of c_{m1} belongs to each category in the tag space. The equations follows:

$$P_r[T_{m1} = T_i(i \in [1, n])] = \frac{\sum_j (T_{m1,j} \in T_a)}{\sum_{i=1}^n T_{m1,i}} \qquad (3)$$

4 The Features to Train ELM

4.1 Brief Introduction of ELM

In this section, we further discuss the issue that the extracting features train extreme
learning machine (ELM) in detail. First, it is brief introduction of ELM.

For N arbitrary distinct samples (x_i, t_i), where $x_i = [x_{i1}, x_{i2}, \dots, x_{in}]^T \in R^n$ and $t_i = [t_{i1}, t_{i2}, \dots, t_{im}]^T \in R^m$, standard SLFNs with $\tilde{N}$ hidden nodes and activation function
$g(x)$ are mathematically modeled in [10] as

$$\sum_{i=1}^{\tilde{N}} \beta_i g_i(x_j) = \sum_{i=1}^{\tilde{N}} \beta_i g(w_i \cdot x_j + b_i) = o_j \quad (j = 1, \dots, N) \qquad (4)$$

where $w_i = [w_{i1}, w_{i2}, \dots, w_{in}]^T$ is the weight vector connecting the ith hidden node
and the input nodes, $\beta_i = [\beta_{i1}, \beta_{i2}, \dots, \beta_{im}]^T$ is the weight vector connecting the ith
hidden node and the output nodes, and b_i is the threshold of the ith hidden node.

That standard SLFNs with $\tilde{N}$ hidden nodes with activation function $g(x)$ can
approximate these N samples with zero error means that $\sum_{j=1}^{\tilde{N}} \|o_j - t_j\| = 0$, i.e.,
there exist β_i, w_i and b_i such that

$$\sum_{i=1}^{\tilde{N}} \beta_i g(w_i \cdot x_j + b_i) = t_j \quad (j = 1, \dots, N) \qquad (5)$$

The above N equations can be written compactly as β_i, w_i and b_i such that

$$H\beta = T \qquad (6)$$

where

$$
\begin{aligned}
&H(w_1, \ldots, w_{\widetilde{N}}, b_1, \ldots, b_{\widetilde{N}}, x_1, \ldots, x_{\widetilde{N}}) \\
&= \begin{bmatrix}
g(w_1 \cdot x_1 + b_1) & \cdots & g(w_l \cdot x_L + b_L) \\
\vdots & \ddots & \vdots \\
g(w_1 \cdot x_N + b_1) & \cdots & g(w_L \cdot x_N + b_L)
\end{bmatrix}_{N \times L}
\end{aligned} \qquad (7)
$$

where H is called the hidden layer output matrix of the neural network; the ith hidden node output with respect to inputs $\{x_1, x_2, \ldots, x_N\}$. T is the target matrix of the output layer.

4.2 Multi-label Classification of ELM

The article is less for multi-label classification in ELM. Most multi-label classification learning methods still are based on transformation of the multi-label classification problem to a binary classification problem and then use binary classification learning methods to classify the input data. Because of the location semantic annotation as a multi-label classification problem, we also proposed to address the location semantic annotation problem by learning a binary ELM for each tag in the tag space in order to support the multi-label classification. We give the training stage algorithm of multi-label classification of ELM. It is illustrated in Algorithm 1.

Algorithm 1: Multi-label classification of ELM Algorithm

Input: Input N training samples (x_i, t_i),
hidden node number L,
the whole tag space $T(T_1, T_2, \ldots, T_n)$
Output: Output the tag space element number corresponding to β
1 **for** $j = 1$ *to number of elements in the tag space* **do**
2   **for** $i = 1$ *to L* **do**
3   randomly assign input weight w_i;
4   randomly assign bias b_i; end for;
5   Calculate H_j;
6   Calculate $\beta_j = H_j^+ T$

---

5 Experiments

In this section, we conduct a series of experiments to verify the effectiveness of our proposed similar user pattern (SUP) method and find the most effective feature combinations to annotate semantic tags for the locations from all extraction methods by using extreme learning machine (ELM). In the following, we first describe the real world datasets used in the experiments, introduce the metrics employed to evaluate the performance, and finally analyze the experiment results.

5.1 Dataset Setting and Performance Metrics

We collect three datasets from Foursquare, a very popular LBSN service all around the world. The one dataset consists of 5,687 users and their check-in records in UK. The second dataset contains 5842 users and their check-in records in USA. The third dataset consists of 4300 users and their check-in records in Ireland between July and August 2013. Among all those locations of check-in records in three datasets, 19 % of them are not specified with any semantic tags. We group all the tags of locations into 13 general categories.

In order to conduct the experiments, we pre-process this raw dataset to obtain a ground-truth dataset for performance evaluation. First, the locations in the ground-truth dataset should have category tags, so we filter out those locations without category tags. Then, we select the 20 % over the ground-truth dataset as the test samples and remove category tags. We use the features of locations to train ELM in order to recover the category tags for those the test samples.

Multi-label classification requires a different set of performance metrics to evaluate the effectiveness and efficiency of the training method. We evaluate the performance by the following four metrics: Hamming Loss, Average Precision, Recall and F1-Measure which they are widely employed in previous multi-label classification studies [12, 16–18]. Our experiments mainly evaluate the following two aspects: (1) the effect of parameters on the performance; (2) Evaluating our SUP algorithm; (3) Finding the most effective feature combinations for annotating semantic tags. For initial input of ELM, we use the sigmoidal additive activation function in the simulations and set the number of hidden node to 180.

5.2 Experimental Result

First, we give the effect of parameters on the performance by the experiment. In Sect. 3, we have introduced feature extraction method SUP which contains three parameters: (1) when calculating the Similar Users. based on the check-in activities of human have time distributed regularity, we divided 24 h one day into m time periods; (2) we need to calculate the top-k Similar Users for SUP method which involves how to select parameters k; (3) for calculating the tag probability of SUP Algorithm, when we give the check-in activity c_{m1} to annotate tags, we use parameters $[h_{m1} - i, h_{m1} + i]$ to define a certain time period.

Because of the article length limit, we only give experiment comparison chart by the third parameter. For the first and second parameters, we only give experiment results. When the second parameter k value is 5 and third parameter is $[h_{m1} - 1, h_{m1} + 1]$, by the experiment, we finally find that dividing 24 h one day into eight time periods including 8:00–11:00, 11:00–14:00, 14:00–17:00, 17:00–20:00, 20:00–23:00, 23:00–2:00, 2:00–5:00, 5:00–8:00 is best for the SUP algorithm performance. When the first parameter is eight time periods and third parameter is $[h_{m1} - 1, h_{m1} + 1]$, by the experiment, the results show that when k select 4–6, the SUP algorithm performance is best. Now we focus on the results of the third parameter experiments. As mentioned earlier, it assumes that we give the check-in activity c_{m1} to annotate tags. When the first parameter is eight time periods and k value is 5, we respectively use Hamming Loss, Average Precision, Recall and F1-Measure to evaluate the performance of three datasets of UK, USA, Ireland within those time period $([h_{m1} - 0.5, h_{m1} + 0.5], [h_{m1} - 1, h_{m1} + 1], [h_{m1} - 1.5, h_{m1} + 1.5], [h_{m1} - 2, h_{m1} + 2])$. As shown in Fig. 2, we can conclude that the selection of parameter i inside

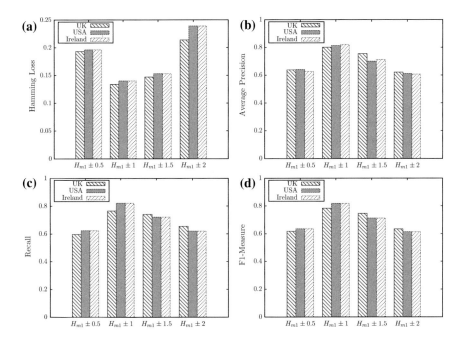

Fig. 2 Selecting a appropriate certain period. **a** Hamming loss. **b** Average precision. **c** Recall. **d** F1-Measure

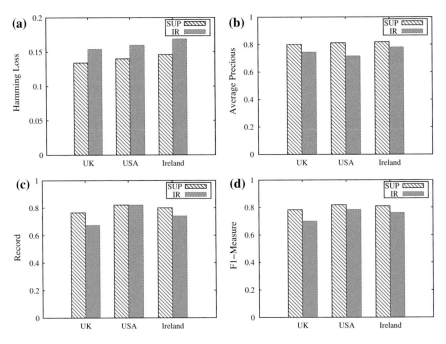

Fig. 3 Performance comparison of *SUP* and *IR*. **a** Hamming loss. **b** Average precision. **c** Recall.
d F1-Measure

$[h_{m1} - i, h_{m1} + i]$ is very important for the performance of the SUP method. Whatever
the performance evaluation in three datasets in UK, USA, Ireland, the period of
$[h_{m1} - 1, h_{m1} + 1]$ and $[h_{m1} - 1.5, h_{m1} + 1.5]$ are relatively better. The reason is that
the time period chosen too short may lead to little check-in activities conform to the
rules, and it is not easy to do probability. But if the time period chosen too long, it
is likely that many check-in activities conform ro rules and lead calculating the tag
probability is inaccurate. So the experiment result is reasonable.

In Fig. 3, we have made a comparative experiment between our proposed method
SUP and Ye et al. proposed method of IR [17]. According to the above the experiment
result, when the first parameter is eight time periods, the second parameter k value
is 5 and third parameter is $[h_{m1} - 1, h_{m1} + 1]$, we respectively use Hamming Loss,
Average Precision, Recall and F1-Measure to evaluate the performance of SUP and
IR in three datasets of UK, USA, Ireland. We can conclude that our proposed method
SUP outperform IR method in all the four metrics. The reason is that IR method is
very good algorithm, but, there are some limitations that IR algorithm just use a
single user check-in regularity for annotation location tags. Although the check-in
activities of users do exhibits a general regularity, there must be some special case.
If the user' check-in record is messy and has no daily regularity, it is bound to affect
the results of the annotation location tags. Our SUP method do daily statistical for
the top-k Similar Users in the period of time, so the affected degree is small.

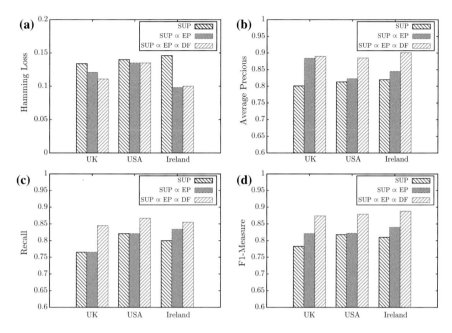

Fig. 4 Performance comparison of *SUP*, *SUP* $\propto$ *EP* and *SUP* $\propto$ *EP* $\propto$ *DF*. **a** Hamming loss. **b** Average precision. **c** Recall. **d** F1-Measure

In Fig. 4, our proposed method SUP combined with the other feature extraction methods to train ELM in order to find the most effective feature combinations from all extraction methods. According to the above the experiment result, when the first parameter is eight time periods, the second parameter k value is 5 and third parameter is $[h_{m1} - 1, h_{m1} + 1]$, we respectively use Hamming Loss, Average Precision, Recall and F1-Measure to evaluate the performance of *SUP*, *SUP* $\propto$ *EP*, and *SUP* $\propto$ *EP* $\propto$ *DF* in three datasets of UK, USA, Ireland where the EP is proposed by Ye et al. [17] and DF is proposed by Ou et al. [15]. We can conclude that the combination of feature extraction methods *SUP* $\propto$ *EP* $\propto$ *DF* is better than other in the four metrics. The reason is that our SUP do not consider features in each individual location. It is just one feature from all extraction features. *SUP* $\propto$ *EP* $\propto$ *DF* includes multiple features which consider not only features of individual locations but also capture the Similar Users by exploiting the regularity of user check-in activities to determine the probability that the location tag belongs to each category in the tag space.

6 Conclusions

In this paper, we propose a new extracting location feature method similar user pattern (SUP) by users' check-in records to annotate category tags for all locations lacking tags and improve the semantic annotation technique. Besides the SUP considers that the check-in activities of users do exhibits a strong regularity, some users daily behaviors have similarities. We also combine with other existing semantic annotation of locations algorithms to learn a set of binary extreme learning machine (ELM) classifiers in order to find the most effective feature combinations to annotate semantic tags. Finally, we use the proposed features with some existing method in our experiment and result shows our proposed method SUP have a satisfactory improvement in performance metrics and the most effective feature combinations is $SUP \propto EP \propto DF$.

Acknowledgments This research is partially supported by the National Natural Science Foundation of China under Grant Nos. 61272181 and 61173030; the National Basic Research Program of China under Grant No. 2011CB302200-G; the 863 Program under Grant No. 2012AA011004; and the Fundamental Research Funds for the Central Universities under Grant No. N120404006.

References

1. Adomavicius, G., Tuzhilin, A.: Toward the next generation of recommender systems: a survey of the state-of-the-art and possible extensions. IEEE Trans. Knowl. Data Eng. **17**(6), 734–749 (2005)
2. Bao, J., Zheng, Y., Wilkie, D., Mokbel, M.F.: A survey on recommendations in location-based social networks. In: ACM Transaction on Intelligent Systems and Technology (2013)
3. Boutell, M.R., Luo, J., Shen, X., Brown, C.M.: Learning multi-label scene classification. Pattern Recogn. **37**(9), 1757–1771 (2004)
4. Chowdhury, G.: Introduction to Modern Information Retrieval, 3rd edn. Facet Publishing, London (2010)
5. Clare, A., King, R.D.: Knowledge discovery in multi-label phenotype data. In: Principles of Data Mining and Knowledge Discovery, pp. 42–53. Springer, Berlin (2001)
6. Cortes, C., Vapnik, V.: Support-vector networks. Mach. Learn. **20**(3), 273–297 (1995)
7. Huang, G.B., Siew, C.K.: Extreme learning machine: RBF network case. In: Control, Automation, Robotics and Vision Conference, 2004. ICARCV 2004 8th. vol. 2, pp. 1029–1036. IEEE (2004)
8. Huang, G.B., Siew, C.K.: Extreme learning machine with randomly assigned RBF kernels. Int. J. Inf. Technol. **11**(1), 16–24 (2005)
9. Huang, G.B., Zhu, Q.Y., Siew, C.K.: Extreme learning machine: a new learning scheme of feedforward neural networks. In: 2004 IEEE International Joint Conference on Neural Networks, 2004. Proceedings. vol. 2, pp. 985–990, July 2004
10. Huang, G.B., Zhu, Q.Y., Siew, C.K.: Extreme learning machine: theory and applications. Neurocomputing **70**(1), 489–501 (2006)
11. Joachims, T.: Text Categorization with Support Vector Machines: Learning with Many Relevant Features. Springer, Heidelberg (1998)
12. Kongsorot, Y., Horata, P.: Multi-label classification with extreme learning machine. In: 2014 6th International Conference on Knowledge and Smart Technology (KST), pp. 81–86. IEEE (2014)
13. Liao, L., Patterson, D.J., Fox, D., Kautz, H.: Learning and inferring transportation routines. Artif. Intell. **171**(5), 311–331 (2007)
14. McCallum, A.: Multi-label text classification with a mixture model trained by EM. In: AAAI99 Workshop on Text Learning, pp. 1–7 (1999)
15. Ou, W.: Extracting user interests from graph connections for machine learning in location-based social networks. In: Proceedings of the MLSDA 2014 2nd Workshop on Machine Learning for Sensory Data Analysis, p. 41. ACM (2014)

16. Schapire, R.E., Singer, Y.: Boostexter: a boosting-based system for text categorization. Mach. Learn. **39**(2), 135–168 (2000)
17. Ye, M., Shou, D., Lee, W.C., Yin, P., Janowicz, K.: On the semantic annotation of places in location-based social networks. In: Proceedings of the 17th ACM SIGKDD International Conference on Knowledge Discovery and Data Mining, pp. 520–528. ACM (2011)
18. Zhang, M.L., Zhou, Z.H.: A k-nearest neighbor based algorithm for multi-label classification. In: 2005 IEEE International Conference on Granular Computing, vol. 2, pp. 718–721. IEEE (2005)

Feature Extraction of Motor Imagery EEG Based on Extreme Learning Machine Auto-encoder

Lijuan Duan, Yanhui Xu, Song Cui, Juncheng Chen and Menghu Bao

Abstract Feature extraction plays an important role in brain computer interface system that significantly affects the success of brain signal classification. In this paper, a feature extraction method of electroencephalographic (EEG) signals based on Extreme Learning Machine auto-encoder (ELM-AE) is applied. Firstly, the original data is classified by Extreme Learning Machine (ELM) and the number of hidden layer's neuron with the highest accuracy is selected as the dimension of feature extraction. Then, ELM-AE's output weight learns to represent the features of the original data. Finally, the features are classified by Support Vector Machine (SVM) classifier. Experiment result shows the efficiency of our method for both the speed of feature extraction and the accuracy of the classification for data set la, which is a typical representative of one kind of BCI competition 2003 data.

Keywords Brain computer interface · Electroencephalogram · Feature extraction · Extreme learning machine auto-encoder

L. Duan (✉) · Y. Xu · S. Cui · M. Bao
Beijing Key Laboratory of Trusted Computing, College of Computer Science and Technology, Beijing University of Technology, Beijing 100124, China
e-mail: ljduan@bjut.edu.cn

Y. Xu
e-mail: xu.xiao.chang@163.com

S. Cui
e-mail: cuisong@emails.bjut.edu.cn

M. Bao
e-mail: irisbao@163.com

L. Duan · J. Chen
Beijing Key Laboratory on Integration and Analysis of Large-Scale Stream Data, College of Computer Science and Technology, Beijing University of Technology, Beijing 100124, China
e-mail: juncheng@bjut.edu.cn

© Springer International Publishing Switzerland 2016
J. Cao et al. (eds.), *Proceedings of ELM-2015 Volume 1*,
Proceedings in Adaptation, Learning and Optimization 6,
DOI 10.1007/978-3-319-28397-5_28

1 Introduction

An electroencephalogram-based brain computer interface is a growing research field which allows users to control computers and other external devices by brain activities rather than depend on the normal output pathways of peripheral nerves and muscles [1]. We focus on motor imagery, which is the mental rehearsal of a motor act, such as movements of hands, limbs, tongue and fingers without any overt motor activities [2]. EEG is most commonly used for capturing motor imaginary brain activities in BCI systems, because of its fine temporal resolution, non-invasiveness, easy implementation and low set-up costs [3].

The accurate classification is a key issue for efficient EEG-based communication and control, which is dependent on extracting relevant features and developing a classification algorithm which is suitable for the features [4–7]. A number of features have been extracted to design BCI such as amplitude values of EEG signals [8], band powers (BP) [9], power spectral density (PSD) values [10], autoregressive (AR) and adaptive autoregressive (AAR) parameters [11] and so on.

A method which can balance classification accuracy of the extracted feature and consuming time while extracting features should be used for the EEG data. Huang et al. proposes a new learning algorithm called extreme learning machine (ELM) for single-hidden layer feed forward neural networks (SLFNs) with a fast learning speed and good generalization [12–14]. Based on ELM, new method called ELM auto-encoder is proposed by Chamara Kasun et al. [15]. The features can be represented ELM-AE's output weight.

In this paper, the extraction method is based on ELM auto-encoder. At first, the input data is classified by Extreme Learning Machine and the number of hidden layer's neuron with the highest accuracy is selected as the dimension of feature extraction. Then, ELM-AE's output weight learns to represent the features of the input data. At last, the features are classified by Support Vector Machine (SVM) classifier. Experiment result shows not only the features extracted by our method obtained the highest classification accuracy but also the speed of feature extraction is fastest in comparison with other common EEG feature extraction methods.

The rest of the paper is organized as follows. The methods are demonstrated in Sect. 2. The experiments are displayed in Sect. 3. The conclusion is drawn in the last section.

2 Method

In this section, we first introduce the framework of the algorithm. And then we describe the method of Extreme Learning Machine auto-encoder. At last, we illustrate feature extraction based on Extreme Learning Machine auto-encoder in details.

2.1 Algorithm Description

The framework of our method is drawn in Fig. 1. Firstly, feature vectors of original EEG signals are extracted by the method based on ELM-AE. Then, in order to verify the efficiency of the feature vectors, SVM classifier is used to obtain the classification accuracy which is regarded as the criteria to evaluate the extracted features.

2.2 Extreme Learning Machine Auto-encoder

Extreme Learning Machine auto-encoder is based on the network of ELM and the theory of auto-encoder. According to ELM theory, ELMs are universal approximators [16], hence ELM-AE is as well. ELM-AE's main objective is to represent the input features meaningfully in three different representations as drawn in Fig. 2 [15]. ELM-AE's output weight is responsible for representing the features of the input data.

As shown in Fig. 2, when repenting the input features, ELM-AE has the following three different representations:

- Compressed. Represent features from a higher dimensional input data space to a lower dimensional feature space;
- Sparse. Represent features from a lower dimensional input data space to a higher dimensional feature space;
- Equal. Represent features from an input data space dimension equal to feature space dimension.

The method of ELM-AE can be summarized as Table 1.

Fig. 1 The framework of the algorithm

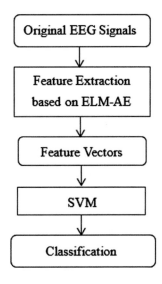

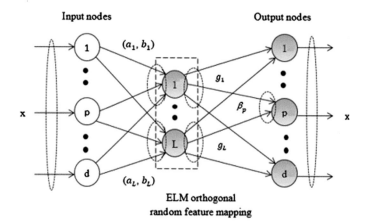

Fig. 2 ELM-AE's network structure

Table 1 Extreme learning machine auto-encoder	Algorithm 1: Extreme learning machine auto-encoder
	Input: input data $\aleph = \{(x_i, y_i)\}_{i=1}^{N} \in R^d \times R^m$, number of hidden neurons L, activation
	01: Step 1. Generate weight a and bias b of hidden nodes randomly, then orthogonalize them $a^T a = I, b^T b = 1$
	02: Step 2. Calculate the output matrix of hidden nodes $H = \begin{bmatrix} G(a_1,b_1,x_1) & \ldots & G(a_L,b_L,x_1) \\ \vdots & \ldots & \vdots \\ G(a_1,b_1,x_N) & \ldots & G(a_L,b_L,x_N) \end{bmatrix}_{N \times L}$
	03: Step 3. Calculate output weights β
	04: If d ! = L,
	05: If N $\geq$ L, $\beta = (\frac{I}{C} + H^T H)^{-1} H^T X$
	06: Else $\beta = H^T (\frac{I}{C} + HH^T)^{-1} X$
	07: Else $\beta = H^{-1} Y, \beta^T \beta = I$
	Output: code: a, b Encode: β

2.3 Feature Extraction Based on Extreme Learning Machine Auto-encoder

As ELM-AE has the same solution as the original extreme learning machine except that its target output is the same as input x, and the hidden node parameters (a_i, b_i) are made orthogonal after being randomly generated [15], the number of hidden nodes of ELM-AE is selected according to the accuracy of original ELM. Before

Fig. 3 Procedure of feature
extraction with the method of
ELM-AE

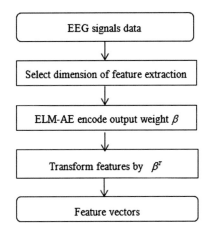

extracting features, we use original ELM to classify EEG data with different numbers of hidden nodes. Then, the numbers of hidden nodes of ELM with the highest classification accuracy is chosen as the numbers of hidden nodes of ELM-AE, namely, the features' dimension after extracting by the feature extraction based on ELM-AE. The procedure of feature extraction method based on ELM-AE is shown in Fig. 3.

As shown in Fig. 3, in the beginning, to select dimension of feature extraction, the EEG signals data is classified by original ELM with different numbers of hidden nodes and then the number of hidden nodes of ELM with the highest classification accuracy is considered as the number of hidden nodes of ELM-AE. After that, we use ELM-AE with the selected number of hidden nodes to calculate the output weight β for representing features. In the end, to obtain the feature vectors, β is transposed and the feature vector is the input data multiplied by β^T.

3 Experiments

In this section, firstly, we introduce the EEG dataset used in the experiment. Then, we present the parameters selection of feature extraction. Finally, we compare performance of our method with other common methods in the aspects of speed of feature extraction and classification accuracy of the features.

3.1 Description of EEG Data

The data used in this paper comes from the BCI competition 2003 data set Ia, which is a batch of high quality of the data set provided by University of Tübingen,

Institute of Medical Psychology and Behavioral Neurobiology, Niels Birbaumer [17, 18]. The dataset was taken from a healthy subject. The subject was asked to move a cursor up and down on a computer screen, while his cortical potentials were taken [19]. During the recording, the subject received visual feedback of his slow cortical potentials (Cz-Mastoids). Cortical positivity leads to a downward movement of the cursor on the screen. Cortical negativity leads to an upward movement of the cursor. All the trails are composed by training set (268 trials, 135 for class 0, 133 for class 1) and testing set (293 trials, 147 for class 0, 146 for class 1). Each trial lasted 6 s. During every trial, the task was visually presented by a highlighted goal at either the top or bottom of the screen to indicate negativity or positivity from second 0.5 until the end of the trial. The visual feedback was presented from second 2 to 5.5. Only this 3.5 s interval of every trial was provided for training and testing. The sampling rate of 256 Hz and the recording length of 3.5 s resulted in 896 samples per channel for every trial.

Six EEG electrodes were located according to the International 10–20 system as shown in Fig. 3 [20] and referenced to the vertex electrode Cz as follows: Channel 1: A1 (left mastoid); Channel 2: A2 (right mastoid); Channel 3: F3 (2 cm frontal of C3); Channel 4: P3 (2 cm parietal of C3); Channel 5: F4 (2 cm frontal of C4); and Channel 6: P4 (2 cm parietal of C4) (Fig. 4).

3.2 Parameters Selection of Feature Extraction

According to our previous research on BCI competition 2003 data set Ia [21], the optimal electrodes of A1 and A2 are selected in the experiment. Compute the classification accuracy of A1 and A2 respectively in ELM classifier with the number of hidden nodes ranging from 20 to 900 with the step of 20. Then select the number of hidden nodes with the highest accuracy. Finally using the best hidden nodes, extract the feature of electrode A1 and A2 respectively with the method of ELM-AE. The result is shown in Fig. 5.

In Fig. 5, the two red pentagrams are the highest accuracy, namely 0.8381 and 0.7814, and their corresponding number of hidden nodes, namely 440 and 200,

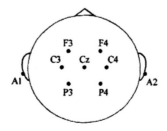

Fig. 4 Distribution of EEG electrodes for 6 channels

Fig. 5 Classification accuracy of electrodes A1 and A2 by the classifier of ELM with different numbers of hidden layer nodes

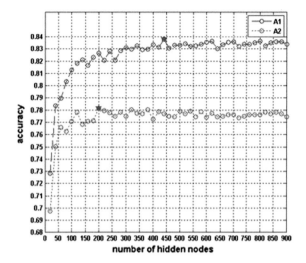

which are the numbers of hidden nodes of ELM-AE about electrode A1 and A2 respectively. Thus, after being extracted features the dimensions of the features of electrode A1 and A2 are 440 and 200 respectively.

3.3   Comparison Performance with Other Methods

To verify the efficiency of feature extraction method based on ELM-AE, we compare our method with other related methods which are commonly used in feature extraction of EEG signals data, such as Principle Component Analysis (PCA) [22], Linear Discriminative Analysis (LDA) [22] and Waveform Packet Decomposition (WPD) [20]. In this section, we compare the above methods from the aspect of classification accuracy of extracted features and the speed of extracting features.

Comparison Performance about Classification Accuracy of Extracted Features. To test the classification accuracy of different methods, SVM classifier which is widely used in the field of EEG classification is adopted in this experiment. The results are shown in Table 2.

Table 2 Comparison of accuracy of two features by ELM-AE and other related methods

Method	Dimension of electrode A1	Dimension of electrode A2	Accuracy (%)
PCA [22]	19	13	76.45
LDA [22]	1	1	50.17
WPD [20]	272	272	64.85
ELM-AE	440	200	**86.69**

Bold value indicates the best result

Table 3 Speed comparison of two features by our method and other methods

Method	Dimension of electrode A1	Dimension of electrode A2	Speed (s)
PCA [22]	19	13	0.2436
LDA [22]	1	1	3.5209
WPD [20]	272	272	98.2906
ELM-AE	440	200	**0.2054**

Bold value indicates the best result

From the Table 2, the method of feature extraction based on ELM-AE is proved effectively although the feature's dimension is a bit large.

Comparison Performance about the Speed of Extracting Features . In this subsection, we compare performance about the speed of extracting features. All the experiments are performed on a PC with an Intel Core i5-4570 processor at 3.2 GHz, 8G RAM and coded with MATLAB 2012b.

In Table 3, the fastest feature extraction method is ELM-AE. In other words, even though the dimension of features' extracted by our method is large, the speed of feature extraction is still fast. This advantage of speed is in accordance with the characteristic of basic ELM.

All in all, from the analysis of Tables 2 and 3, the efficiency of feature extraction method based on ELM-AE can be verified from the aspect of classification accuracy of extracted features and the speed of extracting features.

4   Discussion and Conclusion

In this paper, we presented a feature extraction method based on ELM-AE. We test the efficiency of the method on BCI EEG signal for identifying motor imagery in the aspect of classification accuracy of extracted features and the speed of extracting features. The main conclusions are summarized as follows:

- The method is suitable for the high-dimensional EEG data of motor imagery task. From the better accuracy of the features extracted by our method, the method can extract the discriminative information of original EEG data efficiently.
- In the view of speed of feature, our method can maintain the advantage of fast learning speed, which is the obvious characteristic of ELM. Although the experiments are all offline, we believe the merit of our method may be prospective to handle online data which is strict with speed.

In this paper, the dimension of the features extracted by our method is a bit large. In the process of EEG features, feature selection is a common technique. Therefore, we believe our method may be suitable for combination with related feature selection as well.

Acknowledgements This research is partially sponsored by Natural Science Foundation of China (Nos. 61175115, 61370113 and 61272320), Beijing Municipal Natural Science Foundation (4152005 and 4152006), the Importation and Development of High-Caliber Talents Project of Beijing Municipal Institutions (CIT&TCD201304035), Jing-Hua Talents Project of Beijing University of Technology (2014-JH-L06), and Ri-Xin Talents Project of Beijing University of Technology (2014-RX-L06), the Research Fund of Beijing Municipal Commission of Education (PXM2015_014204_500221) and the International Communication Ability Development Plan for Young Teachers of Beijing University of Technology (No. 2014-16).

References

1. Vaughan, T.M., Wolpaw, J.R.: The third international meeting on brain-computer interface technology: making a difference. IEEE Trans. Neural Syst. Rehabil. Eng. A Publ. IEEE Eng. Med. Biol. Soc. **14**(2), 126–127 (2006)
2. Wolpaw, J.R., et al.: Brain-computer interface technology: a review of the first international meeting. IEEE Trans. Rehabil. Eng. **8**(2), 164–173 (2000)
3. Kayikcioglu, T., Aydemir, O.: A polynomial fitting and KNN based approach for improving classification of motor imagery BCI data. Pattern Recogn. Lett. **31**(11), 1207–1215 (2010)
4. Wang, T., Deng, J., He, B.: Classifying EEG-based motor imagery tasks by means of time–frequency synthesized spatial patterns. Clin. Neurophysiol. **115**(12), 2744–2753 (2004)
5. Vallabhaneni, A., He, B.: Motor imagery task classification for brain computer interface applications using spatiotemporal principle component analysis. Neurol. Res. **26**(3), 282–287 (2004)
6. Vallabhaneni A, Wang T, He B.: Brain-Computer Interface. Neural Engineering, Springer US, pp. 85–121 (2005)
7. Garrett, D., Peterson, D.A., Anderson, C.W., et al.: Comparison of linear, nonlinear, and feature selection methods for EEG signal classification. IEEE Trans. Neural Syst. Rehabil. Eng. **11**(2), 141–144 (2003)
8. Kaper, M., Meinicke, P., Grossekathoefer, U., Lingner, T., Ritter, H.: BCI competition 2003–data set iib: support vector machines for the p300 speller paradigm. IEEE Trans. Biomed. Eng. **51**(6), 1073–1076 (2004)
9. Pfurtscheller, G., Neuper, C., Flotzinger, D., Pregenzer, M.: EEG-based discrimination between imagination of right and left hand movement. Electroencephalogr. Clin. Neurophysiol. **103**(6), 642–651 (1997)
10. Chiappa, S. et al.: HMM and IOHMM modeling of EEG rhythms for asynchronous BCI systems. ESANN, pp. 193–204 (2004)
11. Pfurtscheller, G., Neuper, C., Schlogl, A., Lugger, K.: Separability of EEG signals recorded during right and left motor imagery using adaptive autoregressive parameters. IEEE Trans. Rehabil. Eng. **6**(3), 316–325 (1998)
12. Huang, G.B., Zhu, Q.Y., Siew, C.K.: Extreme learning machine: a new learning scheme of feedforward neural networks. In: Proceedings of 2004 IEEE International Joint Conference on Neural Networks, vol. 2, pp. 985–990 (2004)
13. Huang, G.B., Zhu, Q.Y., Siew, C.K.: Extreme learning machine: theory and applications. Neurocomputing. **70**(1), pp. 489–501. [Code:http://www.ntu.edu.sg/home/egbhuang/elm_random_hidden_nodes.html]
14. Huang, G.B.: What are extreme learning machines? filling the gap between Frank Rosenblatt's dream and John von Neumann's puzzle. Cogn. Comput. **7**, 263–278 (2015)
15. Kasun, L.L.C., Zhou, H.M., Huang, G.B.: Representational learning with ELMs for big data. IEEE Intell. Syst. **28**(6), 31–34 (2013)

16. Huang, G.B., Chen, L., Siew, C.K.: Universal approximation using incremental constructive feedforward networks with random hidden node. IEEE Trans. Neural Networks **17**(4), 879–892 (2006)
17. Birbaumer, N.: DataSets Ia for the BCI competition II. http://www.bbci.de/competition/ii/#datasets
18. Blankertz, B., Müller, K., Curio, G., et al.: The BCI competition 2003: progress and perspectives in detection and discrimination of EEG single trials. IEEE Trans. Biomed. Eng. **51**(6), 1044–1051 (2004)
19. Hinterberger, T., Schmidt, S., Neumann, N., et al.: Brain-computer communication and slow cortical potentials. IEEE Trans. Biomed. Eng. **51**(6), 1011–1018 (2004)
20. Ting, W., Guo, Z.Y., Hua, H.Y., et al.: EEG feature extraction based on wavelet packet decomposition for brain computer interface. Measurement **41**(6), 618–625 (2008)
21. Duan, L.J., Zhang, Q., Yang, Z., Miao, J.: Research on heuristic feature extraction and classification of EEG signal based on BCI data set. Res. J. Appl. Sci. Eng. Technol. **5**(3), 1008–1014 (2013)
22. Panahi, N., Shayesteh, M.G., Mihandoost, S., et al.: Recognition of different datasets using PCA, LDA, and various classifiers. In: 5th International Conference on Application of Information and Communication Technologies (AICT 2011). Baku, October 12–14, United States, IEEE Computer Society, pp. 1–5 (2011)

Multimodal Fusion Using Kernel-Based ELM for Video Emotion Recognition

Lijuan Duan, Hui Ge, Zhen Yang and Juncheng Chen

Abstract This paper presents a multimodal fusion approach using kernel-based Extreme Learning Machine (ELM) for video emotion recognition by combing video content and electroencephalogram (EEG) signals. Firstly, several audio-based features and visual-based features are extracted from video clips and EEG features are obtained by using Wavelet Packet Decomposition (WPD). Secondly, video features are selected using Double Input Symmetrical Relevance (DISR) and EEG features are selected by Decision Tree (DT). Thirdly, multimodal fusion using kernel-based ELM is adopted for classification by combing video and EEG features at decision-level. In order to test the validity of the proposed method, we design and conduct the EEG experiment to collect data that consisted of video clips and EEG signals of subjects. We compare our method separately with single mode methods of using video content only and EEG signals only on classification accuracy. The experimental results show that the proposed fusion method produces better classification performance than those of the video emotion recognition methods which use either video content or EEG signals alone.

Keywords Multimodal fusion · Kernel-based ELM · Video emotion recognition · EEG

L. Duan · H. Ge · Z. Yang (✉)
Beijing Key Laboratory of Trusted Computing, College of Computer Science and Technology, Beijing University of Technology, Beijing 100124, China
e-mail: yangzhen@bjut.edu.cn

L. Duan
e-mail: ljduan@bjut.edu.cn

H. Ge
e-mail: imissyoutata@emails.bjut.edu.cn

L. Duan · H. Ge · J. Chen
Beijing Key Laboratory on Integration and Analysis of Large-Scale Stream Data, College of Computer Science and Technology, Beijing University of Technology, Beijing 100124, China
e-mail: ustcjuncheng@foxmail.com

© Springer International Publishing Switzerland 2016
J. Cao et al. (eds.), *Proceedings of ELM-2015 Volume 1*,
Proceedings in Adaptation, Learning and Optimization 6,
DOI 10.1007/978-3-319-28397-5_29

1 Introduction

With the rapid development of multimedia technology, various digital videos emerge in large numbers. However, some contents of the video are not suitable for juveniles to watch, such as contents with violence and eroticism. For the healthy growth of juveniles, establishing a good network environment is necessary. Recognizing different video emotions accurately is a basis of establishing a good network environment. Recently, how to recognize different video emotions accurately is a challenging and important issue for many researchers [1, 2].

Multimodality can represent video content more comprehensively and recognize video emotion more accurately than single modality, so multimodal fusion approaches for video emotion recognition are becoming increasingly popular. In current multimodal fusion approaches, face, speech, video, text and physiological signal are the most common combined modalities. Bailenson et al. [3] uses facial features and physiological signal to carry out real-time classification of evoked emotions. Mansoorizadeh et al. [4] combines facial features with speech features to construct multimodal information fusion application of human emotion recognition. Koelstra [5] uses facial expressions and electroencephalogram (EEG) signal for affect recognition and implicit tagging of videos. Ye et al. [6] presents an approach combing speech signal with textual content for emotion recognition. Wang et al. [7] combines video content with EEG signal at two different levels (feature-level and decision-level) to annotate videos' emotional tags for the first time and proves that fusion accuracy of video content and EEG signal is higher than video only or EEG only. In previous studies, various classifiers are adopted, such as support vector machine (SVM) [8], neural network [9], hidden Markov models [10]. In Wang's method [7], three Bayesian Networks are adopted for fusion.

In this study, we propose a novel multimodal fusion method using kernel-based Extreme Learning Machine (ELM) for video emotion recognition. In the proposed method, the fusion of video content and EEG is adopted at decision-level. The contributions of this paper mainly include: (1) to the best of our knowledge, kernel-based ELM is applied to implement the fusion of video content and EEG signal for the first time; (2) a multimodal fusion method using kernel-based ELM is proposed for video emotion recognition; (3) the fusion method is shown experimentally to be more accurate than single mode method of using video content or EEG signal separately.

The remainder of this paper is organized as follows. In Sect. 2, the proposed multimodal fusion method is presented particularly. In Sect. 3, experimental materials and EEG experiment protocol are illustrated. In Sect. 4, the parameter selection that used in experiment and performance evaluation of the classification method are described, respectively. Finally, the "Conclusions" is given.

2 Methods

The proposed multimodal fusion method involves three steps: (1) obtain video features, (2) obtain EEG features, (3) carry out classification fusion using kernel-based ELM and obtain the final decision. The framework of the proposed multimodal fusion method is shown as Fig. 1.

2.1 Feature Extraction

Content-based video features. In order to describe video clips accurately, content-based video features consist of audio-based features and visual-based features in our research.

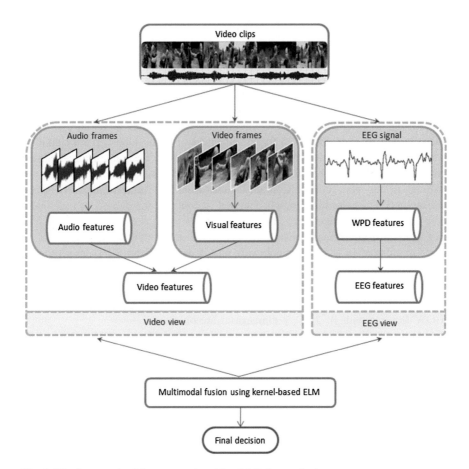

Fig. 1 The framework of the proposed multimodal fusion method

Table 1 Audio-based features

Domain	Feature list	Statistics	Dimension
Time domain	Zero crossing rate	$(\text{Std/mean})^2$	1
	Short time energy	$(\text{Std/mean})^2$	1
Frequency domain	Spectral flux	Std	1
	Spectral rolloff	Std	1
	Spectral centroid	Std	1
Time-frequency domain	MFCCs	Std	20

Audio-based features. Actually, the digital audio signal is time-varying signal. For purpose of analyzing audio signal in a traditional way, it is assumed that the audio signal is stable in a few milliseconds. To obtain short-time audio signal, window operation is adopted. That is to say, each audio signal is divided to frames and one audio signal corresponds to with one video clip. For each audio frame, low-level features of 25 dimensions are extracted, shown as Table 1. These features are widely used in audio and speech processing and audio classification. In conclusion, 25 audio-based features are got in each video clip.

Visual-based features. Digital video is a serial of images composed of frames, which contains rich information. In order to remove redundant information and reduce calculation amount, key frames are extracted and then visual features of key frames are extracted. In this paper, the hierarchical clustering approach based on color histogram is adopted to extract key frames. Assume that the duration of each video clip is t seconds, the video frame sequences of each video clip are classified by clustering into $3 * t$ clusters. Then from every cluster, the frame nearest cluster center is selected as one key frame. Finally, $3 * t$ key frames of each video clip are obtained. In this work, color histogram features from key frames in the HSV space are extracted. The color histogram is obtained by counting the number of times each color occurs in the image array [11]. We divide H to 2^4, S to 2^2, V to 2^2, so 256 HSV color histogram features in each key frame are obtained. After a frame of image is sampled in a dense sample way and is described by SIFT descriptors, a descriptor set is obtained. Then the descriptor set is quantized into W visual words by using K-means clustering. After that, the spatial pyramid descriptor of entire image is formed, which has Q level and $2^{2(Q-1)}$ cells at level Q. For one key frame, PHOW features with $W^*2^{2(Q-1)}$ dimensions are obtained.

EEG features. The EEG signal is down-sampled to 500 Hz and EOG is subtracted from EEG data. 8–30 Hz band-pass filter is used to reduce artifacts. Wavelet packet decomposition (WPD) is used in our research. WPD features have been previously used for EEG signal analysis and worked very well at classification accuracy [12]. In the WPD analysis, signal is decomposed to high frequency component and low frequency component. Accordingly, WPD coefficients are obtained. Then each component is decomposed similarly. For one electrode's EEG signal corresponding to one video clip, one second time window is used to process signal and the signal is divided into t segments. For signal of each window, the level of decomposition is set to J and db6 is selected as wavelet basis to obtain 2^J WPD features. In this way, $2^J * t$ WPD features in each electrode's signal are got. For E electrodes' EEG signals corresponding to one video clip, $E * 2^J * t$ WPD features are obtained.

2.2 Feature Selection

Video feature selection. In this paper, Double Input Symmetrical Relevance (DISR) is adopted to select video features. The DISR has two properties: first, a combination of features can return more information on the output class than the sum of the information that is returned by each of the features individually; secondly, it is intuitive to assume a combination of the best performing subsets of $d - 1$ features as the most promising set. In video features, audio-based features just have 25 dimensions, so only features on visual-based features are selected. Considering the balance between audio-based features and visual-based features, the dimension of DISR features is the same as the dimension of audio-based features, namely, 25 DISR features are obtained. Then 50 video features are got.

EEG feature selection. EEG signal has redundant information in high dimensional space. Therefore, a method to reduce the dimensionality should be chosen to remove redundant information. Decision tree (DT) has been widely used in the classification because of its fast speed and high precision. However, the main factor that influences the performance of decision tree classification is the selection problem [13]. For EEG signal, attribute selection problem is spatial feature selection problem. So DT is selected as EEG features selection approach. C4.5 is one of the widely used DT algorithms. In this paper, decision tree algorithm C4.5 is adopted to select EEG features. EEG data's each feature is regarded as one attribute. Simplified tree's all attributes are used as EEG features selected. Finally, Z EEG features are obtained from $E * 2^J * t$ WPD features.

2.3 Classification

A brief review of kernel-based ELM. ELM is proposed for "generalized" single hidden layer feed forward networks (SLFN) by Huang et al. [14]. It is shown that the learning speed of ELM is much faster than other learning algorithms such as SVM. The essence of ELM is that the hidden layer of SLFNs need not be tuned. The output function of ELM with L hidden nodes for generalized SLFNs can be expressed by

$$f_L(\mathbf{x}_i) = \sum_{j=1}^{L} \boldsymbol{\beta}_i h_i(\mathbf{x}_i) = \mathbf{h}(\mathbf{x}_i)\boldsymbol{\beta} \qquad (1)$$

where $\mathbf{x}_i$ is the input sample vector, and the output weight vector between the hidden layer of L nodes to the c ($c \geq 1$) output nodes is denoted as $\boldsymbol{\beta} = [\boldsymbol{\beta}_1, \ldots, \boldsymbol{\beta}_L]^T$, and ELM nonlinear feature mapping is denoted as $\mathbf{h}(\mathbf{x}_i) = [h_1(\mathbf{x}_i), \ldots, h_L(\mathbf{x}_i)]$. In real applications, $h_i(\mathbf{x}_i)$, the output of the jth hidden node, can be $h_i(\mathbf{x}_i) = G(\mathbf{w}_j, b_j, \mathbf{x}_i)$, where $G(\cdot)$ is the activation function of the hidden nodes, and $(\mathbf{w}_j, b_j)$ are hidden node parameters. Given N input vectors, Eq. (1) can be written in matrix form as $\mathbf{H}\boldsymbol{\beta} = \mathbf{T}$, where the hidden layer output matrix is

$$\mathbf{H} = \begin{bmatrix} \mathbf{h}(\mathbf{x}_1) \\ \vdots \\ \mathbf{h}(\mathbf{x}_N) \end{bmatrix} = \begin{bmatrix} G(w_1, b_1, \mathbf{x}_1) & \cdots & G(w_L, b_L, \mathbf{x}_1) \\ \vdots & \ddots & \vdots \\ G(w_1, b_1, \mathbf{x}_1) & \cdots & G(w_L, b_L, \mathbf{x}_1) \end{bmatrix} \quad (2)$$

and the training data label matrix is

$$\mathbf{T} = \begin{bmatrix} \mathbf{t}_1^T \\ \vdots \\ \mathbf{t}_N^T \end{bmatrix} = \begin{bmatrix} t_{11} & \cdots & t_{1c} \\ \vdots & \ddots & \vdots \\ t_{N1} & \cdots & t_{Nc} \end{bmatrix}. \quad (3)$$

In ELM, the input weights $\mathbf{w}_j$ and hidden biases are randomly generated so that $\mathbf{H}$ does not need to be tuned. $\boldsymbol{\beta}$ is solved by $\min_{\boldsymbol{\beta} \ni \mathbf{R}^{L \times c}} \|\mathbf{H}\boldsymbol{\beta} - \mathbf{T}\|^2$. To train ELM, a least-square solution must be found by using Moore-Penrose generalized inverse. The optimal solution is given by

$$\beta^* = \mathbf{H}^\dagger \mathbf{T} \quad (4)$$

where $\mathbf{H}^\dagger$ is the Moore-Penrose generalized inverse of $\mathbf{H}$, and $\mathbf{H}^\dagger = (\mathbf{H}^T\mathbf{H})^{-1}\mathbf{H}^T$ or $\mathbf{H}^T(\mathbf{H}\mathbf{H}^T)^{-1}$.

In order to make learning system more stable and generalization performance better, kernel-based ELM [15] introduces a positive regularization coefficient into ELM. If $N > L$, we have

$$\beta^* = (\mathbf{H}^T\mathbf{H} + \mathbf{I}/\gamma)^{-1}\mathbf{H}^T\mathbf{T} \quad (5)$$

where $\mathbf{I}$ is an identity matrix of dimension L, and $\mathbf{H}^T\mathbf{H}$ is called "ELM kernel matrix". If $N < L$, we get

$$\beta^* = \mathbf{H}^T(\mathbf{H}\mathbf{H}^T + \mathbf{I}/\gamma)^{-1}\mathbf{T} \quad (6)$$

where $\mathbf{I}$ is an identity matrix of dimension N, and $\mathbf{H}\mathbf{H}^T$ is called "ELM kernel matrix".

Classification fusion using Kernel-based ELM. In this work, decision-level fusion is adopted. Z EEG features and 50 video features are respectively input into kernel-based ELM. The actual outputs ($\mathbf{T}$ in Eq. 3) of kernel ELM are regarded as decision features. So EEG decision features and video decision features are

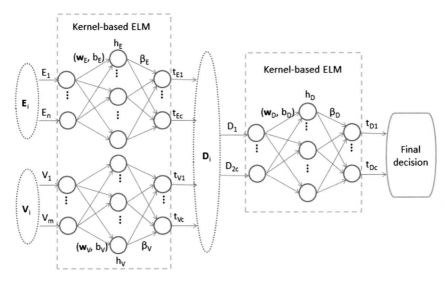

Fig. 2 Schematic diagram of the proposed classification fusion system

obtained. Then EEG decision features and video decision features are combined together to form new decision features. Finally, the new decision features are input into kernel-based ELM and final decision can be obtained. The proposed classification fusion system is depicted graphically by Fig. 2. E_i denotes EEG features corresponding with one video clip. The dimension of E_i is n and $n = Z$. V_i denotes video features of the same video clip. The dimension of V_i is m and $m = 50$. If the actual outputs $\mathbf{T}_E = \begin{bmatrix} t_{E1} & \cdots & t_{Ec} \end{bmatrix}$ and $\mathbf{T}_D = \begin{bmatrix} t_{D1} & \cdots & t_{Dc} \end{bmatrix}$, decision features D_i is given by

$$\mathbf{D}_i = \begin{bmatrix} \mathbf{T}_E & \mathbf{T}_D \end{bmatrix} \tag{7}$$

where c denotes the number of video classes, and both $\mathbf{T}_E$ and $\mathbf{T}_D$ denote the actual outputs of kernel-based ELM. In addition, t_{Dj} ($j = 1, \ldots c$) also denotes the actual outputs of kernel-based ELM.

3 Materials and Experiment

3.1 Materials

The video dataset is created for the experiment. 90 video clips are extracted from different famous movies and television program. The movies include two major genres, which are action and drama. These video clips include three emotional classes: violence, neutral, eroticism. In addition, the clips in television program

Human and Nature were selected as neutral clips, and the clips in various movies were included as the other two classes' clips. The number of video clips in each class is the same. Each video clip only includes one emotional event, and has duration of approximately six seconds. The video dataset were built by six students (3 males and 3 females).

3.2 EEG Experiment Protocol

13 healthy subjects (7 males and 6 females, from 24 to 28 years old) participated in the experiment. At the start of the experiment, the screen displayed the experimental instruction and the announcements in the experiment. After watching them, the user pressed the space key to start the experiment. For one subject, 30 clips are selected from the video dataset and each 10 clips belonged to the same class. For avoiding subject to form the memory of inertia, the clips were played randomly. Before playing each clip, prompt displayed in the screen to attract the attention of the subject. When 30 clips were finished playing, the experiment ended. During the experiment, the subject's EEG signals were recorded. The experiment repeated for each subject. Finally, 17 subjects' EEG signals are obtained.

4   Experimental Results and Analyses

4.1   Result of EEG Feature Selection

In this paper, decision tree algorithm C4.5 is adopted to select EEG features. EEG data's each feature is regarded as one attribute. Simplified tree's all attributes are used as EEG features selected. In the experiment, simplified tree contained 14 attributes. Therefore, 14 EEG features are obtained from 3072 WPD features. The position of the selected features is C6, C5, M1, PO3, FP1, FT7, F3, POZ and AF4. In addition, FP1 includes five features, and C5 includes two features and other electrodes only have one feature. AF4 and F3 correspond to brain's frontal area. FT7 corresponds to frontal- temporal area. C5 and C6 correspond to brain's central area. PO3 and POZ correspond to brain's parietal-occipital area. According to findings in brain research, frontal lobe is associated with attention, short memory, and planning; parietal lobe is associated with movement; occipital lobe is associated with vision; temporal lobe is associated with sensory input processing, language comprehension, and visual memory retention. The selected features demonstrate audio-visual task stimulate brain's frontal area, central area and parietal-occipital area. This result is consistent with findings in brain research and confirms that applying decision tree to EEG feature selection is reasonable.

Table 2 Classification results comparing the proposed method with single mode methods

Method	Optimal parameters	Accuracy (%)
Video only	Regularization_coefficient = 2^1; 'RBF_kernel'; Kernel_para = 1	73.33
EEG only	Regularization_coefficient = 2^{10}; 'RBF_kernel'; Kernel_para = 10^{17}	55.56
Fusion	(V)Regularization_coefficient = 2^4; 'RBF_kernel'; Kernel_para = 10^2	76.67
	(E)Regularization_coefficient = 2^1; 'RBF_kernel'; Kernel_para = 10^3	
	(D)Regularization_coefficient = 2^4; 'RBF_kernel'; Kernel_para = 1	

4.2 Evaluation of Classification Fusion Using Kernel-Based ELM

In order to test the validity of the proposed method, we compare our method separately with single mode methods of using video content only and using EEG signals only on classification accuracy. Classification accuracy is obtained by using the averaged tenfold cross-validation. To select optimal parameters that make classification performance optimal, regularization coefficient of kernel-based ELM ranges from 2^{-15} to 2^{15} and kernel_para ranges from 10^{-15} to 10^{20}, and step sizes are 2 and 10 respectively. The results of classification are shown in Table 2. In particular, the first row of Fusion in Optimal parameters denotes the optimal parameters of kernel-based ELM whose inputs are video features. The second row of Fusion in Optimal parameters denotes the optimal parameters of kernel-based ELM whose inputs are EEG features. The last row of Fusion in Optimal parameters denotes the optimal parameters of kernel-based ELM whose inputs are decision features. From the last column of the Table 2, we can see that the accuracy of the proposed fusion method is higher than that of method using video only and EEG only. The classification accuracy obtained by using the proposed method is about 3.34 % higher than using video content only and approximately 21.11 % higher than using EEG signals only.

5 Conclusions

In this paper, we proposed a multimodal fusion method using kernel-based ELM to recognize video emotion by combing video content and EEG signals. Then we design and conduct the EEG experiment to collect data. At last, the proposed fusion method is applied to our collected data. The main conclusions are summarized as follows:

1. The method is suitable for three-class recognition of video emotion. The validity of our approach has been demonstrated with our data set.
2. We compare the proposed method separately with single mode methods of using video content only and EEG signals only on classification accuracy. The experimental results show that the proposed fusion method produces better classification performance than those of the video emotion recognition methods which use either video content or EEG signals alone. In addition, the classification accuracy obtained by using the proposed method is about 3.34 % higher than using video content only and approximately 21.11 % higher than using EEG signals only.

Acknowledgements This research is partially sponsored by Natural Science Foundation of China (Nos. 61175115, 61370113 and 61272320), Beijing Municipal Natural Science Foundation (4152005 and 4152006), the Importation and Development of High-Caliber Talents Project of Beijing Municipal Institutions (CIT&TCD201304035), Jing-Hua Talents Project of Beijing University of Technology (2014-JH-L06), Ri-Xin Talents Project of Beijing University of Technology (2014-RX-L06), the Research Fund of Beijing Municipal Commission of Education (PXM2015_014204_500221) and the International Communication Ability Development Plan for Young Teachers of Beijing University of Technology (No. 2014-16).

References

1. Lin, J., Sun, Y., Wang, W.: Violence detection in movies with auditory and visual cues. In: Proceedings of the 2010 International Conference on Computational Intelligence and Security. IEEE Computer Society, pp. 561–565 (2010)
2. Nie, D., Wang, X.W., Shi, L.C., et al.: EEG-based emotion recognition during watching movies. Int. IEEE/EMBS Conf. Neural Eng. **1359**, 667–670 (2011)
3. Bailenson, J.N., et al.: Real-time classification of evoked emotions using facial feature tracking and physiological responses. Int J Hum Mach Stud **66**(5), 303–317 (2008)
4. Mansoorizadeh, M.: Multimodal information fusion application to human emotion recognition from face and speech. Multimed. Tools Appl. **49**(2), 277–297 (2010)
5. Koelstra, R.A.L.S.: Affective and implicit tagging using facial expressions and electroencephalography. Queen Mary University of London (2012)
6. Ye, W., Fan, X.: Bimodal emotion recogition from speech and text. Int. J. Adv. Comput. Sci. Appl. **5**(2), 26–29 (2014)
7. Wang, S., Zhu, Y., Wu, G., et al.: Hybrid video emotional tagging using users' EEG and video content. Multimed. Tools Appl. **72**(2), 1257–1283 (2014)
8. Chuang, Z.J., Wu, C.H.: Multi-modal emotion recognition from speech and text. Int. J. Comput. Linguist. Chin. Lang. Process. **1**, 779–783 (2004)
9. Pantic, M., Caridakis, G., André, E., et al.: Multimodal emotion recognition from low-level cues. Cognit. Technol. 115–132 (2011)
10. Sun, K., Yu, J.: Video affective content representation and recognition using video affective tree and hidden markov models. Lecture Notes Comput. Sci. 594–605 (2007)
11. Jasmine, K.P., Kumar, P.R.: Integration of HSV color histogram and LMEBP joint histogram for multimedia image retrieval. Adv. Intell. Syst. Comput. (2014)
12. Wu, T., Yan, G.Z., Yang, B.H., et al.: EEG feature extraction based on wavelet packet decomposition for brain computer interface. Measurement **41**(6), 618–625 (2008)

13. Chen, X., Wu, J., Cai, Z.: Learning the attribute selection measures for decision tree. In: Fifth international conference on machine vision (ICMV 2012): algorithms, pattern recognition, and basic technologies, **8784**(2), 257–259 (2013)
14. Huang, G.B., Zhu, Q.Y., Siew, C.K.: Extreme learning machine: a new learning scheme of feedforward neural networks. In: 2004 International Joint Conference on Neural Networks (IJCNN'2004), (Budapest, Hungary), July 25–29 (2004)
15. Huang, G.B., Zhou, H., Ding, X., Zhang, R.: Extreme learning machine for regression and multiclass classification. IEEE Trans. Syst. Man Cybernet. Part B: Cybernet. **42**(2), 513–529 (2012). (This paper shows that ELM generally outperforms SVM/LS-SVM in various kinds of cases.)

Equality Constrained-Optimization-Based Semi-supervised ELM for Modeling Signal Strength Temporal Variation in Indoor Location Estimation

Felis Dwiyasa, Meng-Hiot Lim, Yew-Soon Ong and Bijaya Panigrahi

Abstract Signal strength can be used to estimate location of a wireless device. As compared to other signal measures such as time-based and angle-based metrics, signal strength is normally embedded in wireless transceivers. This allows us to add location estimation feature on top of any wireless systems without requiring hardware modification. However, signal strength is affected by many environmental factors which cause temporal and spatial variation that could degrade the accuracy of location estimation system if not handled properly. In this paper, we focus on the temporal variation effect which is inevitable in dynamic environments where people and surrounding objects are typically not stationary. We try to improve the Location Estimation using Model Trees (LEMT) algorithm, a previous work that uses M5 model tree, by proposing that the calibration of the radio map over time can be done using Equality Constrained-optimization-based Semi-Supervised Extreme Learning Machine (ECSS-ELM). By using continuous signal strength readings collected from reference tags and tracking tag of a 2.4-GHz Radio Frequency Identification (RFID) system, we found that the algorithm can achieve comparable performance with much faster training time and testing time as compared to the M5 model tree.

Keywords Semi-supervised · ELM · Temporal variation · Signal strength

F. Dwiyasa (✉) · M.-H. Lim
School of Electrical and Electronic Engineering, Nanyang Technological University,
Singapore 639798, Singapore
e-mail: felis001@e.ntu.edu.sg

Y.-S. Ong
School of Computer Engineering, Nanyang Technological University, Singapore
639798, Singapore

B. Panigrahi
Department of Electrical Engineering, Indian Institute of Technology Delhi,
New Delhi 110016, India

© Springer International Publishing Switzerland 2016
J. Cao et al. (eds.), *Proceedings of ELM-2015 Volume 1*,
Proceedings in Adaptation, Learning and Optimization 6,
DOI 10.1007/978-3-319-28397-5_30

383

1 Introduction

A wide range of sensors such as infrared [1], ultrasonic [2] and inertial sensor [3] have been proposed to offer a solution for indoor navigation system. There are also numerous research works on wireless-based positioning based on Ultra Wideband (UWB) [4], Wi-Fi [5–7], and Radio Frequency Identification (RFID) [6, 7]. In wireless communication system, signal strength is one of the most popular techniques for positioning.

Multiple devices are typically required to provide accurate positioning of a tracked device. One common approach is to use several measurement units [5–8]. Some methods use several reference devices places in static positions [6, 7]. In cooperative learning, a tracked device may also use the location information gathered from other tracked devices nearby [8].

Various building layout and materials, furniture placements, human activity within the building, and interferences from other wireless devices may strengthen or weaken a signal. Fingerprint-based technique could adapt to different test-site layouts by doing offline training phase before the real testing phase is conducted. However, when there is any change in environmental condition, the radio map pattern may change and the data collected during training phase may not correlate well with the data at testing phase.

As repeated training data collection can be a tedious manual process, automatic real-time calibration has been considered to provide an efficient way in capturing temporal variation effect. In LANDMARC algorithm [6], reference tags are installed in a grid of fixed positions to capture the temporal variation. LANDMARC assumes that functional relationship between tags is inversely proportional to the distance between them. It also requires that all tags are nearly identical and that reference tags are placed in a way that the k-nearest reference tags are not blocked in an unbalanced manner [6].

Location Estimation using Model Trees (LEMT) [7] proposes that functional relationship between tags can be learned during training phase. By doing so, LEMT is found to be not much affected by non-uniformity of device hardware and propagation path.

Despite its better robustness, LEMT has a potential complexity issue, particularly because the depth of its tree depends on the training data and therefore the complexity of the algorithm may grow unpredictably. To solve uncertain training time experienced by the model tree, we would like to explore Extreme Learning Machine (ELM) approach which has a very quick training time as compared to other neural network approaches. To our knowledge, there has been no previous work that explores the use of ELM to replace tree-based model. We apply ELM algorithm to learn the functional relationship between the reference tags and tracking tag when the signal strength is experiencing temporal variation due to environmental changes.

Our work extends the work presented in [9], in which semi-supervised ELM is used to learn spatial variation. Instead of spatial variation, our work uses the semi-supervised ELM for learning temporal variation. The semi-supervised ELM allows us to use unlabeled data, which, in our location system, are the signal strength of the reference tags. Unlabeled data are easy to collect because we do not need to manually label the location of the tracking tag during the data collection. We also merge the concept of equality constrained-optimization-based from [10] into the semi-supervised ELM [9] to improve accuracy, and then apply the model to replace M5 model tree in LEMT algorithm [7]. The ability of the model to provide a good approximation to adapt a changing radio map will be evaluated in terms of speed and accuracy.

2 Learning Temporal Variation

2.1 LEMT Algorithm

Because signal strength is affected by many environmental factors, it is inevitable that signal strength value may change over time due to environmental dynamics. This phenomenon is often referred to as temporal variation. It is one of the most important factors to consider when developing indoor location estimation system. A calibrated location estimation system which does not adapt to temporal variation would have its radio map obsolete and would certainly experience performance degradation over time.

LEMT algorithm [7] is developed based on the assumption that tags at static locations have functional relationship that does not change over time. The algorithm consists of two phases: offline training phase and online testing phase. The functional relationship between the reference tags and the tracking tag is learned during offline training phase. Later on, the functional relationship is used to estimate the signal strength of the tracking tag during online testing phase.

Consider a 2-dimensional coordinate of physical location $l_i = (x_i, y_i)$ for $1 \leq i \leq n$, where n is the number of possible physical locations. For each location i, the signal strength of m reference tags and a tracking tag are measured by p readers. During training phase, the objective is to find the functional relationship which is defined as

$$ s_j^{(i)} = f_{ij}\left(r_{1j}^{(i)}, r_{2j}^{(i)}, \ldots, r_{mj}^{(i)} \right) \tag{1} $$

where $r_{kj}^{(i)}$ is the signal strength of reference tag k measured by reader j at location i, $f_{ij}(.)$ is the functional relationship of reader j at location i, and $s_j^{(i)}$ is the signal strength of tracking tag measured by reader j at location i.

As mentioned previously, LEMT assumes that $f_{ij}(.)$ does not change over time and therefore the functional relationship that is learned from training data can be applied on testing data. The functional relationship $f_{ij}(.)$ is learned using M5 model tree algorithm [11] which constructs a decision tree that contains piecewise linear functions. The tree is built by recursively splitting the training data into subsets with the objective that the subsets maximize the expected reduction in error. A detailed explanation on how an M5 model tree is formed can be found in [7, 11].

During online testing phase, the signal strength of the tracking tag is estimated based on the signal strength of reference tags and the functional relationship that has been learned during the training phase:

$$\sigma_j^{(i)} = f_{ij}\left(r_{1j}^{(i)}, r_{2j}^{(i)}, \ldots, r_{mj}^{(i)}\right) \tag{2}$$

The most likely location of the tracking tag, L, is the location l_i that minimizes the Euclidean distance D_i between the estimated signal strength $\sigma_j^{(i)}$ and the tracking tag signal strength $s_j^{(i)}$ for all readers, as defined by:

$$D_i = \sqrt{\sum_{j=1}^{p}\left(\sigma_j^{(i)} - s_j^{(i)}\right)^2} \tag{3}$$

$$L = \arg\min_{l_i} D_i \tag{4}$$

As presented in [7], the time complexity of LEMT algorithm is $O(m'np)$, where m' is the average depth of the model trees. The number of independent model trees that must be built and stored is $O(np)$. This complexity analysis shows that the time and space requirements linearly increase when we use more readers, more reference tags, and more locations to distinguish.

2.2 Extreme Learning Machine

ELM is a single hidden-layer feedforward network which has its hidden layer neurons randomized instead of optimized [12]. Because the output layer is the only layer optimized during training phase, ELM has a very fast training time as compared to conventional neural network algorithms which use backpropagation training. Various regression and classification problems can be solved by ELM with reasonable accuracy and much faster training time [10].

Similar with other neural network approaches, each neuron in ELM performs a nonlinear function such as sigmoid, Gaussian, hard-limit, or various types of kernel methods. Therefore, ELM can be used to approximate nonlinear functions. Using the same notations used in Sect. 2.1, our objective is to use ELM to solve the functional relationship $f_{ij}(.)$ during the training phase. We chose ELM which has a very fast training time in lieu of M5 model tree used in LEMT algorithm.

To achieve this, we build a training dataset which consists of rows of input vector and target output. As shown in Eqs. (5) and (6), we define the input vector **u** as the signal strength of all reference tags, and the target output v as the signal strength of the tracking tag.

$$\mathbf{u} = \{r_{1j}^{(i)}, r_{2j}^{(i)}, \ldots, r_{mj}^{(i)}\} \tag{5}$$

$$v = s_j^{(i)} \tag{6}$$

During training phase, the ELM learns the relationship between the input vector and target output by randomizing the weights of hidden neurons and solving the weights of output neurons. During testing phase, the weights of hidden neurons and output neurons are used to estimate the signal strength of the tracking tag. The signal strength estimates obtained from ELM are then used to substitute $\sigma_j^{(i)}$ in LEMT equation in Eq. (3).

In this work, we focus on several types of ELM, which are: (1) Original ELM (2) Equality Constrained-optimization-based ELM (EC-ELM) (3) Semi-Supervised ELM (SS-ELM) and (4) Equality Constrained-optimization-based Semi-Supervised ELM (ECSS-ELM). The main difference of those algorithms is on how the weights of output neurons are solved.

The original ELM algorithm [12] solves the weights of output neurons β as:

$$\beta = \mathbf{H}^+\mathbf{T} \tag{7}$$

where $\mathbf{H}$ is a weight matrix of output neurons, $+$ is Moore-Penrose pseudoinverse operator and $\mathbf{T}$ is the training data output.

In EC-ELM [10], the pseudoinverse in Eq. (7) is replaced by matrix regularization and matrix inverse as follow:

$$\beta = \left(\frac{\mathbf{I}}{C} + \mathbf{H}^T\mathbf{H}\right)^{-1}\mathbf{H}^T\mathbf{T} \tag{8}$$

In SS-ELM [9], additional unlabeled training samples are used as a constraint for the target regression. Unlabeled training samples are training data that are constructed by training input only, whereas labeled training samples are training data that are constructed by training input and training output. Unlabeled data are easier to collect than labeled data because they do not contain training output which is usually obtained from manual labeling process. SS-ELM adds smoothness constraint and calculates the weights of output neurons as:

$$\beta = \left(\left(\mathbf{J} + \lambda\tilde{\mathbf{L}}^T\right)\mathbf{H}\right)^+\mathbf{J}\mathbf{T} \tag{9}$$

where $\mathbf{J}$ is a diagonal matrix which has its diagonal elements set to 1 for labeled training data and 0 for unlabeled training data, $\tilde{\mathbf{L}}$ is a Laplacian matrix that is aimed to include smoothness constraint and λ is a constraint weight that controls how much the smoothness factor affects the weights of output neurons.

The Laplacian matrix is defined as:

$$\tilde{\mathbf{L}} = \tilde{\mathbf{D}} - \tilde{\mathbf{W}} \tag{10}$$

$$\tilde{W}_{ij} = \exp\left(\frac{-\left|\mathbf{X}_i - \mathbf{X}_j\right|^2}{2\delta^2}\right) \tag{11}$$

$$\tilde{D}_{ii} = \sum_{j=1}^{m} \tilde{W}_{ij} \tag{12}$$

where $\mathbf{X}_t$ is the training input data at time t, and δ is the standard deviation of Gaussian process. For the rest of the paper we assume that δ is calculated as the standard deviation of $\left|\mathbf{X}_i - \mathbf{X}_j\right|^2$.

In addition to the ELM algorithms presented above, we propose ECSS-ELM which merges the equality constrained-optimization concept in Eq. (8) and semi supervised concept in Eq. (9). It calculates the weights of output neurons as:

$$\beta = \left(\frac{\mathbf{I}}{C} + \mathbf{Q}^T\mathbf{Q}\right)^T \mathbf{Q}^{-1}\mathbf{T} \tag{13}$$

$$\mathbf{Q} = \left(\mathbf{J} + \lambda\tilde{\mathbf{L}}^T\right)\mathbf{H} \tag{14}$$

3 Experimental Testing

3.1 Test Scenario

To analyze the temporal variation phenomenon, we conducted signal strength measurement for 50 min using a 2.4-GHz active RFID system. The system consists of 2 readers, 8 reference tags and 1 tracking tag in an indoor environment shown in Fig. 1. The tags broadcast a ping signal every 3 s and the readers measure the signal strength of every incoming ping signals.

We placed 4 reference tags in Room A and the other 4 tags in Room B. All reference tags were placed on the floor at fixed positions throughout the experiment. The tracking tag was placed in Room A for the first 24 min and then moved to Room B during a 1-min break. The tracking tag was then left in Room B for the following 24 min, followed by a 1-min break.

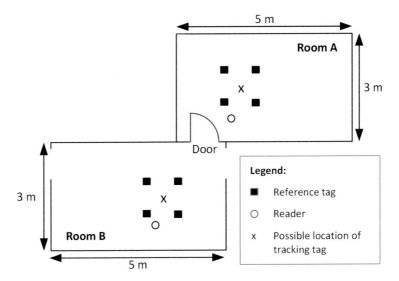

Fig. 1 Test site layout

Because ping signals may not always be received well by readers due to unexpected anomalies such as signal collision, we deal with missing data by taking the average of the signal strength received during 15 s. In total we get a dataset that contains 48 min × 4 samples/min = 192 data samples.

Temporal variation was deliberately introduced by changing the state of a door that separates Room A and Room B every 3 min. The door was either fully closed or fully opened.

Training data is built by taking the first 12-min data of all tags measured by each reader at each location. The last 12-min data are used as testing data. We took consecutive T_r samples of training data as labeled data, where T_r represents the training data length that varies from 3, 6, 9 and 12 min. The remaining training data are unlabeled and used for SS-ELM and ECSS-ELM.

Simulation was repeated for 50 times to cover random statistical error. For each trial, we used fixed time-width sliding window for selecting training data.

We ran Matlab simulation by using the source codes of the original ELM and M5 model tree taken from [13] and [14] respectively. ELM parameter configurations used for the experiment are as shown in Table 1.

Table 1 Parameter configurations for the ELM algorithms

(1) Original ELM	(2) EC-ELM	(3) SS-ELM	(4) ECSS-ELM
Regression mode	Regression mode	Regression mode	Regression mode
Sigmoid activation	Sigmoid activation	Sigmoid activation	Sigmoid activation
100 hidden nodes	100 hidden nodes	100 hidden nodes	100 hidden nodes
	$C = 2^{24}$	$\lambda = 6$	$\lambda = 6 \ C = 2^{24}$

3.2 Data

The signal strength of four reference tags placed in Room A, four reference tags placed in Room B, and the tracking tag are as shown in Fig. 2a, b. Clearly, the signal strength fluctuate even when tags are not physically moved.

Fluctuations are more significant for pairs of reference tag and reader that are blocked by the door. In Fig. 2a, Reader A which was placed in Room A observes more fluctuations on the signal strength of the reference tags in Room B, whereas in Fig. 2b Reader B observes more fluctuations on the signal strength of the reference tags in Room A.

Tags that are located in the same room tend to have closer signal strength values as compared with tags that are located in separated rooms. In the first 24-min data of Fig. 2a, which correspond to the data measured by Reader A when the tracking tag was in room A, the signal strength data of the tracking tag range from −30 to

Fig. 2 Signal strength of reference tags and tracking tag. **a** Signal strength measured by Reader A. **b** Signal strength measured by Reader B

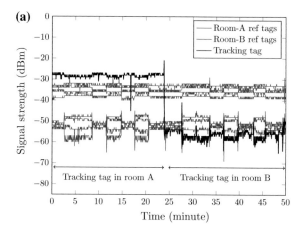

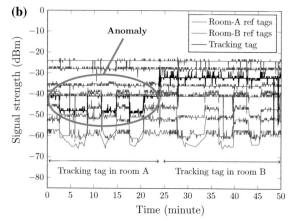

−27 dBm for most of the time. It is closer to the signal strength data of the reference tags in Room A which range from −40 to −32 dBm, rather than the signal strength data of the reference tags in Room B which range from −60 to −48 dBm. Similarly, as shown in the last 24-min data of Fig. 2a, b, the signal strength data of the tracking tag when the tracking tag was in Room B are closer to the signal strength data of reference tags in Room B.

However, an anomaly can be observed in the first 24-min data of Fig. 2b which correspond to the signal strength measured by Reader B when the tracking tag is in Room A. Here the signal strength data of the tracking tag range from −48 to −40 dBm, which are in between the signal strength of the reference tags located in Room A and Room B. Therefore, reference tags that are physically closer to a tracking tag are not always the ones that have closer signal strength fingerprints.

This anomaly would cause difficulties for the LANDMARC algorithm to decide the k-nearest reference tags because the k reference tags that have closer fingerprints for Reader A would differ from those for Reader B. Reader B would not have k-nearest fingerprints from one room only, whereas reader A could easily pick the k-nearest fingerprint from Room A. It is highly possible that this anomaly is caused by the indoor propagation behavior which can be unpredictable due to the complexity of indoor environment. Selecting wrong k-nearest reference tags would deteriorate the performance of LANDMARC [6].

In contrast, LEMT may not be affected by the anomaly because it does not perform reference tags selection as LANDMARC. Furthermore, the patterns on the signal strength values when the door is opened or closed seem to be repeating over time. If the patterns are repeating, the functional relationship could be learned, either by using model tree originally used in LEMT algorithm or by using ELM algorithm we propose.

3.3 Performance Evaluation

For speed comparison, we recorded the training time and testing time of M5 model tree and ELM for each trial. Because LEMT algorithm requires independent functional relationship modeling for each location and each reader, we accumulated the time required to obtain the functional relationship of all locations and all readers so that we could compare the overall performance.

We calculated the Root Mean Square Error (RMSE) to measure the error between the estimated signal strength and the actual signal strength of the tracking tag in Eqs. (1) and (2) for each reader at each location using the following equation:

$$RMSE_j^{(i)} = \sqrt{\frac{1}{T_s} \sum_{t=1}^{T_s} \left(\sigma_j^{(i)} - s_j^{(i)} \right)^2} \tag{15}$$

We also measured the localization accuracy, which is defined as the correctness percentage in pinpointing the most likely location of the tracking tag, whether it is in Room A or Room B in our test case.

The training time, testing time, RMSE, and localization accuracy of the repeated experiments for all readers at all locations are then aggregated and presented as statistical mean and standard deviation.

As shown in Tables 2 and 3, the training time and testing time of ELM algorithms are much shorter than M5 model tree. Moreover, the training time and testing time of ELM algorithms are relatively constant regardless of the length of training data. As for M5 model tree, the training time and testing time increase significantly when more training data are used. The training time of M5 model tree ranges from 71.2 to 443.2 ms, whereas the testing time of M5 model tree ranges from 36.2 to 62.2 ms.

Furthermore, ELM algorithms are much faster than M5 model tree in terms of training time and testing time. For 12-min training data, the training time of ELM is between 14 and 35 times faster than M5 model tree, whereas its testing time is between 10 and 14 times faster.

Although the training time and testing time required by M5 model tree is under 1 s, which is still reasonable, we should note that the scale of our experiment is still very small. In real deployment, the training time and testing time can be significant when hundreds or thousands of devices are involved. Speed improvement is important when we want to achieve a large-scale real-time localization system.

Table 2 Comparison of training time (mean ± standard deviation)

Algorithms	Training data length (min)			
	3	6	9	12
	Training time (ms)			
M5 model tree	71.2 ± 65.2	186.1 ± 19.4	322.7 ± 27.9	443.2 ± 20.1
Original ELM	8.7 ± 6.9	12.1 ± 6.0	20.6 ± 7.6	27.2 ± 10.2
EC-ELM	19.2 ± 78.7	11.6 ± 6.8	11.9 ± 3.2	12.4 ± 3.3
SS-ELM	30.3 ± 4.5	30.5 ± 1.7	30.7 ± 1.3	30.9 ± 1.3
ECSS-ELM	24.9 ± 21.5	22.8 ± 8.4	28.1 ± 20.3	22.0 ± 5.2

Table 3 Comparison of testing time (mean ± standard deviation)

Algorithms	Training data length (min)			
	3	6	9	12
	Testing time (ms)			
M5 model tree	36.2 ± 9.7	49.0 ± 7.9	57.4 ± 7.7	62.2 ± 2.3
Original ELM	7.8 ± 6.1	6.0 ± 2.3	6.3 ± 4.1	6.3 ± 2.0
EC-ELM	4.9 ± 1.3	6.7 ± 4.8	6.4 ± 2.9	6.1 ± 3.3
SS-ELM	4.7 ± 1.9	4.3 ± 0.5	4.4 ± 0.7	4.3 ± 0.5
ECSS-ELM	5.8 ± 3.6	5.9 ± 4.1	6.5 ± 6.1	5.3 ± 2.5

Table 4 Comparison of root mean square error (mean ± standard deviation)

Algorithms	Training data length (min)			
	3	6	9	12
	RMSE (dB)			
M5 model tree	2.20 ± 0.81	1.76 ± 0.76	1.48 ± 0.12	1.48 ± 0.00
Original ELM	49.52 ± 89.29	52.29 ± 155.93	26.20 ± 23.03	46.87 ± 18.20
EC-ELM	4.53 ± 2.16	2.94 ± 3.98	1.64 ± 0.20	1.54 ± 0.10
SS-ELM	6.61 ± 3.87	8.35 ± 4.41	11.10 ± 4.72	16.42 ± 3.93
ECSS-ELM	2.16 ± 0.46	1.70 ± 0.28	1.53 ± 0.18	1.47 ± 0.08

Table 5 Comparison of location estimation accuracy (mean ± standard deviation)

Algorithms	Training data length (min)			
	3	6	9	12
	Localization accuracy (%)			
M5 model tree	99.93 ± 0.26	99.96 ± 0.22	100.00 ± 0.00	100.00 ± 0.00
Original ELM	85.89 ± 19.19	89.54 ± 13.74	85.65 ± 8.76	78.63 ± 4.50
EC-ELM	99.11 ± 3.19	99.54 ± 2.93	100.00 ± 0.00	100.00 ± 0.00
SS-ELM	97.98 ± 4.67	95.46 ± 6.92	90.96 ± 8.60	82.83 ± 5.64
ECSS-ELM	100.00 ± 0.00	100.00 ± 0.00	100.00 ± 0.00	100.00 ± 0.00

Table 4 shows that the best RMSE is achieved by M5 model tree and ECSS-ELM. As for the remaining ELM algorithms, their RMSE are significantly higher.

In Table 5, it is clear that ECSS-ELM obtains the best localization accuracy. Except for the original ELM, the localization accuracy of the remaining algorithms is not too far from 100 % in most cases. This could have happened due to the simplicity of our experimental setup. In an environment with more complex temporal variation with higher number of locations to decide, accuracy degradation is expected.

4 Conclusions and Future Work

In this paper, we have presented several ELM algorithms to learn indoor radio map. As compared to M5 model tree, which is the original method proposed in LEMT algorithm, the training and testing speed of ELM can be much faster. The speed improvement would be useful if the localization system needs to fulfill a strict time constraint and be deployed in a larger scale. Among the four ELM algorithms tested, ECSS-ELM achieves the best accuracy in terms of RMSE and localization accuracy with a reasonable training time and testing time.

As temporal variation introduced in this work may only represent a simple type of signal fluctuation, the first research direction we want to pursue is to conduct the temporal variation observation in a more realistic indoor environment where more complex human activities are present on the test site. Second, we will increase the number of readers and reference tags to observe the performance of our proposed method when applied on a large-scale system. Third, we will also consider various methods to deal with missing data such as presented in [15] because some signal strength data might be missing in a large-scale indoor environment due to the limited range of wireless transceivers. Finally, we will explore other regression and prediction methods such as methods presented in [16, 17] in order to find an indoor radio map updating method that has reasonable training time and achieves higher accuracy than what we have obtained in this work.

Acknowledgments This research is supported by the National Research Foundation Singapore under its Interactive Digital Media (IDM) Strategic Research Programme.

References

1. Want, R., Hopper, A., Falcao, V., Gibbons, J.: The active badge location system. In: ACM Transactions on Information Systems (TOIS), vol. 10, no. 1, pp. 91–102. ACM (1992)
2. Randell, C., Muller, H.: Low cost indoor positioning system. In: Ubiquitous Computing (Ubicomp 2001), pp. 42–48. Springer, Heidelberg (2001)
3. Coronel, P., Furrer, S., Schott, W., Weiss, B.: Indoor location tracking using inertial navigation sensors and radio beacons. In: The Internet of Things, vol. 4952, pp. 325–340. Springer, Heidelberg (2008)
4. Guvenc, I., Sahinoglu, Z., Orlik, P.V.: TOA Estimation for IR-UWB systems with different transceiver types. IEEE Trans. Microw. Theory Tech. **54**(4), 1876–1886 (2006)
5. Haeberlen, A., Flannery, E., Ladd, A.M.: Practical robust localization over large-scale 802.11 wireless network. In: Proceedings of the 10th Annual International Conference on Mobile Computing and Networking, pp. 70–84. ACM (2004)
6. Ni, L.M., Liu, Y., Lao, Y.C., Patil, A.P.: LANDMARC: indoor location sensing using active RFID. In: Wireless Networks, vol. 10. no. 6, pp. 701–710. Springer, Heidelberg (2004)
7. Yin, J., Yang, Q., Ni, L.M.: Learning adaptive temporal radio maps for signal-strength-based location estimation. In: IEEE Transactions on Mobile Computing, vol. 7, no. 7, pp. 869–883. IEEE Press, New York (2008)
8. Satizábal, H.F., Upegui, A., Perez-Uribe, A., Rétornaz, P., Mondada, F.: A social approach for target localization: simulation and implementation in the marXbot robot. In: Memetic Computing, vol. 3, no. 4, pp. 245–259. Springer, Heidelberg (2011)
9. Liu, J., Chen, Y., Liu, M., Zhao, Z.: SELM: semi-supervised ELM with application in sparse calibrated location estimation. In: Neurocomputing, vol. 74, pp. 2566–2572. Elsevier, Amsterdam (2011)
10. Huang, G.B., Zhou, H., Ding, X., Zhang, R.: Extreme learning machine for regression and multiclass classification. IEEE Trans. Syst. Man Cybern. Part B Cybern. **42**(2), 513–529 (2012)
11. Quinlan, J. R.: Learning with continuous classes. In: 5th Australian Joint Conference on Artificial Intelligence, vol. 92, pp. 343–348. World Scientific (1992)
12. Huang, G.B., Zhu, Q.Y., Siew, C.K.: Extreme learning machine: a new learning scheme of feedforward neural networks. In: IEEE International Joint Conference on Neural Networks, vol. 2, pp. 985–990. IEEE Press, New York (2004)

13. Matlab Codes of ELM Algorithm (for ELM with random hidden nodes and random hidden neurons). http://www.ntu.edu.sg/home/egbhuang/elm_random_hidden_nodes.html
14. M5PrimeLab: M5 Regression Tree and Model Tree Toolbox, version 1.0.2. http://www.cs.rtu.lv/jekabsons/Files/M5PrimeLab.zip. Accessed 16 May 2015
15. Zhang, J., Song, S., Zhang, X.: Sparse Bayesian ELM handling with missing data for multi-class classification. In: Proceedings of ELM-2014, vol. 1, pp. 1–13. Springer, Heidelberg (2015)
16. Kraipeerapun, P., Nakkrasae, S., Fung, C.C., Amornsamankul, S.: Solving regression problem with complementary neural networks and an adjusted averaging technique. Memetic Comput. 2(4), 249–257 (2010)
17. Miranian, A., Abdollahzade, M.: Developing a local least-squares support vector machines-based neuro-fuzzy model for nonlinear and chaotic time series prediction. IEEE Trans. Neural Netw. Learn. Syst. 24(2), 207–218 (2013)

Extreme Learning Machine with Gaussian Kernel Based Relevance Feedback Scheme for Image Retrieval

Lijuan Duan, Shuai Dong, Song Cui and Wei Ma

Abstract As for the huge gap between the low-level image features and high-level semantics, content-based image retrieval still could not receive a satisfying result by now. Since the special request of the relevance feedback, making full use of the rare number of labeled data and numerous unlabeled data is an ideal way. Because ELM has excellent classification accuracy and processing time, and high accuracy and fast speed are the key factors to evaluate the relevance feedback performances. In this paper, we proposed an Extreme learning Machine with Gaussian kernel Based Relevance Feedback scheme for image retrieval, to overcome the above limitations, our method uses three component classifiers to form a strong learner by learning different features extracted from the hand-marking data, then we use it to label the image database automatically. From the experiments we can see the use of the ELM with kernel have high classification accuracy, the processing time get largely decreased at the same time. Thus, it improves the efficiency of entire relevance feedback system. The experiments results show that the proposed algorithm is significantly effective.

Keywords Image retrieval · Relevance feedback · Extreme learning machine · Unlabeled data

L. Duan (✉) · S. Dong · S. Cui · W. Ma
Beijing Key Laboratory of Trusted Computing, College of Computer Science
and Technology, Beijing University of Technology, Beijing 100124, China
e-mail: ljduan@bjut.edu.cn

S. Dong
e-mail: dongs@emails.bjut.edu.cn

S. Cui
e-mail: cuisong@emails.bjut.edu.cn

W. Ma
e-mail: mawei@bjut.edu.cn

L. Duan · S. Dong · S. Cui · W. Ma
Beijing Key Laboratory on Integration and Analysis of Large-Scale Stream Data,
College of Computer Science and Technology, Beijing University of Technology,
Beijing 100124, China

J. Cao et al. (eds.), *Proceedings of ELM-2015 Volume 1*,
Proceedings in Adaptation, Learning and Optimization 6,
DOI 10.1007/978-3-319-28397-5_31

1   Introduction

In recent years, with the explosive increase in volume of digital images, content-based image retrieval technique is getting popular, lots of systems had been developed in the decades, including QBIC, Photobook, MARS, PicHunter and others [1, 2]. In a CBIR system, the system fails to be so perceptive to the user's intention that cannot return the satisfactory results, which mostly owning to the huge gap between the high-level semantic concepts and the low-level image features. In that way, relevance feedback methods were proposed [3]. And this technique had widely applied in vary content-based image retrieval systems [4, 5].

In this paper, we proposed an Extreme learning Machine (with Gaussian kernel) based relevance feedback approach. Firstly, we extract the features of labeled data and use them to train the component ELM classifiers. Secondly, we predict the whole database based on the tri-training method and vote for the result. Thirdly, the new labeled data are used to retrain the component classifiers and get the final learner group.

Cox et al. [6] proposed an interactive relevance feedback scheme based on Bayesian model by optimizing the features' probability distribution. Rui et al. [3] proposed an optimizing scheme for the relevance feedback performance by analyzing the features of positive samples. Zhou et al. [7] combined the semi-supervised method with active learning method by labeling the uncertain samples, which could have the unlabeled-data used.

The rest of this paper is organized as follows. Section 2 describe the method we proposed. Section 2.1 contains the description of our ELM with kernel based relevance feedback system. Section 3 describes the experiments and the performance of our scheme. Section 4 concludes this paper.

2   Method

2.1   *Extreme Learning Machine with Gaussian Kernel*

Extreme learning machine (ELM) was first proposed by Huang et al. [8]. ELM works for generalized single-hidden layer feed forward networks (SLFNs) [9, 10]. ELM algorithm tends to provide better generalization performance at extremely fast learning speed. Structure of the SLFNs [10] is shown as Fig. 1.

With L hidden nodes in output layer, the output function of SLFNs can be expressed by:

$$f_L(\mathbf{x}) = \sum_{i=1}^{L} \boldsymbol{\beta}_i g_i(\mathbf{x}) = \sum_{i=1}^{L} {}_i\boldsymbol{\beta}_i G(a_i, b_i, \mathbf{x}). \tag{1}$$

The function also can be written as $f(\mathbf{x}) = \mathbf{h}(\mathbf{x})\boldsymbol{\beta}$ where $\boldsymbol{\beta} = [\boldsymbol{\beta}1, \boldsymbol{\beta}2, \dots, \boldsymbol{\beta}L]$ is the vector of the output weights between the hidden layer of L neurons and the

Fig. 1 Single-hidden layer
feed forward networks

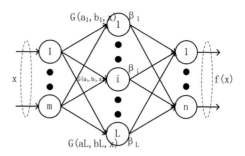

output neuron and $h(x) = [h1(x), h2(x), \ldots, hL(x)]$ is the output vector of the hidden layer with respect to the input x, which maps the data from input space to the ELM feature space [11].

In particular, L equals to N, which is rare condition because L is far smaller than N in actual problem, that is to say, that there is error between the output value and the actual value. So, the most important thing is to find least-squares solution $\boldsymbol{\beta}$ of the linear system.

$$\mathbf{H}\boldsymbol{\beta} = \mathbf{T}, \quad \boldsymbol{\beta} = \mathbf{H}^{\dagger}\mathbf{T}. \tag{2}$$

where $\mathbf{H}^{\dagger}$ is the Moore-Penrose Generalized inverse of matrix $\mathbf{H}$ [12, 13], $\mathbf{H}^{\dagger} = (\mathbf{H}'\mathbf{H})^{-1}\mathbf{H}'$ or $\mathbf{H}'(\mathbf{H}\mathbf{H}')^{-1}$, depending on the singularity of $\mathbf{H}'\mathbf{H}$ or $\mathbf{H}\mathbf{H}'$.

In the newly developed kernel ELM, it's getting more stable to introduce a positive coefficient into the learning system. If $\mathbf{H}'\mathbf{H}$ is nonsingular, the coefficient $1/\lambda$ is added to the diagonal of $\mathbf{H}'\mathbf{H}$ in the calculation of the output weights $\boldsymbol{\beta}$ After that, $\boldsymbol{\beta} = \mathbf{H}'(\mathbf{I}/\lambda + \mathbf{H}\mathbf{H}')^{-1}$, the corresponding function of the regularized ELM is:

$$f(\mathbf{x}) = \mathbf{h}(\mathbf{x})\boldsymbol{\beta} = \mathbf{h}(\mathbf{x})\mathbf{H}'\left(\frac{1}{\lambda}\mathbf{I} + \mathbf{H}\mathbf{H}'\right)^{-1}\mathbf{T}. \tag{3}$$

Huang et al. [11] shown that ELM with a kernel matrix can be defined as follows. Let $\boldsymbol{\Omega}_{ELM} = \mathbf{H}\mathbf{H}' : \boldsymbol{\Omega}_{ELM_{i,j}} = \mathbf{h}(\mathbf{x}i)\mathbf{h}(\mathbf{x}j) = K(\mathbf{x}i, \mathbf{x}j)$. The output function can be written as:

$$f(\mathbf{x}) = \mathbf{h}(\mathbf{x})\mathbf{H}'\left(\frac{1}{\lambda}\mathbf{I} + \mathbf{H}\mathbf{H}'\right)^{-1}\mathbf{T} = \begin{bmatrix} K(\mathbf{x}, \mathbf{x}_1) \\ \vdots \\ K(\mathbf{x}, \mathbf{x}_N) \end{bmatrix}' \left(\frac{1}{\lambda}\mathbf{I} + \boldsymbol{\Omega}_{ELM}\right)^{-1}\mathbf{T}. \tag{4}$$

The hidden layer feature mapping $\mathbf{h}(\mathbf{x})$ need not to be known, and instead its corresponding kernel $K(\mathbf{u}, \mathbf{v})$ can be computed. In this way, the Gaussian kernel is used, $K(\mathbf{u}, \mathbf{v}) = \exp(-\gamma||\mathbf{u} - \mathbf{v}||2)$ [14].

2.2 Algorithm Description

Due to the special requirement of relevance feedback, short processing time and high accuracy are the key points. Our method shown the promising effect.

Step 1: Label the data.

In the image retrieval system we proposed, let L denote the labeled data set, U denote the unlabeled data set. While $L = P \cup N$, where P denote the labeled positive samples and N denote the labeled negative samples.

Step 2: Feature extraction and train the component classifiers.

Inspired by the co-training paradigm [15], the labeled data used as the data of first train, and extract three different features to train the component classifier to reduce the variance of unstable procedures, which leading to improved prediction [16].

Step 3: Vote.

The whole database as known as U is put into the combined-classifier-group. From the Fig. 2 we can see the framework of voting procedure. Each component classifier will give its predicted label of ith trail of U data set, and the output is set as Y_{in}. All the trails are fed into m number of sub-ELM. Finally, a class label set ψ can be got.

Step 4: Label the unlabeled data automatically.

Let L_temp denote the whole database with labels. L_temp is used to retrain the component classifiers in order to level up the classification ability. From the Fig. 3, the dataset U has been classified by updating the classifier group.

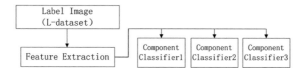

Fig. 2 The framework of voting

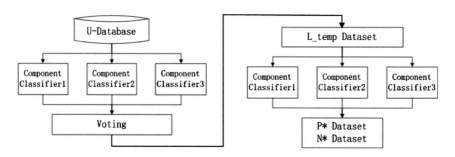

Fig. 3 The framework of labeling the unlabeled data

For the *i*th trail, P_i is the total number of the component classifiers which predict the class of *i*th trail to be 0; Q_i is the total number of the component classifiers which predict the class of *i*th trail to be 1. The sum of P_i and Q_i is 3. The vote function can be shown as bellow:

$$\text{vote}(\Psi) = \begin{bmatrix} Y_1 \\ \vdots \\ Y_M \end{bmatrix}_{M \times 1}, Y_i = \lfloor \frac{Q_i}{P_i} \rfloor \qquad (5)$$

Step 5: Update the classifiers.

As the result, let P^* denote the positive dataset, N^* denote the negative dataset. In each iteration of the feedback, the whole result list is returned by the ascend order of the similarity rank.

3 Experiments

In this section, we firstly introduce the image database our experiments performed on. Secondly we describe the methods of the feature extraction. Then we illustrate the structure of our relevance feedback system. Finally, we compare the performance of our methods with other methods in the aspect of the average-AP values and the aspect of processing time.

3.1 Image Database

We perform our experiments on the COREL photo gallery, which contains 1000 images into 10 categories, and 6000 images into 60 categories. The size of every image in the database is 256×384 or 384×256. A ground truth of the image database is needed to evaluate the performance of our experiment. Thus, the natural categories of the COREL photo gallery are used as the semantic categories, and we define that images belong to the same category are relevant, otherwise, are irrelevant.

3.2 Feature Selection

For the image retrieval system with relevance feedback, use 3 image features are used: SIFT, color and LBP.

Based on the former work in our laboratory, using Bag-of-Features model to extract two image features which based on SIFT and color, respectively. The color information is the most informative feature because of its robustness with respect to scaling, rotation, perspective, and occlusion.

Local Binary Pattern as known as LBP which is an effective image descriptor, it has been used as the texture feature of the image.

3.3 Image Retrieval System

For the image relevance feedback, the image retrieval system is required to return the images which are the most semantically relevant. Figure 4 shows our system has three main parts. The feedback part is the key function in the system.

- Retrieval part: Extract three features which mentioned above of every image in the image database. Computing the Euclidean distance as the similarity between the query and each image of the database, and return the top 20 images as the retrieval result.
- Feedback part: Let the user choose and label the negative images, extract the image features, the images that displayed and without labeled are set positive. Training classifier groups for the each kind of features respectively, to label the unlabeled database automatically. Updating the classifier groups by using the new-labeled data, as the result, the final classifier is using for classifying the unlabeled database.
- Display part: Displaying the retrieval result for the retrieval operation, displaying the feedback result of the each iteration for the relevance feedback operation.

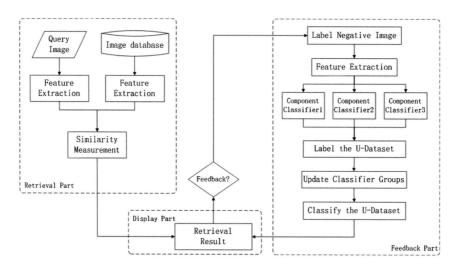

Fig. 4 The framework of image retrieval system with relevance feedback

3.4 Experiment Results

Before the relevance feedback procedure, the top 20 images are result which displayed on the panel. The user interface is shown on the Fig. 5. The image that at the top of left is the query image and the other images on the 'result panel' are the retrieval result. User label the 'negative', the rest images are labeled 'positive' automatically.

From Fig. 6 has shown only 4 images of the retrieval result meet our request. After the first iteration of relevance feedback, the result had notable improvement, the result shown on Figs. 7, 8 and 9 show the feedback results by iteration 1, 2, 3, respectively. The 'right' images are getting more, semantically, the result is more and more meets our request. It took us only 3 iterations to have all the results right.

We use the Average Precision (AP) measure. The retrieval result has been optimized by performing our relevance feedback scheme. At the each iteration of relevance feedback, The AP value can be obtained, which is defined as the average of precision value obtained after each relevant image is retrieved. Let $\bar{P}$ denotes the average precision which is obtained at the current iteration, and it is computed by:

$$\bar{P} = \sum_{E_i \in R} \frac{P_i}{|S|} \tag{6}$$

where P_i denotes the precision value obtained after the system retrieves i top-ranked images, E_i is the element of the relevant images set, S is the set of all relevant images that belong to the same category as the query, and $|S|$ denotes the cardinality of S. The AP calculated over all the relevant images can avoid the fluctuation of precision that is usually encountered by the traditional precision measurement.

Fig. 5 The user interface and the retrieval result

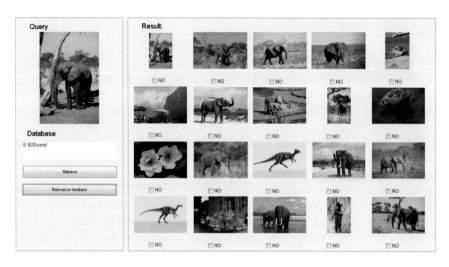

Fig. 6 The result of first relevance feedback iteration

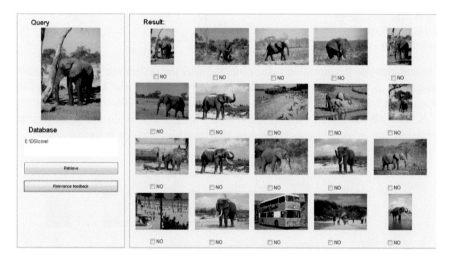

Fig. 7 The result of second relevance feedback iteration

We report the feasibility and utility of our algorithm, and compare it with five feedback methods. Figures 9 and 10 show the average-AP value of different method.

According to the uncertainty of the ELM which brought by the random parameters, 3 ELM are used classifiers to get a stable component-classifier, respectively. Thus, 9 elm classifiers are used to compose the component-classifiers. Because of the randomness of the ELM, the experiments performed 20 times to get the stable data. From the figures we can see that the result of 9-elm-classifiers

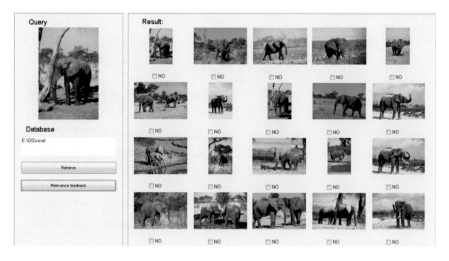

Fig. 8 The result of third relevance feedback iteration

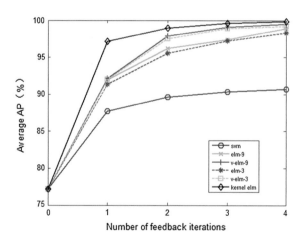

Fig. 9 The average-AP values of different method performed on 10 categories

method is very close to the result of 3-elm-classifiers method, so 3-elm-classifiers method is better in order to reduce the unnecessary processing time. The average-AP values of the kernel ELM method are superior for the 4 iterations. And the final iteration our method perform the most perfect result than the other methods.

Figures 11 and 12 show the processing time of various methods getting short as the number of feedback iterations increases. And they show our methods based on ELM with Gaussian kernel have enormous advantage in each iteration whatever on 10 categories or 60 categories which is the most important factor to the relevance feedback. Eventually, from the result of accuracy and time, the effectiveness of the ELM with Gaussian kernel based relevance feedback scheme is utilized.

Fig. 10 The average-AP
values of different method
performed on 60 categories

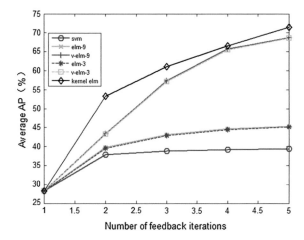

Fig. 11 The processing times
of various methods performed
on 10 categories

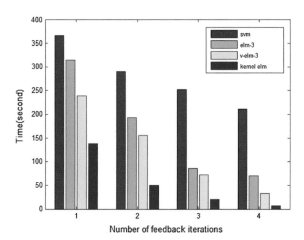

Fig. 12 The processing times
of various methods performed
on 60 categories

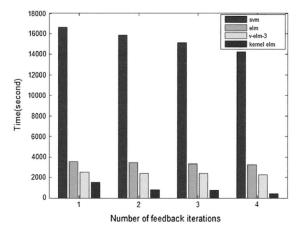

4 Conclusions

Recently, Extreme Learning Machine has been widely applied in relevance feed-back. It's an important and efficient way to improve the performance of image retrieval. Most of the advantages of ELM are suitable for the relevance feedback, such as: short processing time, high accuracy of classification, and good generalize ability, etc. However, the conventional feedback relevance schemes could not give considerations to the both accuracy and speed. To combine ELM with the semi-supervised method, we can overcome the limitation of the conventional problems. The experiments performed on the Corel-Photo gallery shows that our new method can have an excellent performance.

Acknowledgement This research is partially sponsored by Natural Science Foundation of China (Nos. 61003105, 61175115 and 61370113), the Importation and Development of High-Caliber Talents Project of Beijing Municipal Institutions (CIT&TCD201304035), Jing-Hua Talents Project of Beijing University of Technology (2014-JH-L06), and Ri-Xin Talents Project of Beijing University of Technology (2014-RX-L06), and the International Communication Ability Development Plan for Young Teachers of Beijing University of Technology (No. 2014-16).

References

1. Meng, W., Hao, L., Tao, D., et al.: Multimodal graph-based reranking for web image search. IEEE Trans. Image Process. **21**(11), 4649–4661 (2012)
2. Hu, W., Xie, N., Li, L., et al.: A survey on visual content-based video indexing and retrieval. IEEE Trans. Syst. Man Cybernet. Part C Appl. Rev. **41**(6), 797–819 (2011)
3. Rui, Y., Huang, T.S., Ortega, M., et al.: Relevance feedback: a power tool for interactive content-based image retrieval. IEEE Trans. Circuits Syst. Video Technol. **8**(5), 644–655 (1998)
4. Auer, P., Leung, A.P.: Relevance feedback models for content-based image retrieval. Stud. Comput. Intell. **346**, 59–79 (2011)
5. Wang, M., Ni, B., Hua, X.S., et al.: Assistive tagging: a survey of multimedia tagging with human-computer joint exploration. ACM Comput. Surv. **44**(4), 1173–1184 (2012)
6. Cox, I.J., Miller, M.L., Minka, T.P., et al.: An optimized interaction strategy for bayesian relevance feedback. In: IEEE Conference on Computer Vision and Pattern Recognition (CVPR'98), pp. 553–558 (1998)
7. Zhi-Hua, Z., Ke-Jia, C., Yuan J.: Exploiting unlabeled data in content-based image retrieval. In: Proceedings of ECML-04, 15th European Conference on Machine Learning, pp. 525–536 (2004)
8. Huang, G.B., Zhu, Q.Y., Siew, C.K.: Extreme learning machine: a new learning scheme of feedforward neural networks. Proc. Int. Joint Conf. Neural Netw. **2**, 985–990
9. Huang, G.B., Zhu, Q.Y., Siew, C.K.: Extreme learning machine: theory and applications. Neurocomputing **70**, 489–501 (2006)
10. Huang, G.B., Wang, D.H., Lan, Y.: Extreme learning machines: a survey. Int. J. Mach. Learn. Cybernet. **2**(2), 107–122 (2011)
11. Huang, G.B., Zhou, H., Ding, X., et al.: Extreme learning machine for regression and multiclass classification. IEEE Trans. Syst. Man Cybernet. Part B Cybernet. **42**(2), 513–529 (2012). A Publication of the IEEE Systems Man and Cybernetics Society
12. Serre, D.: Matrices: theory and applications. Mathematics **32**, xvi, 221 (2002)

13. Dwyer, P.S., Rao, C.R., Mitra, S.K., et al.: Generalized inverse of matrices and its applications. J. Am. Stat. Assoc. (1971)
14. Huang, G.B.: An insight into extreme learning machines: random neurons, random features and kernels. Cognit. Comput. **6**(3), 376–390 (2014)
15. Blum, A., Mitchell, T.: Combining labeled and unlabeled data with co-training. In: Colt Proceedings of the Workshop on Computational Learning Theory, pp. 92–100 (1998)
16. Breiman, L.: Bagging predictors. Mach. Learn. **24**(2), 123–140 (1996)

Routing Tree Maintenance Based on Trajectory Prediction in Mobile Sensor Networks

Junchang Xin, Teng Li, Pei Wang and Zhiqiong Wang

Abstract With the wireless sensor networks (WSNs) becoming extremely widely used, mobile sensor networks (MSNs) have recently attracted more and more researchers' attention. Existing routing tree maintenance methods used for query processing are based on static WSNs, most of that are not directly applicable to MSNs due to the unique characteristic of mobility. In particular, sensor nodes are always moving in real world, which seriously affects the stability of the routing tree. Therefore, in this paper, we propose a novel method, named routing tree maintenance based on trajectory prediction in mobile sensor networks (RTTP), to guarantee a long term stability of routing tree. At first, we establish a trajectory prediction model based on extreme learning machine (ELM). And then, we predict sensor node's trajectory through the proposed ELM based trajectory prediction model. Next, according to the predicted trajectory, an appropriate parent nodes are chose for each non-effective node to prolong the connection time as much as possible, and reduce the instability of the routing tree as a result. Finally, extensive experimental results show that RTTP can effectively improve the stability of routing tree and greatly reduce energy consumption of mobile sensor nodes.

Keywords Mobile sensor networks · Extreme learning machine · Routing tree · Trajectory prediction

J. Xin (✉) · T. Li · P. Wang
College of Information Science & Engineering, Northeastern University,
Shenyang, China
e-mail: xinjunchang@ise.neu.edu.cn

T. Li
e-mail: leeteng@163.com

P. Wang
e-mail: wangpearll@163.com

Z. Wang
Sino-Dutch Biomedical & Information Engineering School,
Northeastern University, Shenyang, China
e-mail: wangzq@bmie.neu.edu.cn

© Springer International Publishing Switzerland 2016   409
J. Cao et al. (eds.), *Proceedings of ELM-2015 Volume 1*,
Proceedings in Adaptation, Learning and Optimization 6,
DOI 10.1007/978-3-319-28397-5_32

1   Introduction

In recent years, WSNs have been broadly used in various fields, such as national defense, national economy and environment monitoring, with many existing maintenance algorithms applied to make the routing tree work regularly. As a more complex situation in WSNs, MSNs have attracted more and more researchers' attention. Routing tree of the MSNs, as the basis of sensing data queries, plays an important role in all kinds of sensing applications. However, existing algorithm for routing tree are not applicable to MSNs directly because they cannot keep routing tree structure stable if plenty of sensor nodes are outing from routing tree when sensor nodes move frequently. Therefore, we need to study routing tree maintenance methods to improve stability of the routing tree.

In the MSNs, the geographical location of all sensor nodes will be changed by anytime and anywhere due to the uncertain surrounding environment. Sometimes, one node may move towards the direction far from its parent nodes relatively. If the distance between the node and its parent node exceeds the communication radius, it cannot transmit its sensoring data to the base station. As a result, the routing tree structure that nodes have been established in MSNs will be destroyed. And if the node is out of the routing tree frequently, it will waste a lot of precious energy to reconstruct the local routing tree. Therefore, the longer connection time between the node and it parent node is, the more stable the routing tree structure is, and the more energy is saved.

While a sensor node cannot keep connection with its parent node due to mobility, the key is to select a available node as its new parent node. In this paper, trajectory prediction by extreme learning machine (ELM) is researched firstly. According to the trajectory predicted, the connection time between two sensor nodes in a period time in the future can be calculated. Then, the problem that the new parent node's choice is studied. When a sensor node dropped from the routing tree, the sensor node that has the longest connection time with it can be set as its new parent node. The contributions of this paper can be summarized as follows.

1. A trajectory prediction model is built based on ELM, which can be used to predict a sensor node's trajectory.
2. Basing on the trajectory predicted, a novel method called routing tree maintenance based on trajectory prediction in mobile sensor networks is proposed to improve the stability of routing tree in MSNs.
3. Last but not least, our extensive experimental studies using synthetic data show that the proposed approach can largely improve the stability of the whole routing tree in MSNs.

The remainder of the paper is organized as follows. Section 2 briefly introduces the related works that researched by other academicians. The problem statement is introduced in Sect. 3. Routing tree maintenance based on trajectory prediction in mobile sensor networks is described in Sect. 4. The experimental results are reported in Sect. 5 to show the effectiveness of our approach. Finally, we conclude this paper in Sect. 6.

2 Related Works

Routing tree, as the basis of sensing data queries, plays an important role in all kinds of WSNs applications. Madden et al. [1] propose TAG, a typical in network aggregation approach for routing tree in WSNs. Manjhi et al. [2] propose the combination of two routing methods, tributary and delta structure, which fully plays the advantages of the two routing structures. In addition to that, Sharaf et al. [3] combine the construction of the routing tree with the Group by clause in the query to improve the computational efficiency in the net of aggregation query. Branislav et al. [4] present a method that using a sensor node closer to mobile sink as the relay node. Its routing algorithm updates information potentials for both the current and predicted relay node. Chang et al. [5] have proposed a routing algorithm where all the nodes are mobile. A node on arriving at a new location sends an anchor information request to neighbor nodes. Nguyen et al. [6] dive all nodes into clusters according to the predicted distance between cluster head and node, which allows node to choose a cluster to join with the lowest cost.

Trajectory prediction of moving objects is gradually becoming an active research area. In literature [7], a dynamic adaptive probabilistic suffix tree (PST) prediction method was put forward, which can achieve a better prediction results. Qiao et al. [8] introduced an uncertain trajectory prediction algorithm based on trajectory continuous time Bayesian networks (CTBN). In [9], the definition of path probability and an uncertain path prefix tree are used to generate the uncertain trajectory. Zhang et al. [10] proposed a bus-based ad hoc routing mechanism Vela. Feng et al. [11] studied non-conflict data aggregation scheduling problem in MSN. Then an algorithm based on dynamic programming is proposed. In [12], a hidden Markov model (HMM)-based trajectory prediction algorithm is proposed, called hidden Markov model-based trajectory prediction (HMTP). As traditional neural networks have been widely studied and used in system modeling and perdition, ELM plays an important role for developing efficient and accurate model for these applications [13].

3 Problem Statement

3.1 Routing Tree in MSNs

In MSNs, the geographical location of all sensor nodes will be changed by anytime and anywhere due to the surrounding environment. As a result, the routing tree structure that nodes have been established will be changed, which will lead to the delayed reception or lost of uncertain sensed data collected by sensor nodes.

Definition 1 *(Effective node and non-effective node).* In MSNs, a sensor node S on the routing tree is an effective node if it can transmit its sensed data to the base station within a finite number of hops, if not, sensor node S is called non-effective node.

Definition 2 *(Neighbor node).* In MSNs, all sensor nodes who can communicate with S are the neighbor nodes of S.

Definition 3 *(Ancestor node and descendant node).* In MSNs, if sensor node S is an effective node, then all the sensor nodes on the way from S to base station are ancestor nodes of S, and S is the descendant node of them.

Suppose that Fig. 1 is the initial state of the mobile sensor network ($t = 0$) and Fig. 2 is the structure of the routing tree in 4 consecutive time. In Fig. 2, red lines indicate the nodes who have moved but can still keep communication with their neighbor nodes. Similarly, red dash lines indicate the nodes who are not connected with their neighbor nodes after moving.

Figure 2a is the structure of routing tree at time $t = t_1$. After moving, sensor nodes S_2, S_6 and S_{16} can still keep communication with their parent nodes. So we can say, S_2, S_6 and S_{16} are still effective nodes. But the location volatility of nodes S_7 and S_{15} is relatively large. Node S_{15} and node S_7 are no longer connected with their own parent nodes. And node S_7 is no longer connected with its child node S_{12}. Therefore, sensor nodes like S_{15}, S_7 and S_{12} are non-effective nodes. At the moment $t = t_2$, $t = t_3$ as well as $t = t_4$, the structure of the routing tree is shown in Fig. 2b–d. The transformation of sensor nodes like node S_2 between effective node and non-effective frequently leads to instability of the whole MSN. As a result, it will seriously affect the quality of the transmission of the sensed data between the sensor nodes.

In a word, the most important problem we faced in MSNs is to construct a stable routing tree, and to guarantee the quality of the sensed data transmission among nodes. Most of related academic achievements build routing tree only based on location model which take the nearest sensor node as the new parent node. As we known,

Fig. 1 The initial state of routing tree in mobile WSN

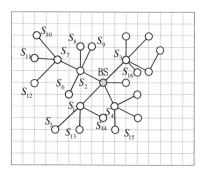

(a) (b)

(c) (d)

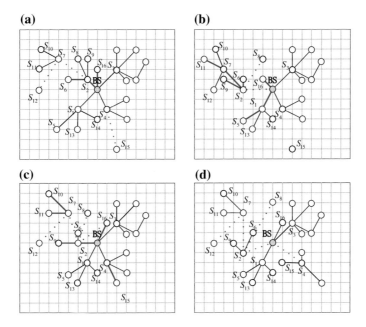

Fig. 2 Example for mobile WSN

location-based model can only guarantee that two sensor nodes keep connection at current moment, which would consume too much energy to check and reconstruct the routing tree. Therefore, this method is defective.

3.2 Location-Based Routing Tree Maintenance in MSNs

Location-based routing tree maintenance in MSNs (LRTM) is a naive method widely used to maintain a routing tree, which will be introduced in this section.

The process of LRTM on sensor node is shown in Algorithm 1 in detail. Sensor node S is a non-effective node that needs a new parent node. Then, S will broadcast messages to all of its neighbor nodes (Lines 1–2). After the neighbor nodes of S receiving the message from S, they will send their locations to S that will choose the nearest one to be its new parent node (Lines 7–11). Finally, return the new parent of S (Line 12).

In the steps of S's new parent node choice, a variable *min_dist* representing the minimal distance is set at first (Line 6). Then, for each sensor node in set L_S, the distance between it and S is calculated (Line 7); If the distance is less than *min_dist*, the new minimal distance should be reset and the parent node of S should be chosen (Lines 9–11).

Algorithm 1 LRTM on sensor nodes

1: **If** sensor node S is a non-effective **then**
2:   Sensor node S broadcast messages to its all neighbor nodes;
3: $L_S = \{\}$;
4: **While** S receives a reply from its neighbor node N **do**
5:   $L_S.add(N)$;
6: $min\_dist = Inf.$;
7: **For** each sensor node N in L_S **do**
8:   $dist = Distance(S, N)$;
9:   **If** $dist < min\_dist$ **then**
10:     $min\_dist = dist$;
11:     $S.parent = N$;
12: **Return** $S.parent$;

As we know, though the naive method LRTM can choose a new parent for non-effective nodes, it cannot maintain the stability of the routing tree, which leads to more energy consumption. If we choose the one that has a longer time to keep connection with non-effective node, energy consumption of routing tree maintenance will be reduce greatly.

4 Routing Tree Maintenance Based on Trajectory Prediction in MSNs

In this section, we firstly introduce the trajectory prediction model based on ELM, in detail. Then, the routing tree maintenance based on trajectory predicted in MSNs is illustrated.

4.1 Trajectory Prediction Model Based on ELM

ELM [14, 15] has been originally developed for single hidden-layer feedforward neural networks (SLFNs) and then extended to the "generalized" SLFNs where the hidden layer need not be neuron alike [16, 17]. ELM first randomly assigns the input weights and the hidden layer biases, and then analytically determines the output weights of SLFNs. Besides, ELM is less sensitive to user-specified parameters and can be deployed faster and more conveniently [18, 19]. And ELM prediction is proved accurate and efficient. As a result, a trajectory prediction model based ELM is first built.

Definition 4 *(Historical trajectory and future trajectory).* In MSNs, given a sensor node S's trajectory $L = \{l_a, \ldots, l_b\}$. For $\forall i \in [a, b]$, $L_f = \{l_a, \ldots, l_i\}$ is the future trajectory of $L_h = \{l_{i+1}, \ldots, l_b\}$, and L_h is the historical trajectory of L_f. The length of L_f is $i - a + 1$ and the length of L_h is $b - i$.

The specific process of trajectory prediction model based on ELM is shown in Algorithm 2. At first, the trajectory of sensor node S is collected by S and stored (Line 1). Then, it cuts the trajectory into two segmentations including historical trajectory and future trajectory. The historical trajectory can be used as the input feature vectors $\mathbf{F}$ of the model. The future trajectory can be used as the actual regression results $\mathbf{T}$ of the model (Lines 2–3). When the number of hidden layer node ranges from 1 to L, the input weights $\mathbf{w}_i$ and biases b_i of hidden neurons are generated randomly (Lines 4–5). Next, the output matrix $\mathbf{H}$ is obtained by $\mathbf{w}_i$, b_i and the feature vector $\mathbf{F}$ of the sensor node's historical trajectory (Line 6). And then, get the parameter β according to $\mathbf{H}^\dagger$ and $\mathbf{T}$ (Line 7). Finally, return parameters $\mathbf{w}$, b and β (Line 8).

Algorithm 2 Trajectory Prediction Model based on ELM

1: Get the trajectory L of sensor node S;
2: $\mathbf{F} = getPositionFeatureVectors(L)$;
3: $\mathbf{T} = getActualRegressionResults(L)$;
4: **For** $i = 1$ to L **do**
5:   Randomly generate input weights $\mathbf{w}_i$ and biases b_i of hidden neurons;
6: Calculate the output matrix $\mathbf{H}$ of hidden layer nodes by $\mathbf{F}$;
7: Calculate $\beta = \mathbf{H}^\dagger\mathbf{T}$;
8: **Return** $< \mathbf{w}, b, \beta >$;

4.2 Routing Tree Maintenance Based on Trajectory Prediction in MSNs

By Sect. 3.2, we know that LRTM can find a new parent node for the non-effective sensor nodes, which can guarantee the connectivity of the whole routing tree. But it is the truth that LRTM can only keep the whole routing tree's connectivity at a moment. In the case that sensor nodes in MSNs move more frequently, LRTM needs to reset parent node for the sensor node which is out of the routing tree. As a result, the more burden of sensor nodes, the more frequent construction of routing tree, the more consumption that unnecessary, all of them will harm the long-term using of MSNs.

Definition 5 *(Isolated loop).* In MSNs, if a sensor node S can transmit it sensed data to itself within a finite number of hops, then we can say there exists a isolated loop in routing tree, represented by $R^C = < S, S >$. The length of R satisfies that $L_{RC} \geq 2$.

Theorem 1 *Given two sensor nodes S_1 and S_2 in a WSN, if S_1, S_2 satisfies that S_1 is the ancestor node of S_2 and S_2 is also the ancestor node of S_1. Then it can be proved that there exists an isolated loop in the routing tree, which is $R^C = < S, S >$ $(S = S_1, S_2)$.*

Proof It can be proved that sensor node S_1 can transmit its sensed data to S_2 because S_1 is the ancestor node of S_2. Similarly, S_2 can transmit its sensed data to S_1. Then, the sensed data collected by S_1 will transmit from S_1 to S_1. That is to say $\exists R^C = < S_1, S_1 >$.

When choose a new parent node for the non-effective node S, there will be isolated loops generate if mistakenly set the descendants of S as its parent node, which will cause serious consequences. According to Throrem 1, it is necessary to classify the candidate parent nodes into three sets including set C_h, set C_{lnd} and set C_d. C_h stores sensor nodes whose level is no less than S's level. The set C_{lnd} stores sensor nodes whose level is less than S's level but they are not the descendants of S. All the sensor nodes in candidate parent nodes belonging to descendants of S are stored in set C_d. S will select new parent node from C_h at first; if failed, then S will find a sensor node in C_{lnd}; if still can not find its appropriate new parent, S will find a sensor node in C_d. When S find its new parent node in set C_d, disconnect all its child nodes in C_d from S and reset its child nodes' parent node except S. As a result, S can find a new parent in C_d.

Algorithm 3 RTTP on sensor nodes

1: **If** sensor node S is a non-effective **then**
2:    Sensor node S broadcast messages to its all neighbor nodes;
3: $C_h = \{\}$;
4: $C_{lnd} = \{\}$;
5: $C_d = \{\}$;
6: **While** S receives a reply from its neighbor node N **do**
7:    **If** $N.level \geq S.level$ **then**
8:       $C_h.add(N)$;
9:    **Else if** $S! = N.ancestor$ **then**
10:       $C_{lnd}.add(N)$;
11:    **Else**
12:       $C_d.add(N)$;
13: **If** C_h is not *NULL* **then**
14:    $S.parent = getParentNode(Set_{higher})$;
15: **Else if** C_{lnd} is not *NULL* **then**
16:    $S.parent = getParentNode(Set_non - d)$;
17: **Else**
18: **For** each sensor node M in C_d **do**
19:    **If** $M.parent = S$ **then**
20:       $M.parent = RTTP(noS)$;
21:       $S.parent = getParentNode(C_d)$;
22: **Return** $S.parent$;

The process of routing tree maintenance by RTTP is shown in Algorithm 3 in detail. First of all, a new parent node for non-effective node S should be chosen. Then, S will broadcast messages to all of its neighbor nodes (Lines 1–2). After receiving a reply from its neighbor nodes N, sensor node S will put N into a corresponding set (Lines 6–12). If there are sensor nodes in C_h, S chooses its new parent node from

C_d (Lines 13–14). If there is not any sensor node in C_h with set C_{lnd} not empty, S chooses its new parent node from C_{lnd} (Lines 15–16). Otherwise, reset parent nodes for the child nodes of S in C_d except itself (Lines 17–21). Finally, return the new parent of S (Line 22).

Definition 6 *(Expected connection time, ECT).* In MSNs, given sensor nodes S_1, S_2, S_1's trajectory $L_{S_1} = \{l_a, l_{a+1}, \ldots, l_b\}(a \le b)$ and S_2's trajectory $L_{S_2} = \{l'_a, l'_{a+1}, \ldots, l'_b\}(a \le b)$, $\exists i \in [a, b]$ and $\forall t \in [a, i]$, S_1 at location l_t and S_2 at location l'_t can keep connection with each other, and at time $t = i + 1$, S_1 is disconnect with S_2, then the ECT between S_1 and S_2 is i.

When choose the appropriate parent node in C_d for non-effective node S, the future trajectory of S is obtained first of all, and future trajectory of all nodes in C_d is obtained subsequently. And, choose the node that has maximum ECT with S as its new parent node. The specific process of the algorithm is shown in 4. Initializing the trajectory L_S empty (Line 1) at first. Then, the future trajectory is predicted by trajectory prediction model based on ELM (Line 2). After that, initializing the variable *max_ECT* zero. Next, for each N node in *CandidateSet* get the future trajectory future trajectory of N by trajectory prediction model based on ELM and calculate the ECT between S and N (Lines 5–6). If the ECT is bigger than *max_ECT*, make *max_ECT* equal ECT and set the parent node of S as N (Lines 7–9). Finally, return the parent node of S.

Algorithm 4 getParentNode()

1: $L_S = \{\}$;
2: $L_S = S.getFurureTrajectory()$;
3: $max\_ECT = 0$;
4: **For** each sensor node N in *CandidateSet* **do**
5:     $L_N = N.getFutureTrajectory()$;
6:     $ECT = getECT(L_N, L_S)$;
7:     **If** $ECT > max\_ECT$ **then**
8:         $max\_ECT = ECT$;
9:         $S.parent = N$;
10: **Return** $S.parent$;

---

5 Performance Evaluations

In this section, the experimental settings are introduced at first, then, the experimental results are illustrated in detail.

5.1 Experimental Settings

All the experiments were conducted on a compute with Inter(R) Core(TM) i7-3770 CPU 3.40 GHz and 8.00 GB RAM. The experimental data is processed basing on

Table 1 Experimental parameters

Parameters' name	Parameters' value
The number of hidden nodes	700
Radious	$2\sqrt{2}, 3, 4, 5, 6$
Time	**200**, 300, 400, 500, 600
The number of sensor nodes	100, 200, 300, 400, **536**

the origin data. The origin data includes the information of 536 taxis in yellow in San Francisco collected by GPS from May 2008 to June 2008. The origin data has four attributes including dimension, longitude, whether there are passengers and time stamp.

As the reason that the moving law of taxis are different from sensor nodes', the dimension, longitude and time stamp in origin data are token in experiments. The dimension is regarded as horizontal coordinates, the longitude is regarded as longitudinal coordinate. In experiments, the data are mapped into a smaller map.

The specific settings of the experimental parameters are shown in Table 1, which includes hidden node number, communication radius and the length of time and setting range of sensor nodes number. Among them, bold font is the default value. In order to improve the computational efficiency of the trajectory prediction model, ELM training function is proceed offline.

5.2 Experimental Results

The root-mean-square error (RMSE) of ELM prediction is shown in Table 2. In the experiments, we take the average value of the RMSE of 536 sensor nodes future trajectory prediction to evaluate the experiments. In Table 2, we compare the average RMSE value of the predicted future trajectory whose length is 1, 5 and 10. Obviously, the longer future trajectroy, the bigger average RMSE. But the RMSE is relatively small while the length of future trajectory is 10, which proved that the trajectory prediction by ELM is better than other methods.

The experiments are shown in Fig. 3. As we can see, both the curves are smooth with the increase of time interval. The gap between results illustrate the better performance of RTTP. Because while a sensor node is out of the routing tree, its new parent node was selected the nearest one by LRTM. But the one who has the longest connection time with it was selected by RTTP that can reduce the average number of dropped nodes of the routing tree.

As the number of the nodes changes, the number of non-effective nodes per minute generated by the RTTP and LRTM for the maintenance of the routing tree is described in Fig. 4. As we can see, the bigger the total number of the sensor nodes is, the bigger the number of invalid nodes per minute generated by the RTTP and LRTM

Table 2 Different length of future trajectory's RMSE

Future trajectory's length	RMSE (%)	Historical trajectory's length
1	0.035	30
5	0.080	35
10	0.125	40

Fig. 3 The average number of non-effective versus time

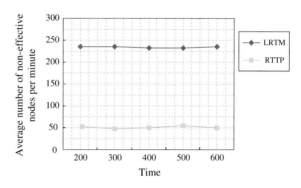

Fig. 4 The average number of non-effective versus the number of sensor nodes

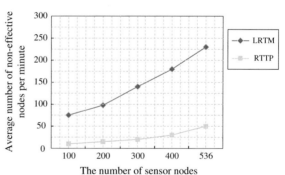

is. But it is clear that there is a significant improvement of the RTTP compared with LRTM, which means RTTP has a better effect on the stability of the routing tree in MSNs.

6 Conclusions

Routing tree plays a vital role in the query processing in WSNs, especially in MSNs. How to improve the stability and minimize energy consumption becomes an essential problem of routing tree maintenance. In this paper, a novel method called RTTP is proposed to keep the whole routing tree of MSNs in a more stable state. Different from the naive methods like LRTM, in RTTP non-effective nodes choose a node as its parent node who can keep connection with it in a longer period. As a result, the average number of non-effective nodes needed to be processed is reduced largely, which can decrease energy consumption. Furthermore, a large number of experiment results prove that compared with LRTM, RTTP can maintain a more stable routing tree with less energy consumption.

Acknowledgments This research was partially supported by the National Natural Science Foundation of China under Grant Nos. 61472069 and 61402089; And the Fundamental Research Funds for the Central Universities under Grant Nos. N130404014.

References

1. Madden, S., Franklin, M.J., Hellerstein, J.M., Hong, W.: TAG: a tiny aggregation service for ad-hoc sensor networks. In: Proceedings of 5th Symposium Operating Systems Design and Implementation (OSDI'02), pp. 131–146 (2002)
2. Manjhi, A., Nath, S., Gibbons, P.B.: Tributaries and deltas: efficient and robust aggregation in sensor network streams. In: Proceedings of the 2005 ACM SIGMOD International Conference on Management of Data (SIGMOD'05), pp. 287–298 (2005)
3. Sharaf, M.A., Beaver, J., Labrinidis, A., Chrysanthis, P.K.: Balancing energy efficiency and quality of aggregate data in sensor networks. VLDB J. (VLDBJ) **13**(4), 384–403 (2004)
4. Kusy, B., Lee, H.J., Wicke, M., Milosavljevic, N., Guibas, L.: Predictive QOS routing to mobile sinks in wireless sensor networks. In: Proceedings of International Conference on Information Processing in Sensor Networks (IPSN'09), pp. 109–120 (2009)
5. Chang, T.J., Wang, K., Hsieh, Y.L.: A color theorybased energy efficient routing algorithm for mobile wireless sensor networks. Int. J. Comput. Netw. Commun. **52**, 531–541 (2008)
6. Nguyen, L.T., Defago, X., Beuran, R., Shinoda, Y.: Energy efficient routing scheme for mobile wireless sensor networks. In: Proceedings of IEEE International Symposium on Wirelesscommunication Systems (ISWCS'08), pp. 568–572 (2008)
7. Wang, X., Jiang, X.H., Lin, J., Xiong, J.B.: Prediction of moving object trajectory based on probabilistic suffix tree. J. Comput. Appl. **33**, 3119–3122 (2013)
8. Qiao, S.J., Peng, J., Li, T.R., Zhu, Y., Liu, L.X.: Uncertain trajectory prediction of moving objects based on CTBN. J. Univ. Electron. Sci. Technol. China **41**(5), 759–763 (2012)
9. Guo, L., Ding, Z.M., Hu, Z.L., Chen, C.: Uncertain path prediction of moving objects on road networks. J. Comput. Res. Dev. **47**, 104–112 (2010)
10. Zhang, F.S., Jin, B.H., Wang, Z.Y., Hu, J.F., Zhang, L.F.: A routing mechanism over bus-based VANETs by mining trajectories. Chin. J. Comput. **38**(3), 648–662 (2015)
11. Feng, C., Li, A.J., Jiang, S.X.: Data aggregation scheduling on wireless mobile sensor networks. Chin. J. Comput. **38**(3), 685–700 (2015)
12. Qiao, S., Shen, D., Wang, X., Han, N., Zhu, W.: A self-adaptive parameter selection trajectory prediction approach via hidden Markov models. IEEE Trans. Intell. Transp. Syst. **16**(1), 284–296 (2015)
13. Huang, G., Huang, G.B., Song, S.J., You, K.Y.: Trends in extreme learning machines: a review. Neural Netw. **61**, 32–48 (2015)
14. Huang, G.B., Zhu, Q.Y., Siew, C.K.: Extreme learning machine: theory and applications. Neurocomputing **70**(1–3), 489–501 (2006)
15. Huang, G.B., Chen, L., Siew, C.K.: Universal approximation usingincremental constructive feedforward networks with random hidden nodes. IEEE Trans. Neural Netw. **17**(4), 879–892 (2006)
16. Huang, G.B., Chen, L.: Convex incremental extreme learning machine. Neurocomputing **70**(16–18), 3056–3062 (2007)

17. Huang, G.B., Chen, L.: Enhanced random search based incremental extreme learning machine. Neurocomputing **71**(16–18), 3460–3468 (2008)
18. Huang, G.B., Ding, X., Zhou, H.: Optimization method based extreme learning machine for classification. Neurocomputing **74**(1–3), 155–163 (2010)
19. Huang, G.B., Zhou, H., Ding, X., Zhang, R.: Extreme learning machine for regression and multiclass classification. IEEE Trans. Syst. Man Cybern. Part B: Cybern. **42**(2), 513–529 (2012)

Two-Stage Hybrid Extreme Learning Machine for Sequential Imbalanced Data

Wentao Mao, Jinwan Wang, Ling He and Yangyang Tian

Abstract In many practical engineering applications, data tend to be collected in online sequential way with imbalanced class. Many traditional machine learning methods such as support vector machine and so on generally get biased classifier which leads to lower classification precision for minor class than major class. To get fast and efficient classification, a new online sequential extreme learning machine method with two-stage hybrid strategy is proposed. In offline stage, data-based strategy is employed, and the principal curve is introduced to model the distribution of minority class data. In online stage, algorithm-based strategy is employed, and a new leave-one-out cross-validation method using Sherman-Morrison matrix inversion lemma is proposed to tackle online imbalance data, meanwhile, with add-delete mechanism for updating network weights. The proposed method is evaluated on the real-world Macau air pollutant forecasting dataset. The experimental results show that, the proposed method outperforms the classical ELM, OS-ELM and meta-cognitive OS-ELM in terms of generalization performance and numerical stability.

Keywords Extreme learning machine · Imbalance problem · Principal curve · Leave-one-out cross-validation · Online sequential learning

W. Mao (✉) · J. Wang · L. He · Y. Tian
School of Computer and Information Engineering, Henan Normal University, Xinxiang, China
e-mail: maowt.mail@gmail.com

W. Mao
Engineering Technology Research Center for Computing Intelligence & Data Mining, Xinxiang, Henan, China

W. Mao
School of Mechanics and Civil & Architecture, Northwestern Polytechnical University, Xián 710129, Shanxi, People's Republic of China

© Springer International Publishing Switzerland 2016   423
J. Cao et al. (eds.), *Proceedings of ELM-2015 Volume 1*,
Proceedings in Adaptation, Learning and Optimization 6,
DOI 10.1007/978-3-319-28397-5_33

1 Introduction

Data imbalance problem occurs commonly in real applications [1]. Many batch
learning algorithms suffer from this problem that has an "undo" effect to minority
class [2]. To improve the generalization of imbalance data, two kinds of strategies
are widely applied. Data-based strategy [3, 4] only focus on sampling method that
adjust the size of training data, with increasing data for minority class or decreasing
data for majority class. Algorithm-based strategy [5, 6] mainly involves introduction
of the cost-sensitive information in a classification algorithm to handle data imbalance. These two strategies and their extensions have been proved effective for many
data imbalance problems [7, 8].

In this paper, we focus on one special imbalance problem where data are collected
in online sequential way with imbalanced class. We call it as *online sequential data
imbalance problem*. There are two main challenges to improve the generalization for
online sequential data imbalance problem. One is how to choose a proper baseline
algorithm for online sequential setting, and another is how to make the majority and
minority classes balanced while obeying the original distribution of online sequential
data. As a extension form of single-hidden layer feedforward neural network(SLFN),
extreme learning machines(ELMs), introduced by Huang [9], have shown its very
high learning speed and good generalization performance in solving many problems
of regression estimate and pattern recognition [10, 11]. As a sequential modification of ELM, online sequential ELM(OS-ELM) proposed by Liang [12] can learn
data one-by-one or chunk-by-chunk. In many applications such as time-series forecasting, OS-ELMs also show good generalization at extremely fast learning speed.
Threfore, OS-ELM is a proper solution for the first challenge. To solve the second
challenge, Vong [13] introduced prior duplication strategy to generate more minority
class data, and utilized OS-ELM to train an online sequential prediction model. The
experimental results on air pollutants forecasting data from Macau also show this
method has higher generalization than many classical batch learning algorithms in
online setting. To our best knowledge, there are very few other researches concerning
this topic.

However, although OS-ELM in [13] works effectively on online sequential data,
it still not yet applies better imbalance strategy in online stage. Specifically speaking,
as the imbalance strategy in [13] is only duplicating minority class observations, it
couldn't explore the real data distribution of minority class. On the other hand, in the
level of algorithm, the method in [13] lacks an efficient mechanism to exclude the
redundant and harmful samples including new generated virtual samples. Therefore,
to solve this problem, this paper blends the data-based strategy and algorithm-based
strategy, and proposes a new two-stage hybrid strategy. In offline stage, the principal
curve is used to explore the data distribution and further establish an initial model
on imbalance data. In online stage, some virtual samples are generated according
to the principal curve, and a new leave-one-out(LOO) cross-validation method is
proposed to tackle online unbalanced data with add-delete mechanism for updating
network weights. This LOO cross-validation method uses Sherman-Morrison matrix

inversion lemma, and can effectively reduce the redundant computation of matrix inversion. Experimental results demonstrate the proposed method outperforms the traditional ELMs in generalization performance.

2 Background

2.1 Review of ELM and OS-ELM

ELM proposed by Huang [14] is a learning method for generalized SLFNs where all the hidden node parameters are randomly generated and the output weights of SLFNs are analytically determined. Furthermore, ELM tends to provide good generalization at extremely fast learning speed because of its simple and efficient learning algorithm in which iterative tuning is not required in the hidden layers [15].

Like ELM, all the hidden node parameters in OS-ELM are randomly generated, and the output weights are analytically determined based on the sequentially arrived data. OS-ELM process is divided into two steps: initialization phase and sequential learning phase [12].

Step 1. Initialization phase: choose a small chunk $M_0 = \{(x_i, t_i), i = 1, 2, \ldots, N_0\}$ of initial training data, where $N_0 \geq \tilde{N}$.

(1) Randomly generate the input weight $\mathbf{w}_i$ and bias $b_i, i = 1, 2, \ldots, \tilde{N}$. Calculate the initial hidden layer output matrix $\mathbf{H_0}$.
(2) Calculate the output weight vector:

$$\beta^0 = \mathbf{D_0}\mathbf{H_0}^T\mathbf{T_0} \tag{1}$$

where $\mathbf{D_0} = (\mathbf{H_0}^T\mathbf{H_0})^{-1}$, $\mathbf{T_0} = [t_1, t_2, \ldots, t_{N_0}]^T$.
(3) Set $k = 0$

Step 2. Sequential learning phase

(1) Learn the $(k + 1)$th training data: $d_{k+1} = (\mathbf{x}_{N_0+k+1}, t_{N_0+k+1})$
(2) Calculate the partial hidden layer output matrix:

$$\mathbf{H}_{k+1} = [g(\mathbf{w}_1 \cdot \mathbf{x}_{N_0+k+1} + b_1) \quad \cdots \quad g(\mathbf{w}_L \cdot \mathbf{x}_{N_0+k+1} + b_L)]_{1 \times L} \tag{2}$$

Set $\mathbf{T}_{k+1} = [t_{N_0+k+1}]^T$.
(3) Calculate the output weight vector

$$\mathbf{D}_{k+1} = \mathbf{D}_k - \mathbf{D}_k\mathbf{H}_{k+1}^T(\mathbf{I} + \mathbf{H}_{k+1}\mathbf{D}_k\mathbf{H}_{k+1}^T)^{-1}\mathbf{H}_{k+1}\mathbf{D}_k \tag{3}$$

$$\beta^{k+1} = \beta^k + \mathbf{D}_{k+1}\mathbf{H}_{k+1}^T(\mathbf{T}_{k+1} - \mathbf{H}_{k+1}\beta^k) \tag{4}$$

(4) Set $k = k + 1$. Go to step 2(1).

2.2 Review of Principal Curve

In 1983, Hastie [16] firstly introduced the theory of principal curve. Afterwards, this theory were successfully applied to solve practical problems, like data visualisation [17] and ecology analysis [18], etc. Principal curve is extension of principal component analysis and its basic idea is to find a continuous one-dimensional manifold that approximate the data in the sense of "self-consistency", i.e. the curve should coincide at each position with the expected value of the data projecting to that position [18]. Intuitively, this curve passes through the "middle" of a high-dimensional data set. In this paper, we choose k-curve for its good practicability. The definition of k-curve is listed as follows [19].

Definition 1 *(K-principal curve [19])* For a data set $X = \{x_1, x_2, \ldots, x_n\} \subset \mathbb{R}^d$, a curve f^* is called a K principal curve of length L for X if f^* minimizes $\triangle(f)$ over all curves of length less than or equal to L, where f is a continuous function $f : I \to \mathbb{R}^d$, $\triangle(f)$ is the expected squared distance between X and f and defined as:

$$\triangle(f) = E[\triangle(X, f)] = E[\inf_{\lambda} \|X - f(\lambda)\|^2] = E[\|X - f(\lambda_f(X))\|^2]$$

where $\lambda_f(x) = \sup \{\lambda : \|X - f(\lambda)\| = \inf_{\tau} \|x - f(\tau)\|\}$ is called projection index.

The goal of K curve is to find a set of polygonal lines with K-segments and with a given length to approximate the principal curve. The algorithm is based on a common model about complexity in statistical learning theory [19]. The framework of this algorithm can be summarized as follows. At the beginning, $f_{1,n}$ is initialized by the first principal component line and in each iteration step, a new vertex is added on $f_{i-1,n}$ which is obtained in $i-1$ step, to increase the number of segments. According to the principle of minimizing the projection distance, the position of vertexes are optimized to construct a new curve $f_{i,n}$. Kégl [19] gave a detailed description of this algorithm.

3 OS-ELM with Two-Stage Hybrid Strategy

3.1 Offline Stage

As principal curve can truly reflect the shape of data set, we employ the principal curve in offline stage to balance the samples in minority and majority classes. Then the initial model is established using the obtained dataset.

The concrete process can be described as follows:

Step 1. Plot the principal curve of minority or majority samples $D = \{(Dx_i, Dy_i)\}$ using k-curve [17] presented in Sect. 2.2. Get the samples $S = \{(Sx_i, Sy_i)\}$ on principal curve using polynomial interpolation.

Step 2. Generate virtual samples for minority class by adding Gaussian white noise on original samples $Z = \{(Zx_i, Zy_i)\}$. Keep the majority class sample set unchanged.

Step 3. Filter samples for two classes using the following formulation:

$$
\begin{cases}
\left| Sx_i - Dx_j \right| \leq \delta_x \\
\left| Sy_i - Dy_j \right| \leq \delta_y
\end{cases}
\tag{5}
$$

where δ_x, δ_y are pre-defined threshold. The samples not meeting Eq. (10) will be excluded from the minority or majority class sample set.

Define the obtained sample set filtered by principal curve as $D = \{(\mathbf{x}_i, t_i) | i = 1, 2, \ldots, N\}$. Given activation function $g(x)$ and the number of hidden neurons L, choose input weight $\mathbf{w}_i$ and bias b_i, $i = 1, 2, \ldots, L$ randomly and calculate the input matrix $\mathbf{H}_1$:

$$
\mathbf{H}_1 = \begin{bmatrix} \mathbf{h}(\mathbf{x}_1) \\ \mathbf{h}(\mathbf{x}_2) \\ \vdots \\ \mathbf{h}(\mathbf{x}_N) \end{bmatrix} = \begin{bmatrix} g(\mathbf{w}_1 \cdot \mathbf{x}_1 + b_1) & \cdots & g(\mathbf{w}_L \cdot \mathbf{x}_1 + b_L) \\ g(\mathbf{w}_1 \cdot \mathbf{x}_2 + b_1) & \cdots & g(\mathbf{w}_L \cdot \mathbf{x}_2 + b_L) \\ & \cdots & \\ & \cdots & \\ & \cdots & \\ g(\mathbf{w}_1 \cdot \mathbf{x}_N + b_1) & \cdots & g(\mathbf{w}_L \cdot \mathbf{x}_N + b_L) \end{bmatrix}_{N \times L}
\tag{6}
$$

Here the output vector is $\mathbf{T}_1 = [t_1, t_2, \ldots, t_N]^T$, and the output weight is:

$$
\beta_1 = \mathbf{H}_1{}^+ \mathbf{T}_1
\tag{7}
$$

where

$$
\mathbf{H}_1{}^+ = (\mathbf{H}_1{}^T \mathbf{H}_1)^{-1} \mathbf{H}_1{}^T
\tag{8}
$$

Let $\mathbf{M}_1 = (\mathbf{H}_1{}^T \mathbf{H}_1)^{-1}$, Eq. (8) can be rewritten as $\mathbf{H}_1{}^+ = \mathbf{M}_1 \mathbf{H}_1{}^T$.

3.2 Online Stage

In this stage, we employ leave-one-out (LOO) cross validation to choose more valuable samples. However, the traditional LOO cross validation is computationally expensive. In this section, we will derive a new fast LOO error estimation method. Meanwhile, we introduce a add-delete mechanism to update output weights in order to highlight the value of new arrived sample and keep the model simple.

3.2.1 Add New Sample

Add the new arrived sample $(\mathbf{x}_{N+1}, t_{N+1})$ into training set. The output vector becomes $\mathbf{T}_2 = [t_1, t_2, \ldots, t_N, t_{N+1}]^T = [\mathbf{T}_1^T \ t_{N+1}]^T$, and the hidden layer matrix becomes $\mathbf{H}_2 = [\mathbf{h}_1^T, \mathbf{h}_2^T, \ldots, \mathbf{h}_N^T, \mathbf{h}_{N+1}^T]^T = [\mathbf{H}_1^T \ \mathbf{h}_{N+1}^T]^T$. Then we have:

$$\mathbf{H}_2^+ = (\mathbf{H}_2^T\mathbf{H}_2)^{-1}\mathbf{H}_2^T \tag{9}$$

Let $\mathbf{M}_2 = (\mathbf{H}_2^T\mathbf{H}_2)^{-1}$, then Eq. (9) becomes:

$$\mathbf{H}_2^+ = \mathbf{M}_2\mathbf{H}_2^T \tag{10}$$

Because

$$\mathbf{H}_2^T\mathbf{H}_2 = [\mathbf{H}_1^T \ \mathbf{h}_{N+1}^T][\mathbf{H}_1^T \ \mathbf{h}_{N+1}^T]^T = \mathbf{H}_1^T\mathbf{H}_1 + \mathbf{h}_{N+1}^T\mathbf{h}_{N+1} \tag{11}$$

we have

$$\mathbf{M}_2^{-1} = \mathbf{M}_1^{-1} + \mathbf{h}_{N+1}^T\mathbf{h}_{N+1} \tag{12}$$

Calculate the inversion of Eq. (12), and according to Sherman-Morrison matrix inversion lemma, we have:

$$\mathbf{M}_2 = (\mathbf{M}_1^{-1} + \mathbf{h}_{N+1}^T\mathbf{h}_{N+1})^{-1} = \mathbf{M}_1 - \frac{\mathbf{M}_1\mathbf{h}_{N+1}^T\mathbf{h}_{N+1}\mathbf{M}_1}{1 + \mathbf{h}_{N+1}\mathbf{M}_1\mathbf{h}_{N+1}^T} \tag{13}$$

As shown in Eq. (13), M_2 can be calculated based on $\mathbf{M}_1$, which reduces computational cost largely. Then we have $\mathbf{H}_2^+$ by substituting Eq. (13) into Eq. (11).

3.2.2 Delete Old Sample

After adding new sample $(\mathbf{x}_{N+1}, t_{N+1})$, to reduce the negative effect of old sample and make the model simple, we need to exclude the oldest sample $(\mathbf{x}_1, t_1)$. After excluding $(\mathbf{x}_1, t_1)$, the output vector becomes $\mathbf{T}_3 = [t_2, t_3, \ldots, t_N, t_{N+1}]^T$, and the hidden layer matrix becomes $\mathbf{H}_3 = [\mathbf{h}_2^T, \mathbf{h}_3^T, \ldots, \mathbf{h}_N^T, \mathbf{h}_{N+1}^T]^T$. We have:

$$\mathbf{H}_3^+ = (\mathbf{H}_3^T\mathbf{H}_3)^{-1}\mathbf{H}_3^T \tag{14}$$

Let $\mathbf{M}_3 = (\mathbf{H}_3^T\mathbf{H}_3)^{-1}$, then we have:

$$\mathbf{H}_3^+ = \mathbf{M}_3\mathbf{H}_3^T \tag{15}$$

Because

$$\mathbf{H}_2^T\mathbf{H}_2 = [\mathbf{h}_1^T \ \mathbf{H}_3^T][\mathbf{h}_1^T \ \mathbf{H}_3^T]^T = \mathbf{h}_1^T\mathbf{h}_1 + \mathbf{H}_3^T\mathbf{H}_3 \tag{16}$$

we have:

$$\mathbf{M}_3^{-1} = \mathbf{M}_2^{-1} - \mathbf{h}_1{}^T\mathbf{h}_1 \tag{17}$$

We use Sherman-Morrison matrix inversion lemma again, and have:

$$\mathbf{M}_3 = (\mathbf{M}_2^{-1} - \mathbf{h}_1{}^T\mathbf{h}_1)^{-1} = \mathbf{M}_2 + \frac{\mathbf{M}_2\mathbf{h}_1{}^T\mathbf{h}_1\mathbf{M}_2}{1 - \mathbf{h}_1\mathbf{M}_2\mathbf{h}_1{}^T} \tag{18}$$

Similar to Eq. (13), $\mathbf{M}_3$ can be obtained directly from $\mathbf{M}_2$. Then we have $\mathbf{H}_3{}^+$ by substituting Eq. (18) into Eq. (15).

3.2.3 Fast Online LOO Error Estimation

In [20], Liu et al. derived a fast LOO error estimation of ELM. The generalization error in ith LOO iteration can be expressed as:

$$r_i = t_i - f_i(\mathbf{x}_i) = \frac{t_i - \mathbf{H}_{\mathbf{x}_i}\mathbf{H}^+\mathbf{T}}{1 - (\mathbf{H}_{\mathbf{x}_i}\mathbf{H}^+)_i} \tag{19}$$

where $(\cdot)_i$ means the ith element, $\mathbf{H}$ is hidden layer matrix, and $\mathbf{H}_{\mathbf{x}_i}$ means the row about the sample $\mathbf{x}_i$ in $\mathbf{H}$.

However, this LOO estimation cannot be directly applied to online sequential scenario. We observe in Eq. (19), when adding a sample and delete another sample, the only affected element is $\mathbf{H}$. So, we simply set $\mathbf{H}^+ = \mathbf{H}_3{}^+$ which is calculated from Eq. (15), and the generalization error in ith LOO iteration can be expressed as:

$$r_i = t_i - f_i(\mathbf{x}_i) = \frac{t_i - \mathbf{Hx}_i\mathbf{H}^+\mathbf{T}}{1 - (\mathbf{Hx}_i\mathbf{H}^+)_i} \tag{20}$$

After introducing PRESS statistic, we have the LOO error estimation:

$$LOO = \frac{1}{N}\sum_{i=1}^{N}r_i^2 \tag{21}$$

4 Experimental Results

In this section, we examine one typical imbalanced real-world data set, i.e., air pollutants forecasting in Macau [13]. Our goal is to demonstrate that the proposed algorithm can efficiently improve the generalization performance of OS-ELM in data imbalance problem. For comparison, we choose three baselines. The first is classical

ELM [14]. The second is OS-ELM [12]. The third is meta-cognitive OS-ELM(MC-OSELM) proposed in [13, 21]. In this approach, minority class samples are dupli-cated directly in online phase. Moreover, the proposed OS-ELM algorithm based on principal curve and fast LOO cross-validation is named PL-OSELM.

As shown in [13], air pollutants data is a typical time-series imbalance data. We use the data collected from 2010 to 2013 year to conduct experiment. Specifically, the data in 2010 are used for initial offline training, the data in 2011 are used for online training, the data in 2012 are used for validation, and the data in 2013 are used for test. The description of data set is shown in Table 1.

It is clear that the data of every year is severely imbalanced. Meanwhile, it is still time series. Hence we use these data to examine the performance of PL-OSELM. First, we use principal curve-based method described in Sect. 3.1 to expand minority sample and cut down redundant majority sample in initial offline stage. We use k-principal curve [17] to plot the principal curve of majority and minority data in 2010, respectively, as shown in Fig. 1.

Obviously, the principal curves have revealed the variation tendency. Based on the obtained principal curve, we exclude the redundant majority samples, and generate virtual minority samples by adding white noise whose intensity is set 10 dB. With the threshold $\delta_x = 0.002$, $\delta_y = 0.005$, we get the new training set described in Table 2.

After preprocessing using principal curve, the ratio between majority and minor-ity classes becomes almost 1:1. The data imbalance problem has been released to some extents.

Table 1 Description of original data set from 2010 to 2013

	2010	2011	2012	2013
Minority sample	31	30	29	51
Majority sample	334	335	337	313
Ratio of minority (%)	8.49	8.22	7.92	14.01
Ratio of majority (%)	91.51	91.78	92.08	85.99

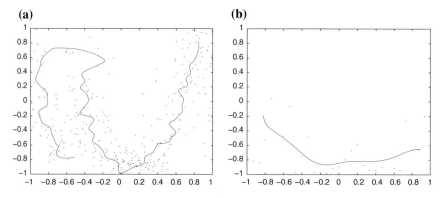

Fig. 1 Principal curve of PM_{10} data in 2010 with **a** majority class and **b** minority class

Table 2 Description of new data set based on principal curve in 2010

	Majority sample	Minority sample	Ratio of majority (%)	Ratio of minority (%)
Orignal data set	334	31	91.51	8.49
New data set	154	131	**54.03**	**45.97**

Table 3 Classification accuracy on minority class of four algorithms on air pollutant data set

	PLOSELM	OSELM	ELM	MCOSELM
Training time (s)	14.7473	0.4940	0.0468	0.2979
Test time (s)	0.0052	0.0052	0.0260	0.3093
Minority training accuracy (%)	**80.31**	20.22	21.31	79.22
Majority training accuracy (%)	93.52	99.30	99.05	95.60
Minority test accuracy (%)	**53.44**	21.84	18.49	50.34
Majority test accuracy (%)	96.34	99.80	99.31	97.73
Whole training accuracy (%)	88.04	92.69	92.56	91.65
Whole test accuracy (%)	89.71	92.26	92.08	90.80

Table 3 provides the comparative results of four algorithms. We mainly focus on the accuracy on minority class. Here the hidden neurons are set 50.

From Table 3, although PL-OSELM get relative low accuracy on majority class and whole data, but it still obtains the highest training and test accuracy on minority class, which is precisely our algorithm's value. Similar to the results in [13], MC-OSELM also gets much higher accuracy on minority class than OS-ELM and ELM, which also demonstrates the necessity of solving online sequential data imbalance problem. OS-ELM gets highest accuracy on whole data, but as stated in section Introduction, this value is meaningless because the correct prediction gathers nearly in majority class.

We also examine the effect of hidden neurons. Figure 2 shows the accuracy of four algorithms with different number of hidden neurons. Note that the slight fluctuation of results are caused by the randomness of ELM.

From Fig. 2, PL-OSELM and MC-OSELM both get satisfying results on minority class, with aggravated performance on majority class, which demonstrate their good ability to handle online sequential imbalance data. In comparison, PL-OSELM behaves in more stable way. Although OS-ELM and ELM get much better accuracy in Fig. 2b, the results have no reference value because almost all minority class samples are incorrectly predicted.

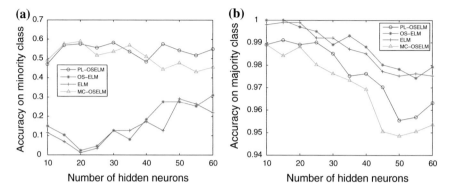

Fig. 2 Classification accuracy of Macau air pollutant data with different number of hidden neurons on **a** minority class and **b** majority class

5   Conclusion and Future Work

In this paper, a kind of data imbalance problem, online sequential data imbalance problem, is addressed. The key idea is to reduce the data imbalance phenomenon in online sequential learning process. Data-based strategy and algorithm-based strategy are integrated into a two-stage hybrid strategy. To realize this strategy, this paper utilizes principal curve in offline stage to reduce the gap between majority and minority classes, and proposes a fast leave-one-out cross-validation error estimation to reduce data imbalance in online stage. Following this strategy, a new OS-ELM algorithm based on principal curve and fast leave-one-out error estimation is proposed. This algorithm can adjust training samples in more efficient way, and update output weights automatically with "add-delete" mechanism. The experimental results on the real-world data set demonstrate the effectiveness of the proposed approach.

Acknowledgments We wish to thank the author C.M. Vong of [13] for useful discussion and instruction. This work was supported by the National Natural Science Foundation of China (No. U1204609,U1404602,61173071) and Postdoctoral Science Foundation of China (No. 2014M550508).

References

1. Murphey, Y.L., Guo, H., Feldkamp, L.A.: Neural learning from unbalanced data. Appl. Intell. **21**, 117–128 (2004)
2. Lu, W.-Z., Wang, D.: Ground-level ozone prediction by support vector machine approach with a cost-sensitive classification scheme. Sci. Total Environ **395**, 109–116 (2008)
3. Blagus, R., Lusa, L.: Class prediction for high-dimensional class-imbalanced data. BMC Bioinf. **11**, 523 (2010)
4. Batista, G.E., Prati, R.C., Monard, M.C.: A study of the behavior of several methods for balancing machine learning training data. SIGKDD Explor. Newsl. **6**, 20–29 (2004)

5. Ducange, P., Lazzerini, B., Marcelloni, F.: Multi-objective genetic fuzzy classifiers for imbalanced and cost-sensitive datasets. Soft Comput.—Fusion Found. Methodol. Appl. **14**, 713–728 (2010)
6. Pang, S., Zhu, L., Chen, G.: Dynamic class imbalance learning for incremental LPSVM. Neural Netw. **44**, 87–100 (2013)
7. Estabrooks, A., Jo, T., Japkowicz, N.: A multiple resampling method for learning from imbalanced datasets. Comput. Intell. **20**, 18–36 (2004)
8. Tang, Y., Zhang, Y.-Q., Chawla, N.V., Krasser, S.: SVMs modeling for highly imbalanced classification. IEEE Trans. Syst. Man Cybern. Part B: Cybern. **39**, 281–288 (2009)
9. Huang, G.-B., Zhu, Q.-Y., Siew, C.-K.: Extreme learning machine: A new learning scheme of feedforward neural networks. In: Proceedings of International Joint Conference on Neural Networks, Budapest, Hungary, vol. 2, pp. 985–990 (2004)
10. Liu, X., Gao, C., Li, P.: A comparative analysis of support vector machines and extreme learning machines. Neural Netw. **33**, 58–66 (2012)
11. Deng, W., Zheng, Q., Wang, Z.: Cross-person activity recognition using reduced kernel extreme learning machine. Neural Netw. **53**, 1–7 (2014)
12. Liang, N.Y., Huang, G.-B.: A fast accurate online sequential learning algorithm for feedforword networks. IEEE Trans. Neural Netw. **17**, 1411–1423 (2006)
13. Vong, C.M., Ip, W.F., Wong, P.K., Chiu, C.C.: Predicting minority class for suspended particulate matters level by extreme learning machine. Neurocomputing **128**, 136–144 (2014)
14. Huang, G.-B., Zhu, Q.-Y., Siew, C.-K.: Extreme learning machine: theory and applications. Neurocomputing **70**, 489–501 (2006)
15. Deng, C.-W., Huang, G.-B., Xu, J., Tang, J.-X.: Extreme learning machines: new trends and applications. Sci. China Inf. Sci. **58**(2), 1–16 (2015)
16. Hastie, T.: Principal curves and surfaces. Stanford University, Department of Statistics, Technical Report 11 (1984)
17. Hermann, T., Meinicke, P., Ritter, H.: Principal curve sonification. In: Proceedings of International Conference on Auditory Display, pp. 81–86. Atlanda, USA (2000)
18. Déath, G.: Principal curves: a new technique for indirect and direct gradient analysis. Ecology **80**(7), 2237–2253 (1999)
19. Kégl, B., Krzyzak, A., Linder, T., Zeger, K.: Learning and design of principal curves. IEEE Trans. Pattern Recogn. Mach. Intell. **22**(3), 281–297 (2000)
20. Liu, X., Li, P., Gao, C.: Fast leave-one-out cross-validation algorithm for extreme learning machine. J. Shanghai Jiaotong Univ. **45**(8), 6–11 (2011)
21. Chong, C.C.: Online Sequential Prediction of Minority Class of Suspended Particulate Matters by Meta-Cognitive OS-ELM. University of Macau (2013)

Feature Selection and Modelling of a Steam Turbine from a Combined Heat and Power Plant Using ELM

Sandra Seijo, Victoria Martínez, Inés del Campo,
Javier Echanobe and Javier García-Sedano

Abstract The modelling of complex industrial processes is a hard task due to the complexity, uncertainties, high dimensionality, non-linearity and time delays. To model these processes, mathematical models with a large amount of assumptions are necessary, many times this is either almost impossible or it takes too much computational time and effort. Combined Heat and Power (CHP) processes are a proper example of this kind of complex industrial processes. In this work, an optimized model of a steam turbine of a real CHP process using Extreme Learning Machine (ELM) is proposed. Previously, with the aim of reducing the dimensionality of the system without losing prediction capability, a hybrid feature selection method that combines a clustering filter with ELM as wrapper is applied. Experimental results using a reduced set of features are very encouraging. Using a set of only three input variables to predict the power generated by the steam turbine, the optimal number of hidden nodes are only eight, and a model with RMSE less than 1 % is obtained.

Keywords Extreme learning machine · Combined heat and power · Feature selection · Nonlinear system modelling

1 Introduction

Modelling the behaviour of complex industrial processes is a hard task for many reasons: uncertainties, high dimensionality, non-linearity and time delays involved in the process behaviour. To model these processes, mathematical models with many assumptions are necessary [1–3]. Most of the times modelling this processes is either almost impossible or it takes too much computational time and effort. Also, to obtain

S. Seijo (✉) · V. Martínez · I. del Campo · J. Echanobe
Department of Electricity and Electronics, University of the Basque Country,
Leioa, Spain
e-mail: sandra.seijo@ehu.es

J. García-Sedano
Optimitive, Avda Los Huetos 79, Vitoria, Spain

© Springer International Publishing Switzerland 2016
J. Cao et al. (eds.), *Proceedings of ELM-2015 Volume 1*,
Proceedings in Adaptation, Learning and Optimization 6,
DOI 10.1007/978-3-319-28397-5_34

results from such physical models is usually time-consuming and not useful for 'real-time' applications as long as they use iterative methods for final solutions.

Combined Heat and Power (CHP) processes is a proper example of complex industrial process. CHP or cogeneration, is the simultaneous production of electricity and heat, both of which are used [4]. In CHP plants the flue gases are used to generate more electricity or in other process. This implies cost saving, because the amount of fuel is reduced. In addition, this fuel saving results in a reduction of pollution.

Computational Intelligence (CI) techniques are a set of methodologies, inspired in nature, able to deal with problems which are very hard to solve with standard methods [5]. One of the most widely used CI technique are Artificial Neural Networks (ANNs). ANN have been used in CHP process, as for example to predict the baseline energy consumption of a cogeneration plant due to its high level of robustness against uncertainty affecting measured values of input variables [6]. In addition, ANNs have demonstrated that they are a useful tool to predict the power generated in a cogeneration plant with simple models where the number of necessary variables is not high [7, 8]. A method to predict the power output based on ANN is carried out in [9], where a study of the relationships between the electricity produced in a cogeneration power plant and the properties of the fuel is performed. Power prediction is the objective in [10] where the process is divided into two submodules and each one has its own ANN model.

Although ANNs have been successfully applied to solve numerous problems, they present some drawbacks that make them unsuitable for an increasing number of cutting-edge applications. It is well known that the design of Back Propagation (BP) based ANNs is a time-consuming task that depends on the skills of the designer to obtain effective solutions. The designer has to select the most suitable network parameters, optimize the parameters to avoid overfitting, and be aware of local minima. As a consequence, applications requiring autonomy (i.e. no human intervention) are difficult to manage using this approach.

Extreme Learning Machines (ELM) have attracted increasing attention recently because they outperform conventional Artificial Neural Networks in some aspects. ELM is a learning algorithm for training a particularly type of single hidden-layer feedforward neural networks (SLFNs) [11, 12]. ELM provides a robust learning algorithm, free of local minima, without overfitting problems and less dependent on human intervention than the ANN. ELM is appropriate for the implementation of intelligent autonomous systems with real-time learning capability.

In this work, an optimized model of a steam turbine of a real CHP process using ELM is proposed. Previously, with the aim of reducing the dimensionality of the system without losing prediction capability, a hybrid feature selection method that combines a clustering filter with ELM as wrapper, is applied.

The rest of the paper is organized as follows: Sect. 2 introduces the extreme learning machines algorithm. In Sect. 3, the combined heat and power plant used in this work is presented. Section 4 explains the data collection and the feature selection method for the steam turbine. Section 5 presents the experimental results for the modelling of the steam turbine: Finally, in Sect. 6 some concluding remarks are noted.

2 Extreme Learning Machine

Extreme learning machines were originally proposed by Huang et al. [13] for the single hidden-layer feedforward neural networks and then extended to the generalized single hidden-layer feedforward networks where the hidden layer needs not to be neuron alike [14].

Suppose a SLFN with n inputs, m outputs and l nodes in the hidden layer (see Fig. 1). The output j of the SLFNs can be written as:

$$y_j = \beta_j h(x) \tag{1}$$

where $\beta_j = [\beta_{j1}, \ldots, \beta_{jl}]^T$ is the weight vector connecting the hidden layer and the jth output node, $h = [h_1, h_i, \ldots, hl]$ is the vector formed by the values $h_i = g(a_i x + b_i)$ being $g()$ the activation function, $a_i = [a_{i1}, \ldots, a_{in}]^T$ the vector connecting the input $x = [x_1, \ldots, x_n]$ with the ith hidden node and b_i the bias of the ith hidden node.

The main difference between ELM and traditional learning approaches is that the hidden layer needs not to be tuned; it is a randomized layer. That is to say, the set of parameters of the hidden nodes (a_i, b_i), $1 \leq i \leq l$, are randomly generated. Therefore, they are independent of the application and of the training samples. Learning in ELM is a straightforward procedure that aims at computing the vector of output weights, β_j in (1), for each output node.

For k arbitrary distinct samples (x^k, t^k), where $x^k = [x_1^k, \ldots, x_n^k]^T \in R^n$ are the input data and $t^k = [t_1^k, t_2^k, \ldots, t_m^k]^T \in R^m$ are the target data, the above linear equations can be written in the matrix form:

$$H(x)\beta = T \tag{2}$$

Fig. 1 The topological structure of the SLFN

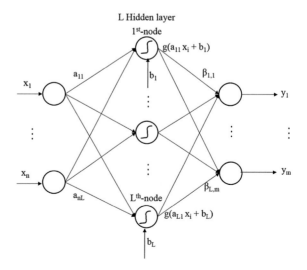

where $H(a_1, \ldots, a_n, b_1, \ldots, b_l, x^1, \ldots, x^k)$

$$H = \begin{bmatrix} g(a_1 \cdot x^1 + b_1) & \ldots & g(a_l \cdot x^1 + b_l) \\ \vdots & \ldots & \vdots \\ g(a_1 \cdot x^k + b_1) & \ldots & g(a_l \cdot x^k + b_l) \end{bmatrix}_{K \times L} \tag{3}$$

$T = [t_1, \ldots, t_K]'_{K \times m}$ is a vector of target labels and $\beta = [\beta_1, \ldots, \beta_m]'_{L \times m}$. The solution of above equation is given as: $\beta = H^{\dagger}T$, where $H^{\dagger}$ is the Moore-Penrose generalized inverse of matrix H [15].

3 Combined Heat and Power Plant

The CHP plant being evaluated in this work is located in Monzón (Huesca), in the North of Spain [16]. The plant generates electricity with four internal combustion engines fed with natural gas and a steam turbine. The four engines are identical,

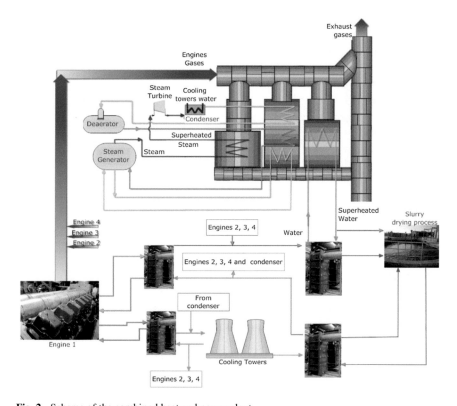

Fig. 2 Scheme of the combined heat and power plant

with the same characteristics and the nominal power of each being 3700 kW. The engines are refrigerated with two refrigeration circuits that use water from the cooling towers as Fig. 2 shows. Therefore, the engines generate electrical energy and high temperature gases. Subsequently, the electrical power generated is sold and the high temperature gases go to an exhaust steam boiler. Moreover, the steam generator creates steam using the heat from the exhaust steam boiler. This steam is used in a steam turbine to generate more electricity, with 1000 kW of nominal power, that is also sold. The heat from exhaust steam boiler is also used in a slurry drying process that uses the slurry from nearby farms. After being processed by the plant it becomes fertilizer and clean water for irrigation.

4 Data Collection and Feature Selection

The data collection used in this work is obtained from the combined heat and power plant described in the previous section. The data collection contains variables from the four engines, the refrigeration circuits, the exhaust steam boiler, the steam turbine, and the slurry drying process. It was collected over a one-year period from the whole cogeneration process with a sample time of 1 min. The data collection starts in December 2012 and finishes in November of 2013. The non informative variables, i.e. variables that are always constant, are removed from the dataset. Subsequently, the Lower Heating Value (LHV) of the natural gas used to feed the engines is added to the dataset with a sample time of one value per day. This variable is provided by the supplying company Enagas [17]. Besides, the ambient temperature and humidity are added too, with a sample time of one hour, provided by the State Meteorological Agency (AEMET) of Spain [18]. Finally, a dataset with 200 signals is available.

In order to have a proper model of the steam turbine, feature selection techniques are important to select a series of features that are particularly informative for the steam turbine behaviour. There are two main categories of feature selection techniques: filter methods and wrapper methods. The former select a reduced subset of variables by evaluating general characteristics of the data (i.e. the selected learning algorithm is not involved in the selection process), while the latter use the performance of the selected machine-learning algorithm to evaluate each subset of variables. Filters measure the relevance of different subsets of features. Usually they order features individually or as nested subsets of features, while the filter assessment is done by means of statistical tests. They are robust against overfitting, but may fail to select the most useful features for a given classifier. On the other hand, wrappers measure the usefulness of feature subsets. They perform an exhaustive search of the space of all feature subsets and use cross validation to evaluate the performance of the classifier. Wrappers are able to find the most useful features, but they favour overfitting and are very time-consuming. As will be seen, a combination of both, filter and wrapper, provides a trade-off between usefulness and robustness.

4.1 Correlational Relationships Between Signals

First, a simple correlation study is performed with the aim of detecting strong lineal relationships between pairs of signals. Signals with correlation coefficients higher than $\|r\| > 0.9801$ have been removed from the original dataset. Taking into account the correlation threshold, a total of 12 signals are eliminated. Henceforth a dataset with 188 signals is available to apply hybrid feature selection techniques.

4.2 Hybrid Feature Selection

The reduced set of 188 signals is examined in order to obtain a collection of informative features. A hybrid feature selection method is then applied to reduce the system dimensionality without losing significant identification capability. The proposal combines a clustering filter based on the nearest shrunken centroids (NSC) procedure for feature selection in high-dimensional problems [19], with a wrapper around the extreme learning machine algorithm. The NSC method is a modification of the nearest centroid classifier that considers denoised versions of the centroids as the class prototypes. The class centroids are increasingly shrunken towards the overall centroid, and as they are shrunk, some features from the initial set no longer contribute to the classification.

Let x_{ij} be the values for the variables $i = 1, \dots, p$, and for the samples $j = 1, \dots, n$. And let C_k be the set of the indices of the n_k samples in class $k = 1, \dots, K$. A t-statistic d_{ik} is used to compare each class k to the overall centroid for each variable i:

$$d_{ik} = \frac{\bar{x}_{ik} - \bar{x}_i}{m_k(s_i + s)} \tag{4}$$

with s_i being the pooled within-class standard deviation for each variable:

$$s_i^2 = \frac{1}{n - K} \sum_k \sum_{j \epsilon C_k} (x_{ij} - \bar{x}_{ij})^2 \tag{5}$$

s is the median value of the s_i over the set of variables, and $m_k = \sqrt{1/n_k + 1/n}$. In the shrinkage, d_{ik} is the reduced by soft-thresholding:

$$d'_{ik} = sign(d_{ik})(|d_{ik} - \Delta|)_+ \tag{6}$$

where $t_+ = t$ if $t > 0$ and zero otherwise. And according to Eq. 4 the shrunken centroids are calculated as follows:

$$\bar{x}'_{ik} = \bar{x}_i + m_k(s_i + s)\, d'_{ik} \tag{7}$$

As the parameter Δ increases, d_{ik} for some variables are reduced to zero for all classes, and all centroids $\bar{x}_{ik}$ are shrunk to $\bar{x}_i$. Those variables are therefore effectively eliminated from the class prediction.

In order to implement the above of the distances of elements (steam turbine power data) to the gravity centers of their intervals. A total of 20 classes have been generated in this case.

The shrinkage is applied from $\triangle = 0$, that is to say, no shrinkage and no signals eliminated, up to $\triangle = 40$, where only one variable is left. The training and testing root mean squared error (RMSE) are computed for different $\triangle$ values in this range. The collection of samples is randomly divided into a training set consisting of the 75 % of data, and a test set with the last 25 %. Training and prediction are evaluated both by the nearest shrunken centroids clustering procedure, and by the extreme learning machine method.

The nearest shrunken centroids classification of a test sample $x^* = (x_1^*, x_2^*, \ldots, x_p^*)$ is done by calculating the standardized distances of sample x^* to each shrunken centroid or prototype of class k:

$$\delta_k(x) = \sum_{i=1}^{p} \frac{(x_i^* - \bar{x}_{ik}')^2}{(s_i + s)^2} \tag{8}$$

The prediction for sample x^* is then carried out by a "winner-takes-all" rule that chooses the class for which the distance δ_k is the smallest.

The extreme learning machine training and prediction are performed at each $\triangle$ value over the subset of active variables. In every case, the average accuracy over 100 trails of ELM has been computed to provide more stable results and minimize the effect of randomness. Figure 3 shows the shrinkage results where good

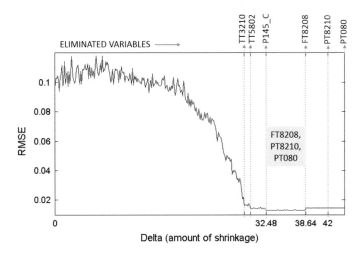

Fig. 3 ELM testing accuracy as a function of the shrinkage parameter $\triangle$, and the relevant variables during the hybrid feature selection method

Table 1 Relevant variables throughout the hybrid feature selection method

Variable TAG	Description of the variable and units
TT3210	Water temperature refrigeration engine B (°C)
TT5802	Gas temperature engines room (°C)
P145_C	Crankshaft pressure of engine C (bar)
FT8208	Steam flow to the steam turbine (kg/h)
PT8210	Inlet pressure steam to steam turbine (bar)
PT080	Steam generator pressure (bar)

performance of the extreme learning machine is noticeable, and some relevant variables are pointed out (Table 1). The minimum ELM test error reached for a shrinkage $\triangle = 32.48$, with a subset of only 3 variables (FT8210, PT8210, PT080) from the initial 188 variables. The zone with shrinkage between $\triangle = 32.48$ and $\triangle = 38.64$ is very stable. Then, variable PT8210 is eliminated and the error increases slightly. We can conclude that the subset with lowest testing error is the most suitable to generate a model to predict the power generated by the steam turbine.

5 Experimental Results

In this section, the optimized model for the steam turbine is developed. For this purpose, the set of 3 input variables selected in the previous section with the hybrid feature selection method from the initial 188 variables, is used to predict the power generated by the steam turbine. Firstly, the optimal number of hidden nodes is calculated using cross validation. The original one year dataset is separated into 12 subsets, each with data relating to one month, except the last 5 days of each month (these data are reserved for later use). From these 12 subsets, fourfold cross validation is applied, using each time 9 months for training and 3 months for testing. For each of the four folds, the hidden nodes are increased from 1 up to 100, and 10 trials for each fold are realized. The testing Root Mean Squared Error (RMSE) average and the standard deviation (STD) average are calculated for the same number of hidden nodes in all the trials done for the four folds.

Figure 4 shows the RMSE average and the STD as a function of the number of hidden nodes. From the results it can be concluded that the zone with lowest RMSE and STD is between 3 and 9 hidden nodes. The minimum testing average error is RMSE = 0.0164 with 8 hidden nodes. Then, it can be concluded that the optimal number of hidden nodes for the steam turbine system is 8. For this number of hidden nodes the standard deviation average is only STD = 0.0045. Then, this means that the number of eight hidden nodes is the optimal number regardless of the random assignment of values to the hidden layer.

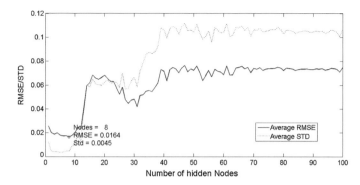

Fig. 4 Testing RMSE and STD according to the number of hidden nodes

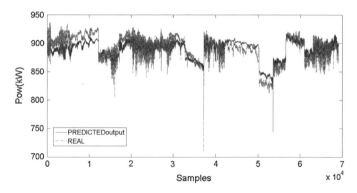

Fig. 5 Real and predicted values of the power generated by the steam turbine using the best model for the testing dataset

After calculating the optimal number of hidden nodes, 100 trials with 8 hidden nodes are obtained and the model with best performance is selected. The training dataset used for this tests is a dataset with all data used in the cross validation to calculate the optimum number of hidden nodes. The testing dataset is composed of the last 5 days of each month. The best model testing error is RMSE = 0.0099. Figure 5 shows, the real values and the predicted values for the power generated by the steam turbine in the testing dataset for the best model. As can be seen, predicted power is able to follow the trend of the real power quite well. However, in the beginning and around 40000–55000 samples, the predictions are not as good as in the rest of the testing dataset.

The model was tested with the last 5 days of each month separately, the testing errors are as shown in Fig. 6. It can be seen that the highest testing errors are for the months with the most extreme environmental conditions, i.e. December, January, July and August. In Fig. 5, the periods with lowest prediction correspond to these months. Despite of this, the optimal model obtained has good performance and its testing accuracy is less than 0.01.

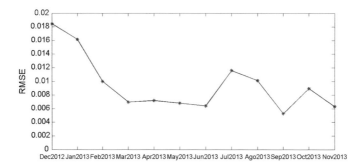

Fig. 6 RMSE testing the last 5 days of each month separately

6 Conclusions

In this work, an optimized model of a steam turbine of a real CHP process using Extreme Learning Machine (ELM) is proposed. In addition, with the aim of reducing the dimensionality of the system without losing prediction capability, a hybrid feature selection method that combines a clustering filter with ELM as wrapper is applied.

First, a correlation study is performed to detect strong lineal relationships between pairs of signals and the signals with higher correlation coefficients than a threshold are removed. Then, the feature selection results reveal that a subset of only three variables, (FT8208, PT8210, PT080), is the most suitable to generate a model to predict the power generated by the steam turbine.

After the feature selection, the optimum number of hidden nodes were found and subsequently the best steam turbine model was determined. Experimental results reveal that ELM with only 3 input variables and 8 hidden nodes is able to predict the power generated by the steam turbine. This straightforward topology provides a high accuracy model with RMSE less than 1 %.

Acknowledgments This work was supported in part by the Basque Country Government under Grants IT733-13, and IG2012/221 (ICOGME), and the Zabalduz Program of the University of the Basque Country (Spain).

References

1. Zare, V., Mahmoudi, S.M.S., Yari, M., Amidpour, M.: Thermoeconomic analysis and optimization of an ammonia-water power/cooling cogeneration cycle. Energy **47**, 271–283 (2012)
2. Feidt, M., Costea, M.: Energy and exergy analysis and optimization of combined heat and power systems. Comparison of various systems. Energies **5**, 3701–3722 (2012)
3. Sen, D., Panua, R., Sen, P., Das, D.: Thermodynamic analysis and cogeneration of a cement plant in india-a case study. In: 2013 International Conference on Energy Efficient Technologies for Sustainability (ICEETS 2013), pp. 641–646 (2013)

4. http://www.cogeneurope.eu/
5. Rutkowski, L.: Computational Intelligence: Methods and Techniques (2008). Cited By 36
6. Rossi, F., Velázquez, D., Monedero, I., Biscarri, F.: Artificial neural networks and physical modeling for determination of baseline consumption of CHP plants. Expert Syst. Appl. **41**(10), 4658–4669 (2014)
7. Nikpey, H., Assadi, M., Breuhaus, P.: Development of an optimized artificial neural network model for combined heat and power micro gas turbines. Appl. Energy **108**, 137–148 (2013)
8. Sisworahardjo, N., El-Sharkh, M.Y.: Validation of artificial neural network based model of microturbine power plant. In: Proceedings of 2013 IEEE Industry Applications Society annual Meeting (2013)
9. Ozel, Y., Guney, I., Arca, E.: Neural network solution to the cogeneration system by using coal. Int. J. Energy **1**, 105–112 (2007)
10. De, S., Kaiadi, M., Fast, M., Assadi, M.: Development of an artificial neural network model for the steam process of a coal biomass cofired combined heat and power (CHP) plant in Sweden. Energy **32**, 2099–2109 (2007)
11. Huang, G.-B., Wang, D., Lan, Y.: Extreme learning machines: a survey. Int. J. Mach. Learn. Cybern. **2**(2), 107–122 (2011). Cited By 323
12. Huang, G.-B., Zhou, H., Ding, X., Zhang, R.: Extreme learning machine for regression and multiclass classification. IEEE Trans. Syst. Man Cybern. Part B: Cybern. **42**(2), 513–529 (2012). Cited By 340
13. Huang, G.-B., Zhu, Q.-Y., Siew, C.-K.: Extreme learning machine: theory and applications. Neurocomputing **70**(1–3), 489–501 (2006). Cited By 1296
14. Huang, G.-B., Chen, L.: Convex incremental extreme learning machine. Neurocomputing **70**, 3056–3062 (2007)
15. Serre, D.: Matrices: Theory and Applications. Springer, New York
16. EnergyWorks. http://www.energyworks.com/
17. Enagas. http://www.enagas.es/portal/site/enagas
18. de Meteorología, A.-A.E.: http://www.aemet.es/en/portada
19. Tibshirani, R., Hastie, T., Narasimhan, B., Chu, G.: Class prediction by nearest shrunken centroids, with applications to DNA microarrays. Stat. Sci. **18**(1), 104–117 (2003). Cited By 156

On the Construction of Extreme Learning Machine for One Class Classifier

Chandan Gautam and Aruna Tiwari

Abstract One Class Classification (OCC) has been prime concern for researchers
and effectively employed in various disciplines for outlier or novelty detection. But,
traditional methods based one class classifier is very time consuming due to its
iterative process for various parameters tuning. This paper presents four novel
different OCC methods with their ten variants based on extreme Learning Machine
(ELM). As we know, threshold decision is a crucial factor in case of OCC, so, three
different threshold declining criteria have been employed so far. Our proposed
classifiers mainly lie in two categories i.e. out of four proposed one class classifiers,
two classifiers belong to reconstruction based and two belong to boundary based. In
four proposed methods, two methods perform random feature mapping and two
methods perform kernel feature mapping. These methods are tested on three
benchmark datasets and exhibit better performance compared to eleven traditional
one class classifiers.

Keywords One class classification (OCC) · One class extreme learning machine
(OCELM) · Autoassociative ELM (AAELM) · Kernelized extreme learning
machine (KELM)

1 Introduction

Novelty or outlier detection [1] has been always prime attention of researchers in
various disciplines and one class classifier [2] has been broadly applied for this
purpose. One class classification (OCC) was coined by Moya et al. [3]. In case of any
binary or multiclass classification problem, data is available for only one class or

C. Gautam (✉) · A. Tiwari
Computer Science and Engineering Department, Indian Institute of Technology Indore,
Indore, India
e-mail: chandangautam31@gmail.com

A. Tiwari
e-mail: artiwari@iiti.ac.in

© Springer International Publishing Switzerland 2016
J. Cao et al. (eds.), *Proceedings of ELM-2015 Volume 1*,
Proceedings in Adaptation, Learning and Optimization 6,
DOI 10.1007/978-3-319-28397-5_35

other classes' data is not available. In such case, we need to learn with single class data. It becomes necessity when data of only one class is available or other classes' data are very rare. As an example, classifier design between healthy and unhealthy people. In this case, it is possible to define the range of the data for the healthy people but not possible to define such a definite range for unhealthy people. Because it is possible to prepare the data based on only existing diseases but not possible to prepare based on forthcoming diseases. One thing is quite clear by this example that which data is available for training, called as positive or normal or target class and another class, which is not possible to define or very rare or unknown, called as negative or abnormal or outlier class. Reason behind the absence of outlier samples can be the very high measurement costs or the less occurrence of an event, examples are failure of nuclear power plant or a rare medical disease or machine fault detection. Even if outlier examples are available then there is a possibility that those examples are not well distributed or does not characterize all possible issues of outliers, therefore, we can't rely on the outlier examples. Thus, one class classification has been broadly applied in the field of novelty or outlier detection. Overall goal of one class classifier is to accept target samples and reject the outliers. Basic assumption of multi class classification also follows here that examples belong to same class share same pattern among them. So, it can be applied in various type of problems [2] viz., (i) Novelty detection, e.g. Machine fault detection (ii) Outlier detection e.g. Intrusion detection in network (iii) Classification in case of badly balanced data e.g. Classification in medical data where data is not properly balanced.

Various methods have been proposed to resolve the one class classification problem. According to Pimentel et al. [1], these methods can be broadly divided into 5 categories: probabilistic or density based, distance based, Information theoretic techniques, domain or boundary based, reconstruction based. While, Tax [2] divided OCC methods in three parts density based, boundary based and reconstruction based. We will discuss about this in the next section. This paper is mainly focused on last two categories i.e. boundary and reconstruction based. Our literature survey is also primarily focused on these two categories only.

The remaining paper is organized as follows. Literature survey about OCC is discussed in Sect. 2. Section 3 presents motivation of our proposed work. Section 4 provides a brief description of ELM. Section 5 discusses about proposed work with three threshold deciding criteria. Section 6 describes about experimental setup and dataset description. Section 7 provides performance based comparison of existing and proposed classifiers on three benchmark datasets. Conclusion and future direction of work are discussed in Sect. 8.

2 Literature Survey

OCC word is firstly coined by Moya et al. [3] and applied for target recognition application. As far as learning for OCC is concerned, it can be divided into three parts: Learning with (i) positive examples only (ii) positive and small amount of negative examples only (iii) positive and unlabeled data. Japkowicz [4] proposed autoassociative based approach for OCC in the absence of counterexamples. OCC have been widely studied by Tax [2] and he developed various model to handle OCC problem viz., three models based on density methods (i) Gaussian model (ii) mixture of Gaussians and (iii) Parzen density estimator, 3 models based on boundary methods (i) k-centers method (ii) NN-d and (iii) SVDD and reconstruction based models (i) k-mean clustering (ii) self-organizing maps (iii) PCA and mixtures of PCA's and (iv) diabolo networks. Tax and Duin [5] proposed algorithm to handle OCC problem based on support vector machine (SVM) using positive examples only. In text classification community, learning with positive and unlabeled data [6] have received much attention. Most recently, Leng et al. [7] proposed OCC based on Extreme Learning Machine (ELM) where ELM consists of only one node in output layer. Leng et al. [7] tested their model with only one type of threshold deciding criteria i.e. rejection of few percentages of most deviant training samples after completion of training on all samples. So, there is lot of scope open for OCC based ELM as Leng et al. [7] also stated in their paper. Although, ELM has been applied earlier also for anomaly or intrusion detection by using binary classification [8–10]. Xiang et al. [10] proposed map reduce based ELM for intrusion detection in big data environment.

3 Motivation of Our Work

Our proposed methods are based on ELM. Various papers have been presented in the past, which exhibited the superiority of ELM [11–13] over traditional machine learning techniques like Back-propagation (BP), SVM, Probabilistic Neural Network (PNN), Multilayer Perceptron (MLP) etc. in terms of generalization capability and training time for binary and multiclass classification. Since, traditional methods have required tuning of weight and various parameters in each iteration but ELM provides result just in one pass. ELM has been well explored for binary and multiclass classification problem but not well explored in case of OCC problem. Since, existing OCC models are based on traditional methods, so, problem lies with traditional methods also remain with existing OCC models. As an example, backpropagation based OCC require many iteration to stabilize the weight, so it will be very time consuming compare to single hidden layer feed forward network (SLFN). We are going to present 4 methods with their 10 variants of OCC based on ELM. Earlier, ELM had been presented for binary and multiclass classification but we are presenting it for one class classification.

4 A Brief Overview of ELM

ELM is proposed by Huang et al. [11, 12] for addressing the slow learning speed of traditional neural network. We can state basic ELM in three steps. For N given training samples, activation function g(x) and by assuming m number of hidden neurons following steps are taken to perform learning of the neurons,

1. Random initialization of input weights W_i and bias b_i for all m number of nodes/neurons, where i = 1, 2...m.
2. Calculate the hidden layer output matrix H by applying X over all m number of hidden neurons with W and b with activation function g as $g(WX+b)$.
3. Calculate the output weight $\beta = H^{\dagger}T$, where, $H^{\dagger} = (H^T H)^{-1} H^T$ and T = Output layer.

Here, Objective of ELM is to minimize the training error as well as norm of the output weights [11, 12]:

$$\text{Minimize: } \|H\beta - T\|^2 \text{ and } \|\beta\|,$$

5 Proposed Work

In this paper, we proposed methods based on two types of OCC viz., reconstruction based one class ELM i.e. autoassociative ELM (AAELM) and boundary based one class ELM (OCELM). The Kernelized version of both AAELM and OCELM viz., autoassociative Kernelized ELM (AAKELM) and one class Kernelized ELM (OCKELM), is also presented in this paper. In all methods, only positive samples are used for training. Now onwards, positive samples will be called as target data and negative samples as outliers.

5.1 Boundary Based

Assumption "Model is trained by only target data x and endeavored to approximate all data to 1 because training have only one class. Since, weights between layers are trained according to pattern of target data i.e. positive data. But if any pattern other than target data will provide to trained model then it will not be properly approximate to one. Therefore, difference between approximated value and one will be large if pattern is differing from target data. If difference will be more than the threshold then treat that sample as outlier otherwise belongs to target data".

5.1.1 One Class Extreme Learning Machine (OCELM)

By following above assumption, algorithm of OCELM (see Fig. 1) is as follows:

1. Normalize the training dataset 'X' in the range of [0 1] using min-max normalization.
2. Select the number of hidden nodes and activation function.
3. Assign random hidden nodes parameter, input weights W and bias b.
4. Calculate the hidden layer output matrix $H = g(W X + b)$, where g is an activation function.
5. The output nodes contain the identity row or column matrix as the target variables due to only one class is available for training.
6. Calculate the output weight $\beta = H^{\dagger}T$, where **T = identity row or column matrix** i.e. we are approximating all training i.e. normal data to one. Therefore, Objective of AAELM is modified as:

$$\text{Minimize} \|H\beta - T\|^2 \text{and} \|\beta\|,$$

where, **T = identity row or column matrix**
7. Calculate the threshold value using anyone threshold criteria **Thr1 or Thr2 but not Thr3** as mentioned in the Sect. 5.3 by using training data only.
8. Calculate the error i.e. difference between predicted value by above steps and **one** for each input sample in testing and if it is greater than threshold value, which was calculated during training then that sample will be treated as outlier otherwise normal data.

5.1.2 One Class Kernelized Extreme Learning Machine (OCKELM)

Training of OCKELM is different from OCELM in term of feature mapping. In OCKELM, kernelized feature mapping is employed instead of random feature

Fig. 1 Architecture of one class extreme learning machine (OCELM)

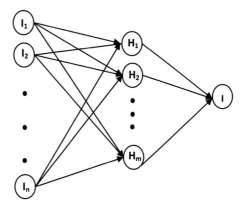

mapping and rest of the procedure is same as OCELM. We have used four different kernels: **polynomial kernel, linear kernel, rbf kernel** and **wave kernel.**

5.2 Reconstruction Based

Assumption "Reconstruct the input vector X at the output layer during training and weights between layers are trained only for target data i.e. positive data. So, if any pattern other than target data passed to the model then that pattern will not be well reconstructed. Therefore, difference between input and output in case of outlier will be more than the difference between input and output of target dataset. So, we can decide threshold in such a way that if dissimilarity is more than this threshold then assume as outlier data otherwise target data".

5.2.1 Autoassociative Extreme Learning Machine (AAELM) Based One Class Classifier

By following above assumption, we modified ELM architecture. Output layer is modified and present input data in the output layer for training i.e. **T = input data**, but those data belongs to only one class (See Fig. 2).

1. Steps 1 to 4 are same as OCELM in Sect. 5.1.1.
2. The output nodes contain input X in output layer.
3. Calculate the output weight $\beta = H^{\dagger}T$, **where T = X (Note:** It is different from **OCELM).** Therefore, Objective of **AAELM** is modified as:
$$\text{Minimize} \|H\beta - T\|^2 \text{and } \|\beta\|,$$
   where, T = Input Data from only one class

Fig. 2 Architecture of autoassociative extreme learning machine (AAELM) for one class classification

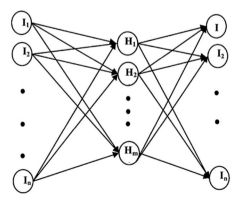

4. Calculate the threshold value using anyone threshold criteria **Thr1, Thr2 or Thr3** as mentioned in the Sect. 5.3.
5. Calculate the error i.e. difference between predicted value by above steps and **Y** for each input sample and if it is greater than threshold value, which was calculated during training then that sample will be treated as outlier otherwise normal data. It is noted that **Y = Testing Data**.

5.2.2 Autoassociative Kernelized Extreme Learning Machine (AAKELM) Based One Class Classifier

Training of AAKELM is different from AAELM in terms of feature mapping. AAKELM use Kernel feature mapping and AAELM use random feature mapping and rest of the procedure is same as AAELM. We have used four different kernels: **polynomial kernel, linear kernel, rbf kernel** and **wave kernel.**

5.3 Threshold Deciding Criteria for Proposed Methods

In One class classification, threshold plays crucial role in distinguishing outlier from target data. We deployed 3 methods to determine the threshold for acceptance of data as target i.e. in other word rejection of data as outlier. One point must be noted that only positive samples are used for deciding the threshold criteria.

1. **Thr1**: Calculate the error using Euclidean distance between actual and predicted on each training data and arrange the error in decreasing order. Afterwards, set the threshold at rejection of 10 % most erroneous data i.e. false negative rate should be expected at the rate of 10 %.
2. **Thr2**: Calculate the error between actual and predicted on training and set the threshold (Thr) using following formula:

$$Thr2 = Error + 0.2 * Std \tag{6}$$

    Error   Mean square error over all the training data
    Std   Standard deviation of error

3. **Thr3**: This threshold criteria is only for reconstruction based methods i.e. not for boundary based methods. Calculate relative error between actual (*act*) and predicted (*pred*) on each attribute of training dataset by using following formula:

$$err_i = abs\left(\frac{act_i - pred_i}{act_i}\right), \text{ where } i = 1, 2 \dots n. \tag{7}$$

```
Well_reconstructed_attribute=0

for i=1 to n

    if err_i < thr_condn1
Well_reconstructed_attribute=
Well_reconstructed_attribute +1;

    end if

end for
```

This error is a deciding factor that whether attribute is well reconstructed in output layer or not. There are 2 conditions require to satisfy for this threshold deciding criteria. We need to satisfy first condition (**thr_condn1**) at this point for deciding how many attributes are well reconstructed at output layer.

Afterwards, we need another **condition (thr_condn2)** for deciding whether data should be treated as genuine or outlier. If number of `Well_recon-structed_attribute` is above **thr_condn2** then it will be treated as target data otherwise outlier. We considered **thr_condn1 = 0.5** and **thr_condn2 = 0.7 * no. of features in dataset** i.e. if more than 70 % attributes are well reconstructed then consider that sample as target sample otherwise outlier sample.

Remark for Thr3: **Thr1** and **Thr2** rejected few percent of data for deciding threshold but **Thr3** doesn't reject any percentage of data for deciding the threshold.

6   Experimental Setup and Dataset Description

Performance of proposed methods has been tested on variety of three benchmark datasets [See Table 1]. These datasets are downloaded from UCI Machine Learning Repository [14] and website of TU Delft [15]. Breast cancer database is obtained

Table 1 Dataset description

Dataset name	Number of attributes	Number of target	Number of outlier	Name of target class
Breast cancer [14, 16]	9	458	241	Benign
Page blocks [14]	10	4913	560	Text
Spambase [14]	57	1813	2788	Not spam

from the University of Wisconsin Hospitals, Madison from Dr. William H. Wolberg [14, 16]. There are 2 classes in this dataset viz., benign and malignant. We used benign class as the target class. Second dataset, Page Blocks dataset has been first employed for classifying all the blocks of the page layout of a document that has been detected by a segmentation process. This is an essential step in document analysis in order to separate text from graphic areas. Indeed, the five classes are: text, horizontal line, picture, vertical line and graphic. We used test class as target class and remaining classes as outlier class. Third dataset, Spambase dataset is created in Hewlett-Packard Labs [14] and consists of 1813 normal samples and 2788 spam samples. Normal samples are used as target class in our experiment. We prepared the dataset for OCC in the same way as prepared by Leng et al. [7]. 50 % data from target class is selected randomly for training and rest of the data used for testing. So, we divided our target class dataset into 50-50. Each classifier has been executed 10 times and calculates the average of 10 runs for each performance evaluation measure viz., Precision, Recall (Sensitivity), F1, Accuracy (Acc) and Area under curve (AUC). But we presented Acc and AUC only due to number page constraint in the conference.

Our proposed methods are compared with the methods proposed by Tax [2]. Tax [2] has done resounding work on OCC. He proposed various one class classifiers as we have discussed in literature. Tax also provided a MATLAB toolbox viz., DD toolbox [17] to employ these classifiers. We deploy below classifiers using consistency based optimization [18]:

- Density based methods: (i) Parzen density data description (parzen_dd) (ii) Naive Parzen density data description description (nparzen_dd) (iii) Gaussian data description (gauss_dd)
- Boundary based methods: (i) K-nearest neighbor data description (knndd) (ii) Minimax probability machine data description (mpm_dd) (iii) Support vector data description (svdd) (iv) Incremental SVDD (incsvdd)
- Reconstruction based methods: (i) Auto-encoder neural network data description (autoenc_dd) (ii) k-means data description (kmeans_dd) (iii) Principal component data description (pca_dd) (iv) Self-Organizing Map data description (som_dd)

Above classifiers are tested with the same benchmark datasets along with the similar criteria (i.e. 10 % rejection (**Thr1**), 50-50 division of dataset etc.) as we applied on our proposed methods. The generated results are presented in tabular form, Tables 2, 3 and 4. All experiments have been executed in MATLAB under Windows 7 environment with Intel Core 5 processor and 4 GB RAM.

Table 2 Average result of Breast cancer dataset over 10 Runs

	AUC	Acc
knndd	94.808	94.936
svdd	74.716	75.362
kmeans_dd	93.321	93.489
parzen_dd	89.061	89.340
nparzen_dd	94.386	94.489
pca_dd	86.593	86.511
mpm_dd	73.952	74.617
incsvdd	93.654	93.681
som_dd	93.958	94.106
autoenc_dd	93.681	93.809
gauss_dd	93.265	93.426
Proposed		
AAELM_thr1 (Sig, 10)	94.502	94.617
AAELM_thr2 (Sig, 10)	94.830	94.936
AAELM_thr3 (Sig, 2)	63.374	63.234
AAKELM_thr1 (Poly, 10^{-6}, [1 1])	95.535	95.638
AAKELM_thr2 (Wav, 10^3, [10^6 1 1])	95.962	96.043
AAKELM_thr3 (RBF, 10^6, [10^{-6}])	94.563	94.596
OCELM_thr1 (Sig, 2)	87.956	87.915
OCELM_thr2 (Sig, 2)	89.217	89.383
OCKELM_thr1 (Wav, 10^3, [10^4 1 1])	95.085	95.170
OCKELM_thr2 (RBF, 1, [10^2])	95.805	95.851

Table 3 Average result of Page blocks dataset over 10 Runs

	AUC	Acc
knndd	83.918	88.077
svdd	74.794	59.622
kmeans_dd	81.326	86.280
parzen_dd	49.604	19.629
nparzen_dd	70.930	82.954
pca_dd	87.816	89.227
mpm_dd	49.331	18.780
incsvdd	68.642	81.698
som_dd	78.153	85.355
autoenc_dd	87.420	89.267
gauss_dd	88.324	89.281
Proposed		
AAELM_thr1 (Sig, 95)	89.590	88.816
AAELM_thr2 (Sig, 100)	89.604	87.885
AAELM_thr3 (Sig, 5)	83.997	81.054
AAKELM_thr1 (RBF, 10^3, [1])	89.857	89.105
AAKELM_thr2 (RBF, 10^3, [1])	90.115	89.536
AAKELM_thr3 (Wav, 10^3, [10^4 1 1])	88.618	88.525
OCELM_thr1 (Sig, 200)	86.087	86.131
OCELM_thr2 (Sig, 250)	85.469	83.654
OCKELM_thr1 (Wav, 1, [10^{-2} 1 1])	88.593	86.844
OCKELM_thr2 (RBF, 1, [10^{-2}])	88.771	85.249

Table 4 Average result of Spambase dataset over 10 Runs

	AUC	Acc
knndd	61.004	46.040
svdd	56.628	76.762
kmeans_dd	56.472	40.189
parzen_dd	55.415	56.194
nparzen_dd	48.117	27.212
pca_dd	61.316	47.737
mpm_dd	53.450	77.122
incsvdd	54.922	37.147
som_dd	52.805	34.434
autoenc_dd	58.261	42.006
gauss_dd	81.431	77.864
Proposed		
AAELM_thr1 (Sig, 30)	81.309	77.753
AAELM_thr2 (Sig, 45)	80.420	78.525
AAELM_thr3 (Sig, 30)	80.759	81.354
AAKELM_thr1 (Lin, 10^6)	82.321	79.686
AAKELM_thr2 (Poly, 10^3, [10^4 1])	82.195	77.515
AAKELM_thr3 (Wav, 10^3, [1 1 1])	81.374	75.731
OCELM_thr1 (Sig, 150)	79.218	77.650
OCELM_thr2 (Sig, 70)	77.386	75.504
OCKELM_thr1 (Wav, 1, [1 1 1])	80.822	77.783
OCKELM_thr2 (Wav, 1, [1 1 1])	80.660	79.112

7 Results and Discussion

The proposed methods are tested on three benchmark datasets and their performances are evaluated by varying numerous variables viz., number of hidden neurons, kernel parameters, regularization parameter, different kernels and three threshold deciding criteria. By employing **Thr1** and **Thr2** on all 4 proposed methods, we generated 8 novel variants of ELM for OCC viz., **OCELM_Thr1, OCELM_Thr2, OCKELM_Thr1, OCKELM_Thr2, AAELM_Thr1, AAELM_Thr2, AAKELM_Thr1, and AAKELM_Thr2**. We proposed 2 novel variants of ELM for OCC by employing **Thr3** with reconstruction based methods **AAELM** and **AAKELM**, viz., **AAELM_Thr3 and AAKELM_Thr3**.

After analyzing Tables 2, 3 and 4, it is quite clear that our proposed approaches outperformed earlier existing approaches by significant margin in terms of AUC and Accuracy. Reconstruction based one class classifier outperforms boundary based classifiers among all the proposed methods. It is also clearly visible that kernel feature mapping performed better compared to random feature mapping for all 3 datasets. Figure 3a–e depict the behavior of the proposed methods viz.,

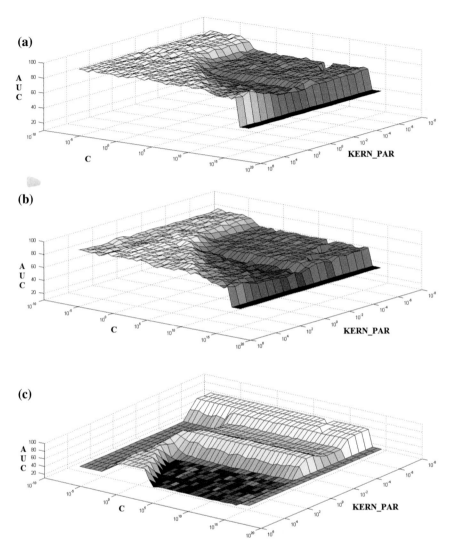

Fig. 3 Above figure from **a–e** represents that how various variants of kernelized based OCC behave after varying the two parameters viz., regularization and Kernel parameter over 3 threshold criteria for RBF kernel on Breast cancer dataset. **a** AAKELM_Thr1. **b** AAKELM_Thr2. **c** AAKELM_Thr3. **d** OCKELM_Thr1. **e** OCKELM_Thr2. *Note* Proposed methods name format in all above Figures and Tables—[AAELM or OCELM]_Thr1, Thr2 or Thr3 (activation function, No. of hidden nodes) [AAKELM or OCKELM]_ Thr1, Thr2 or Thr3 (Kernel name, regularization parameter, Kernel parameter). *Note* All above figure **a–e** contain same axis parameter as **a** (due to space constrains of conference papers, we provided small size of **b–e**)

(d)

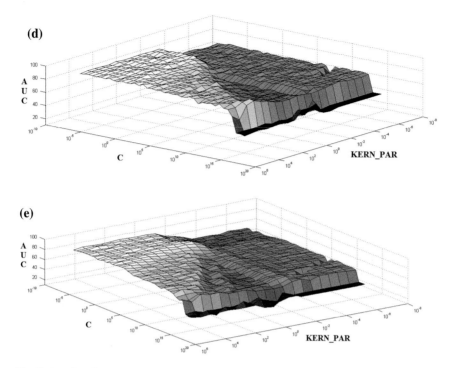

(e)

Fig. 3 (continued)

AAKELM and **OCKELM**, over changing the kernel parameters (KERN_PAR), regularization parameter (C), threshold criteria. Figure 3a–e is depicted for RBF kernel, however we tested with 4 kernels viz., RBF, Polynomial, linear and Wavelet. As you can see in Fig. 3a–e, behavior of both Kernelized methods for **Thr1** and **Thr2** does not change much but it performs totally different for **Thr3**. However, performance of **Thr3** is comparable to other threshold criteria viz., **Thr1** and **Thr2**, except for **AAELM** in the Breast Cancer dataset.

We have tested our function for all 4 kernels and also calculated F1, sensitivity, specificity, standard deviation but didn't present all results due to space constrains.

8 Conclusion and Future Work

In this paper, we proposed ELM based four OCC methods with their ten different variants by using three different threshold deciding criteria. Our proposed approaches are fast and simple compared to earlier OCC methods and also outperformed various existing methods from literature but it needs user intervention for optimal selection of parameters. This verifies a viable and effective alternative for outlier detection using OCC. Although, our **Thr3** performed well but it would fail on those cases where actual value would be zero, so, we need to improve **Thr3**. Our future work will carry this discussion forward for optimal selection of parameter without user intervention and make **Thr3** more robust. These issues need further investigation.

References

1. Pimentel, M.A.F., Clifton, D.A., Clifton, L., Tarassenko, L.: A review of novelty detection. In: Signal Processing, vol. 99, pp. 215–249. Elsevier (2014)
2. Tax, D.L.: One class classification: concept-learning in the absence of counter-examples. PhD thesis, Delft University of Technology (2001)
3. Moya, M., Koch, M., Hostetler, L.: One-class classifier networks for target recognition applications. In: Proceedings of the World Congress on Neural Networks, International Neural Network Society, pp. 797–801 (1993)
4. Japkowicz, N.: Concept-learning in the absence of counterexamples: an autoassociation-based approach to classification. PhD thesis, New Brunswick Rutgers, The State University of New Jersey (1999)
5. Tax, D., Duin, R.: Support vector data description. Mach. Learn. **54**, 45–66 (2004)
6. Lee, W., Liu, B.: Learning with positive and unlabeled examples using weighted logistic regression. In: Proceedings of the 20th International Conference on Machine Learning (ICML) (2003)
7. Leng, Q., Qi, H., Miao, J., Zhu, W., Su, G.: One-Class Classification with Extreme Learning Machine, Mathematical Problems in Engineering. Hindawi Publishing Corporation, Article ID: 412957 (2014)
8. Wang, Y., Li, D., Du, Y., Pan, Z.: Anomaly detection in traffic using L1-norm minimization extreme learning machine. Neurocomputing **149**, 415–425 (2015)
9. Farias, G., Oliveira, A., Cabral, G.: Extreme learning machines for intrusion detection systems, neural information processing. In: Proceedings of the 19th International Conference, ICONIP 2012, Doha, Qatar, pp. 535–543 (2012)
10. Xiang, J., Westerlund, M., Sovilj, D., Pulkkis, G.: Using extreme learning machine for intrusion detection in a big data environment. In: Proceedings of the 2014 Workshop on Artificial Intelligent and Security Workshop (AISec'14), pp. 73–82. ACM, New York, USA (2014)
11. Huang, G.B., Zhu, Q., Siew, C.: Extreme learning machine: a new learning scheme of feedforward neural networks. Int. Joint Conf. Neural Netw. **2**, 985–990 (2004)
12. Huang, G.B., Zhu, Q.C., Siew, C.K.: Extreme learning machine: theory and applications. Neurocomputing, vol. 70, pp. 489–501. Elsevier (2006)
13. Huang, G.B., Zhou, H., Ding, X., Zhang, R.: Extreme learning machine for regression and multiclass classification. IEEE Trans. Syst. Man Cybern. Part B: Cybern. **42**, 513–529 (2012)
14. Lichman, M.: UCI machine learning repository [http://archive.ics.uci.edu/ml]. University of California, School of Information and Computer Science, Irvine, CA (2013)
15. http://homepage.tudelft.nl/n9d04/occ/index.html. Accessed 14 Sept 2015

16. Mangasarian, O.L., Setiono, R., Wolberg, W.H.: Pattern recognition via linear programming: theory and application to medical diagnosis. In: Large-scale numerical optimization, pp. 22–30. SIAM Publications, Philadelphia (1990)
17. Tax, D.M.J.: DD tools 2014, the data description toolbox for MATLAB, version 2.1.1 [http://prlab.tudelft.nl/david-tax/dd_tools.html] (2014)
18. Tax, D.M.J., Muller, K.R.: A consistency-based model selection for one-class classification. In: Proceedings of the 17th International Conference on Pattern Recognition (ICPR'04), IEEE Computer Society, Los Alamitos, Calif, USA, pp. 363–366 (2004)

Record Linkage for Event Identification in XML Feeds Stream Using ELM

Xin Bi, Xiangguo Zhao, Wenhui Ma, Zhen Zhang and Heng Zhan

Abstract Most of the news portals and social media networks are utilizing RSS feeds for information distribution and content sharing. Event identification improves the service quality of feeds providers in the aspect of content distribution and event browsing. However, thriving challenges arise due to representation of structural information and real-time requirement in feeds streams mining. In this paper, we focus on the record linkage problem which classifies stream content into known categories. To realize fast and efficient record linkage over XML feeds stream, we design two classification strategies: a classifier based on ensemble ELMs and an incremental classifier based on OS-ELM. Experimental results show that our solutions provide effective and efficient record linkage for event identification applications.

Keywords XML · Stream · ELM · Classification · Record linkage

1 Introduction

Content providers such as news portals (e.g., CNN, BBC, ABC, etc.) and social media networks (e.g., Twitter, Flickr, etc.) are using RSS (Really Simple Syndication) feeds in the format of XML streams as up-to-date and inclusive content releases. As to content providers, event mining plays a critical role in improving the quality of content distribution and event browsing quality, and attracts wide attention and interest in both academia and industry of information technology.

Event identification problem [1] has been addressed under two different scenarios [2], i.e., known-event and unknown-event identification. In this paper, we only discuss *known-event identification* with priori knowledge of planned events. The identification phase is also referred as *Record Linkage* [3].

X. Bi (✉) · X. Zhao · W. Ma · Z. Zhang · H. Zhan
College of Information Science and Engineering, Northeastern University,
Shenyang 110819, Liaoning, China
e-mail: edijasonbi@gmail.com

© Springer International Publishing Switzerland 2016
J. Cao et al. (eds.), *Proceedings of ELM-2015 Volume 1*,
Proceedings in Adaptation, Learning and Optimization 6,
DOI 10.1007/978-3-319-28397-5_36

463

Since the RSS feeds are XML streams essentially, as news information and the user-generated data volume keep proliferating, record linkage task faces two major *challenges* in RSS feeds stream: (1) the representation of semi-structured feeds formatted in XML containing both semantic and structural information; (2) efficient classification component to handle fast and large-scale feeds stream.

Record linkage task is treated as a problem of *classification over XML stream data* in this paper. We introduce Extreme Learning Machine (ELM) [4, 5] combined with different stream processing strategies to realize real-time identification of fast and large-scale feeds content. In summary, the contributions of this paper are summarized as:

1. Based on the thorough study on XML feeds stream classification, we present a naïve algorithm intuitively as the baseline method;
2. Following the streaming data model, we propose an *ensemble based* algorithm with update strategies of ELM classifiers;
3. Taking all the historical learning experience into consideration, we propose an *online sequential* algorithm based on OS-ELM;
4. Extensive experiments on real-world datasets are conducted to verify the effectiveness and efficiency of our algorithm.

The remainder of the paper is organized as follows. Section 2 gives a survey on the study of event identification and ELM. In Sect. 3 we present the data model and problem definition. In Sect. 4 we introduce a brief of ELM theory, based on which our algorithms are proposed based on different stream strategies for event identification in Sect. 5. We present and discuss our experimental results in Sect. 6, and draw conclusions in Sect. 7.

2   Related Work

Extreme Learning Machine (ELM) [4, 5] achieves extremely fast learning speed and good generalization performance. ELM has been proven to be a powerful learning component in many fields [6–9].

The event detection task [10, 11] was first studied to identify news events on a continuous stream of news documents. Clustering techniques are usually applied to discover event topics [12, 13]. In recent years, most event identification solutions focus on social media streaming data [14–17], which takes all the features of social media into consideration. In this paper, we focus on the record linkage task in feeds stream, which are treated as a semantic classification problem in XML documents stream.

In the aspect of motivation, the most similar work [18] aims at event identification from social media RSS feeds using Naive Bayes Model. However, considering the distinct advantages of ELM, it is a great choice to design ELM based strategies rather than traditional learning algorithms.

In the aspect of techniques, a similar work [19] has proposed UC-ELM and WEC-ELM, which realize classification over uncertain data stream by utilizing weights of instances and classifiers. WEC-ELM also applies classifier update strategies using uncertainties as criteria. Different from UC-ELM and WEC-ELM, in this paper, we (1) focus on XML data classification and utilize semi-structural representation model; (2) design classifier update criteria for certain XML data learning applications other than uncertainty-related threshold. To our best knowledge, there is no existing work on classification over XML stream aiming at record linkage task in event identification.

3 Preliminaries

3.1 Problem Definition

In the event identification problem, each content entry will be identified with an event topic, which is also considered as a classification problem. A feeds stream is essentially a stream of XML documents. Each XML document contains one or more feeds entry. Thus, we consider the record linkage task as an *XML stream classification problem*.

We introduce *sliding window* to denote a finite collection of the XML stream. Note that in most case, each XML document in the RSS feeds contains a variable number of content entries, that is, RSS feeds release up-to-date content in batches. For convenience of definitions and calculations, we denote each element of the stream as an XML document. Thus, we define record linkage task in event identification as a classification problem as Definition 1.

Definition 1 (*Record linkage problem*) Given an RSS feeds stream S, assuming there is a set $S^Y \in S$, each $S_i^Y \in S^Y$ is related to a known event topic $c_i \in C$. For the set $S^N = S \setminus S^Y$, the problem of record linkage is to learn a classification function $\varphi : S^N \to C$ using a learning algorithm, so that each feeds entry $S_i^N \in S^N$ will be assigned with an event topic $c_i \in C$.

3.2 XML Representation

In the feeds stream, each XML document has to be transformed into representation model to be taken as inputs to the classifier component. In this paper, we utilize our proposed Structured Vector Model (DSVM) [20] based on Structured Link Vector Model (SLVM) [21] to further strengthen the ability of representation. DSVM not only inherits the advantages of SVLM in containing structural information, but also benefits in feature subset selection using information gain to achieve improved representation ability. In DSVM, an XML document with m features is represented as

$$\mathbf{d}_{DSVM} = \langle \mathbf{d}^1_{u^j_i}, \ldots, \mathbf{d}^k_{u^j_i}, \ldots, \mathbf{d}^m_{u^j_i} \rangle \qquad (1)$$

where $\mathbf{d}^k_{u^j_i}$ is the kth term feature calculated as

$$\mathbf{d}^k_{u^j_i} = \sum_{j=1}^{m} (\mathrm{TF}(w_k, doc.e_j) \cdot \varepsilon_j) \cdot \mathrm{IDF}_{\mathrm{ex}}(w_k, c) \cdot \rho_{\mathrm{CD}} \qquad (2)$$

where m is the number of elements in XML document doc, $doc.e_j$ is the jth element e_j of doc, ε_j is the dot product of $1 \times s$ unit vector and $1 \times s$ weight vector. $\mathrm{IDF}_{\mathrm{ex}}(w_i, c)$ is the revised IDF. ρ_{CD} is the distribution modifying factor, which is the reciprocal of arithmetic product of WCD and ACD. Due to the space constraint, detailed definition and calculation of DSVM referring in [20] will not be given in this paper.

4 Brief of ELM

In order to achieve extremely fast learning speed and good generalization performance, Extreme Learning Machine (ELM) generates parameters of the single hidden layer randomly to avoid iteratively tuning. Given N arbitrary samples $(\mathbf{x}_i, t_i) \in \mathbf{R}^{n \times m}$, ELM is mathematically modeled as

$$\sum_{i=1}^{L} \beta_i \, G(\mathbf{w}_i, b_i, \mathbf{x}) = \boldsymbol{\beta} \, h(\mathbf{x}) \qquad (3)$$

where L is the number of hidden layer nodes, $\mathbf{w}_i = [w_{i1}, w_{i2}, \ldots, w_{in}]^{\mathrm{T}}$ is the input weight vector from input nodes to the ith hidden node, b_i is the bias of ith hidden node, $\boldsymbol{\beta}_i$ is the output weight from the ith hidden node to the output node. $G(\mathbf{w}_i, b_i, \mathbf{x})$ is the activation function to generate mapping neurons, which can be any nonlinear piecewise continuous functions [22].

The ELM feature mapping denoted as $\mathbf{H}$ is calculated as

$$\mathbf{H} = \begin{bmatrix} h(\mathbf{x}_1) \\ \vdots \\ h(\mathbf{x}_N) \end{bmatrix} = \begin{bmatrix} G(\mathbf{w}_1, b_1, \mathbf{x}_1) & \cdots & G(\mathbf{w}_L, b_L, \mathbf{x}_1) \\ \vdots & \cdots & \vdots \\ G(\mathbf{w}_1, b_1, \mathbf{x}_N) & \cdots & G(\mathbf{w}_L, b_L, \mathbf{x}_N) \end{bmatrix}_{N \times L} \qquad (4)$$

ELM aims to minimize the training error and the norm of output weights:

$$\text{Minimize:} \quad \|\mathbf{H}\boldsymbol{\beta} - \mathbf{T}\|^2 \quad \text{and} \quad \|\boldsymbol{\beta}\| \qquad (5)$$

Therefore, the output weight β can be calculated as

$$\beta = \mathbf{H}^{\dagger}\mathbf{T} \tag{6}$$

where $\mathbf{H}^{\dagger}$ is the Moore-Penrose Inverse of $\mathbf{H}$.

5 Classification over XML Stream

5.1 Naïve Solution

Based on the definitions of stream processing problem, we assume that learning of classifiers focus on the stream elements in the *current* sliding window. That is, the expired elements are out of value immediately. In this case, each time the sliding window slides, we train a new classifier using the current s elements in the sliding window to replace the former one to solve the *concept drift* problem.

Figure 1 shows the flow chart of our naïve solution Baseline Stream Extreme Learning Machine (BS-ELM). The upper arrow with intervals is a feeds stream. The overall framework of BS-ELM (as Algorithm 1) is that: (1) with the current sliding window X_i, BS-ELM use the latest flow-in element S_{i+s-1} to test the former classifier cl_{i-1}; (2) a new classifier is trained using all the elements of X_i; (3) the sliding window continues to slide as the stream flows.

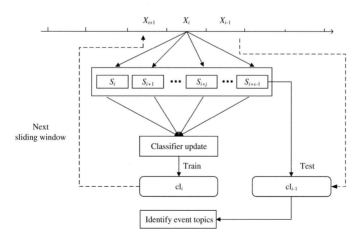

Fig. 1 Flow chart of BS-ELM

Algorithm 1: BS-ELM

Input: Feeds stream S, sliding window X with size s
Output: Classification results
1 $X = \{S_1, \ldots, S_s\}$;
2 Train an initiate classifier using X: $\beta = \mathbf{H}^\dagger\mathbf{T}$;
3 S_{s+1} flows in X and S_1 expires;
4 **while** *!S.end()* **do**
5    $X = \{S_i \mid S_i$ within the current sliding window$\}$;
6   Calculate output of the latest flown-in element X_s: $\mathbf{O}_s = \mathbf{H}\beta$;
7   Identify the event topic with the max index of $\mathbf{O}_i$ to S_s;
8   Update the classifier with the current X: $\beta = \mathbf{H}^\dagger\mathbf{T}$;
9   The sliding windows continues to slide;

5.2 Ensemble Strategy

The naïve algorithm BS-ELM retrains the classifier each time the sliding window slides. In order to avoid *frequent updates* on the premise of concept drift detection, we propose an Ensemble based Stream Extreme Learning Machine (ES-ELM). Compared with BS-ELM, the *improvement* of ES-ELM are:

1. The *sliding step* in ES-ELM is the size of the sliding window, so that more stream elements can participate in both the training and testing procedures;
2. The *ensemble* strategy combined with mechanisms such as voting theory improves the overall classification performance;
3. ES-ELM applies a *lazy strategy* of classifier update, that is, whether the classifiers should be updated is determined by some evaluation criteria of classification performance.

Figure 2 presents the flow chart of ES-ELM. We assume that ES-ELM has already trained m classifiers, namely cl_1, cl_2, ..., cl_m, in the initiate phase. With the current sliding window X_i, ES-ELM tests all the elements with each of the m ensemble classifiers. For each classifier, let's say cl_i, we calculate its error rate. If the error rate is larger than the threshold ε, classifier cl_i will be eliminated. ES-ELM trains a new classifier using all the elements in the current sliding window X_i to replace cl_i. If all the error rate of ensemble classifiers are smaller than ε, we keep all these m classifiers without update. And the sliding window continues to slide as the stream keep flowing.

ES-ELM utilizes *error rate* as the evaluation criterion of classifiers elimination. The formal definition of error rate is given as follows.

Definition 2 (*Error rate*) For an stream element x_i and a classifier cl_j, if x_i is correctly classified by cl_j, we set $flag_{c_j}^{x_i} = 0$, otherwise, $flag_{c_j}^{x_i} = 1$. Thus, for s elements in the current sliding window, the error rate of classifier cl_j is

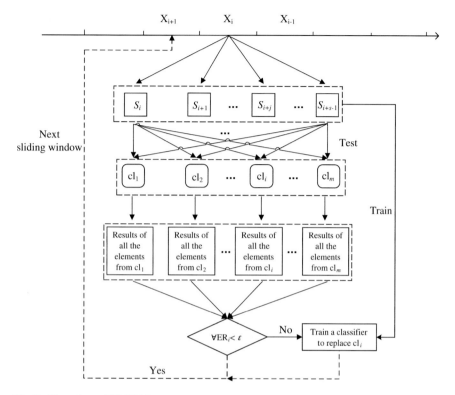

Fig. 2 Flow chart of ES-ELM

$$ER_j = \frac{\sum_{i=1}^{s} flag_{c_j}^{x_i}}{s} \tag{7}$$

The detailed algorithm is described as Algorithm 2.

Given a feeds stream S, the size s of sliding window X, the number of classifiers m and the threshold ε of error rate, ES-ELM initiates m classifiers using the elements in the initiate sliding window (Lines 1, 2). After the first slide (Line 3), for each of the m ensemble classifiers (Lines 6–11), ES-ELM calculates the outputs of each element X_j in the current sliding window (Lines 7, 8). Then the error rate ER_i or each classifier cl_i is calculated (Line 8). If the error rate ER_i of classifier cl_i is larger than threshold ε, ES-ELM eliminates cl_i and trains a new classifier with all the elements in the current sliding window X (Line 11). With all the outputs of the s elements from the m classifiers, ES-ELM utilizes voting mechanism to identify the event topics with max votes to each of the element (Line 12). The procedure will be executed iteratively as the feeds stream flows.

Algorithm 2: ES-ELM

Input: Feeds stream S, sliding window X with size s, number of classifiers m, error rate
        threshold ε
Output: Classification results

1 $X = \{S_1, \ldots, S_s\}$;
2 Train m initiate classifiers using X: $\beta = \mathbf{H}^\dagger \mathbf{T}$;
3 X slides with a step of s elements;
4 **while** *!S.end()* **do**
5       $X = \{S_i \mid S_i$ within the current sliding window$\}$;
6   **for** $i = 1$ *to* m **do**
7   **for** $j=1$ *to* s **do**
8   Calculate output $\mathbf{O}_j^i$ of element X_j using cl_i;
9   Calculate error rate ER_i of classifier cl_i;
10   **if** $ER_i > \varepsilon$ **then**
11   Update cl_i with a new classifier trained with all the elements in X;
12   **for** $i = 1$ *to* s **do**
13   Identify the event topic with the max votes of $\{\mathbf{O}_i^1, \ldots, \mathbf{O}_i^m\}$ to X_i;
14   The sliding windows continues to slide;

5.3 Online Sequential Strategy

Both BS-ELM and ES-ELM strictly follows the definitions and principles of stream processing problem, that is, the out-of-date stream elements expire and do not contribute to the learning of the current classifier. Online Sequential Extreme Learning Machine (OS-ELM) [23] realizes incremental learning, which we believe suites the learning over data stream better. The learning of the stream elements within the current sliding window can be incrementally combined with historical learning experience. In this section, we propose Online Sequential Stream Extreme Learning Machine (OSS-ELM) based on OS-ELM.

In OS-ELM, given a chunk of samples $\mathbf{H}_i$, which will be viewed as the set of the elements in the current sliding window in OSS-ELM, the latest output weight β_{k+1} is calculated by a transition matrix $\mathbf{P}_{k+1}$ as

$$\beta_{k+1} = \beta_k + \mathbf{P}_{k+1}\mathbf{H}_{k+1}^{\mathrm{T}}(\mathbf{T}_{k+1} - \mathbf{H}_{k+1}\beta_k) \tag{8}$$

where

$$\mathbf{P}_{k+1} = \mathbf{P}_k - \mathbf{P}_k\mathbf{H}_{k+1}^{\mathrm{T}}(\mathbf{I} + \mathbf{H}_{k+1}\mathbf{P}_k\mathbf{H}_{k+1}^{\mathrm{T}})^{-1}\mathbf{H}_{k+1}\mathbf{P}_k \tag{9}$$

Figure 3 shows the flow chart of OSS-ELM. After the initiate phase, OSS-ELM first uses the stream elements in the current sliding window X_i to test the former classifier cl_{i-1} and identify event topics to the elements. Then OSS-ELM utilizes incremental calculation to train an up-to-date classifier.

Fig. 3 Flow chart of
OSS-ELM

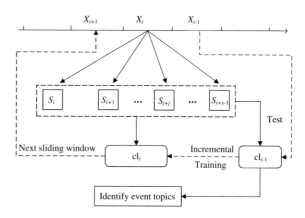

Algorithm 3: OSS-ELM

Input: Feeds stream S, sliding window X with size s, number of classifiers m, error rate
threshold ε

Output: Classification results

1 $X = \{S_1, \ldots, S_s\}$;
2 Train an initiate classifier using X;
3 X slides with a step of s elements;
4 **while** *!S.end()* **do**
5       $X = \{S_i \mid S_i$ within the current sliding window$\}$;
6   **for** $i = 1$ *to* s **do**
7   Calculate output $\mathbf{O}^i$ of element X_i;
8   Identify an event topic to X_i according to $\mathbf{O}^i$;
9   Calculate matrix $\mathbf{H}_{k+1}$ and $\mathbf{P}_{k+1}$;
10   Calculate output weight β_{k+1};
11   The sliding windows continues to slide;

Algorithm 3 describes the detailed procedure of OSS-ELM. After initialize the
sliding window (Line 1) and the classifier (Line 2), the sliding window slides with a
step of the size of the sliding window (Line 3). When processing the rest of the feeds
stream, each time given the updated stream elements set within the current sliding
window (Line 5), OSS-ELM test all the elements in X and identify event topics to
them (Lines 6–8). Then OSS-ELM calculates matrix $\mathbf{H}_{k+1}$ and $\mathbf{P}_{k+1}$ (Line 9), and
then the up-to-date output weight β_{k+1} (Line 10) based on OS-ELM. This procedure
also iterates while the feeds stream continues to flow.

6 Performance Evaluation

6.1 Experiments Setup

All the experiments are conducted on a machine with Intel Core i5 3.50 GHz CPU and 8 GB RAM. We realize our algorithms by MATLAB R2013b on 64-bit Windows 7. Two datasets of XML streams are used in our experiments, namely IBM DeveloperWorks[1] articles and ABC News.[2] We use the feeds channels as known event topics. For each dataset, we choose 6 event topics and 6000 feeds entries under each event topic.

Three parameters have to be set manually in our experiments, which are the number of hidden layer nodes, the size of sliding window, and the number of ensemble classifiers. After a set of preparation experiments, we set the number of hidden layer nodes to 110.

6.2 Evaluation Results

6.2.1 Training Time Comparison with Varied Sliding Window Sizes

Event identification problem requires real-time processing of the fast and large-scale data stream. Thus, we first compare the training time among BS-ELM, ES-ELM, OSS-ELM and SVM. The size of sliding window varies from 20 to 120.

Figure 4 charts the trend of training time with the increasing size of sliding window. It can be seen that when the size of sliding window is relatively small, all the four algorithms have high frequencies of updates. The baseline algorithm has the fast learning speed due to its relatively simple calculation. When the size of sliding window is larger than 60, ES-ELM and OSS-ELM have less training time. As to ES-ELM, the lazy update strategy leads to less times of classifier retraining, and the ensemble strategy improves the overall accuracy of the classifier component. As to OSS-ELM, larger size of sliding window leads to less incremental calculation.

6.2.2 Classificaton Performance Comparison with Varied Sliding Window Sizes

The other important evaluation criterion is classification performance. Figure 5 present the classification performance comparison among BS-ELM, ES-ELM, OSS-ELM and SVM on different datasets.

---

[1]http://www.ibm.com/developerworks.

[2]http://abcnews.go.com.

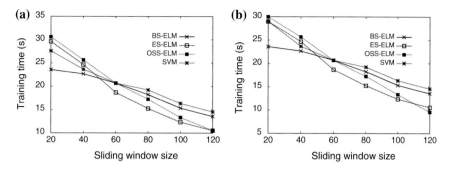

Fig. 4 Training time comparison. **a** Dataset IBM. **b** Dataset ABC News

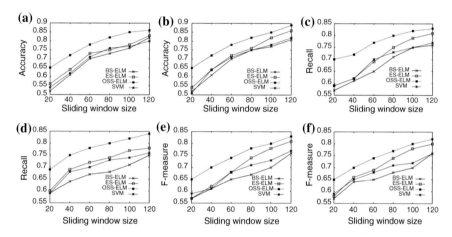

Fig. 5 Classification performance comparison. **a** Accuracy on IBM. **b** Accuracy on ABC. **c** Recall on IBM. **d** Recall on ABC. **e** F1 on IBM. **f** F1 on ABC

From this set of experimental results, we find that BS-ELM has the lowest performance, since BS-ELM utilizes the naïve strategy to learn classifiers with only the stream elements in the sliding window. SVM based methods has the similar performance to BS-ELM. ES-ELM has a higher performance due to its ensemble strategy. OSS-ELM has the highest performance, since it takes all the historical learning experience into consideration without expiring any former elements. On the other hands, all these four algorithms gain higher performance when the size of sliding window increases.

6.2.3 Influence of Ensemble Number on ES-ELM

As to ES-ELM, a parameter of ensemble number has to be set manually. Ensemble number decides the number of classifiers being maintained by ES-ELM. Figure 6 shows the influence of ensemble number on the classification performance of ES-ELM.

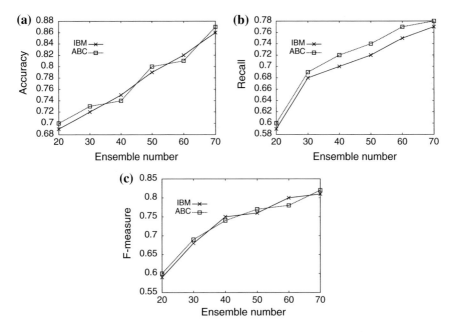

Fig. 6 Classification performance comparison with varied ensemble numbers. **a** Accuracy. **b** Recall. **c** F-measure

Figure 6a shows the trend of accuracy as the ensemble number increases. Figure 6b shows the trend of recall and Fig. 6c shows the trend of F-measure. This set of experiments demonstrate that though training more classifiers cost more learning time, more ensemble classifiers lead to better classification performance.

7 Conclusion

Record linkage task in RSS feeds is treated as classification problem over XML stream in this paper. We propose ELM based algorithms with ensemble strategy and online sequential strategy respectively to realize efficient classification over fast and large-scale XML stream. Experimental results indicate that our algorithms gains effectiveness and efficiency for XML stream problems, which provides practical guidance to event identification applications.

Acknowledgments This research is partially supported by the National Natural Science Foundation of China under Grant Nos. 61272181, 61173029, and 61173030; the National Basic Research Program of China under Grant No. 2011CB302200-G; the 863 Program under Grant No. 2012AA011004; and the Fundamental Research Funds for the Central Universities under Grant No. N120404006.

References

1. Becker, H., Naaman, M., Gravano, L.: Learning similarity metrics for event identification in social media. In: Proceedings of the Third ACM International Conference on Web Search and Data Mining, WSDM '10, pp. 291–300. ACM, New York, NY, USA (2010)
2. Psallidas, F., Becker, H., Naaman, M., Gravano, L.: Effective event identification in social media. IEEE Data Eng. Bull. **36**(3), 42–50 (2013)
3. Reuter, T., Cimiano, P., Drumond, L., Buza, K., Schmidt-Thieme, L.: Scalable event-based clustering of social media via record linkage techniques. In: Proceedings of the Fifth International Conference on Weblogs and Social Media, Barcelona, Catalonia, Spain, 17–21 July 2011
4. Huang, G.-B., Zhu, Q.-Y., Siew, C.-K.: Extreme learning machine: a new learning scheme of feedforward neural networks. In: International Symposium on Neural Networks, vol. 2 (2004)
5. Huang, G.-B., Zhu, Q.-Y., Siew, C.-K.: Extreme learning machine: theory and applications. Neurocomputing **70**, 489–501 (2006)
6. Zong, W., Huang, G.-B.: Face recognition based on extreme learning machine. Neurocomputing **74**, 2541–2551 (2011)
7. Zhao, X., Wang, G., Bi, X., Gong, P., Zhao, Y.: XML document classification based on ELM. Neurocomputing **74**, 2444–2451 (2011)
8. Wang, B., Wang, G., Li, J., Wang, B.: Update strategy based on region classification using elm for mobile object index. Soft Comput. **16**(9), 1607–1615 (2012)
9. Wang, G., Zhao, Y., Wang, D.: A protein secondary structure prediction framework based on the extreme learning machine. Neurocomputing **72**, 262–268 (2008)
10. Allan, J., Papka, R., Lavrenko, V.: On-line new event detection and tracking. In: Proceedings of the 21st Annual International ACM SIGIR Conference on Research and Development in Information Retrieval, pp. 37–45. ACM (1998)
11. Kumaran, G., Allan, J.: Text classification and named entities for new event detection. In: Proceedings of the 27th Annual International ACM SIGIR Conference on Research and Development in Information Retrieval, pp. 297–304 (2004)
12. Dey, L., Mahajan, A., Haque, S.K.M.: Document clustering for event identification and trend analysis in market news. In: Proceedings of the Seventh International Conference on Advances in Pattern Recognition, ICAPR 2009, pp. 103–106. IEEE Computer Society, Kolkata, India, 4–6 Feb 2009
13. Yin, J.: Clustering microtext streams for event identification. In: Sixth International Joint Conference on Natural Language Processing, pp. 719–725 (2013)
14. Becker, H., Naaman, M., Gravano, L.: Event identification in social media. In: 12th International Workshop on the Web and Databases, WebDB 2009, Providence, Rhode Island, USA, 28 June 2009
15. Weiler, Scholl, M.H., Wanner, F., Rohrdantz, C.: Event identification for local areas using social media streaming data. In: Proceedings of the ACM SIGMOD Workshop on Databases and Social Networks, pp. 1–6. ACM (2013)
16. Vavliakis, K.N., Symeonidis, A.L., Mitkas, P.A.: Event identification in web social media through named entity recognition and topic modeling. Data Knowl. Eng. **88**, 1–24 (2013)
17. Weiler, A., Grossniklaus, M., Scholl, M.H.: Event identification and tracking in social media streaming data. In: Proceedings of the Workshops of the EDBT/ICDT 2014 Joint Conference, Athens, Greece, 28 Mar 2014, pp. 282–287 (2014)
18. Trabelsi, C., Yahia, S.: A probabilistic approach for events identification from social media rss feeds. In: Database Systems for Advanced Applications Hong, B., Meng, X., Chen, L., Winiwarter, W., Song, W. (eds.), of Lecture Notes in Computer Science, vol. 7827, pp. 139–152. Springer, Berlin Heidelberg (2013)
19. Cao, K., Wang, G., Han, D., Ning, J., Zhang, X.: Classification of uncertain data streams based on extreme learning machine. Cogn. Comput. **7**(1), 150–160 (2015)
20. Zhao, X., Bi, X., Qiao, B.: Probability based voting extreme learning machine for multiclass xml documents classification. World Wide Web 1–15 (2013)

21. Yang, J., Chen, X.: A semi-structured document model for text mining. J. Comput. Sci. Technol. (2002)
22. Huang, G., Song, S., Gupta, J., Wu, C.: Semi-supervised and unsupervised extreme learning machines. IEEE Trans. Cybern. **99**, 1–1 (2014)
23. Liang, N.-Y., Huang, G.-B., Saratchandran, P., Sundararajan, N.: A fast and accurate online sequential learning algorithm for feedforward networks. IEEE Trans. Neural Netw. **17**, 1411–1423 (2006)

Timeliness Online Regularized Extreme Learning Machine

Xiong Luo, Xiaona Yang, Changwei Jiang and Xiaojuan Ban

Abstract A novel online sequential extreme learning machine (ELM) algorithm with regularization mechanism in a unified framework is proposed in this paper. This algorithm is called timeliness online regularized extreme learning machine (TORELM). Like the timeliness managing extreme learning machine (TMELM) which incorporates timeliness management scheme into ELM approach for the incremental samples, TORELM also processes data one-by-one or chunk-by-chunk under the similar framework, while the newly incremental data could be prior to the historical data by maximizing the contribution of the newly increasing training data. Furthermore, in consideration of the disproportion between empirical risk and structural risk in some traditional learning methods, we add regularization technique to the timeliness scheme of TORELM through the use of a weight factor to balance them to achieve better generalization performance. Therefore, TORELM may has its unique feature of higher generalization capability with a small testing error while implementing online sequential learning. And the simulation results show that TORELM performs better than other ELM algorithms.

Keywords Incremental learning · Extreme learning machine · Timeliness managing extreme learning machine · Timeliness online regularized extreme learning machine

X. Luo (✉) · X. Yang · C. Jiang · X. Ban
School of Computer and Communication Engineering, University of Science and Technology Beijing, Beijing 100083, China
e-mail: xluo@ustb.edu.cn

X. Luo · X. Yang · C. Jiang · X. Ban
Beijing Key Laboratory of Knowledge Engineering for Materials Science,
Beijing 100083, China

© Springer International Publishing Switzerland 2016
J. Cao et al. (eds.), *Proceedings of ELM-2015 Volume 1*,
Proceedings in Adaptation, Learning and Optimization 6,
DOI 10.1007/978-3-319-28397-5_37

1 Introduction

Neural networks (NNs) have been widely used in machine learning due to their ability of solving those problems that classical techniques are not able to deal with. Among the available NN-based machine learning algorithms, extreme learning machine (ELM) for single-hidden layer feedforward network (SLFN) has attracted much attention because of its quickness and simplicity [1–5]. Usually, the traditional ELM algorithm would make full use of training data to establish an ELM model. Therefore, it may be difficult to obtain the whole samples only once time in the real situations. Then, the newly incremental data and historical data need to be separated. To avoid such limitation, an online sequential ELM (OSELM) was proposed to learn data one-by-one or chunk-by-chunk with fixed or varying chunk size [6]. And it will discard those data for which the training has been already done. And only new arrived single or chunk of samples are handled and learned.

Furthermore, some practice experiences show that the newer data has more effective and obvious relevance than it of the older data. Thus, a timeliness should be taken into consideration for the newly incremental data in some actual applications. And those newly incremental data play a critical role in implementing the learning task. With this problem considered, a timeliness managing ELM (TMELM) was proposed through the full use of the incremental data [7]. Through the use of the adaptive timeliness weight and iteration scheme in TMELM, the incremental data can contribute reasonable weight to represent the current situation to ensure the stability of the model.

Although ELM is extremely fast in speed and has good generalization ability, its solution has some drawbacks. First, ELM based on empirical risk minimization (ERM) principle [8] is very likely to result an over-fitting model. Also, it provides weak control capacity since it directly calculates the minimum norm least-squares solution. Finally, it does not consider heteroskedasticity in real applications. In order to address these issues, regularization based on structural risk minimization (SRM) is taken into consideration. According to statistical learning theory [9], the real prediction risk of learning is consisted of empirical risk and structural risk. A model with good generalization ability should have the best tradeoff between the two risks. So, a weight factor for empirical risk is introduced to regular the proportion of the SRM and ERM to achieve better generalization performance [10]. Motivated by it, a novel algorithm named timeless online regularized ELM (TORELM) is proposed in this paper. This approach can overcome singular and ill-posed problems by using regularization mechanism while implementing online sequential learning. And, the generalization performance of this algorithm is improved. Simulation results show that it is faster to achieve the minimize error than the traditional algorithms.

The rest part of this paper is organized as follows: Sect. 2 briefly describes the related works related to ELM and OSELM, Sect. 3 proposes an implementation of TORELM for the prediction, Sect. 4 analyzes the performance of proposed algorithm via simulation, and Sect. 5 provides a conclude for this paper.

2 ELM and OSELM

For N arbitrary distinct training samples $\{(\mathbf{x}_i, \mathbf{t}_i)_{i=1}^N\}$, where $\mathbf{x}_i = [x_{i1}, x_{i2}, \ldots, x_{in}]^T \in \mathbb{R}^n$ and $\mathbf{t}_i = [t_{i1}, t_{i2}, \ldots, t_{im}]^T \in \mathbb{R}^m$, the corresponding output function of ELM with L hidden neurons and activation function $g(\cdot)$ are mathematically modeled as (1) and $\mathbf{w}_j = [w_{j1}, w_{j2}, \ldots, w_{jn}]^T$ is the weight vector connecting the jth hidden neuron, the input neurons $\beta_j = [\beta_{j1}, \beta_{j2}, \ldots, \beta_{jm}]^T$ is the weight vector connecting the jth hidden neuron and the output neurons, and b_j is the threshold of the jth hidden neuron. In addition, $\mathbf{w}_j \cdot \mathbf{x}_i$ denotes the inner product of $\mathbf{w}_j$ and $\mathbf{x}_i$ [11].

$$\mathbf{t}_i = \sum_{j=1}^L \beta_j g(\mathbf{w}_j \cdot \mathbf{x}_i + b_j), \tag{1}$$

The above N equations can be written compactly as:

$$\mathbf{H}\beta = \mathbf{T}, \tag{2}$$

Here, $\mathbf{H}$ is called the hidden layer output matrix of the SLFN. And the ith column of $\mathbf{H}$ is the ith hidden node output with respect to the inputs $\mathbf{x}_1, \mathbf{x}_2, \ldots \mathbf{x}_N$. The ith row of $\mathbf{H}$ is the hidden layer feature mapping with respect to the ith input $\mathbf{x}_i$.

The orthogonal projection method can be efficiently used in ELM: $\mathbf{H}^\dagger = (\mathbf{H}^T\mathbf{H})^{-1}\mathbf{H}^T$ if $\mathbf{H}^T\mathbf{H}$ is nonsingular, where $\mathbf{H}^\dagger$ is the Moore-Penrose generalized inverse of $\mathbf{H}$. Therefore, the solution of β is [12]:

$$\beta = \mathbf{H}^\dagger\mathbf{T} = (\mathbf{H}^T\mathbf{H})^{-1}\mathbf{H}^T\mathbf{T}. \tag{3}$$

Furthermore, OSELM as an incremental learning algorithm does not need to handle all the samples [6]. The effect of incremental data is influenced by the correction $\Delta\beta$, which modifies the historical model β_0 to form a new model β^* based on the following equation:

$$\beta^* = \beta_0 + \Delta\beta(X^*). \tag{4}$$

In [13], a solution to this model was provided. The detailed steps are as follows. Given a chunk of initial training set of safety data $\aleph_0 = \{(\mathbf{x}_i, \mathbf{t}_i)_{i=1}^{N_0}\}$ $(N_0 \geqslant L)$, under the ELM scheme, we can find that

$$\beta_0 = \mathbf{K}_0^{-1}\mathbf{H}_0^T\mathbf{T}_0, \tag{5}$$

where $\mathbf{K}_0 = \mathbf{H}_0^T\mathbf{H}_0$. Then, suppose that we are given another chunk of data set $\aleph_1 = \{(\mathbf{x}_i, \mathbf{t}_i)_{i=N_0+1}^{N_0+N_1}\}$, where N_1 denotes the number of new samples in this data set. Considering both training data sets $\aleph_0$ and $\aleph_1$, the output weight β_1 becomes [6]:

$$\beta_1 = \mathbf{K}_1^{-1}\begin{bmatrix}\mathbf{H}_0\\\mathbf{H}_1\end{bmatrix}^{\mathrm{T}}\begin{bmatrix}\mathbf{T}_0\\\mathbf{T}_1\end{bmatrix} = \beta_0 + \mathbf{K}_1^{-1}\mathbf{H}_1^{\mathrm{T}}(\mathbf{T}_1 - \mathbf{H}_1\beta_0), \tag{6}$$

where

$$\mathbf{K}_1 = \begin{bmatrix}\mathbf{H}_0\\\mathbf{H}_1\end{bmatrix}^{\mathrm{T}}\begin{bmatrix}\mathbf{H}_0\\\mathbf{H}_1\end{bmatrix} = \mathbf{K}_0 + \mathbf{H}_1^{\mathrm{T}}\mathbf{H}_1.$$

As can be seen from (6), the calculation of β_1 is based on β_0, which will improve the computational efforts. Then, $(\mathbf{T}_1 - \mathbf{H}_1\beta_0)$ is obtained by using the old model parameters β_0, where $\mathbf{H}_1\beta_0$ is considered the error of predicting the newly added data. With the increase of incremental number, after $(k+1)$ times incremental learning, the model parameter can be written as follows:

$$\beta_{k+1} = \beta_k + \mathbf{K}_{k+1}^{-1}\mathbf{H}_{k+1}^{\mathrm{T}}(\mathbf{T}_{k+1} - \mathbf{H}_{k+1}\beta_k). \tag{7}$$

3   Timeless Online Regularized Extreme Learning Machine (TORELM)

Different from OSELM, under the adaptive timeliness weight and iteration scheme in TMELM, the incremental data can contribute reasonable weights to represent the current situation [7]. The current collection of data has higher contribution to the system model. Thus, a penalization weight w is designed to adjust the contribution of data. Then, (7) can be rewritten as [7]:

$$\beta_{k+1} = \beta_k + w \cdot \mathbf{K}_{k+1}^{-1}\mathbf{H}_{k+1}^{\mathrm{T}}(\mathbf{T}_{k+1} - \mathbf{H}_{k+1}\beta_k). \tag{8}$$

Here, the penalization weight w reflects the timeliness effect of newly incremental data. Moreover, w can be expressed as follows [14]:

$$w = 1 + 2 \cdot \exp(-|\mathrm{mean}(a_1) - \mathrm{mean}(a_2)|^{|\mathrm{var}(a_1) - \mathrm{var}(a_2)|}), \tag{9}$$

where a_1 is the newly incremental data, a_2 is the history adjacent incremental data, mean is the function used to obtain the mean value and var is the function used to obtain the variance value.

In our TORELM, the model parameter is updated with (8).

3.1 Regularization

ELM is still able to be considered as empirical risk minimization theme and tends to generate over-fitting model. Additionally, since ELM doesnt consider heteroskedasticity in real applications, its performance will be affected seriously when outliers exist. Thus, regularization is being taken into consideration for the calculation of error [10].

Generally, we may wish to obtain specific $\mathbf{w}_i, \beta_i, b_i (i = 1, 2, \ldots, N)$ to meet empirical risk minimization and structural risk minimization. So, (1) should be written as

$$
\begin{aligned}
&\min \|\varepsilon\|^2 \\
&\text{s.t. } \sum_{i=1}^{L} \beta_i g(\mathbf{w}_i \cdot \mathbf{x}_j + b_i) - \mathbf{t}_j = \varepsilon_j, j = 1, 2, \ldots, N
\end{aligned}
\tag{10}
$$

where $\varepsilon_j = [\varepsilon_{j1}, \varepsilon_{j2}, \ldots, \varepsilon_{jm}]$ is the resident between target value and real value of the jth sample, and $\varepsilon = [\varepsilon_1, \varepsilon_2, \ldots, \varepsilon_N]$.

According to statistical learning theory, the real prediction risk of learning is consisted of empirical risk and structural risk. And a weight factor γ is introduced to be regularized for empirical risk. The empirical risk is represented by $\|\varepsilon\|^2$ and structural risk is represented by $\|\tau\|^2$ which is obtained by maximizing the distance of margin separating classes [15]. Moreover, with the purpose of obtaining a robust estimate weakening outlier interference, the error ε_j is weighted by variable v_i. Thus, $\|\varepsilon\|^2$ is going to be extended to $\|\mathbf{D}\varepsilon\|^2$, where $\mathbf{D} = \text{diag}(v_1, v_2, \ldots, v_N)$. Therefore, the proposed regularized mathematic model can be written as follows [10]:

$$
\begin{aligned}
&\min \frac{1}{2}\|\tau\|^2 + \frac{1}{2}\gamma\|\mathbf{D}\varepsilon\|^2 \\
&\text{s.t. } \sum_{i=1}^{L} \beta_i g(\mathbf{w}_i \cdot \mathbf{x}_j + b_i) - \mathbf{t}_j = \varepsilon_j, j = 1, 2, \ldots, N
\end{aligned}
\tag{11}
$$

We can adjust the proportion of empirical risk and structural risk by means of changing γ. The Lagrangian function for (11) can be described as:

$$
\begin{aligned}
L(\tau, \varepsilon, \alpha) &= \frac{\gamma}{2}\|\mathbf{D}\varepsilon\|^2 + \frac{1}{2}\|\tau\|^2 - \sum_{j=1}^{N} \alpha_j (\sum_{i=1}^{L} \beta_i g(\mathbf{w}_i \cdot \mathbf{x}_j + b_i) - \mathbf{t}_j - \varepsilon_j) \\
&= \frac{\gamma}{2}\|\mathbf{D}\varepsilon\|^2 + \frac{1}{2}\|\tau\|^2 - \alpha(\mathbf{H}\beta - \mathbf{T} - \varepsilon),
\end{aligned}
\tag{12}
$$

where $\alpha_j \in \mathbb{R}$ $(j = 1, 2, \ldots, N)$ is the Lagrangian multiplier with the equality constraint of (11) in (12).

Then, the method of solving this equation is described as:

$$
\begin{aligned}
\frac{\partial L}{\partial \beta} &\rightarrow \beta^{\mathrm{T}} = \alpha \mathbf{H}, \\
\frac{\partial L}{\partial \varepsilon} &\rightarrow \gamma \varepsilon^{\mathrm{T}} \mathbf{D}^2 + \alpha = 0, \\
\frac{\partial L}{\partial \alpha} &\rightarrow \mathbf{H}\beta - \mathbf{T} - \varepsilon = 0.
\end{aligned}
\tag{13}
$$

By solving (13), we will obtain the solution of β:

$$\beta = \left(\frac{\mathbf{I}}{\gamma} + \mathbf{H}^T \mathbf{D}^2 \mathbf{H}\right)^{\dagger} \mathbf{H}^T \mathbf{D}^2 \mathbf{T}, \tag{14}$$

where $\mathbf{I}$ is an unit matrix.

There are many kinds of method to compute the weights v_j [16], e.g.

$$v_j = \begin{cases} 1 & |\varepsilon_j/\hat{s}| \leq c_1 \\ \frac{c_2 - |\varepsilon_j/\hat{s}|}{c_2 - c_1} & c_1 \leq |\varepsilon_j/\hat{s}| \leq c_2 \\ 10^{-4} & \text{otherwise} \end{cases} \tag{15}$$

where $\hat{s}$ is robust estimate of the standard deviation of the unweighted regularized ELM error variables ε_j, and

$$\hat{s} = \frac{\text{IQR}}{2 \times 0.6745}. \tag{16}$$

The inter quartile range (IQR) is the difference between the 75th percentile and the 25th percentile. During the process of estimating $\hat{s}$, one takes into account how much the estimates error distribution deviates from a Gaussian distribution. In addition, the constants c_1 and c_2 are set as 2.5 and 3, respectively [17].

3.2 Algorithm Framework

The framework of TORELM is summarized as follows:

(1) Assign the model parameters by N_0 initial samples, such as the number of hidden neurons L, activation function $g(\cdot)$.
(2) Determine arbitrary input weight $\mathbf{w}_i$ and bias b_i, $i = 1, \ldots, L$.
(3) Calculate the initial hidden layer output matrix $\mathbf{H}_0$.
(4) Calculate the α and ε_i respectively based on $\alpha = -\gamma(\mathbf{H}_0\beta_0 - \mathbf{T}_0)^T$ and $\varepsilon_i = \frac{\alpha_i}{\gamma}(i = 1, 2, \ldots, N_0)$. Also, $\hat{s}$ and weights of v_j are supposed to be computed.
(5) Update the initial output weight $\beta_0 = (\frac{\mathbf{I}}{\gamma} + \mathbf{H}_0^T \mathbf{D}^2 \mathbf{H}_0)^{\dagger} \mathbf{H}_0^T \mathbf{D}^2 \mathbf{T}_0$.
(6) Add a group of samples $\aleph_1$ and calculate the penalty weight w using (9).
(7) Calculate $\mathbf{H}_1$ and β_1 using $\aleph_1$ and (6). Here, k is set to 1.
(8) Calculate β_k using (8) and $j = 0$.
(9) If $|\beta_{k(j+1)} - \beta_{k(j)}| < \sigma$, the iteration stops, else
$\beta_{k(j+1)} = \beta_{k(j)} + w \cdot \mathbf{K}_{k+1}^{-1} \mathbf{H}_{k+1}^T (\mathbf{T}_{k+1} - \mathbf{H}_{k+1}\beta_{k(j)})$. Then, $j = j + 1$.
(10) Obtain $\beta_{k(j+1)}$.
(11) Learn new increment data with $k = k + 1$ and skip to step 8.

4 Simulation Results and Discussions

4.1 Simulation Description

The idea behind of TORELM is to strengthen the recent data and weaken the older data with the goal of SRM and ERM. In order to test the effectiveness of this algorithm, we evaluate the performance of TORELM through the successful prediction rate. Meanwhile, the performance of TORELM is also evaluated by comparing different traditional schemes, including OSELM, WOSELM [14], and TMELM. The difference between WOSELM and TMELM is the selection of penalization weight w. The adaptive weight is calculated in TMELM instead of fixed weight used in WOSELM.

In addition, the performance of accuracy is evaluated by prediction error. And in this paper, we use the root mean square error (RMSE) to measure the prediction error between the predicted value and the actual value.

4.2 Results and Discussions

The initial data are randomly generated the same as [14]. Let X be the input data set. It is randomly generated in the range $[-10, 10]$ and the output value Y meets the 'SinC' relationship with X according to (17).

$$X = 20 \cdot \text{rand}(1, Q) - 10, \quad Y = \frac{\sin(X)}{X}, \tag{17}$$

where Q is the number of data.

To make the whole data set more like a real world problem. We add noise in the range $[-0.4, 0.4]$, that is:

$$Y = Y + 0.8 \cdot \text{rand}(1, Q) - 0.4. \tag{18}$$

Taking timeliness into consideration, the distribution of incremental data does not ought to be the same as that of initial data. Therefore, the incremental data X is obtained using (19). And the output Y is adjusted with a scalar and bias as follows:

$$X = 20 \cdot \text{rand}(1, Q) + 10, \quad Y = A \cdot \frac{\sin(X)}{X} + B, \tag{19}$$

where A and B are set 1.5 and 2, receptively.

The testing data are generated the same as incremental data. And, the initial number of data is 1000, the number of hidden neurons is 50, the activate function is radial basis function ('rbf'), the number of increment is 10, and the number of incremental

Fig. 1 Testing errors with
different weights

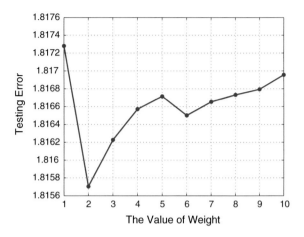

Fig. 2 Testing errors with
different $\log_2 \gamma$

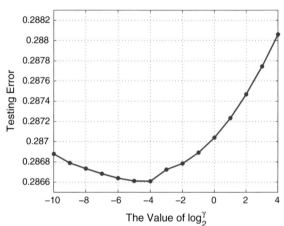

data is 500 once. In our new algorithm, we first study the relationship between different weight and RMSE, which is showed in Fig. 1. If the testing data is not closer to the incremental data, it will produce a large testing error. So, we should take the testing error into consideration. In Fig. 1, the increasing value of weight within [1,10], the testing error first comes down, then goes up. And the weight of w is set as 2 which is better than others.

In our algorithm, a weight factor γ is being trained to balance the empirical risk and structural risk. In order to obtain the relationship between γ and the testing error, we set different value of γ to test. From Fig. 2, we can see that the testing error is relatively small and it reaches the bottom level after a declination. When the $\log_2 \gamma$ is set to -4, the performance is best.

The performance of TORELM and other ELM-based learning algorithms is compared. Figure 3 shows the testing error of different algorithms. During the

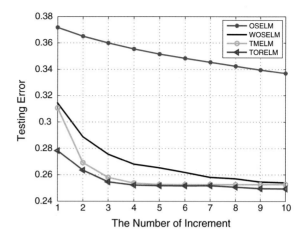

Fig. 3 Performance comparison in testing error

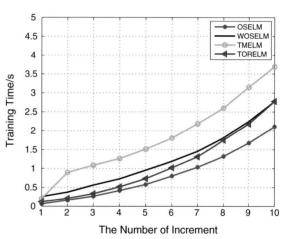

Fig. 4 Performance comparison in training time

computation, with the increase of increment number, the training data will be larger that covers more samples. Thus, the testing error for all algorithms comes down. From Fig. 3, we can see that with the increase of increment number, the testing error shows a decreasing trend. However, the testing error of OSELM is the highest because of its absence of weight scheme. Compared with OSELM, WOSELM using a fixed weight has a smaller testing error on the data. But, a fixed weight is not reasonable for the varied training data. Then, TMELM is being trained to obtain a good testing accuracy and it does achieve a fine result. Furthermore, considering the disproportion between empirical risk and structural risk, regularization is used to develop TORELM based on TMELM. Unsurprisingly, TORELM quickly achieves small testing error and stability due to its anti-noise function and weight scheme. Moreover, TORELM is more stable than other algorithms due to its regulation. Thus, the generalization performance of the proposed algorithm is improved.

Figure 4 shows the training time consumption. Even though TORELM algorithm employs iteration and weight schemes, it just costs a litter more time than OSELM. On the whole, TORELM is more outstanding than others. It achieves good performance both in testing error and time consumption.

5 Conclusion

Motivated by TMELM, a novel online sequential learning algorithm TORELM with regularization mechanism is proposed. This algorithm can implement processing task as OSELM by learning data one-by-one or chunk-by-chunk. After employing the timeliness management and adaptive weight techniques to train the samples under TMELM scheme, our proposed algorithm considers empirical risk and structural risk with a good tradeoff between them. A weight factor for empirical risk is introduced to implement regularization for the proportion of structural risk and empirical risk to achieve better generalization performance and strengthen the control ability. Then this optimal model with the minimum prediction error could be achieved by selecting appropriate weight factor. In addition, the proposed scheme holds strong anti-noise ability when outliers exist in the dataset. Therefore, we can see that the performance of TORELM is better than that of other ELM-based algorithms through the analysis of simulation results. It spends less time to achieve the minimum error when the data has noise samples. In doing so, it may be more feasible that the incremental data can contribute reasonable weight and appropriate proportion of structural risk and empirical risk to obtain an optimal model.

Acknowledgments This work was jointly supported by the National Natural Science Foundation of China under Grants 61174103, 61272357, and 61300074, the National Key Technologies R&D Program of China under Grant 2015BAK38B01, and the Aerospace Science Foundation of China under Grant 2014ZA74001.

References

1. Huang, G., Huang, G.B., Song, S., You, K.: Trends in extreme learning machines. A Rev. Neural Netw. **61**, 32–48 (2015)
2. Luo, X., Chang, X., Liu, H.: A taylor based localization algorithm for wireless sensor network using extreme learning machine. IEICE Trans. Inf. Syst. **E97D**, 2652–2659 (2014)
3. Mirza, B., Lin, Z., Liu, N.: Ensemble of subset online sequential extreme learning machine for class imbalance and concept drift. Neurocomputing **149**, 316–329 (2015)
4. Cambria, E., Gastaldo, P., Bisio, F., Zunino, R.: An ELM-Based model for affective analogical reasoning. Neurocomputing **149**, 443–455 (2015)
5. Huang, G.B., Bai, Z., Kasun, L., Vong, C.: Local receptive fields based extreme learning machine. IEEE Comput. Intell. Mag. **10**, 18–29 (2015)
6. Huang, G.B., Liang, N.Y., Rong, H.J., Saratchandran, P., Sundararajan, N.: On-Line sequential extreme learning machine. In: IASTED International Conference on Computational Intelligence, pp. 232–237. Acta Press, Calgary, Canada (2005)

7. Liu, J.F., Gu, Y., Chen, Y.Q., Cao, Y.S.: Incremental localization in WLAN environment with timeliness management. Chinese J. Comput. **36**, 1448–1455 (2013)
8. Vapnik, V.N.: The nature of statistical learning theory. IEEE Trans. Neural Netw. **10**, 988–999 (1995)
9. Zhang, D.J., Mei, C.L., Liu, G.H.: Soft sensor modeling based on improved extreme learning machine algorithm. Comput. Eng. Appl. **48**, 51–54 (2012)
10. Deng, W.Y., Zheng, Q.H., Chen, L.: Regularized extreme learning machine. In: IEEE Symposium on Computational Intelligence and Data Mining, pp. 389-395. IEEE Press, New York (2009)
11. Huang, G.B., Ding, X., Zhou, H.: Optimization method based extreme learning machine for classification. Neurocomputing **74**, 155–163 (2010)
12. Rao, C.R., Mitra, S.K.: Generalized Inverse of Matrices and Its Applications. Wiley, New York (1971)
13. Liang, N.Y., Huang, G.B., Saratchandran, P.: A fast and accurate online sequential learning algorithm for feedforward networks. IEEE Trans. Neural Netw. **17**, 1411–1423 (2006)
14. Gu, Y., Liu, J.F., Chen, Y.Q., Jiang, X.L., Yu, H.C.: TOSELM: timeliness online sequential extreme learning machine. Neurocomputing **128**, 119–127 (2014)
15. Glenn, F., Mangasarian, O.L.: Proximal support vector machine classifiers. In: ACM SIGKDD International Conference on Knowledge Discovery and Data Mining, pp. 77–86. ACM Press, New York (2001)
16. David, H.A.: Early sample measures of variability. Stat. Sci. **13**, 368–377 (1998)
17. Rousseeuw, P.J., Leroy, A.: Robut Regression and Outlier Detection. Wiley, New York (1987)

An Efficient High-Dimensional Big Data Storage Structure Based on US-ELM

Linlin Ding, Yu Liu, Baoyan Song and Junchang Xin

Abstract With the rapid development of computer and the Internet techniques, the amount of data in all walks of life increases sharply, especially accumulating numerous high-dimensional big data such as the network transactions data, the user reviews data and the multimedia data. The storing structure of high-dimensional big data is a critical factor that can affect the processing performance in a fundamental way. However, due to the huge dimensionality feature of high-dimensional data, the existing data storage techniques, such as row-store and column-store, are not very suitable for high-dimensional and large scale data. Therefore, in this paper, we present an efficient high-dimensional big data storage structure based on US-ELM, *H*igh-dimensional *B*ig Data *File*, named *HB-File*, which is a hybrid storage model of row-store and column-store. With the intensive experiments, we show the effectiveness of *HB-File* for storing the high-dimensional big data.

Keywords US-ELM · HDFS · Big data · High-dimensional data

1 Introduction

With the rapid development of computer and the improvement of human cognitive abilities, the understanding view and depth of things by human also continues extending and deepening. Many attributes are derived to describe the things and entities, so the high-dimensional data is generated, such as the network transactions data, the user reviews data and the multimedia data. Especially when the era of

L. Ding (✉) · Y. Liu · B. Song
School of Information, Liaoning University, Shenyang 110036, Liaoning, China
e-mail: dinglinlin@lnu.edu.cn

B. Song
e-mail: bysong@lnu.edu.cn

J. Xin
College of Information Science & Engineering, Northeastern University, Shenyang
110819, Liaoning, China

© Springer International Publishing Switzerland 2016
J. Cao et al. (eds.), *Proceedings of ELM-2015 Volume 1*,
Proceedings in Adaptation, Learning and Optimization 6,
DOI 10.1007/978-3-319-28397-5_38

data explosion comes, many data sets to be processed and analyzed are being the "big data", so more and more high-dimensional data forms the high-dimensional big data. For example, the number of user comments is close to 3.2 billion every day in Facebook. The high-dimensional big data mixes the typical features of both high-dimensional data and big data, which brings the new problems and challenges of the query processing and optimization of high-dimensional big data. In this case, the storing structure of high-dimensional big data is a critical factor that can affect the processing performance in a fundamental way.

However, the existing storage structures of big data are not suitable for storing high-dimensional big data. For example, the column-store structure, typical HBase [1], is very fit for storing the data with sparse columns features. But, due to the large amount and high coherence among dimensions of high-dimensional big data, if we use the pure column-store technology to manage high-dimensional big data, there would be numerous join operations among the dimensions during recovering the data. Instead, if we use row-store structure, typical HDFS [2], to store the high-dimensional big data, the single data record would be very long due to so many data dimensions. So, each data block only has a little high-dimensional big data records, which would reduce the storage efficiency. In a word, it is an urgent need to design efficient storage model for efficient storing high-dimensional big data.

Therefore, in this paper, we present an efficient high-dimensional big data storage model, *H*igh-dimensional *B*ig Data *File*, named *HB-File*. First, a table stored high-dimensional big data in *HB-File* vertically partitioned into two tables, respectively *key dimension* table and *non key dimension* table, which are confirmed by US-ELM [3] and FCM [4] algorithms. Second, the *non key dimension* table is directly stored in HDFS. The *key dimension* table is stored in HBase according to the cluster results of US-ELM and FCM. Each *key dimension* cluster is stored in a column family of HBase. Then, *HB-File* utilizes a column-wise data compression within each *column group* to avoid unnecessary column decompression during query execution. So, according to the characteristics of high-dimensional big data, *HB-File* is a mixed data storage structure combining proper column and row storage based on US-ELM and FCM.

The remainder of this paper is organized as follows. Section 2 briefly introduces the background, containing the ELM, US-ELM and the data placement for big data. Our *HB-File* structure is proposed in Sect. 3. The experimental results to show the performance of *HB-File* are reported in Sect. 4. Finally, we conclude this paper in Sect. 5.

2 Background

2.1 Review of ELM and US-ELM

Nowadays, Extreme Learning Machine (ELM) [5] and its variants [6–8] have the characteristics of excellent generalization performance, rapid training speed and little human intervene, which have attracted increasing attention from more and more researchers. ELM is originally designed for single hidden-layer feedforward neural networks (SLFNs [9]) and is then extended to the "generalized" SLFNs. Though ELMs have become popular in a wide range of domains, ELMs are primarily applied to supervised learning problems such as classification and regression, which greatly limits their applicability. Obtaining labels for fully supervised learning is time consuming and expensive, while a multitude of unlabeled data are easy and cheap to collect. Only a few existing research works based on ELM can process the problem of semi-supervised learning or unsupervised learning. The manifold regularization framework was added into the ELM for processing labeled and unlabeled data [10], which extended ELMs for semi-supervised learning.

The US-ELM extends ELMs to handle unsupervised learning problems and inherits the computational efficiency and the learning capability of traditional ELMs. The US-ELM uses spectral techniques for embedding and clustering by combining Laplacian Eigenmaps (LE) [11] and spectral clustering (SC) [12]. In all these algorithms, an affinity matrix is first built from the input patterns. When LE and SC are used for clustering, then k-means is adopted to cluster the data in the embedded space. The US-ELM consists of two stages: the random feature learning stage and the output weights learning stage. The random feature learning stage can generate the hidden layer, which is the essence of the ELM theory. The output weights learning stage of US-ELM is obtained by solving a generalized eigenvalue problem.

2.2 Data Storage for Big Data

In recent years, MapReduce [13] and its variants [14–16] have become the common methods of processing big data [17] and gained series of the research achievements. For data storage of big data based on MapReduce, there are mainly three data storage structures widely used, horizontal row-store, vertical column-store and hybrid store structure.

For horizontal row-store [18, 19], the row-store structure adopts the one-size-fits-all method to store data. Data records are placed contiguously in a disk page. The major advantage of row-store for a Hadoop-based system is that it has fast data loading and strong adaptive ability to dynamic workloads. But, the row-store cannot provide fast query processing and cannot achieve a high data compression.

For vertical column-store [20, 21], the vertical store scheme is based on a column-oriented store model for read-optimized data warehouse systems. In a vertical storage, a relation is vertically partitioned into several sub-relations. Column-store can avoid reading unnecessary columns during a query execution, and can easily achieve a high compression ratio by compressing each column within the same data domain. However, column-store cannot guarantee that all fields in the same record are located in the same cluster node.

For hybrid store structure, the main representatives are PAX [22], RCFile [23] and their improvements. For a record with multiple fields from different columns, PAX puts them in a single disk page to save additional operations for record reconstructions. Within each disk page, PAX uses a mini-page to store all fields belonging to each column, and uses a page header to store pointers to mini-pages. RCFile applies the concept of *"first horizontally-partition, then vertically-partition"* from PAX. It combines the advantages of both row-store and column-store.

Though each of these structures has its own advantages to store data records in different situations, all the structures are not very fit for storing the high-dimensional big data. The reason is that high-dimensional big data has not only huge data dimensions and data volume, but also the coherence of the dimensions is very high. Pure row-store structure and pure column-store structure are not satisfy the features of high-dimensional big data obviously. The PAX structure use column-store inside each disk page which cannot improve the I/O performance for big data. RCFile first horizontally partitioned the data records using row-store, where the *non key dimension* of high-dimensional big data would waste a lot of computation and transfer. So, it is necessary to design an efficient data structure for high-dimensional big data.

3   The Design of *HB-File* Structure

3.1   Overview of HB-File

In this section, we present *HB-File* (*H*igh-dimensional *B*ig Data *File*), a data placement structure designed for high-dimensional big data storage structure in Hadoop ecosystem. *HB-File* applies the concept of *"first vertically-partition, then horizontally partition, last vertically compression"* according to the characteristics of high-dimensional big data. It combines the advantages of both row-store and column-store. First, as column-store, *HB-File* can store the dimensions of high-dimensional big data separately following our design, so it can skip unnecessary column reads. Second, as row-store, *HB-File* guarantees the similar data in *key dimension* are located in the same node to reduce the network transfer. Last, *HB-File* can exploit a column-wise data compression to skip unnecessary column reads. *HB-File* is designed and implemented on the Hadoop Distributed File System (HDFS). Figure 1 shows the workflow of *HB-File* structure.

First, we divide the whole data dimensions of high-dimensional big data into *key dimension* and *non key dimension* by using US-ELM and FCM algorithm with sampled data. Then, The *non key dimension* of data records are directly stored in HDFS blocks for their features. Because the *key dimension* of data records forms several clusters, each cluster will be stored in one *column group*, and then be partitioned horizontally into blocks according to the design of *HB-File*. Last, the columns in each *column group* of *HB-File* can be compressed.

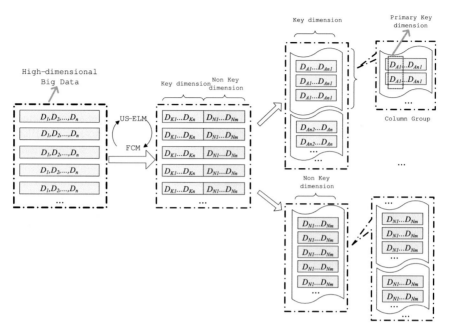

Fig. 1 Overview of *HB-File* structure

3.2 Choose Key and Non Key Dimension

The dimensions of high-dimensional big data are huge and the correlations of these dimensions are very tight, so pure column-store or pure row-store can not be used for efficient storing of high-dimensional big data. The dimension information of high-dimensional big data shows obvious clustering characteristics. That is to say, the huge dimensions can be divided into many clusters by suitable cluster algorithms. The dimensions in the same cluster have similar attributes and information. If the features of one cluster can be found, it can be used for representing the cluster. So, it is necessary to use the cluster information to enhance the storing efficiency of high-dimensional big data. There are many methods to represent a cluster. Because each cluster has its cluster centric, in this paper, we use the cluster centric as the *key dimension*. The definition of *key dimension* is shown in Definition 1.

Definition 1 (*Key Dimension*) The high-dimensional big data can be clustered into many clusters. The clustering centric dimension of each cluster is defined as the **key dimension** of this cluster.

According to the definition of *key dimension*, many cluster algorithms can be used to identify the *key dimension*. As shown above, the US-ELM extends ELMs to handle unsupervised learning problems and inherits the computational efficiency and the learning capability of traditional ELMs. There are two main stages in the training process, the random feature learning stage and the output weights learning stage.

In the first stage, it is the essence of the ELM theory to generate the random feature learning, which is the embedding matrix of the training data. For high-dimensional big data, this stage can be also regarded as the dimensionality reduction. In the second stage, the US-ELM treats each row of the embedding matrix as a point, and clusters the N points into k clusters using k-means algorithm. However, the number of clusters about the high-dimensional big data can not be identified and estimated easily according to the characteristics of high-dimensional big data, so k-means algorithm is not suitable for the clustering of *key dimension*. That is to say, US-ELM algorithm can be used for identifying the embedding matrix of high-dimensional big data.

In this paper, according to the characteristics of high-dimensional big data, we improve the US-ELM algorithm to cluster the dimensions by adding FCM algorithm. The fuzzy c-means algorithm (FCM) is a widely used clustering algorithm, which can cluster the data records according to the features. The clustering course containing US-ELM and FCM is used to cluster the dimensions of high-dimensional big data. There are also two stages in this course, the first course is the same to the first stage of US-ELM, and the embedding matrix can be gained in the first stage. Then, in the second stage, we use FCM algorithm to cluster the embedding matrix of high-dimensional big data. Because the FCM algorithm can identify the cluster centric, the *key dimension* can be gained. The dimensions of *key dimension* and *non key dimension* can be divided. However, due to the huge dimensions of high-dimensional big data, after one course of US-ELM and FCM, the number of *key dimension* is still very large, so it is necessary to design the method of gaining the proper *key dimension*. This course of US-ELM and FCM can be a loop. The loop can find the cluster centric, so it is needed to design a standard. There are two standards for identifying the cluster centric. Once the clustering results remain stable or the number of cluster centrics reach the predefined value, this loop can be stopped.

3.3   *Horizontal Partition and Data Compression*

After the above course, the *key dimension* and *non key dimension* of high-dimensional big data can be gained. Then, the *non key dimension* of data records are stored in HDFS according to the principle of HDFS. The *key dimension* of data records are stored in *column group*s according to the cluster results of data records. One cluster is stored in a *column group*. In each *column group*, the data records are partitioned horizontally. In order to efficiently store the data records, the similar data records should be stored in the same data nodes to minimize network transfer. So, in all clusters of *key dimension*, we use Definition 2 to define the *primary key dimension*. After that, we partition the data records horizontally in each *column group* by the *primary key dimension*. We store high-dimensional big data into *HB-File*, each dimension will be stored into a column, the column stored *key dimension* will be called as *key column*.

Definition 2 (*Primary Key Dimension*) The biggest variance of clustering centric dimension of key dimension is defined as the **primary key dimension** of data records.

After horizontal partition, in each column cluster, the column information can be compressed. Because all the values in the same column are the same type so they can be compressed well. *HB-File* takes two kinds of compression. One is normal Gzip algorithm to compress *non key dimension* and the other one is the RLE (Run Length Encoding) algorithm to compress *key dimension*. Since *key dimension* in *HB-File* will be involved in massive computations and decompressions when processing queries, it is not suitable for Gzip algorithm. We use RLE algorithm to compress *key dimensions* as the RLE algorithm can find the long runs of the same data records. With RLE compression algorithm, we can avoid unnecessary data read and decompression of other dimensions. RLE algorithm gets the most optimal compression effect in sorted sequence, but sorting all the columns will waste a lot of time. So we only use it on the *key dimensions* that will reduce the sorting time and data restructing time. While compressing all the column data, after horizontal partition, according to the concept of "each column cluster stores in a *column group*", we sort the *key column* of the *column group* and store the whole *column group* into *HB-File*. Because of existing the updates, we do not need to guarantee that all the *HB-File*s is in order but we must guarantee that *key column* is in order in each *HB-File*. Then compress *HB-File* by RLE for *key column* and Gzip for *non key column*.

4   Performance Evaluation

4.1   Experimental Setup

The experimental setup is a Hadoop cluster running on 9 nodes in a high speed Gigabit network, with one node as the Master node, the others as the Slave nodes. Each node has an Intel Quad Core 2.66 GHZ CPU, 500 GB disk, 4 GB memory and CentOS Linux 6.4. We use Hadoop 0.20.2 and compile the source codes under JDK 1.7. The US-ELM algorithm is implemented in MATLABR2009a.

We have conducted a comprehensive evaluation of *HB-File* using synthetic data sets. We generate different data sizes and different dimension numbers of our experiment data. The data sizes are from 128 GB to 1 TB, where the default size is 128 GB. The dimension numbers are from 120 to 300, where the default dimension number is 120. We compared *HB-File* with RCFile, row-store and column-store in three aspects, space occupancy rate, data loading time and query execution time.

4.2 Experimental Results

Space Occupancy Rate. We generate different sizes and dimension numbers of raw data to measure the data space occupancy rate. Space occupancy rate is the rate of the actual data space occupancy and real data size.

Figures 2 and 3 show the space occupancy rate of different storage structures with different data sizes and dimension numbers. Figure 2 shows that the space occupancy rate of all the storage structures grow with the increasing of data volume, since the meta data grows while the records of data set is greater. Meta data and real column data are not compressed together so that they will effect the compression efficiency. From Fig. 2 we can also conclude that except for raw data, row-store structure has the worst compression efficiency compared with column-store structure, since a column-wise data compression is much better than a row-wise data compression with mixed data domains.

Fig. 2 Performance of space occupancy rate changing data size

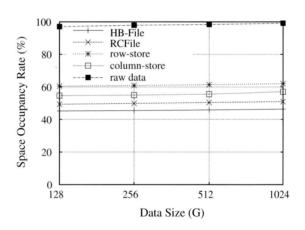

Fig. 3 Performance of space occupancy rate changing data dimension

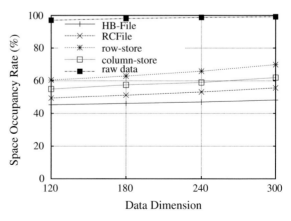

From Fig. 3 we can see that the growth of dimensions effects the rate much more than the growth of data volume. The reason is that the growth of dimensions will lead to the increasing of meta data columns, and it will limit the compression performance. So we can see *HB-File* has slightly better compression than RCFile, because US-ELM has clustered all the similar column previously, and a Column Controller holds all the meta data of a *column group*. While storing column meta data and real column data together, it cannot compress them separately. So, with too many columns in high-dimensional big data, RCFile cannot perform well because each column has a column holder and column holders burden on the compression performance.

Query Execution Time. We designed three groups of experiments to demonstrate the effectiveness of *HB-File* for different kinds of queries, range query, multiple columns query and similarity query.

(1) Range query. We execute range query on the above data set. The query finds all the results that meet the where clause condition, like "where $a<x<b$", where x is an attribute in the table, a and b is the upper and lower bounds of x.

According to Fig. 4, query on row-store is the slowest. While querying on row-store structures, query executor has to read all the unnecessary columns of a record so that it will waste a lot of time. *HB-File* is the fastest among all the column-store structures. Data in *HB-File* is stored in order, so when processing a range query, *HB-File* can avoid unrelated file blocks by using ordered data records and improve query performance.

(2) Mutiple columns query. We also execute multiple columns query on the above table. Compared with range query, multiple columns query includes more than one column. We design a multiple columns query to demonstrate the effectiveness of *HB-File*.

Figure 5 shows the execution times of the multiple columns query with the four data storage structures. RCFile is much slower than *HB-File*. The reason is that when processing related columns query, RCFile needs much network I/O and join operations since RCFile does not store related column together. While our *HB-File* has

Fig. 4 Performance of range query

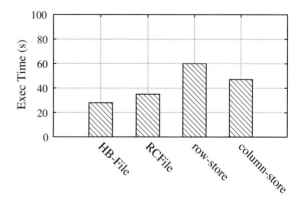

Fig. 5 The performance of mutiple columns query on query A

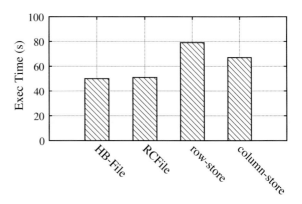

Fig. 6 Performance of similarity query

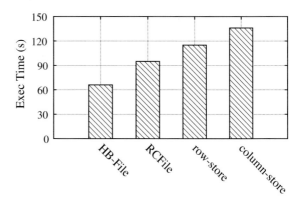

preprocessed the high-dimensional data into *column group* via US-ELM, each *column group* stores on the same node so that it can avoid lots of join operations and network transmission.

(*3*) *Similarity query*. Besides the two queries above mentioned, similarity query is also very common in high-dimensional big data. We define Euclidean distance as the metric of similarity. As Fig. 6 shows, similarity on *HB-File* is faster than any other storage structures. For the reason, *HB-File* can only compute the distance by *key dimensions* that are regarded as the representatives of all the *column groups*. So similarity query can be much faster and more accurate on *HB-File*. While other storage structures do not differentiate *key dimensions* from the all, so they cannot avoid computing all the columns. High-dimensional big data has large amounts of columns so the query costs much more calculations and takes longer executing time.

5   Conclusions

The management of high-dimensional big data is the research hot of database. The storing of high-dimensional big data is important for the query and analysis of high-dimensional big data. In this paper, we present an efficient high-dimensional big data storage model, named *HB-File*. First, a table stored high-dimensional big data in *HB-File* vertically partitioned into two tables, respectively *key dimension* table and *non key dimension* table, which are confirmed by US-ELM and FCM algorithms. Second, the *non key dimension* table is directly stored in HDFS. The *key dimension* table is stored in HBase according to the cluster results of US-ELM and FCM. Each *key dimension* cluster is stored in a column family of HBase. Then, *HB-File* utilizes a column-wise data compression within each Column Family to avoid unnecessary column decompression during query execution. So, according to the characteristics of high-dimensional big data, *HB-File* is a mixed data storage structure combining proper column and row storage for storing high-dimensional big data.

Acknowledgments This work is supported by National Natural Science Foundation of China (NO. 61472169,61472069,61502215,61472089). Science Research Normal Fund of Liaoning Province Education Department (NO. L2015193). Young Research Foundation of Liaoning University. (NO. LDQN201438). Doctoral Scientific Research Start Foundation of Liaoning Province 2015.

References

1. Carstoiu, D., Lepadatu, E., Gaspar, M.: Hbase—non SQL database, performances evaluation. Int. J. Adv. Comput. Technol. **2**(5), 42–52 (2010)
2. Leo, S., Zanetti, G.: Pydoop: a python mapreduce and HDFS API for hadoop. In: Proceedings of the 19th ACM International Symposium on High Performance Distributed Computing, HPDC 2010, pp. 819–825. Chicago, Illinois, USA, 21–25 June 2010
3. Huang, G., Song, S., Gupta, J.N.D., Wu, C.: Semi-supervised and unsupervised extreme learning machines. IEEE Trans. Cybern. **44**(12), 2405–2417 (2014)
4. Cannon, R.L., Dave, J.V., Bezdek, J.C.: Efficient implementation of the fuzzy c-means clustering algorithms. IEEE Trans. Pattern Anal. Mach. Intell. **8**(2), 248–255 (1986)
5. Huang, G.-B., Zhu, Q.-Y., Siew, C.-K.: Extreme learning machine: a new learning scheme of feedforward neural networks. In: Proceedings of the 2004 IEEE International Joint Conference on Neural Networks, vol. 2, pp. 985–990. IEEE (2004)
6. Sun, Y., Yuan, Y., Wang, G.: Extreme learning machine for classification over uncertain data. Neurocomputing **128**, 500–506 (2014)
7. Zong, W., Huang, G.-B.: Learning to rank with extreme learning machine. Neural Process. Lett. **39**(2), 155–166 (2014)
8. Sun, Y., Yuan, Y., Wang, G.: An os-elm based distributed ensemble classification framework in p2p networks. Neurocomputing **74**(16), 2438–2443 (2011)
9. Huang, G.-B., Zhu, Q.-Y., Siew, C.-K.: Extreme learning machine: theory and applications. Neurocomputing **70**(1), 489–501 (2006)
10. Liu, J., Chen, Y., Liu, M., Zhao, Z.: SELM: semi-supervised ELM with application in sparse calibrated location estimation. Neurocomputing **74**(16), 2566–2572 (2011)
11. Belkin, M., Niyogi, P.: Laplacian eigenmaps for dimensionality reduction and data representation. Neural Comput. **15**(6), 1373–1396 (2003)

12. Ng, A.Y., Jordan, M.I., Weiss, Y.: On spectral clustering: analysis and an algorithm. In: Advances in Neural Information Processing Systems 14 Neural Information Processing Systems: Natural and Synthetic, NIPS 2001, pp. 849–856. Vancouver, British Columbia, Canada, Dec 3–8 2001

13. Dean, J., Ghemawat, S.: Mapreduce: simplified data processing on large clusters. Commun. ACM **51**(1), 107–113 (2008)

14. Deng, D., Li, G., Hao, S., Wang, J., Feng, J.: Massjoin: a mapreduce-based method for scalable string similarity joins. In: ICDE, pp. 340–351. IEEE (2014)

15. Qin, L., Yu, J.X., Chang, L., Cheng, H., Zhang, C., Lin, X.: Scalable big graph processing in mapreduce. In: SIGMOD Conference, pp. 827–838. ACM (2014)

16. Zhang, Y., Chen, S., Wang, Q., Yu, G.: i^2 mapreduce: incremental mapreduce for mining evolving big data. IEEE Trans. Knowl. Data Eng. **27**(7), 1906–1919 (2015)

17. Bin Cui, H.M., Ooi, B.C.: Big data: the driver for innovation in databases. Nat. Sci. Rev. **1**(1), 27–30 (2014)

18. Ghemawat, S., Gobioff, H., Leung, S.: The google file system. In: Proceedings of the 19th ACM Symposium on Operating Systems Principles 2003, SOSP 2003, pp. 29–43. Bolton Landing, NY, USA, 19–22 Oct 2003

19. Raman, V., Swart, G., Qiao, L., Reiss, F., Dialani, V., Kossmann, D., Narang, I., Sidle, R.: Constant-time query processing. In: Proceedings of the 24th International Conference on Data Engineering, ICDE 2008, pp. 60–69. Cancún, México, 7–12 April 2008

20. Boncz, P.A., Manegold, S., Kersten, M.L.: Database architecture optimized for the new bottleneck: memory access. In: VLDB'99, Proceedings of 25th International Conference on Very Large Data Bases, pp. 54–65. Edinburgh, Scotland, UK, 7–10 Sep 1999

21. Gates, A., Dai, J., Nair, T.: Apache pig's optimizer. IEEE Data Eng. Bull. **36**(1), 34–45 (2013)

22. Ailamaki, A., DeWitt, D.J., Hill, M.D., Skounakis, M.: Weaving relations for cache performance. In: VLDB 2001, Proceedings of 27th International Conference on Very Large Data Bases, pp. 169–180. Roma, Italy, 11–14 Sep 2001

23. He, Y., Lee, R., Huai, Y., Shao, Z., Jain, N., Zhang, X., Xu, Z.: Rcfile: a fast and space-efficient data placement structure in mapreduce-based warehouse systems. In: Proceedings of the 27th International Conference on Data Engineering, ICDE 2011, pp. 1199–1208. Hannover, Germany, 11–16 April 2011

An Enhanced Extreme Learning Machine for Efficient Small Sample Classification

Ying Yin, Yuhai Zhao, Ming Li and Bin Zhang

Abstract ELM, as an efficient classification technology, is used to many popular application domain. However, ELM has weak generalization performance when the original data set is small related to its feature space. In this paper, aiming to the above problem, an enhanced ELM classification framework is proposed to improve the accuracy of ELM classifier. At first, the method automatically obtains the k discretization intervals for the continuous data and removes the irrelevant features and the redundancy features by mutual information. Further, we only select those features which have high relevance with the object node by an improved Markov Boundary identify algorithm. Finally, Obtaining the enhanced ELM classifier by an efficient weight voting mechanism. The experiments conducted on real-life small sample datasets demonstrate that the proposed framework outperforms the previous methods, especially for small sample data.

Keywords Extreme Learning Machine · Representative features · Small sample data

1 Introduction

In recent years, classification problem has regained extensive research efforts from computer scientists, due to the explosive emergence of new classification applications, especially with the emergence of the big data. It is one of the challenges for the researcher on how to learn a model from the various data and classify the data quickly, such as protein sequences classification in bioinformatics [1, 2], online

Y. Yin · Y. Zhao (✉) · M. Li · B. Zhang
College of Information Science and Engineering, Northeastern University,
Shenyang, China
e-mail: zhaoyuhai@ise.neu.edu.cn

Y. Yin
e-mail: yinying@ise.neu.edu.cn

Y. Zhao
Key Laboratory of Computer Network and Information Integration (Southeast University),
Ministry of Education, Nanjing, China

© Springer International Publishing Switzerland 2016
J. Cao et al. (eds.), *Proceedings of ELM-2015 Volume 1*,
Proceedings in Adaptation, Learning and Optimization 6,
DOI 10.1007/978-3-319-28397-5_39

501

social network prediction [3], XML document classification [4], cloud resource classification [5], online real-time stream data prediction [6], Uncertain Graph [1] and user-generated text documents from the Internet and so on [7]. How to classify the data quickly and correctly is an important thing.

Extreme Learning Machine (ELM) is becoming popular since it generally requires far less training time than the conventional learning machines [8–13]. ELM has originally been developed based on Single-hidden Layer Feedforward Neural Networks (SLFNS) in [14, 15]. In ELM, hidden nodes parameters are chosen randomly. A Survey has been done on ELM and its variants in [7].

ELM has a better classification and prediction performance in many domains. However, it rise new challenges to ELM because of the extremely large dimensionality of the feature space. Despite of the success of ELM on classification and prediction, it remains unclear how to tackle the difficulty on the insufficiency of training samples. That is, ELM has weak generalization performance when the original data set is small related to its feature space. For example, in bioinformatics, some special characteristics of microarray data pose the great challenges to most existing data analysis algorithms. This is because a typical microarray data is often of severely limited number of samples, but of several orders of magnitude more dimensions (genes). According to the traditional learning theory, given n dimensions (genes), the required number of samples m for the reliable classifier learning should be on the scale of $O(2n)$ [16]. However, even the minimum requirement ($m = 10 * n$) as a statistical "rule of thumb" is patently impractical for a real microarray dataset [16]. As such, selecting a small number of representative features (genes) showing distinct profiles in different classes of samples becomes highly necessary.

With the aim of choosing a subset of good features with respect to the target concepts, feature subset selection is an effective way to solve the small sample problem through reducing dimensionality removing irrelevant data, increasing learning accuracy, and improving result comprehensibility. The wrapper and filter techniques are two basic approaches for feature selection [16]. The wrapper methods use the predictive accuracy of a predetermined learning algorithm to determine the goodness of the selected features, the accuracy of the learning algorithms is usually high. However, the generality of the selected features is limited and the computational complexity is high. The filter methods's computational complexity is low and having good generality capability. However, the accuracy of the learning algorithms is not guaranteed [10, 17, 18]. The filter methods are usually a good choice when the number of features is very large. However, the filter methods ignore the dependence relationship of attributes with each other.

Among all the existing solutions, Bayesian graphical model [19] has proved its advantages, especially on robust statistical accuracy and bounded time complexity. Markov Boundary, which is the global optimal feature combination with respect to the class label. Interestingly, Markov Boundary considers the dependency relationship between attributes. Thus, we will adopt a group of conditional independence tests to derive Markov Blanket in an efficient and effective way.

The main contributions in this paper are as follows: (1) A Framework for the whole process of constructing ELM using Markov Boundary; (2) Methods for automatic obtaining k discretization intervals for the continuous data. (2) Proposal of a idea to identify representative feature subsets with markov boundary for different category. (3) Obtain enhanced ELM classifier by training ELM with efficient weight vote mechanism.

The remainder of this paper is organized as follows: Sect. 2 gives a brief overview of ELM. Section 3 presents the classification architecture based on enhanced ELM. Section 4 studies the feature prefiltration mechanism. In Sect. 5, we report the enhanced ELM classification with voting strategy. Finally, in Sect. 6, we summarize the present study and draw some conclusions.

2   Brief Introduction of Extreme Learning Machine

Three common approaches of feedforward networks training were summarized in [20]: (1) gradient-descent based (e.g., back propagation method for multi-layer feedforward neural networks); (2) least square based (e.g., ELM for the generalized single-hidden layer feedforward networks); (3) standard optimization method based (e.g., SVM for a specific type of Single-hidden Layer Feedforward Networks). ELM and its variants [9, 21] based on SLFNs for classification and can achieve better generalization performance than that of conventional learning algorithms. Moreover, ELM is less sensitive to user specified parameters, and can be deployed faster and more conveniently [20].

As mentioned, ELM is based on SLFN type classifiers. Standard SLFNs with N arbitrary samples $(\mathbf{x_i}, \mathbf{t_i}) \in \mathbf{R}^{n \times m}$ and activation function $g(x)$ are modeled in [15] as

$$\sum_{i=1}^{L} \beta_i g_i(\mathbf{x_j}) = \sum_{i=1}^{L} \beta_i g(\mathbf{w_i} \cdot \mathbf{x_j} + b_i) = \mathbf{o_j}, (j = 1, \dots, N) \tag{1}$$

where L is the number of hidden layer nodes, $\mathbf{w_i} = [w_{i1}, w_{i2}, \dots, w_{in}]^T$ is the weight vector between the ith hidden node and the input nodes, $\beta_i = [\beta_{i1}, \beta_{i2}, \dots, \beta_{im}]^T$ is the weight vector between the ith hidden node and the output nodes, and b_i is the threshold of the ith hidden node. Then we have the output of ELM

$$f(x) = \sum_{i=1}^{L} \beta_i g(\mathbf{a_i}, b_i, \mathbf{x}) \tag{2}$$

where

$$H(\mathbf{w_1}, \dots, \mathbf{w_L}, b_1, \dots, b_L, \mathbf{x_1}, \dots, \mathbf{x_L}) = \begin{bmatrix} g(\mathbf{w_1} \cdot \mathbf{x_1} + b_1) & \dots & g(\mathbf{w_L} \cdot \mathbf{x_1} + b_L) \\ \vdots & \dots & \vdots \\ g(\mathbf{w_1} \cdot \mathbf{x_N} + b_1) & \dots & g(\mathbf{w_L} \cdot \mathbf{x_N} + b_L) \end{bmatrix}_{N \times L},$$

$$\beta = \left[\beta_1^T, \dots, \beta_L^T \right]_{m \times L}^T$$

The decision function for binary classification [8] is

$$d(x) = sign(\sum_{i=1}^{L} \beta_i g(\mathbf{a_i}, b_i, \mathbf{x})) = sign(\beta \cdot \mathbf{H}) \tag{3}$$

When g(x) approximates the N samples with zero error that $\Sigma_{j=1}^{L} \|o_j - t_j\| = 0$, their outputs β_i, w_i and b_i such that

$$\sum_{i=1}^{L} \beta_i g(\mathbf{w_i} \cdot \mathbf{x_j} + b_i) = \mathbf{t_j}, j = 1, \dots, N \tag{4}$$

The equation above can be expressed compactly as following

$$\mathbf{H}\beta = \mathbf{T} \tag{5}$$

where $\mathbf{T} = [\mathbf{t}_1^T, \dots, \mathbf{t}_L^T]_{m \times L}^T$.

The ELM algorithm is a relatively fast method as compared to the conventional learning algorithms. ELM not only tends to reach the smallest training error but also the smallest norm of weights [7]. Given a training set $\aleph = \{(\mathbf{x_i}, \mathbf{t_i}) | \mathbf{x_i} \in \mathbf{R^n}, \mathbf{t_i} \in \mathbf{R^m}, i = 1, \dots, N\}$, activation function $g(x)$ and hidden node number L, algorithm ELM is described as following [15].

Algorithm 1 ELM

1: **for** i=1 to L **do**
2:     randomly assign input weight w_i
3:     randomly assign bias b_i
4: **end for**
5: calculate **H**
6: calculate $\beta = \mathbf{H^\dagger T}$

---

3   The Enhanced ELM Classification Framework

In order to immediately comprehend our idea, we illustrate the whole process of constructing ELM using Markov Boundary based the process consists of three major phases: (1) preprocessing, which discretizates automatically the continuous data into the k representative features selection by the Fig. 1. As shown, intervals; (2) the representative features selection, instead of constructing a large Bayesian network by all features, we only select those features which have high relevance with the

Fig. 1 The framework

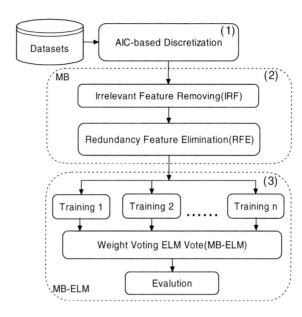

object node; (3) Enhanced ELM classifier construction by training ELM with efficient weight voting mechanism using the extracted feature subsets.

(1) Preprocessing. Most of the data in the real world is continuous. The discretization makes the continuous attributes simple and uniform, since all features would be encoded by the same scheme regardless of their original data type. On the other hand, for continuous data, it is often a trouble matter to automatically determine the number of discrete attributes. It relies on the ability of a discretization algorithm to find good intervals. This part will be discussed in Sect. 4.

(2) The representative features selection. Since not all attributes provide useful information, a dimensionality reduction process is needed to identify and remove the redundant feature as much as possible. Dimensionality reduction is the process of reducing the computing space, and can be performed by means of feature selection. Therefore, the good feature subsets should contain those features with highly correlated with the class [22]. Keeping these in mind, we develop a novel algorithm which can efficiently and effectively deal with both irrelevant and redundant features, and obtain a quick training time and a good classification accuracy. The representative feature selection processing (shown in Fig. 1 (2)) which composed of the two connected steps of irrelevant feature removal and redundant feature elimination. The first step obtains relevant features to the target by eliminating irrelevant ones, and the second step removes redundant features from relevant ones via choosing representatives from different categories, and thus produces the final feature subset. The irrelevant feature removal is relative easy once the relevance measure is defined. We eliminate the redundant feature by an efficient way to derive the Markov Boundary. It involves (i) the construction of Markov Boundary; (ii) pruning; and (iii) the selection of representative features from the learning. Limit by space, we omit the details.

(3) Enhanced ELM classifier. The framework first select a group of representative features (shown in Fig. 1 (3)) which composed of the two connected steps of irrelevant feature removal and redundant feature elimination. Besides, we applied an efficient weight voting mechanism proposed in this paper in classifying samples to achieve a higher accuracy than the original ELM. Then, an efficient postprocessing method is also proposed to further optimize the voting results. This part will be described in Sect. 5.

4 Feature Prefiltration

At first, we adopt a model selection criteria, AIC, to automatically determine the number of discrete attributes. However, for high dimensional and multivariate data, there are some attributes which have little correlation with class labels. All these properties were calculated to learn a model with labels are expensive. In order to reduce the computational cost, this paper only selects those attributes associated with the labels. In order to determine which attributes are associated with the sample label, using mutual information method for initial filtering. Mutual information measures how much the distribution of the feature values and target classes differ from statistical independence. This is a nonlinear estimation of correlation between feature values or feature values and target classes. The mutual information of event X and event Y is defined as:

$$I(X, Y) = H(X) + H(Y) - H(X, Y) \tag{6}$$

where H(X, Y) is the joint entropy, defined as:

$$H(X, Y) = \sum p(x, y) log p(x, y) \tag{7}$$

and p(x, y) is the probability.

By computing the mutual information between a feature value and the target class, we can obtain the correlation degree between a feature value and the target class. Further, we get the candidate boundary nodes by sorting mutual information values.

Feature selection is essentially a task to remove irrelevant and/or redundant features. Irrelevant features severely affect the accuracy of the learning machines. Redundant features are a type of irrelevant feature [16]. The distinction is that a redundant feature implies the co-presence of another feature; individually, each feature is relevant, but the removal of one of them will not affect learning performance. The selection of features can be achieved in two ways: One is to rank features according to some criterion and select the top k features, and the other is to select a minimum subset of features without learning performance deterioration. In other words, subset selection algorithms can automatically determine the number of selected features, while feature ranking algorithms need to rely on some given threshold to select features. Therefore, we will adopt the Markov Boundary [23] to find the representative features for a special object class.

5   ELM Classification with Voting Decision Function

However, Not all ELM training results are of high accuracy due to various reason, such as, data distribution sparsity or the unsuitable hidden nodes setting. Therefore, it is unfair to the classifier when we use all ELM results voting with assigning the same weight. Further, the method affects the real accuracy rate. This is because the training error on every ELM running is different. Due to many application data in reality were composed of multiple labels. The two commonly used methodologies for multi-class classification are one against-all (OAA) and one-against-one (OAO) [24]. We refer the ELM-OAO [24] process. At first, Classifying the m classes into $t(t-1) = 2$ parts. Each part is trained by an ELM classifier $elm(x;y)$. That is, when classifying a sample, the ELM classifier outputs j^+ or j^- if the sample is of class x or y. In this paper, we consider an optimized decision function using different weight to vote ELM for multi-class classification. The voting result is computed by the decision function once the process completes all ELM voting for the output nodes. We only assign the highest number of votes as the final classification result.

Algorithm 2 presents the training process. At first, the data sets D need to be divided into training data sets and validation data sets. For every partitioned training data (in line 1), the phase invokes ELM using the algorithm 1 and the decision function is invoked after all the ELM classifiers complete training phase(in line 2–10).

Algorithm 2 MB-ELM
1: divide the data sets D into training data sets and validation data sets
2: **for** $j = 1$ to $(m\text{-}1)$ **do**
3:    **for** $k = (j+1)$ to m **do**
4:        select samples belonging to class j, k
5:        assign to training subsets $train\_data_{(j,k)}$
6:        train an $ELM_{(j,k)}$ with $train\_data_{(j,k)}$ using Algorithm 1
7:    **end for**
8: **end for**
9: computing with voting decision function(VDF)
10: assign the input test sample to the ith class with VDF

6   Conclusion

In this paper, an enhanced ELM classification framework based on representative features is proposed to improve the accuracy of ELM classifier. At first, the method automatically obtains the k discretization intervals for the continuous data and removes the irrelevant features and the redundancy features by mutual information. Further, we only select those features which have high relevance with the object node by an improved Markov Boundary identify algorithm. Finally, Obtaining the enhanced ELM classifier by training ELM with an efficient weight voting mechanism. A series of experiment results demonstrate that the proposed enhanced ELM classifiers outperforms the other existed methods. A series of experiments show F-ELM has the highest performance and efficiency.

Acknowledgments Project supported by the National Nature Science Foundation of China (No. 61272182, 61100028, 61572117,61173030, 61173029), State Key Program of National Natural Science of China (61332014,U1401256), New Century Excellent Talents (NCET-11-0085) and Fundamental Research Funds for the Central Universities (N130504001).

References

1. Wang, Z., Zhao, Y., Wang, G., Li, Y., Wang, X.: On extending extreme learning machine to non-redundant synergy pattern based graph classification. Neurocomputing (IJON) **149**, 330–339 (2015)
2. Zhao, Y., Wang, G., Yin, Y., Li, Y., Wang, Z.: Improving ELM-based microarray data classification by diversified sequence features selection. Neural Comput. Appl. (2014). doi:10.1007/s00521-014-1571-7
3. Sun, Y., Yuan, Y., Wang, G.: An on-line sequential learning method in social networks for node classification. Neurocomputing (IJON) **149**, 207–214 (2015)
4. Zhao, X., Wang, G., Bi, X., Gong, P., Zhao, Y.: XML document classification based on ELM. Neurocomputing (IJON) **74**(16), 2444–2451 (2011)
5. Xia, M., Weitao, L., Yang, J., Ma, Y., Yao, W., Zheng, Z.: A hybrid method based on extreme learning machine and k-nearest neighbor for cloud classification of ground-based visible cloud image. Neurocomputing (IJON) **160**, 238–249 (2015)
6. Cao, K., Wang, G., Han, D., Ning, J., Zhang, X.: Classification of uncertain data streams based on extreme learning machine. Cogn. Comput. (COGCOM) **7**(1), 150–160 (2015)
7. Huang, G.-B., Wang, D., Lan, Y.: Extreme learning machines: a survey. Int. J. Mach. Learn. Cybern. **2**, 107–122 (2011). doi:10.1007/s13042-011-0019-y
8. Huang, G.-B., Ding, X., Zhou, H.: Optimization method based extreme learning machine for classification. Neurocomputing **74**(1–3), 155–163 (2010)
9. Li, M.-B., Huang, G.-B., Saratchandran, P., Sundararajan, N.: Fully complex extreme learning machine. Neurocomputing **68**, 306–314 (2005)
10. Wei, X., Li, Y., Feng, Y.: Comparative study of extreme learning machine and support vector machine. ISNN **1**, 1089–1095 (2006)
11. Jun, W., Shitong, W., Chung, F.-L.: Positive and negative fuzzy rule system, extreme learning machine and image classification. Int. J. Mach. Learn. Cybern. **2**, 261–271 (2011)
12. Wang, X., Chen, A., Feng, H.: Upper integral network with extreme learning mechanism. Neurocomputing **74**(16), 2520–2525 (2011)
13. Chacko, B., Vimal Krishnan, V., Raju, G., Babu Anto, P.: Handwritten character recognition using wavelet energy and extreme learning machine. Int. J. Mach. Learn. Cybern. **3**(2), 149–161 (2012)
14. Huang, G.-B., Siew, C.K.: Extreme learning machine: Rbf network case. In: ICARCV, pp. 1029–1036 (2004)
15. bin Huang, G., yu Zhu, Q., kheong Siew, C.: Extreme learning machine: theory and applications. Neurocomputing **70**, 489–501 (2006)

16. Liu, H., Motoda, H.: Computational Methods of Feature Selection. Chapman, Hall/CRC, Danvers (2007)
17. Souza, J.: Feature selection with a general hybrid algorithm. Ph.D., University of Ottawa, Ottawa, Ontario, Canada (2004)
18. Langley, P., Selection of relevant features in machine learning. In: Proceedings of the AAAI Fall Symposium on Relevance, pp 1-5 (1994)
19. Banerjee, S., Ghosal, S.: Bayesian structure learning in graphical models. J. Multivar. Anal. (MA) **136**, 147–162 (2015)
20. Cortes, C., Vapnik, V.: Support vector networks. Mach. Learn. **20**, 273–297 (1995)
21. Feng, G., Huang, G.-B., Lin, Q., Gay, R.: Error minimized extreme learning machine with growth of hidden nodes and incremental learning. IEEE Trans. Neural Netw. **20**(8), 1352–1357 (2009)
22. Hall, M.A., Smith, L.A.: Feature selection for machine learning: comparing a correlation-based filter approach to the wrapper. In: Proceedings of the Twelfth International Florida Artificial intelligence Research Society Conference, pp 235–239 (1999)
23. Feng, D., Chen, F., Wenli, X.: Analysis of markov boundary induction in bayesian networks: a new view from matroid theory. Fundam. Inform. (FUIN) **107**(4), 415–434 (2011)
24. Rong, H.-J., Huang, G.-B., Ong, Y.-S.: Extreme learning machine for multi- categories classification applications. IEEE Int. Jt. Conf. Neural Netw. 1709–1713 (2008)

Code Generation Technology of Digital Satellite

Ren Min, Dong Yunfeng and Li Chang

Abstract The digital satellite is a complex system. And it has great development difficulty, low extensibility, great difficulty of debug and test. A theory that complex system is resolved into a set of underlying decisions so that the complex system can be handled by program is advanced based on the theory of dimension decomposition of engineering methodology and cognitive science. A complete digital satellite code generation process is designed according to the theory. The digital satellites configuration and code generation based on several satellites are completed. And digital satellites are used for simulations in the typical condition. The simulations reveal that digital satellite can be assembled automatically using the theory.

Keywords Digital satellite · Satellite simulators · Fractals · Code generation

1 Background

Digitization and informatization is the core of a new generation of intelligent manufacturing. The parallel system is effective way to deal with complex problems. Digital satellite is the parallel system of real satellite and an important tool in the design, development and transit process of satellite.

Research on digital satellite has made some achievements [1–9]. However, the digital satellites in traditional development mode have great development difficulty, a long lead time and huge resource consumption. And the traditional digital

R. Min · D. Yunfeng · L. Chang (✉)
School of Astronautics, Beihang University, Beijing, People's Republic of China
e-mail: 619920580@qq.com

R. Min
e-mail: 286611209@qq.com

D. Yunfeng
e-mail: sinosat@buaa.edu.cn

© Springer International Publishing Switzerland 2016
J. Cao et al. (eds.), *Proceedings of ELM-2015 Volume 1*,
Proceedings in Adaptation, Learning and Optimization 6,
DOI 10.1007/978-3-319-28397-5_40

satellites have low extensibility. The change of design requirements is accompanied by a lot of modification of source code. With the development of digital simulation technology, the traditional development mode of digital satellite is difficult to meet the demand of development, production and simulation of real satellites.

The core of artificial intelligence is that let the machine make the decision and finish the work instead of human. Cognitive science is the basis of artificial intelligence and an important force in promoting the development of artificial intelligence. Cognitive science has become a frontier subject. It has received extensive attention in academic circles and has made progress on several fronts. The method that let the machine do the source code writing instead of human not only improve the efficiency, but also encapsulate design specification inside the tool of satellite design. The method gives a new insight into the aerospace industry informatization.

2   Dimension Decomposition of the Programmer Decision

Human intelligence is reflected in the process to deal with the complicated situation. Human can analyze problem and split it up into several sub problems when human face the complicated situation. The complicated problem can be solved if every sub problems can be solved. Engineering methodology is abstraction and refinement of method which human solve problems use. The program will be intelligent if it is taught to solve problems according to engineering methodology.

The method that problems are decomposed according to the concept of dimension is a not repetitive and missing method of decomposition. The common dimensions are dimensions of system, time and logic. Satellite can be decomposed into payload and service module according to the dimensions of system. Service module can be further decomposed into structure, power, thermal control, telemetry and command, propulsion, attitude and orbit control, on-board data management subsystems. The satellite assembly process can be decomposed into components library establishment, satellite physical components assembly, information transmission assembly, simulation deployment configuration and common code assembly. The method of decomposition according to logic dimension is to decompose problems according to the logic of handling problems process. The process can be decomposed into defining problems and solving problems. Defining problems can be decomposed into defining goals and indexes design. Objectives of satellite should be defined such as working life, observation targets and performance indexes of satellite should be defined such as the coverage performance indexes and control performance indexes in the early stage of digital satellite design. Solving problems can be decomposed into proposing solutions and implementing solutions. Satellite should be designed and the generation program can generate the source code of digital satellite according to the result of design in source code of digital satellite generation process.

Dimensions have fractal properties so that a grid in dimensions can be further decomposed. A gird in Fig. 1 can be further decomposed according to system

Fig. 1 A grid in dimensions

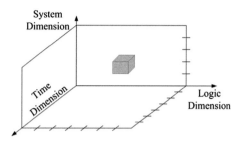

dimension, time dimension and logic dimension. System dimension and time dimension are often reflected in the high level decomposition. The fractal property of logic dimension is more obvious when it is closer to the bottom operation. Each stages and steps of digital satellite are decomposed into data or sources that can be processed by machines so that digital satellite code generation is made feasible.

Each work can be decomposed into several parts which can form a tree structure. The work will be completed if the tree were executed. The tree structure is stored in databases in order to facilitate the execution of the work. The qualitative description can be separated from quantitative analysis when the tree structure is being stored so that the stored data are streamlined and scalability and extensibility of the program are increased. The programs can assemble a complete digital satellite by reading the information from databases and executing the operations according to the information to complete each links.

3 Program Assembly

3.1 Method of Program Generation

The source codes of digital satellite simulator can be decomposed into statements according to the dimensions. Program statements are the basic elements of source codes. The process of statements generation is to read variable information from databases, to piece the information and fixed codes together into a string according to a standard format and to print the string in a source file. Statements can be divided into variable definition, variable assignment, function call and condition judgment statements.

A variable is a string. The variable type need be declared and the variable value need be initialized to meet the needs of programming, such as

```
double GyroMeasureValue[1] = {0};
```

The parts before the equal sign are the type, name and dimension of the variable. The parts after the equal sign are fixed variable value initialization. The program can read the information that the type of the variable is "double", the name of the variable is "GyroMeasureValue" and the dimension of the variable is 1 from database. And the program piece the information together into a string according to the format as "type" + "name" + "[dimension]" + "={0};". And the program prints the string in a source file. The generation of variable definition statement is completed.

The variable assignment statement is similar to the variable definition statement, such as

```
GyroMeasureValue[0] = 0.1;
```

The parts before the equal sign are the name and dimension of the variable. The parts after the equal sign are the value of the variable. The program can read the information that the name of the variable is "GyroMeasureValue", the dimension of the variable is 0 and the value of the variable is 0.1 from database. And the program piece the information together into a string according to the format as "name" + "[dimension]" + "=" + "value;". And the program prints the string in a source file. The generation of variable assignment statement is completed.

The function call statement is composed by the name and the parameters of the function. The name of the function is generally fixed. And it can also be read from databases. The parameters of the function can be fixed or be read from databases or be calculated by program, such as

```
SendUARTPackage(ulUARTGyroChannelNoforGyro, chrRawValue, 17);
```

The name of the function is fixed. The first parameter is read from database. The second parameter is fixed. And the third parameter is calculated by program. The program can piece the information together into a string according to the format as "name of function" + "(" + "name of first parameter" + ", name of second parameter," + "value of third parameter" + ")". And the program prints the string in a source file. The generation of function call statement is completed.

The condition judgment statement is composed by the fixed frame and the variable condition. The condition can be read from databases completely. It can also be constructed of the name of variable, the logic of judgment and the value of the condition which are read from databases. Such as

```
if(ucThrusterIsUpdated == 1)
```

The program can read the information that the name of the variable is "ucThrusterIsUpdated", the logic of the condition is equation and the value of the condition is 1. The program can piece the information together into a string according to the format as "if(" + "name of variable" + "logic of judgment" + "value of condition" + ")". And the program prints the string in a source file. The generation of condition judgment statement is completed.

Several parts of source codes in program are fixed and selectively enabled according to the requirements of satellite assemble, such as functional models of devices, algorithms of the satellite, etc. These parts do not need to generate each statement. These parts can be stored in source files and the names and paths of files are stored in databases. Program can read the information of needed models and algorithms from databases, find the specified files and print the source codes in a source file completely. The assembly of parts of source codes is completed.

The value of several parameters of source codes will change according to the requirements when the source codes are being assembled. The parameters of fixed source codes can be assigned in two ways. One of them is to generate the variable assignment statements in head files. The names and values of variables are stored in databases. The program can generate the head files according to the process of generation of the variable assignment statement. And the source codes include the head file to assign the variables. Another way is to generate the values of parameters in XML files. The satellite simulator can read the XML files to assign the variables.

3.2   Static Assembly

Static refers to that data do not change over time and the data and the data is updated via information transmission. So that the main work of static assembly is information flow assembly. The static assembly can be decomposed into device assembly and satellite assembly.

The source codes of device information transmission are highly generic. The sources of different devices have the same format. So that the flow of information transmission is defined and the program can piece the variable information together into source codes according to a standard format.

The core of source codes of device information transmission is the process of packing and unpacking of information package. The process of packing information package is introduced because the process of packing and unpacking of information package is similar. The source codes of packing information package are composed by the packing package statements of each data in the package. Such as

```
memcpy(chrRawValue+6, &TimeNow, 8);
```

The statement is a function call statement. The name of the function is fixed. The first half part of the first parameter is fixed name of variable. The second half part of the first parameter is the position of the data which can be calculated by program in the process of packing package. The second parameter is fixed. The third parameter is the length of variable which can be read form databases. The information and ordering of variable in package are stored in databases so that program can query the name and length of each variable in the package and generate the statements of packing package of each variable in the package in order to complete the generation of whole source codes of packing package.

The information flow assembly of satellite can be completed by assembling the information flow between devices and algorithm module of satellite on the basis of information flow assembly of devices. The content of the information package is fixed so that the source codes of information flow between devices and algorithm module of satellite can be decomposed into several fixed parts of source codes. The package that the digital satellite need is stored in databases. And the program can generate the whole source codes of information flow according to the process of parts of source codes assembly.

3.3 Dynamic Assembly

The dynamic assembly of digital satellite can be decomposed into devices assembly, onboard algorithm assembly and satellite assembly.

The functional models of devices can be decomposed into the principle model, error model and failure model. The source codes of the functional models of each device are fixed so that the source codes can be stored in classified files according to the model of devices. The information of which device is needed by digital satellite is stored in databases. And the program can assemble the models of devices according to the process of parts of source codes assembly.

The onboard algorithm assembly primarily includes attitude control system assembly and data management system assembly. The attitude control system primarily includes outlier elimination, filtering, attitude determination and attitude control algorithm. Data management system primarily includes CCSDS telemetry processing, data compression, video and image mosaicking algorithm etc. Each algorithm is fixed. And the values of parameters of each algorithm need to be modified to meet the different needs of the design. So the source codes of the algorithms are stored in several files. The program can assemble algorithms according to the process of parts of source codes assembly and assign the parameters of algorithms according to the method of generating the head files.

The work of attitude control system can be divided into different modes for different tasks. The different modes can enable different algorithms and the parameters of the algorithms can be different. The flight modes interrelate through various switching conditions and form a flight flow. The execution process of a flight mode includes parameters input, algorithm execution, condition judgment and mode switching. The source codes of parameters input are variable assignment statements. The source codes of algorithm execution are function call statements. The source codes of condition judgment are condition judgment statements. The source codes of mode switching are function call statements. The names and values of variables, the names of algorithms, the conditions of switching and the name of target mode are stored in databases so that the program can generate the source codes of execution process of flight modes according to the process of generation of the corresponding statements.

The satellite assembly can be decomposed into assemblies of dynamic algorithm, thermodynamic algorithm, energy flow algorithm, fuel flow algorithm, radio transmission loss algorithm and optical imaging algorithm.

The dynamic algorithm can be decomposed into the dynamic modules of rigid body, flexible construction, liquid sloshing etc. The source codes of algorithms of each dynamic module are fixed and there is no coupling between modules. The source codes of dynamic algorithm are stored in files. So the program can assemble algorithms according to the process of parts of source codes assembly and generate the values of dynamic parameters in XML files which are read by digital satellite simulator to complete the assembly of the dynamic algorithm.

The thermodynamic algorithm is to calculate the satellite temperature field according to the thermal network method. The source codes of algorithm are fixed. Coefficients of network of thermal conduction and radiation of different satellites are different. The source codes of thermodynamic algorithm are stored in files. So the program can assemble algorithms according to the process of parts of source codes assembly. The structure and material of satellite is stored in databases. The program can read the information and partition thermal unit by using ANSYS software automatically and generate the values of coefficients in files which are read by digital satellite simulator to complete the assembly of the thermodynamic algorithm.

The energy flow algorithm, fuel flow algorithm, radio transmission loss algorithm and optical imaging algorithm are composed by fixed algorithm and functional models of related devices. The energy flow algorithm relates to battery and solar array. The fuel flow algorithm is relates to tank, gas cylinder and valve. The radio transmission loss algorithm is relates to antenna and amplifier. The optical imaging algorithm is relates to camera. The fixed source codes of algorithm are stored in files and the information of which device is needed by digital satellite is stored in databases. So the program can assemble the fixed parts of source codes according to the process of parts of source codes assembly. And the program can further assemble the whole algorithms by cooperating with the functional models of devices.

4 Cases of Digital Satellite Assembly

Three different satellites have been assembled for validating the feasibility, effectiveness and generality of code generation of digital satellite. The digital satellites were assembled as follows.

The devices that satellites need were defined. The parameters of devices were entered in databases such as size, mass, electric power, thermal power etc. And the principle models, error models and failure models of devices were stored in files.

The main load-carrying structure was chosen. The devices that the satellite uses were chosen. And the number and the installation information of devices were configured. The first satellite used gyro, sun sensor, earth sensor, star sensor, GPS, thruster, wheel, heater, battery, solar array, oxidizer tank, fuel tank, gas cylinder, valve, antenna, amplifier and camera. The second satellite used the same type of devices with the first satellite, but the models and installation information of devices were different. The third satellite used CMG additionally on the basis of the first satellite. The overall constitution of the satellites was gotten after the process above.

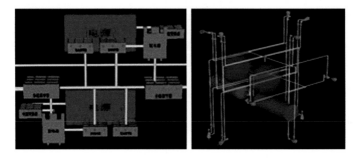

Fig. 2 Three-dimension displays of power and propulsion subsystems

The constitution of power and propulsion subsystem of one satellite is showed in Fig. 2.

The processes of information transmission between devices of three satellites were defined respectively. The processes of information transmission between devices and algorithm module were defined and were stored in files.

The execution processes of flight modes and the switching conditions between flight modes were defined. The flight modes of different satellites were different. The parameters of data management algorithms were defined.

The dynamic algorithm of the first satellite only included the part of rigid body. The dynamic algorithm of the second satellite included the part of rigid body and flexible construction and the parameters of flexible construction were inputted. The dynamic algorithm of the third satellite included the part of rigid body, flexible construction and liquid sloshing and the parameters of flexible construction and liquid sloshing were inputted.

The program of generation was run to read the results of configuration of three satellites respectively and generate the simulators of three satellites successfully. The three simulators were run in the typical working conditions. The results of simulation referred to that the simulators which were generated by program were consistent with the requirements. The three-dimension display of one satellite is showed in Fig. 3.

Fig. 3 Three-dimension display of the satellite

5 Conclusion

A theory that complex system is resolved into a set of underlying decisions so that the complex system can be handled by program is advanced based on the theory of dimension decomposition of engineering methodology and cognitive science. The theory is used for development of digital satellite and the digital satellite simulator is decomposed into several logic and regular modules. The method of generation of statements is advanced based on the results of decomposing. And a complete digital satellite code generation process is designed according to the theory. The digital satellites configuration and code generation based on several satellites are completed. And digital satellites are used for simulations in the typical condition. The simulations reveal that digital satellite can be assembled automatically using the theory. The code generation technology of digital satellite based on the theory greatly simplifies the development of digital satellite, shortens the lead time of digital satellite and makes the mass production of digital satellite possible.

The theory above has strong generality. The theory can be expanded to development of ground testing program and give references for code generation of several kinds of program.

References

1. J, Zhou: Digitization and intellectualization for manufacturing industries. China Mech. Eng. **23**(20), 2395–2400 (2012)
2. Jianying, Z.: The significance, technologies and implementation of intelligent manufacturing. Mech. Manuf. Autom. Major **42**(3), 1–6 (2013)
3. Feiyue, W.: Parallel system methods for management and control of complex systems. Control Decis. **19**(5), 485–489 (2004)
4. Lee, S., Cho, S., Lee, B.S., et al.: Design, implementation, and validation of KOMPSAT-2 software simulator. ETRI J. **27**(2), 140–152 (2005)
5. Yunfeng, D., Shimin, C., Jianmin, S., et al.: Dynamic Simulation Technology of Satellite Attitude Control. Science Press, Beijing (2010)
6. Zhongzhi, S.: Intelligence Science. Tsinghua University Press, Beijing (2006)
7. Thagard, P.: Theory and experiment in cognitive science. Artif. Intell. **171**(18), 1104–1106 (2007)
8. Li, X., Gui, X.L.: Cognitive model of dynamic trust forecasting. J. Softw. **21**(1), 163–176 (2010)
9. van Ditmarsch, H.P., Ruan, J., Verbrugge, R.: Sum and product in dynamic epistemic logic. J. Logic Comput. **18**(4), 563–588 (2008)

Class-Constrained Extreme Learning Machine

Xiao Liu, Jun Miao, Laiyun Qing and Baoxiang Cao

Abstract In this paper, we have proposed a new algorithm to train neural network, called Class-Constrained Extreme Learning Machine (C^2ELM), which is based on Extreme Learning Machine (ELM). In C^2ELM, we use class information to constrain different parts of connection weights between input layer and hidden layer using Extreme Learning Machine Auto Encoder (ELM-AE). In this way, we add class information to the connection weights and make the features in the hidden layer which are learned from input space be more discriminative than other methods based on ELM. Meanwhile, C^2ELM can retain the advantages of ELM. The experiments shown that C^2ELM is effective and efficient and can achieve a higher performance in contrast to other ELM based methods.

Keywords Extreme Learning Machine · Class-Constrained · Discriminative · Extreme Learning Machine Auto Encoder

X. Liu · B. Cao
School of Information Science and Engineering, Qufu Normal University,
Rizhao 276826, Shandong, China
e-mail: xiao.liu13@vipl.ict.ac.cn

B. Cao
e-mail: bxcao@126.com

X. Liu · J. Miao (✉)
Key Lab of Intelligent Information Processing of Chinese Academy
of Sciences (CAS), Institute of Computing Technology, CAS, Beijing 100190, China
e-mail: jmiao@ict.ac.cn

L. Qing
School of Computer and Control Engineering, University of Chinese Academy
of Sciences, Beijing 100049, China
e-mail: lyqing@ucas.ac.cn

© Springer International Publishing Switzerland 2016
J. Cao et al. (eds.), *Proceedings of ELM-2015 Volume 1*,
Proceedings in Adaptation, Learning and Optimization 6,
DOI 10.1007/978-3-319-28397-5_41

1 Introduction

Extreme Learning Machine (ELM) [1–5] is firstly proposed to train "generalized" single-hidden layer feedforward neural networks (SLFNs). In ELM, the connection weights between input and hidden layer are called input weights. The input weights and hidden biases are chosen randomly. The connection weights between hidden layer and output layer, which are called output weights, are analytically determined [4], i.e. the only free parameters which need to be learned are the output weights [6]. The learning process is without iteratively tuning and with a fast learning speed. In this way, ELM can be regarded as a linear system [2] and the objective is to minimize the training error and the norm of output weights at the same time. Thus, ELM has a good generalization performance according to the work of Bartlett [7]. Because of the fast learning speed and good generalization performance, ELM has been widely used in many aspects, such as regression [5], classification [8], clustering [8, 9] and feature learning [10, 11].

In order to achieve a desirable performance, ELM usually uses a large number of hidden nodes which increases the computation cost of training and testing process and easily leads the model to be over-fitting. Besides, as the randomly chosen input weights and hidden biases determine the computation of the output weights, the input weights and hidden biases may exist a more compact and discriminative parameter set which can contribute to improve the performance of ELM [12]. So as to get a more compact and discriminative parameter set, several methods are proposed. Yu et al. [13] used back-propagation method to train the SLFNs and then used the learned connection weights between input layer and hidden layer and the hidden biases of SLFNs to initialize the input weights and hidden biases of ELM, which can achieve a better performance than randomly chosen input weights and hidden biases. Kasun et al. [11] learned the feature mapping matrix using ELM-AE, which is useful to initialize the input weights of deep ELM, and also improved the performance of ELM. Zhu et al. [14] proposed constrained ELM (C-ELM), which constrain the input weights to a set of difference vectors of between-class samples in training data. McDonnell et al. [15] proposed shaped input-weights and combined different approaches of parameters initialization to explore the performance of ELM.

In this paper, we propose a method based on ELM-AE. We reconstruct each class of training samples and all the training samples using ELM-AE to get the feature mapping matrices respectively, i.e. output weights of ELM-AE. Then we combine all the feature mapping matrices to form the input weights of ELM, where we can not only get the feature mapping related to all the samples, but also use class information to constrain the feature mapping. In this way, we get a more discriminative input weights of ELM and we call this model Class-Constrained ELM (C^2ELM). Experimental results show that C^2ELM can achieve a better performance than ELM and other methods which are based on ELM. Besides, the results show that our method can have a better generalization ability with less number of hidden nodes.

The remaining of this paper is organized as follows: Sect. 2 briefly describes the ELM and ELM-AE; the proposed C^2ELM is described in Sect. 3; experiments are presented in Sect. 4; in Sect. 5, we conclude our work.

2 Preliminaries

2.1 Extreme Learning Machine

Extreme Learning Machine (ELM) replaces universal but slow learning methods with randomly chosen input weights and hidden biases and analytically determines output weights.

Generally, for N arbitrary distinct samples $\{(\mathbf{x}_i, \mathbf{t}_i) | \mathbf{x}_i \in R^k, \mathbf{t}_i \in R^m, i = 1, \dots, N\}$, the number of hidden nodes L and activation function $\{G(\mathbf{w}, b, \mathbf{x}) | \mathbf{x} \in R^n, b \in R, \mathbf{x} \in R^n, i = 1, \dots, N\}$, where $\mathbf{w}$ is input weights and b is the biases of hidden nodes, firstly, we can get the mapping from input data to the random feature space:

$$\mathbf{H} = \begin{bmatrix} \mathbf{h}(\mathbf{x}_1) \\ \vdots \\ \mathbf{h}(\mathbf{x}_N) \end{bmatrix} = \begin{bmatrix} G(\mathbf{w}_1, b_1, \mathbf{x}_1) & \cdots & G(\mathbf{w}_L, b_L, \mathbf{x}_1) \\ \vdots & \cdots & \vdots \\ G(\mathbf{w}_1, b_1, \mathbf{x}_N) & \cdots & G(\mathbf{w}_L, b_L, \mathbf{x}_N) \end{bmatrix}_{N \times L} \tag{1}$$

Then the output of ELM is given by Eq. 2:

$$\mathbf{f}(\mathbf{x}) = \sum_{i=1}^{L} h_i(\mathbf{x})\beta_i = \mathbf{h}(\mathbf{x})\beta \tag{2}$$

where $\beta = \begin{bmatrix} \beta_1, \dots, \beta_L \end{bmatrix}$ is the output weights and $\mathbf{h}(\mathbf{x}) = \begin{bmatrix} h_1(\mathbf{x}), \dots, h_L(\mathbf{x}) \end{bmatrix}$ is the feature space for an input sample $\mathbf{x}$, where $\{h_i(\mathbf{x}) = G(\mathbf{w}_i, b_i, \mathbf{x}), i = 1, \dots, L\}$. So for N input samples, we can get $\mathbf{H} = \begin{bmatrix} \mathbf{h}(\mathbf{x}_1), \dots, \mathbf{h}(\mathbf{x}_N) \end{bmatrix}^T$. The output weights β can be calculated by Eq. 3:

$$\beta = \mathbf{H}^\dagger \mathbf{T} \tag{3}$$

where $\mathbf{H}^\dagger$ is the Moore-Penrose generalized inverse [16] of matrix $\mathbf{H}$ and $\mathbf{T} = \begin{bmatrix} \mathbf{t}_1^T, \dots, \mathbf{t}_N^T \end{bmatrix}^T$ is the label vector or ground truth.

In general, we could solve the learning problem according to Eq. 4:

$$\min \|\beta\| + C\|\mathbf{H}\beta - \mathbf{T}\| \tag{4}$$

where C is the regularization factor and it means that we not only calculate the smallest norm of β but also minimize training error at the same time. In this way we can get a more robust solution. Then the output weights β can be calculated as

$$\boldsymbol{\beta} = \mathbf{H}^T \left(\frac{\mathbf{I}}{C} + \mathbf{HH}^T \right)^{-1} \mathbf{T} \qquad (5)$$

or

$$\boldsymbol{\beta} = \left(\frac{\mathbf{I}}{C} + \mathbf{H}^T \mathbf{H} \right)^{-1} \mathbf{H}^T \mathbf{T} \qquad (6)$$

where $\mathbf{I}$ is the identity matrix.

2.2 Extreme Learning Machine Based AutoEncoder

Similar with AutoEncoder [17], if we let the output space of ELM reconstruct the input space in a unsupervised way, i.e. we set $\mathbf{t} = \mathbf{x}$, ELM can learn the feature mapping matrix of input data by reconstruction. In this way, the output weights $\boldsymbol{\beta}^T$ can be considered as the feature mapping matrix from the input space to the hidden space and this method is called Extreme Learning Machine AutoEncoder (ELM-AE) [11].

The output weights $\boldsymbol{\beta}_{ELM-AE}$ of ELM-AE is the feature mapping matrix of input space and can be calculated by Eq. 7:

$$\boldsymbol{\beta}_{ELM-AE} = \mathbf{H}^{\dagger} \mathbf{X} \qquad (7)$$

Also, we can get a more robust solution by Eqs. 8 and 9:

$$\boldsymbol{\beta}_{ELM-AE} = \mathbf{H}^T \left(\frac{\mathbf{I}}{C} + \mathbf{HH}^T \right)^{-1} \mathbf{X} \qquad (8)$$

or

$$\boldsymbol{\beta}_{ELM-AE} = \left(\frac{\mathbf{I}}{C} + \mathbf{H}^T \mathbf{H} \right)^{-1} \mathbf{H}^T \mathbf{X} \qquad (9)$$

3   Class-Constrained Extreme Learning Machine

The output weights $\boldsymbol{\beta}_{ELM-AE}$ learn the variance information [11] by reconstructing the input data. In fact, feature mapping matrix $\boldsymbol{\beta}_{ELM-AE}$ maps the input data to a reconstruction space which can represent all the input data. In our algorithm, we try to add the class information to constrain the feature mapping matrix.

In ELM-AE, the output weights $\boldsymbol{\beta}_{ELM-AE}$ learn the variance information of all sample data. This is a generative learning method. It ignores the class information, which is discriminative and may be helpful for classification. So in our algorithm, we try to use ELM-AE to learn variance information of each class to constrain the feature mapping matrix.

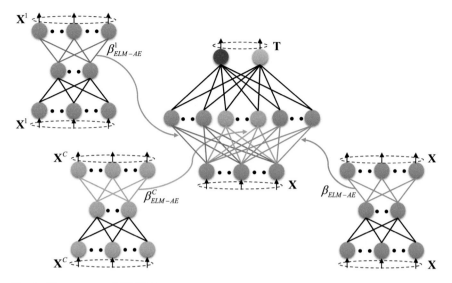

Fig. 1 The structure of C^{2}ELM

For input data of each class $\mathbf{X}^c = \{(\mathbf{x}_j^c)|\mathbf{x}_j^c \in R^k, c = 1, \dots, C, j = 1, \dots, n^c\}$, where C represent the number of class and n^c is the number of samples in each class and $N = n^1 + \cdots + n^C$, we use samples of each class to train an ELM-AE to get the output weights $\boldsymbol{\beta}_{ELM-AE}^c$ using Eq. 7. In each ELM-AE, the number of hidden nodes is l^c, which is related to class c. Then we use all the samples to train the $\boldsymbol{\beta}_{ELM-AE}$ with the number of hidden nodes l^{all}. We get the feature mapping matrix, i.e. input weights of ELM, $\mathbf{W}_{ELM} = \left[\boldsymbol{\beta}_{ELM-AE}^1, \dots, \boldsymbol{\beta}_{ELM-AE}^C, \boldsymbol{\beta}_{ELM-AE}\right]^T$, with the number of the hidden nodes $L = l^1 + \cdots + l^C + l^{all}$. The structure of C^{2}ELM is shown in Fig. 1.

In our proposed algorithm, in fact, we add the class information to constrain the ELM feature mapping matrix and we call it Class Constrained Extreme Learning Machine (C^{2}ELM). From the above discussion, the training algorithm for C^{2}ELM can be concluded in the Algorithm 1.

4 Performance Evaluation

In this section, we evaluate our C^{2}ELM and compare it with some other methods which are used to initialize ELM. These methods are evaluated on large scale datasets. We compare C^{2}ELM with the baseline Extreme Learning Machine (ELM), Extreme Learning Machine Auto Encoder (ELM-AE) [11], Constrained Extreme Learning Machine (CELM) [14], Computed Input Weights Extreme Learning Machine (CIW-ELM) [15] and Receptive Field Extreme Learning Machine (RF-ELM) [15]. Note that, we use randomly input weights and biases instead of orthogonalization of the input weights and bias to initialize the ELM-AE [11].

Algorithm 1 Class-Constrained Extreme Learning Machine (C^2ELM)

Input: The training samples $\{(\mathbf{x}_i, \mathbf{t}_i) | \mathbf{x}_i \in R^k, \mathbf{t}_i \in R^m, i = 1, \dots, N\}$, the number of hidden nodes
   L of ELM, the number of hidden nodes l^c, l^{all} of ELM-AE;
Output: The input weights $\mathbf{W}_{ELM}$, hidden layer biases $\mathbf{b}_{ELM}$ and output weights β_{ELM} of ELM;
1: Obtain samples of each class $\mathbf{X}^c = \{(\mathbf{x}_j^c) | \mathbf{x}_j^c \in R^k, c = 1, \dots, C, j = 1, \dots, n^c\}$ from all samples
   $\mathbf{X}$;
2: Calculate ELM-AE output weights β_{ELM-AE}^c with the number of hidden nodes l^c for each class
   input data $\mathbf{X}^c$;
3: Calculate ELM-AE output weights β_{ELM-AE} for all the input data $\mathbf{X}$ with the number of hidden
   nodes l^{all} which satisfies $L = l^1 + \dots + l^C + l^{all}$;
4: Combine all the ELM-AE output weights to form the input weights of ELM: $\mathbf{W}_{ELM} =$
   $\left[\beta_{ELM-AE}^1, \dots, \beta_{ELM-AE}^C, \beta_{ELM-AE} \right]^T$;
5: Generate bias vectors $\mathbf{b}_{ELM}$ randomly;
6: Calculate the hidden layer output matrix $\mathbf{H}$ by Eq. 1;
7: Calculate the output weight β_{ELM} by Eq. 3;

The experiments were carried out in a desktop computer with a core i7-470@3.4 GHz processor and 32GB RAM runing in MATLAB R2014a.

We have evaluated our C^2ELM on two large size of databases, which are MNIST [19] and CIFAR-10 [20]. The MNIST is a commonly used dataset to test the performance of ELM, which contains 60,000 training samples and 10,000 testing samples and each sample is real-value image with the size of 28×28 pixels. The CIFAR-10 is also used in classification task which is with the size of $32 \times 32 \times 3$ and contrains 50,000 training samples and 10,000 testing samples. In our experiments, the values of the original samples of these two datasets are normalized in a range between 0 and 1.

In this part, we compare C^2ELM with baseline ELM, ELM-AE, CELM on all the two datasets, while CIW-ELM and RF-ELM datasets are just used on MNIST, where these two algorithms do not give the configurations on CIFAR-10 [15]. Ten rounds are conducted at intervals of 100 nodes of hidden layer. In experiments, we set $L = C \times l^c + l^{all}$, where L is the number of hidden nodes in ELM and l^c, l^{all} are the numbers of hidden nodes in ELM-AE as described in Sect. 3 and $l^c \approx l^{all}$.

4.1 Experiments Results on MNIST

In this experiment, these methods are implemented without regularized terms. The performance on MNIST is illustrated in Fig. 2. We can see that the performance of C^2ELM is the best one, where the curve of the testing accuracy is above other algorithms based on ELM, when the number of hidden nodes are more than 100 nodes.

When the number of hidden nodes is less than 2,000 nodes, the advantage of C^2ELM is very obvious. When more than 2,000 nodes, the performance tends to be stable, but C^2ELM is still better than others. Table 1 shows the testing accuracy

Fig. 2 Experiments results
on MNIST dataset

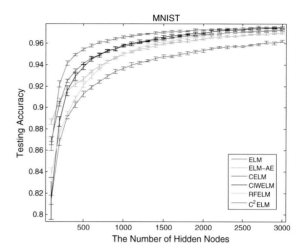

Table 1 Test accuracy on MNIST

Algorithm	The number of hidden nodes					
	500	1000	1500	2000	2500	3000
Baseline ELM	0.9125	0.9365	0.9477	0.9531	0.9571	0.9622
ELM-AE	0.9265	0.9495	0.9596	0.9638	0.9676	0.9697
CELM	0.9412	0.9572	0.9640	0.9685	0.9707	0.9725
CIW-ELM	0.9378	0.9585	0.9661	0.9694	0.9729	0.9750
RF-ELM	0.9208	0.9498	0.9617	0.9668	0.9709	0.9737
C^2ELM	**0.9541**	**0.9660**	**0.9704**	**0.9724**	**0.9737**	**0.9753**

at intervals of 500 nodes of hidden layer. From Table 1, we also can see that the performance of C^2ELM is better than the other algorithms.

Besides, we also calculate the time consumption and compare C^2ELM with baseline ELM, ELM-AE and CELM, where the Matlab codes have the similar code structure, which is shown in Table 2. We can see that, even though C^2ELM is not as faster as baseline ELM, the time consumption is close to each other.

4.2   Experiments Results on CIFAR-10

Figure 3 shows the performance on CIFAR-10. We can see that C^2ELM also achieves the best performance in comparison with the other ELM based algorithms. When the number of hidden nodes are less than 1,500 nodes, the advantage of C^2ELM is also very obvious. The curve of the testing accuracy of C^2ELM is always above

Table 2 Time consumption on MNIST (s)

Algorithm	The number of hidden nodes					
	500	1000	1500	2000	2500	3000
Baseline ELM	4.48	10.22	17.34	27.44	40.61	56.51
ELM-AE	8.90	21.07	36.37	58.18	86.59	121.61
CELM	4.54	10.58	18.33	28.97	43.09	60.53
C^2ELM	5.68	12.47	20.42	32.01	47.01	64.85

Fig. 3 Experiments results on CIFAR dataset

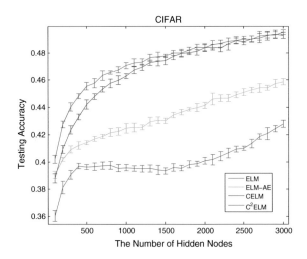

Table 3 Test accuracy on CIFAR-10

Algorithm	The number of hidden nodes					
	500	1000	1500	2000	2500	3000
Baseline ELM	0.3962	0.3957	0.3935	0.4008	0.4099	0.4279
ELM-AE	0.4142	0.4247	0.4302	0.4415	0.4510	0.4586
CELM	0.4422	0.4629	0.4744	**0.4839**	0.4882	0.4928
C^2ELM	**0.4555**	**0.4707**	**0.4777**	0.4837	**0.4901**	**0.4951**

other algorithms, which shows that when we use the class information to constrain the ELM feature mapping matrix is efficient. Table 3 shows the testing accuracy at intervals of 500 nodes of hidden layer and time consumption is shown in Table 4.

Table 4 Time consumption on CIFAR-10 (s)

Algorithm	The number of hidden nodes					
	500	1000	1500	2000	2500	3000
Baseline ELM	5.86	12.18	19.60	29.34	41.88	57.51
ELM-AE	12.60	26.07	41.39	60.67	85.98	117.76
CELM	5.98	12.50	18.21	27.24	36.92	51.00
C^2ELM	9.13	15.78	23.7	35.18	49.55	63.39

5 Conclusion

In this paper, we initialize the input weights of ELM using ELM-AE, where we not only reconstruct input data of all samples but also reconstruct each class. In this way, we add the class information, which is helpful for classification, to the feature mapping matrix. The results have shown that C^2ELM can achieve a better performance in comparison with ELM based algorithms on the large scale datasets, such as MNIST and CIFAR-10.

Acknowledgments This work is supported in part by the National Natural Science Foundation of China (Nos. 61272320 and 61472387) and the Beijing Natural Science Foundation (No. 4152005).

References

1. Huang, G.-B., Chen, L., Siew, C.-K.: Universal approximation using incremental constructive feedforward networks with random hidden nodes. IEEE Trans. Neural Netw. **17**(4), 879–892 (2006)
2. Huang, G.-B., Zhu, Q.-Y., Siew, C.-K.: Extreme Learning Machine: a new learning scheme of feedforward neural networks. In: IEEE International Joint Conference on Neural Networks, vol. 2, pp. 985–990 (2004)
3. Zhang, R., Lan, Y., Huang, G.-B., Xu, Z.-B.: Universal approximation of Extreme Learning Machine with adaptive growth of hidden nodes. IEEE Trans. Neural Netw. Learn. Syst. **23**(2), 365–371 (2012)
4. Huang, G.-B., Zhu, Q.-Y., Siew, C.-K.: Extreme Learning Machine: theory and applications. Neurocomputing **70**, 489–501 (2006)
5. Huang, G.-B., Zhou, H., Ding, X., Zhang, R.: Extreme Learning Machine for regression and multiclass classification. IEEE Trans. Syst. Man Cybern. Part B **42**(2), 513–529 (2012)
6. Huang, G., Huang, G.-B., Song, S.-J., You, K.-Y.: Trends in Extreme Learning Machine: a review. Neural Netw. **61**, 32–48 (2015)
7. Bartlett, P.-L.: The sample complexity of pattern classification with neural networks: the size of the weights is more important than the size of the network. IEEE Trans. Inf. Theory **44**(2), 525–536 (1998)
8. Huang, G., Song, S., Gupta, J.N.D., Wu, C.: Semi-supervised and unsupervised Extreme Learning Machines. IEEE Trans. Cybern. **99**, 1 (2014)
9. Kasun, L.L.C., Liu, T.-C, Yang, Y., Lin, Z.-P., Huang, G.-B.: Extreme Learning Machine for clustering. In: Proceedings of ELM-2014, pp. 435–444 (2014)

10. Huang, G.-B.: An insight into Extreme Learning Machines: random neurons, random features and kernels. Cogn. Comput. 1C15 (2014)
11. Kasun, L.L.C., Zhou, H., Huang, G.-B., Vong, C.M.: Representational learning with Extreme Learning Machines for big data. IEEE Intell. Syst. **28**(6), 31–34 (2013)
12. Zhu, Q.-Y., Qin, A.K., Suganthan, P.N., Huang, G.-B.: Evolutionary Extreme Learning Machine. Pattern Recogn. **38**(10), 1759–1763 (2005)
13. Yu, D., Deng, L.: Efficient and effective algorithms for training single-hidden-layer neural networks. Pattern Recogn. **33**(5), 554–558 (2012)
14. Zhu, W.-T., Miao, J., Qing, L.-Y.: Constrained extreme learning machine: a novel highly discriminative randlom feedforward neural network. In: IEEE International Joint Conference on Neural Networks, pp. 800–807 (2014)
15. McDonnell, M.-D., Tissera, M.-D., van Schaik, A., Tapson, J.: Fast, simple and accurate handwritten digit classification using extreme learning machines with shaped input-weights. arXiv:1412.8307 (2014)
16. Rao, C.R., Mitra, S.K.: Generalized Inverse of Matrices and its Applications. Wiley, New York (1971)
17. Bengio, Y., Lambin, P., Popovici, D., Larochelle, H.: Greedy layer-wise training of deep networks. In: Advances in Neural Information Processing Systems, pp. 153–160 (2007)
18. Hoerl, A.E., Kennard, R.W.: Ridge regression: biased estimation for nonorthogonal problems. Technometrics **12**(1), 55–67 (1970)
19. LeCun, Y., Bottou, L., Bengio, Y., Haffner, P.: Gradient-based learning applied to document recognition. Proc. IEEE **86**(22), 2278–2324 (1998)
20. Krizhevsky, A.: Learning multiple layers of features from tiny images (2009)

Author Index

B

Bai, Mei, 107, 135
Ban, Xiaojuan, 227, 293, 477
Bao, Menghu, 361
Bi, Xin, 347, 463
Bisio, Federica, 265

C

Cao, Baoxiang, 521
Cao, Jiuwen, 169
Cao, Yuping, 65
Chang, Li, 511
Chen, Huajun, 51
Chen, Huiling, 237
Chen, Jiaoyan, 51
Chen, Juncheng, 361, 371
Chen, Xi, 51
Chen, Yuemei, 13
Chu, Zhongyi, 211
Cui, Song, 361, 397

D

del Campo, Inés, 435
Diao, Liang, 237
Ding, Linlin, 489
Dong, Fang, 159, 169
Dong, Fei, 39
Dong, Shuai, 397
Duan, Lijuan, 361, 371, 397
Dwiyasa, Felis, 383

E

Echanobe, Javier, 435

G

García-Sedano, Javier, 435
Gastaldo, Paolo, 265
Gautam, Chandan, 447
Ge, Hui, 371

Gianoglio, Christian, 265
Gong, Yong, 193
Gu, Yu, 77
Guo, Chong, 227, 293

H

Hao, Liling, 147
He, Liang, 159
He, Ling, 423
Hu, Guyu, 179
Hu, Jun, 211
Hu, Xiaohui, 159
Huang, Shan, 13

J

Jiang, Changwei, 477
Jiang, Jingfei, 27
Jiang, Juping, 27
Jiang, Mingchu, 121
Jin, Xinyu, 169

K

Kok, Stanley, 39
Kong, Xiaowang, 77
Kou, Yue, 135

L

Li, Guohui, 293
Li, Ming, 501
Li, Na, 121
Li, Teng, 409
Liao, Shouyi, 307
Lim, Meng-Hiot, 383
Liu, Bing, 279
Liu, Hong, 159
Liu, Huaping, 307
Liu, Jun, 147
Liu, Junbiao, 159, 169
Liu, Mingming, 279

Liu, Xiao, 521
Liu, Yu, 489
Liu, Zeshen, 193
Liu, Zhiqiang, 27
Long, Jingtao, 347
Luo, Xiong, 477
Luo, Yang, 147

M
Ma, Wei, 397
Ma, Wenhui, 463
Mao, Wentao, 423
Martínez, Victoria, 435
Miao, Jun, 521
Min, Ren, 511
Mirza, Bilal, 39

N
Nie, Tiezheng, 135

O
Ong, Yew-Soon, 383

P
Pan, Zhisong, 121, 179, 193
Pang, Jun, 77
Panigrahi, Bijaya, 383
Parth, Yogesh, 1

Q
Qing, Laiyun, 521

R
Ragusa, Edoardo, 265

S
Seijo, Sandra, 435
Shen, Derong, 135
Shen, Qing, 227
Shuai, Liguo, 237
Song, Baoyan, 489

T
Tian, Fengchun, 249
Tian, Shuo, 319, 333
Tian, Xuemin, 65
Tian, Yangyang, 423
Tiwari, Aruna, 447

V
van de Vosse, Frans, 147

W
Wang, Botao, 13
Wang, Cong, 227
Wang, Guoren, 13, 93, 107
Wang, Jinwan, 423
Wang, Pei, 409
Wang, Shicheng, 307
Wang, Weidong, 279
Wang, Xiaohui, 65
Wang, Xite, 107, 135
Wang, Zhanghui, 93
Wang, Zhaoxia, 1
Wang, Zhiqiong, 319, 333, 409
Wang, Zhongyang, 333

X
Xin, Junchang, 107, 319, 333, 409, 489
Xu, Jia, 77
Xu, Jinwei, 27
Xu, Lisheng, 147
Xu, Yanhui, 361
Xu, Yingnan, 27
Xue, Jiao, 193

Y
Yang, Benqiang, 147
Yang, Chunwei, 307
Yang, Xiaona, 477
Yang, Zhen, 371
Yao, Yang, 147
Yin, Ying, 501
Yu, Ge, 13, 77, 135, 319
Yu, Xin, 347
Yu, Yajun, 179
Yunfeng, Dong, 511

Z
Zhan, Heng, 463
Zhang, Bin, 501
Zhang, Chen, 279
Zhang, David, 249
Zhang, Hanyuan, 65
Zhang, Lei, 249
Zhang, Ningyu, 51
Zhang, Zhen, 347, 463
Zhao, Xiangguo, 347, 463
Zhao, Yuhai, 93, 501
Zhu, Weihang, 237
Zunino, Rodolfo, 265